WorldMinds: Geographical Perspectives on 100

# WorldMinds:
# Geographical Perspectives on 100 Problems

Commemorating the 100th Anniversary of
the Association of American Geographers
1904–2004

*Celebrating Geography – the Next 100 Years*

*Edited by*

DONALD G. JANELLE
BARNEY WARF
KATHY HANSEN

*Editorial Assistants*

BENJAMIN SPRAGUE
GAMAIEL ZAVALA

 Springer

A C.I.P. Catalogue record for this book is available from the Library of Congress

ISBN-10 1-4020-1612-3 (HB)
ISBN-13 978-1-4020-1612-7 (HB)
ISBN-10 1-4020-1613-1 (PB)
ISBN-13 978-1-4020-1613-4 (PB)

Published by Springer,
P.O. Box 17, 3300 AA Dordrecht, The Netherlands.

*www.springer.com*

*Printed on acid-free paper*

Printed in the Netherlands.

# FOREWORD

Geography today is a vibrant amalgam of theories, methods, and data about past, current, and emerging worlds. Geography and the geographers who produce it dwell at the intersection of the natural sciences, the social sciences, and the humanities, and various admixtures of their theories, methods, and data constitute the 100 chapters included in *WorldMinds*. Arrayed under the rubrics of politics and power, human wellbeing, cities, livelihood, ecosystems, human-environment interactions, hazards, natural systems, new methods, and human perceptions, these 100 short essays reveal and exemplify the conceptual and topical richness of contemporary North American geography.

As is evident in these rubrics and essays, geography today is a many-splendored enterprise ranging, as the editors note, "from feminist deconstruction to fluvial geomorphology." Geographers have something strikingly valuable to say about many, if not most of the problems that confront individuals and groups in locales and regions ranging from the plots of smallholders to the entire globe. The diverse chapters of *WorldMinds* well illustrate some of the key geographical perspectives that contribute usefully to the broader understanding of common problems.

- Geographers, more than those who profess or practice any other discipline, focus on places as phenomena of intrinsic interest or as the stages on which events take place and processes unfold. The majority of the chapters in *WorldMinds* are rooted in places. The new recognition of place-based analysis in the natural and social sciences is a tribute by other scientists to the enduring power of the geographical perspective.

- Geographers have always attended to the *connections* that tie places to each other, and to the complex flows of commodities, people, and information that move over those connections. How events that occur or decisions that are made at one place affect other places is another fundamental ingredient of geographical analysis. Many of the chapters in *WorldMinds* explore such linkages, either as the focus of the contribution or as critical aspects that foster explanation and understanding of an event or process at a place.

- *Human-environment interactions* have been central to geographical thinking for centuries, finding early modern expression in George Perkins Marsh's 1864 volume entitled *Man and Nature: Or, Physical Geography as Modified by Human Action*, revised and republished in 1874 as *The Earth as Modified by Human Action*. More vital than ever in an era of widespread concern about global climate change and its consequences for peoples and the places they occupy, the human ecology tradition finds rich expression in many of the contributions to *WorldMinds*, especially in sections VI and VII. The value of the "integrated analysis" and "coupled systems" perspectives in global change research comes as no revelation to geographers.

- *Scale* is an abiding component of geographic description and analysis, and few of these essays miss an opportunity to illustrate the profound ways processes vary with geographic scale and the reciprocal ways scale of analysis affects conclusions about geographic processes of all kinds.
- *Maps*, of course, are as critical to geography as the skeleton is to anatomy. It's difficult even to conceive of the arrangements of places on the earth's surface and the connections among them at various scales without maps. Mapping has been at the forefront of geographic description and analysis since the origins of the discipline in Ionia 2,400 years ago. At times mapping has defined the discipline and has indeed often been the definition of "geography" in the popular imagination. Contemporary mapping is powerful and flexible beyond the wildest imaginings of Claudius Ptolemy or Gerard Mercator. It is no less useful than mapping was at the dawn of Western science and during the age of Western discovery, however. Section IX on Mobilizing Geographic Technologies gives appropriate emphasis to this essential geographic method.

The publication of *WorldMinds* is particularly propitious at this time of pressing needs for understanding our world and addressing the challenges posed by issues such globalization, regional and ethnic conflict, environmental hazards, terrorism, poverty, and sustainable development. As the AAG celebrates its centennial – reflecting on its past and examining its future – it is most appropriate that we consider how geography as a discipline can and will contribute to the needs of society and the world at large.

Geographers are nothing if not eclectic, and we can think of no better way for an organization of geographers to mark its centennial than by rejoicing in the diversity that has animated *WorldMinds* and that is evident in every section of the book. We are pleased to have played small parts in advancing geography through our membership in the Association of American Geographers and our service to the organization as recent and current executive directors. We commend the editors and authors of *WorldMinds* for conceiving and producing this most suitable celebration of the discipline's value in today's world, and are pleased on behalf of the Association to make it available to a wide audience as a memento of the AAG's centennial year.

Ronald F. Abler, Executive Director *Emeritus*, AAG
Douglas B. Richardson, Executive Director, AAG

# ACKNOWLEDGEMENTS

*WorldMinds* is a gift from geographers to the broader community in commemoration of the 100th anniversary of the Association of American Geographers. This 100-chapter volume has involved geographers from academia, business, government, and non-government organizations; authors include students, practitioners, and senior scholars. Their contributions highlight the breadth of what Geography renders to society, science, and scholarship. The editors thank the 151 authors and co-authors for volunteering talent to provide insight on how geographical thinking and methodologies help to interpret and to resolve the many problems that confront society and environment.

The Center for Spatially Integrated Social Science (Michael Goodchild, Director) at the University of California Santa Barbara provided logistical support for managing the project. We extend special thanks to Benjamin Sprague and Gamaiel Zavala. Ben assisted with copyediting and file management. Gamaiel set up an interface on the World Wide Web to coordinate the editing and communication among editors and provided technical assistance for graphics. We also thank the staff of Kluwer Academic Publishers – Myriam Poort, Susan Jones, and André Tournois – for being so supportive under a tight publication schedule.

As editors, it has been a pleasure to work with so many dedicated geographers and to help oversee the production of a commemorative volume of this scope. We thank the Association's executive officers and the AAG Council for this privilege and honor.

Donald G. Janelle, Santa Barbara, California
Barney Warf, Tallahassee, Florida
Kathy Hansen, Bozeman, Montana

# CONTENTS

## Part III
## ENHANCING LIFE IN CITIES

## Part IV
## INTEGRATING LOCAL AND GLOBAL ECONOMIES

## Part V
## MAINTAINING HEALTHY ECOSYSTEMS

**Part VI**

**BALANCING ENVIRONMENT WITH ECONOMY**

## Part VII
## UNMASKING DANGEROUS ENVIRONMENTS

## Part VIII
## ACTIVATING PHYSICAL GEOGRAPHY

## Part IX
## MOBILIZING GEOGRAPHIC TECHNOLOGIES

## Part X
## (RE)ASSESSING CULTURE AND IDENTITY

# INTRODUCING *WORLDMINDS*

BARNEY WARF
*Florida State University*

DONALD G. JANELLE
*University of California, Santa Barbara*

KATHY HANSEN
*Montana State University*

Academic work is often popularly satirized for its lack of practical import beyond the comforting walls of the ivory tower; indeed, the very term "academic" often connotes well-intentioned, perhaps amusing intellectuals who make no substantive contribution to the "real world." The image of absent-minded professors smoking pipes in tweed jackets is an obsolete representation of the occupation, but a persistent and popular one nonetheless. Geography in particular, at least in the historical context in which it developed in the United States, has long suffered for a reputation as a reservoir of the trivial and insignificant, a view unfortunately manifested in the simplistic popular images of a "capes-and-bays" approach that privileged memorizing capitals and exports above the critical understanding of pressing social and environmental predicaments.

Yet, increasingly over the past century, the momentum of contributions by geographical scholars has created a remarkable effervescence of conceptual and methodological practices that have enhanced the discipline's relevance in scientific and social discourse. Environmental issues – informed by political ecology, fears of global warming, and a worldwide ecological crisis – have taken on new conceptual sophistication and policy relevance. Human geography has witnessed one field after another refashioned by various forms of social theory, political economy, and philosophically self-conscious inquiry. Methodologically, geographers today have an arsenal of tools ranging from GIS to remote sensing to complex mathematical algorithms that were unavailable to earlier generations. And other concerns, such as how to teach geography to a woefully geographically illiterate public, remain unsullied in their importance. Over the past century, geographers have increasingly received attention as generators of original concepts and tools for interpreting the world and informing its inhabitants. Their work has advanced knowledge and science to help humankind understand and resolve many critical issues.

Geography has become increasingly attuned to a variety of social and environmental problems that range greatly in size, scope, and intensity. It has examined the origins, nature, and potential solutions to issues such as soil erosion, pollution, climate change, natural and human-made hazards, poverty and hunger, homelessness, water quality, ethnic strife, war, unemployment, social inequality, and the extinction of species. Geographers have long sought to address these predicaments using many different conceptual approaches and methodologies. At times their work involves input into public policy making; at others, the output of geographers is used informally by individuals and groups to attack pressing issues in their communities and daily lives, including toxic waste dumps and traffic congestion.

As part of its centennial celebration in 2004, the Association of American Geographers (AAG) wished to showcase geographers' contributions to resolving critical social and environmental issues. Starting in 1904, the organization, like the discipline, has grown rapidly in size and sophistication. Today, geographers embrace an extraordinarily diverse constellation of topics and approaches in their research and teaching, from feminist deconstruction to fluvial geomorphology. This collection of essays surveys the types of social, ecological, and methodological problems that geographers study and the contributions they make to resolve them. The book is designed to reveal the utility of geography at many intellectual levels: as a conceptual discipline that contributes theoretically; as an applied practice that informs policy-making; and as a coherent set of methodologies to gather and analyze data about Earth and its occupants.

This book is dedicated to the idea, naïve perhaps to some, that academics does matter, that geographers can make a substantive difference to the lives of people and the well being of the planet that gives the discipline its name.


## ORGANIZATION OF THE VOLUME

In the spirit of the AAG's centennial celebration, this volume contains 100 essays, each focused on a particular problem. Defining a "problem," of course, is a tricky issue. A matter of pressing urgency to some may appear trivial to others. Some problems are straightforward and obvious: few, for example, would disagree that hunger is a predicament demanding attention and intervention. Others are less clear, particularly when multiple causes are at work, when the lines between cause and effect become blurry, when significant time lapses or geographic distances separate origins and impacts. Geographers' contributions to understanding and alleviating these problems are many and varied. In some cases, no immediate solution may present itself, and the best we can do is to accept the limitations of the situation and hope future strategies are revealed in due course. Rarely are simple and obvious answers available to difficult and complicated predicaments. Ideally, problems can be identified and aborted before they begin, although obviously this rarely happens in practice. Typically, problems that are addressed comparatively earlier in their unfolding can be mitigated more easily than those that have appeared as crises or emergencies. Often problems require political interventions that do not find support among those most able to allocate resources.

The short space allocated to each author in this volume does not lend itself to lengthy, in-depth analyses; rather, this work is designed to showcase, however briefly, a particular issue whose resolution improves people's lives. Lending coherence to the diverse litany of problems and issues found in this text is no small feat. The very range of subjects, from AIDS to deforestation to feng shui to the modifiable area unit problem, testifies to the enormous scope of geography as an intellectual discipline. It may be difficult for the reader to find common themes throughout, yet the volume is more than a display of controlled chaos.

The volume is divided into ten groups of essays. Part I, "Mixing Geography and Politics," points to some ways in which the struggle for, and escape from, power poses predicaments at several spatial scales. These issues include, to take but a sample, ways of establishing a coherent system of property rights in Mozambique in the aftermath of the devastating conflict there, the social origins and strategies used by opposition to globalization, finding and demining hidden landmines, one of the military's most insidious weapons, and contestations over the public memory and representation of place.

In the second Part, "Addressing Human Needs: Health and Education," geographers point to numerous issues that revolve around two fundamental human requirements. One study offers a tantalizing view of the AIDS crisis and the virus's susceptibility to the mineral selenium. Others write on spatial patterns of disease and mortality, ways to convey geographic information to persons without sight, or how the Finns cope with the prolonged darkness of their winters. Relatedly, others write about the spatiality of education, including the problematic geographies of educational services, illiteracy, geographic education, and ways of fomenting university-community collaboration to overcome the "town-gown" dichotomy.

Part III, "Enhancing Life in Cities," focuses on urban areas, where half the human race now lives. Given the dense concentrations of people they harbor, their complex infrastructures and multitude of economic and social activities, it is little wonder that many geographic problems are located there. The essays here concern ways of measuring and alleviating the particular impediments faced by women and minorities, minimizing traffic congestion, identifying and combating urban environmental injustice, and cleaning up the streets, water, and most polluted sites of these locales.

In "Integrating Local and Global Economies" (Part IV), nine selections explore different scalar manifestations of problems associated with the production process, noting how the local and the global are shot through with each other. These range from approaches to sensitize global corporations to the importance of local cultures to the unfolding of the Asian economic crisis that wrought devastation in the late 1990s. Several authors put their thumbs on the mutating geographies unleashed by the contemporary round of restructuring, including the surfacing of high technology centers in China, spicy offshore financial centers on the periphery of the global economy, and the network of call centers that has sprung up around one of the most important but overlooked industries in the United States, telecommunications. Predicaments associated with our neighbors to the south and north, Mexico and Canada, are also evaluated.

Human impacts on ecosystems have long been a central topic of geographic concern. Given the momentous changes underway at present, fueled by economic change, population growth, globalization, and other forces, innumerable species are at risk of extinction. Eleven essays in Part V, "Maintaining Healthy Ecosystems," chart various facets of the threat to biodiversity. Four of the essays are concerned with the rich forests of Central and South America. Others examine particular ecosystems in the United States, including wetlands. Given that oceans cover 71 percent of the world's surface, it is surprising that more geographers don't study marine habitats, as two of these papers do. Furthermore, these essays serve to refute long-standing assumptions that nature lies "outside" the domain of human affairs. By enfolding nature within social relations of material practice and discourse, the authors demonstrate that the biophysical environment is shaped, molded, and even created through human action. "Nature," it seems, is not "natural" any more.

The protection of ecosystems is often counterpoised to the needs associated with economic development, the focus of Part VI, "Balancing Environment With Economy." These essays shed light on the diverse ways in which geographers have advocated sustainable development, energy efficiency, and resource conservation. For example, China's growing economy has rapidly escalated its use of coal, with major environmental consequences. The so-called war on drugs has impacted land use change in the cocoa fields of Amazonia, as two essays document. American agricultural overproduction and overgrazing worldwide are significant modifiers of the national and global environment. Protecting forests and satisfying human needs for the space and resources they offer are two, often conflicting goals, and the analysis and resolution of these issues is a fundamentally geographic practice.

Geographical landscapes can be hazardous, as Part VII, "Unmasking Dangerous Environments," demonstrates. The threats to social and natural environments include hazardous wastes and fuels, pesticides, acid rain, hurricanes and blizzards, and those in coastal regions. Minimizing the risks that such hazards present requires a combination of appropriate research techniques (in these cases including spatial analysis and GIS), an adequate understanding of the behavior of the dynamics of the biophysical environment, and an in-depth appreciation of the social origins, consequences, and policy options that dangerous phenomena present. Hazards thus offer an ideal opportunity to meld human and physical geography, theory and methods.

Physical geographers are intimately involved in addressing human problems as well. In Part VIII, "Activating Physical Geography," the particular galaxy of issues emanating from such processes as global climate change, water resource conservation, and coal extraction is given its due. Two essays focus on urban climatology, while others address the interface between land use and the hydrosphere, as in the application of geomorphology to protect horseshoe crabs or to assess the efficiency of levees. These examples serve to remind those of us who may have forgotten that the natural environment still matters profoundly in the construction of human geographies, that the gulf between "human" and "physical" domains is an artificial and analytically misleading one, and that no accounting of geographers' contributions to the human condition can exclude that segment of the discipline most concerned with climate, landforms, and other facets of the earth's physical landscape.

The explosion of geographical information systems over the last two decades has unfurled a host of new topics and approaches within the discipline, as well documented by Part IX, "Mobilizing Geographic Technologies." GIS is simultaneously a tool and a language, a set of software programs and a worldview, and offers enormous possibilities for tackling complex problems by integrating and analyzing diverse and hitherto incompatible datasets. Half of the essays in this section are concerned with making GIS more accessible to communities in various ways. Others apply GIS to a variety of local circumstances, including the Wheat Belt of the northern United States, watershed modeling, and national land cover change. GIS has come into its own as a formidable and respectable part of the discipline, and its integration into geographic problem solving is a welcome step forward in geography's attempts to attain improved analytical sophistication.

Matters of culture, perception, and meaning may not, at first glance, appear to constitute "predicaments" or problems to those accustomed to thinking of social, environmental, or public policy matters in a traditional sense. Yet as the essays in the final section, Part X, entitled "(Re)Assessing Culture and Identity" indicate, questions of individual and collective identity are inextricably sutured to space and place. This group includes such issues as the geographies of children's lives and their representations of space, the legitimacy (or lack thereof) of immigrants, the relevance (or lack thereof) of religion to contemporary geographical concerns, and the ways in which natural and cultural landscapes interact in our representations and constructions of them. Far from being outside the range of problems addressed by geographers, therefore, these topics are located close to the very heart of the act of "geo-graphing" or "earth writing," geography's original (and yet contemporary) definition.

## ADOPTING THE SPIRIT OF *WORLDMINDS*

These essays, ranging in scale from those concerned with the global economy to very specific case studies of unique habitats, stretch from one end of the discipline to another and push those borders at the same time. If any theme can be discerned that runs throughout, it is geographers' long-standing and ever-present concern that the world be made more habitable, that the quality of life for the planet's peoples be enhanced, and that academic knowledge be harnessed as a tool to improve public and private decision making and resource allocation.

*WorldMinds* is designed to show off the ways in which geographers approach problems of the world. The editors' intent is that readers adopt the spirit of *WorldMinds* to use the theories and tools of geography to address the problems and predicaments that confound the human and physical conjunction of life on earth. Now it is in your hands.

## PART I

# MIXING GEOGRAPHY WITH POLITICS

Land, water, human rights, governance, symbolic practices, and other factors are often the subjects of political conflict. Politics – the exertion of power – underlies all human relations, even when it is invisible. Political struggles are inevitably spatial, structured in their severity and significance by the underlying geographies of class, gender, religion, ethnicities, demographic compositions, past conflicts, and other cleavages that mark the human condition. The perspectives of geography are essential to applications of insight that bring understanding and resolution to some of the most vexing social problems and to the allocation of authority and justice.

*Barney Warf*

# ENVIRONMENTAL PROBLEMS AND INTERNATIONAL DEMOCRACY

RON JOHNSTON
*University of Bristol*

International environmental problems – such as global warming, holes in the ozone layer, biodiversity loss, and the management of common resources (especially oceanic and atmospheric) – raise major problems for democratic practice. Within individual countries, the state regulates environmental use, and its exercise of that power is generally accepted, especially within representative democracies. It is trusted to act for the general good of all, future as well as present.

Environmental processes are unconstrained by state boundaries, however. Many of the associated problems – many of them caused, or at least exacerbated, by human actions – cannot be controlled by individual states. Since nobody can claim, let alone exercise, power at larger scales than individual states – save where it has been partially yielded, as in the European Union (EU) – the prospects for tackling such problems are unpromising.

Hardin's (1968) classic paper on the tragedy of the commons – in which individual rational behavior proves to be detrimental to the general good – is widely deployed as a metaphor for contemporary global environmental issues. In responding to it, writers have identified three potential solutions to situations where individual behavior fails to deliver the best collective outcome. (For an excellent introduction to these, see Vogler 2000.) How do they apply to global environmental problems? For a problem to be tackled effectively internationally there must be broad agreement on not only its existence and causes (on which see Hajer 1995) but also on the means for tackling it – a substantial set of barriers to environmental action (Trudgill 1990).

## PRIVATIZATION

The first potential resolution is to *privatize the commons* (as with British nineteenth century enclosures). Individuals or corporate bodies own the resources, which it is argued they will manage in their own interests and

*D. G. Janelle et al. (eds.), WorldMinds: Geographical Perspectives on 100 Problems*, 3–7.

hence the general long-term good: resource privatization will ensure its sustainability.

Market logic implies allocating resources to the highest bidders (which will benefit the already well-endowed – whether individuals, corporate bodies, or states), but the new owners may exploit them at unsustainable rates to obtain profits (or repay loans). Thus privatization may benefit sellers in the short term, but not meet long-term interests.

But who could sell common global resources? How would a collective sales agreement be reached and the proceeds distributed? Would each resident get an equal share of the proceeds, even influence over the negotiations? Could states buy common resources? No auctions have occurred, but some collective agreements allow states to claim as their own what were formerly considered common resources, as under The Law of the Sea agreements (Steinberg 2001) – an unequal allocation that benefits large states with long coastlines and disadvantages landlocked states.

## COLLECTIVE ACTION

The second potential mechanism for global resource governance is *collective agreement over their use with associated collective management*. Individual states negotiate on behalf of their residents (including corporate bodies) to promote the common good. Or do they? Representative governments rarely survive for long if they fail to deliver favorable conditions for their electorates, and – as negotiations at the various Earth Summit conferences illustrate – many governments are unwilling to act for a perceived general good if they believe this might be detrimental to the short-term interests of (some at least of) their citizens. For example, some states have accordingly opted out of the Moon Treaty and the Biodiversity Convention and resisted CFC reduction protocols.

Some international environmental agreements may succeed, especially those involving only a small number of (usually neighboring) states. Others may work too, when either the benefits of conforming, or the disadvantages of not doing so, are apparent (see Laver 1984, on inexhaustible resources). If these conditions do not apply, however, agreements may fail. The International Convention on Whaling's ban was only briefly adhered to by several countries with whaling fleets that ensured that other states could do little in response if they abrogated the agreed upon terms.

## EXTERNAL REGULATION

Neither privatization nor collective agreement is therefore likely on its own to provide lasting, viable resolutions to global environmental problems involving exhaustible resources. The third option – *regulation by an external body* – may be more viable. If such a body operates in the general good, its regulation of environmental use (as in local land use planning) should benefit all more equitably than the other two approaches.

Therein lies a problem: how can such a body be successfully established? Territorially bounded nation-states can resolve problems that are largely contained within their borders (i.e., with few spillovers), but they are poorly placed to resolve many global environmental problems that currently threaten the quality of life. There is no equivalent of the nation-state at the supra-national scale, though there is shared sovereignty in a few cases, notably the European Union.

Is an accountable super-state to govern the global commons a feasible solution? Representative democracies are based on equality – not only "one person, one vote" but also "one vote, one value," hence the importance of equal-sized constituencies in many liberal democracies. Should this be so in a global democracy, such as giving equal weight to voters in all states or, more realistically, weighting each state's voting power according to its population? If it were, a coalition comprising only a few states – including China and India – could dominate the other roughly 180. Power in such a forum would be very unequally distributed: a large literature shows that an equitable distribution of votes ("one person, one vote") rarely, if ever, produces an equitable distribution of power ("one vote, one value") (see Johnston 1996).

Resolution of this problem of unequal voting power is extremely difficult. It is almost impossible to design a system involving a large number of weighted voters with equal voting power. The European Union institutions, for example, have always involved over-representation for the smallest country (Luxembourg) and under-representation for the largest (Germany) and usually several other states: thus, "one person, one vote" is absent in practice and the distribution of weighted voting power means that "one vote, one value" is too. As the number of member states changes, there are prolonged negotiations over the number of votes for each country, and over the number needed for a proposal to pass, but this approach never equalizes voting power.

The only situation involving equal power is unanimity, i.e., giving every member a veto. In large bodies, this is almost certainly a recipe for stalemate, and for nothing being done. So how can international institutions be constructed? In the United Nations, power is very unequally allocated; a few countries not only have permanent places (though not a majority) on the Security Council, but also the power of veto there. In other institutions, votes are weighted according to various criteria: in the World Bank they are related to the size of a country's deposits: the more its of money that is used, the more say it has on how. And in bodies established to oversee international environmental resolutions – as with the Montreal Protocol on CFC Reductions (Rowlands 1995) and the Deep Sea Bed Authority – stronger and wealthier countries have insisted on voting systems that give them power which is proportionate to some criterion (such as their volume of CFC production) but not others (such as their populations).

The construction of democratically accountable international regimes to resolve environmental problems is thus fraught with difficulty (see chapters by Castree, Dalby, and Litfin in Agnew et al. 2003). The allocation of both votes and power within such bodies is highly problematic, and unlikely to conform to what is considered best practice within individual democratic countries – equality.

There is a further problem. How is conformity with agreed policies to be monitored and, if necessary, ensured? Sovereign nation-states can require compliance with laws and can sanction offenders. Such powers are rarely ceded to supra-national institutions, as debates about the International Criminal Court illustrate (van der Wusten 2002). States are very reluctant to allow external monitoring of their actions and supra-national bodies, lacking the essential characteristic of sovereign states (a monopoly over violence), cannot insist on compliance and/or sanctions other than through moral pressure. States may agree to do more in some circumstances, including economic sanctions, or even armed intervention. But this tactic has been very rare. It is difficult to get initial agreement for such actions, and they are unlikely to be deployed save in the most extreme circumstances, which are unlikely to include contributing to global environmental problems.

## CONCLUSION

In true Hobbesian style, Hardin (1968: 1247) concluded that the only viable solution to the tragedy of the commons is "mutual coercion, mutually agreed on," which is a widely accepted rationale for territorially bounded states. No alternative is on offer for international environmental regimes, but agreement on mutual coercion in an unequal world is far from realized: will it be – until, perhaps, it is too late? Most environmental problems need collective action if they are to be resolved, and yet few of those involved are likely to participate in such actions voluntarily. Within individual countries, the state has the power to enforce such action for the general good. But most of the major contemporary environmental problems are international – indeed global – in their scope, and effective collective action needs international agreement regarding the nature of the problems, how they can be resolved, who will pay for their resolution and how conformity to agreed rules is to be policed. Until regimes covering those issues can be created, in forms that are democratically acceptable, resolution of major problems is unlikely. Will there be enough political will to create such regimes – before the problems become irresolvable?

## REFERENCES

Agnew, J., K. Mitchell, and G. Toal, eds. 2003. *A Companion to Political Geography*. Oxford: Blackwell.

Hajer, M. 1995. *The Politics of Environmental Discourse: Ecological Modernization and the Policy Process*. Oxford: Oxford University Press.

Hardin, G. 1968. The Tragedy of the Commons. *Science* 162: 1243–1248.

Johnston, R. 1996. *Nature, State and Economy: A Political Economy of the Environment*. Chichester: John Wiley.

Laver, M. 1984. The Politics of Inner Space: Tragedies of Three Commons. *European Journal of Political Research* 12: 59–71.

Rowlands, I. 1995. *The Politics of Global Atmospheric Change*. Manchester: Manchester University Press.

Steinberg, P. 2001. *The Social Construction of the Ocean*. Cambridge: Cambridge University Press.

Trudgill, S. 1990. *Barriers to a Better Environment*. London: Belhaven.

Van der Wusten, H. 2002. New Law in Fresh Courts. *Progress in Human Geography* 26: 151–154.

Vogler, J. 2000. *The Global Commons: Environmental and Technological Governance* (second edition). Chichester: John Wiley.

# 2

# GEOGRAPHIES OF POWER IN THE POST-COLD WAR WORLD-SYSTEM

## THOMAS KLAK
*Miami University*

For a half-century the United States and the USSR oversaw a bi-polar world neatly divided into their spheres of influence. But the Cold War division abruptly ended with the Soviet Bloc's collapse (1989–1991). How can we understand the present geographical organization of the international system of states? This is an important question: we must understand the sources and dynamics of international power if we are to effectively participate in creating a more just world order.

This chapter argues that to understand the sources and dynamics of international power we need to examine the structure of the world-system (Wade 1996). World-Systems Theory (WST) helps to explain the geography of the global hierarchy by distinguishing between countries of the core (U.S.), semi-periphery (most of Latin America), and periphery (Central America and the Caribbean; Figure 2.1). This chapter introduces WST and then uses it to understand economic and geopolitical relationships between countries of the world, and in particular in the American Hemisphere.

## WORLD SYSTEMS THEORY

World Systems Theory takes a global geographical and historical perspective to analyze how these richer and poorer countries interrelate in the world economy. WST argues that any particular country's development conditions and prospects are primarily shaped by economic processes and interrelationships operating at the global scale (Shannon 1996). Each country is deeply ingrained in a hierarchically arranged world-system, with a single international division of labor, and wedded by commodity chains that stretch from raw materials and industrial components to consumer goods (Gereffi and Korzeniewicz 1994).

While economic processes play a lead role, the world-system is at once a world-economy, a world-polity, and a world-culture (Wallerstein 1991). WST identifies regularly occurring historical cycles of various durations associated with the level and quality of business activity. The main economic periods

*D. G. Janelle et al. (eds.), WorldMinds: Geographical Perspectives on 100 Problems, 9–14.*
© 2004 *Springer. Printed in the Netherlands.*

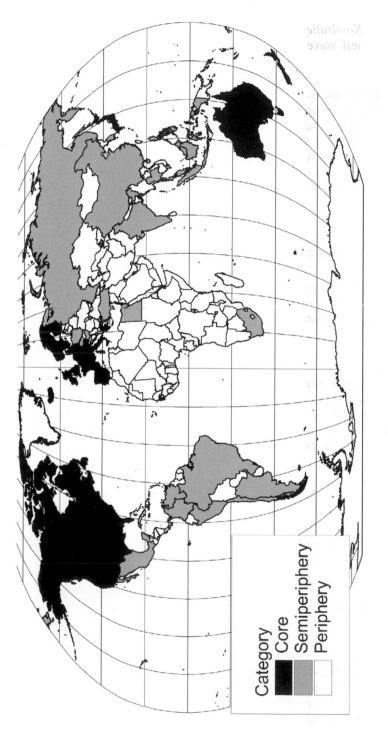

*Figure 2.1.* The World-System's major regions at the dawn of the Twenty-First Century.

for WST are *Kondratieff cycles* or *long waves* that last about 50–60 years. Each new Kondratieff wave represents a major upsurge in innovation and institution-building. The previous long wave ran from the 1940s through the 1980s, coincident with the bi-polar Cold War world (Gwynne et al. 2003).

The dawn of the twenty-first century finds all countries united within one U.S.-led capitalist world-system that is taking shape within a new long wave cycle. Institutions and organizing principles are being contested, fixed, or expanded. Neoliberal free trade policies and trading blocs such as the North American Free Trade Association (NAFTA) are designed to stabilize and ensure continued profitability and global power for the United States and other core countries and their transnational corporations in this new cycle (Anderson 2001).

## GEOGRAPHIES OF POWER AND WEALTH

Today, the world-system is enormously unequal. Despite (or, world-system theorists argue, *because of*) several centuries of worldwide economic integration and trade, global inequalities are stark and continue to worsen. The difference in per capita income separating the richest and poorest country was 3:1 in 1820, 35:1 in 1950, and 72:1 in 1992. In 1998, the world's three richest people were from the United States (Microsoft's Bill Gates and Paul Allen, and Warren Buffett of the insurance and investment firm Berkshire Hathaway) and had assets worth $156 billion. This equals the combined GDPs of the world's 43 poorest countries, including Haiti, Honduras and Nicaragua, home to 600 million people. Core countries have 20 percent of global population but 86 percent of global GDP. At the other extreme, about half of the world lives on less than two dollars a day (UN 2002).

Within this highly unequal global hierarchy are some important place-specific dynamics. In recent times, individual countries have risen (e.g., Chile thanks to its non-traditional export boom) or fallen (e.g., Honduras thanks to Hurricane Mitch in 1998) in the hierarchy of power, development, and economic potential. Most of Latin America has declined economically and socially relative to East Asia, which has amassed industrial export power (Figure 2.1; Gwynne and Kay 2003).

*Core* countries such as the United States are the principal sites of global economic (and especially industrial) power and diversity, technical sophistication, the creation of and control over innovations, wealth, and mass consumerism. Regarding control over innovations, core countries (primarily the United States) hold 97 percent of the world's patents. Following from this economic prowess are the associated military power and political influence. Core countries collectively establish and enforce the rules of the global order. The United States has recently been expanding its economic influence over Latin America, first through NAFTA, and now through a proposed Free

Trade Association of the Americas (FTAA), which Secretary of State Colin Powell says will provide "free access, over the entire hemisphere, without any difficulty or obstacle, for our products, services, technology and capital."

While the United States continues in this new cycle to be the leader of the capitalist global economy, it no longer dominates the world as it did during the Cold War cycle. Western Europe and Japan have grown many globally influential corporations and are now members of a core triad. Latin American and Caribbean countries are seeking to assuage U.S. hemispheric domination by negotiating trade pacts with the European Union.

That the world-system has fundamentally changed since the Soviet collapse is evidenced by the new terms used to describe it. During the previous long wave, the terms "Second World" and "Third World" were used commonly to describe countries such as Cuba and Colombia, respectively. But after the Cold War and since the rise of global neoliberal free trade policies, it is now common to hear some of those same non-core countries, including some former Soviet Bloc countries, referred to as "emerging markets," a grouping that corresponds with WST's semiperiphery (Figure 2.1). Countries that have not earned "emerging market" classification are relegated to the periphery (Gwynne et al. 2003).

The *Semiperiphery* mixes characteristics of the core (e.g., industry, capital, export power, prosperity) and the periphery (e.g., poverty, primary product reliance, vulnerability to outside decision making). These features correlate with size. Semiperipheral countries tend to be larger in population, natural resources, industry, and economic output than peripheral countries, and they attract more investment from the core. Mexico, Brazil, Argentina, and Chile are among the world's top ten semiperipheral countries, which together receive 75 percent of all Foreign Direct Investment (FDI) outside the core.

The semiperiphery is also a turbulent category, in that its members frequently rise or fall in the global hierarchy, as Argentina's precipitous collapse and massive capital flight since 2001 illustrates. In semiperipheral countries such as Brazil under its new socialist president known as "Lula," there is much hope for development. Because of its economic power, Brazil has a narrow window of opportunity to achieve it. But there are also intense interactions with the United States and other core countries bent on fostering their own development while maintaining the hierarchical status quo. These hemispheric interactions are illustrated by the contrasting visions of the FTAA from Colin Powell (quoted above) and Lula, who said the U.S. plan "isn't a policy of integration, it is a policy of political annexation."

Countries of the *Periphery* such as in Central America and the Caribbean tend to be smaller and more marginal in the world-system. In international trade, they do little but provide raw materials for industries elsewhere. The region's tenuous garment industry has recently declined as investors have moved to Mexico and more recently China. Peripheral countries have poor living conditions and bleak development prospects. The United States marginalizes peripheral countries because they are insufficiently profitable or, as Cuba's four decades of socialist defiance illustrates, they present challenges to the capitalist world order. Peripheral status applies to about half of all countries, which together receive only eight percent of the FDI outside the core.

Note the geographical clustering of countries in the three categories in Figure 2.1. At present the core is mainly North America, Western Europe, and Japan. The semiperiphery is mainly the larger countries of Latin America, East Asia, Russia, and key Eastern European countries. The periphery is mainly Central America, the Caribbean, Africa, and Southwest Asia.

The semiperiphery versus periphery distinction is important for conceptualizing the world as a relational, dynamic, evolving, and unpredictable system. A country's position in the world-system is historically *path dependent* (Grabher and Stark 1998) but not in a narrowly deterministic fashion. Countries can move between categories over time, depending on their development success, international aid and alliances, and the nature of the current development policies (Gwynne et al. 2003).

## CONCLUSION

World-Systems Theory, with its keen sense of historical, cyclical, technological, and geographical patterns, is a satisfying antidote to the superficiality in most popular interpretations of global economic change. WST provides a useful way to organize our thinking about evolving economic and political relations among the regions and countries of the American hemisphere. WST cannot be expected to account for all of the important trends and local impacts. It is important to be sensitive to economic, social, and policy trends, at all geographical scales from global to local, that follow a different logic from that emanating from the capitalist world-system (Taylor and Flint 2000). With such a flexible framework for understanding the dynamics of international power in mind, geographers and others can more effectively participate in creating a future world order that is more just and equitable.

## REFERENCES

Anderson, S. 2001. *Seven Years Under NAFTA. Institute for Policy Studies.* See www.ips-dc.org.
Gereffi, G. and M. Korzeniewicz, eds. 1994. *Commodity Chains and Global Capitalism.* Westport: Praeger.
Grabher, G. and D. Stark. 1998. Organizing Diversity: Evolutionary Theory, Network Analysis and Post-Socialism. In J. Pickles and A. Smith, eds. *Theorising Transition*, 54–75. London: Routledge.
Gwynne, R., T. Klak, and D. Shaw. 2003. *Alternative Capitalisms: Geographies of Emerging Regions.* London: Edward Arnold Publishers.

Gwynne, R. and C. Kay, eds. 2003. *Latin America Transformed: Globalization and Modernity* (2nd edition). London: Edward Arnold Publishers and NY: Oxford University Press.

Shannon, T., 1996. *An Introduction to the World-System Perspective*, 2nd edition. Boulder: Westview Press.

Taylor, P. and C. Flint. 2000. *Political Geography: World-Economy, Nation-State and Locality*, 4th edition. NY: Prentice Hall.

United Nations. 2002. *Johannesburg Summit 2002*. See http://www.johannesburgsummit.org/index.html.

Wade, R. 1996. Japan, the World Bank, and the Art of Paradigm Maintenance: The East Asian Miracle in Political Perspective. *New Left Review* 217: 3–36.

Wallerstein, I. 1991. *Geopolitics and Geoculture: Essays on the Changing World-System*. Cambridge: Cambridge University Press.

# 3

# RURAL PROPERTY RIGHTS IN A PEACE PROCESS: LESSONS FROM MOZAMBIQUE

JON D. UNRUH

*Indiana University*

Land rights issues have proven to be one of the most vexing problems in a peace process. In the developing world and the Middle East, the disintegration of land and property rights institutions during armed conflict, and yet the importance of land, homeland, and territory to the cause and conduct of conflict presents particular dilemmas for recovery. Such a process must attempt to both address issues fundamental to the cause and maintenance of the conflict, as well as land-related problems important to recovery.

## LAND TENURE AND CIVIL CONFLICT

The importance of land tenure (rural property rights) issues during and subsequent to civil conflict is reflected in the significant role that agrarian reform has played in many insurgent and revolutionary agendas. As Shipton (1994: 347) observes, "nothing evokes deeper passions or gives rise to more bloodshed than do disagreements about territory, boundaries, or access to land resources." Managing such issues in a peace process is not only important to avoiding disenfranchisement of local populations from land rights – a primary factor contributing to conflict – but also to the secure re-engagement of populations in familiar land uses, agricultural production, food security, and trade opportunities important to recovery.

Subsequent to the end of a conflict, land tenure issues are thrust to the fore over large geographic areas in a short period of time for considerable numbers of people. And like the complex histories involving property, land, and territory that play a role in conflict scenarios, postwar re-establishment of ownership, use, and access rights for individuals, communities and peoples will likewise be complicated and problematic, providing significant potential for renewed confrontation (Unruh 2002a). Because of the spatial nature of both armed conflict and land tenure, geographers are uniquely suited to addressing the problem of land tenure during postwar recovery. Such a topic resides within the fields of political ecology (Bryant 1992) and political economy (Emel and Peet 1989; Marston 1983) within the discipline.

*D. G. Janelle et al. (eds.), WorldMinds: Geographical Perspectives on 100 Problems*, 15–19.
© 2004 *Springer. Printed in the Netherlands.*

## THE PEACE PROCESS IN MOZAMBIQUE

The recent peace process in Mozambique provides an important example. The civil war ending in the early 1990s (Sidaway 1992: 239) initiated a process that needed to manage the reintegration of over six million dislocatees and refugees (approximately 40 percent of the national population) back into primarily agricultural pursuits. Land disputes were very frequent and acute, especially between commercial land interests operating from the formal national land tenure system, and returnees operating largely from traditional customary tenure systems. With weapons still in wide circulation and land issues becoming increasingly fraught, there was significant concern that the numerous land disputes might provide a flashpoint for a return to armed conflict. Because the national land law held documented title – which almost all returnees did not have – to be the only legitimate evidence in land disputes, there existed a very problematic disconnect between opportunities to resolve disputes peaceably and postwar reality (Unruh 2001: 4). Geographical research and subsequent work by a geographer with both the Mozambican Inter-ministerial Land Commission and a group of international donors led to more formalized definition of traditional customary evidence (by necessity spatially referenced) useful in land dispute resolution; and the inclusion of this evidence in national land policy reform, including a new land law.

The problems were numerous. The formal land tenure system was crippled, rendering itself open to abuses and non-compliance for important steps within formal law. The postwar reduction of resources, personnel, and institutions responsible for executing and enforcing formal legal procedures, together with continued insecurity in numerous parts of the country, combined to significantly reduce the capacity and legitimacy of the formal system. At the same time the customary land tenure system underwent substantial change during the conflict, especially for those for whom dislocation and migration to new areas was a significant experience (Unruh in press).

The resulting postwar land tenure situation, especially in agronomically favorable or important areas, was one where the formal tenure system was used by commercial interests to gain access to lands that were also allocated under customary tenure regimes to small-scale agriculturalists. Because these different groups appealed to different sets of evidence that resided within different, and often incompatible or opposing notions of legitimacy (often due to the war) for claim or rights to land, there were no conflict resolution institutions able to legitimately consider the different forms of evidence.

## THE ROLE OF SPATIALLY EXPLICIT EVIDENCE FOR LAND RIGHTS

The use of formal land dispute resolution by the state favored claimants in possession of some form of documentation as evidence for a claim. Those not participating in the state land tenure system used (and continue to use) an array of customary spatially referenced evidence that connects them to a community, and to community land, with history of occupation and physical signs of occupation playing a significant role in this connection. Customary institutions for land dispute resolution held membership in local lineages and community, and testimony from lineage and community members regarding history of land use and occupation, to be legitimate evidence. Commercial or "outside" land interests did not have such evidence. While documents were admissible forms of evidence in formal Mozambican law, oral testimony and corroboration were not. Thus, based on admissible forms of evidence, formal dispute resolution decisions were made in favor of documentation. Such an inequitable arrangement, operating in aggregate, carried serious risks toward instability, impoverization, land degradation, and rural exodus (Unruh 2001: 6).

The problem, more generally, became one of defining what was regarded as legitimate evidence. Within the domain of adjudication, the question of who controls the "language," and the "translations," of reality into evidence for use in adjudication, mapping, and demarcations, becomes critically important (Shipton 1994: 348; Murphy 1990: 545). This control legitimizes or de-legitimizes spatial units of aggregation, kinds of rights, or ways of land use, or they justify appropriations and expropriations (Shipton 1994: 349). Such an evidentiary problem in a postwar context becomes particularly difficult because the prevalence of weapons can quickly lead to violence in land disputes.

## GEOGRAPHICAL RESEARCH ON THE EVIDENCE PROBLEM

The research on the spatio-evidence problem examined customary evidence according to their social and cultural-ecological character. Social evidence is largely oral or testimonial, and is provided or confirmed by members of a community or lineage. This type of evidence relates to historical occupation, and ties individuals, households, and land to local communities. Cultural-ecological evidence is defined as that which exists due to smallholder activity on the landscape, such as the presence of economically valuable trees, current and historical field boundaries, tombs, etc. This type of evidence best demonstrates occupation. Cultural ecological evidence however is problematic on its own, and to a significant degree needs corroborative social evidence for meaning. In this regard, testimony from neighbors, relatives, and the customary leadership regarding boundaries, land occupation, land and tree tenure, land inheritance and the history of these, will be much more valuable in a land claim if they are all linked. Social evidence ties individuals to communities, and cultural-ecological evidence corroborated by social evidence constitutes the connection between the physical signs of human occupation of land and the social aspects that play a large role in creating cultural-ecological evidence.


## POLICY CONTRIBUTIONS

The results of the research were incorporated into the Land Commission's deliberations on land policy reform for Mozambique. On 31 July, 1997, after two weeks of parliamentary debate, the National Assembly approved a new land law. The key changes regarding conflict resolution that were adopted as articles in the revised law indicate that:

- use of nonwritten forms of customary evidence, such as oral testimony, to defend claims to land is permitted;
- rural smallholders are explicitly granted land use rights through "occupation," and such rights are not to be prejudiced by or inferior to rights received through a formal written title;
- local community "participation" is required in the formal titling process; and,
- registering of land in the name of the local community is permitted (Unruh 2002b).

Efforts are underway in Mozambique to encourage domestic and international nongovernmental organizations to play a role in bringing about local understanding of the revised land law. Ultimately the inclusion of customary evidence in the national land law and subsequent communication of the revised law to the provincial, district, and village levels encourages the evolution of land dispute resolution institutions by expanding the menu of legitimate evidence.

Because all societies experience land conflict, what is important to a peace process is equitable access to legitimate land tenure dispute resolution institutions between groups who may view land resources very differently, possess profoundly different evidence with which to pursue claims, and may have participated or sympathized with different sides in the conflict. The Mozambican peace process is widely heralded as a success, in part by the way the land issue was handled. And, much attention is currently focused on the Mozambican case so that lessons learned can be tailored to the specifics of other peace process efforts.


## REFERENCES

Bryant, R. 1992. Political Ecology: An Emerging Research Agenda in Third-World Studies. *Political Geography* 11: 12–36.

Emel, J. and R. Peet. 1989. Resource Management and Natural Hazards. In R. Peet and N. Thrift, eds. *New Models in Geography: The Political Economy Perspective*, Volume 1, 49–76. London: Unwin Hyman.

Marston, S. 1983. Natural Hazards Research: Towards a Political Economy Perspective. *Political Geography Quarterly* 2: 339–348.

Murphy, A. 1990. Historical Justifications for Territorial Claims. *Annals of the Association of American Geographers* 80: 531–548.

Shipton, P. 1994. Land and Culture in Tropical Africa: Soils, Symbols, and the Metaphysics of the Mundane. *Annual Review of Anthropology* 23: 347–377.

Sidaway, J. 1992. Mozambique: Destabilization, State, Society, and Space. *Political Geography* 11: 239–258.

Unruh, J. 2001. Postwar Land Dispute Resolution: Land Tenure and the Peacekeeping Process in Mozambique. *International Journal of World Peace* 18: 3–30.

Unruh J. 2002a. Local Land Tenure in the Peace Process. *Peace Review* 14: 337–342.

Unruh J. 2002b. Land Dispute Resolution in Mozambique: Evidence and Institutions of Agroforestry Technology Adoption. In R. Meinzen-Dick, A. Mcculloch, F. Place, and B. Swallow, eds. *Innovation in Natural Resource Management: The Role of Property Rights and Collective Action in Developing Countries*, London: Johns Hopkins University Press.

Unruh, J. in press. Land Tenure and Legal Pluralism in the Peace Process. *Peace and Change: A Journal of Peace Research.*

# 4

# GLOBALIZATION AND PROTEST:
# SEATTLE AND BEYOND

BRUCE D'ARCUS
*Miami University*

> "Americans may never again think the same way
> about free trade and what it costs"
>
> (Lacayo 1999: 35).

On 30 November 1999, officials from more than 130 countries assembled in Seattle, Washington for meetings of the World Trade Organization (WTO). The meetings were to set the agenda for future discussions about the rules and procedures that regulate how money and goods move across national borders. In a world marked by increasing economic integration of national economies, most of the trade delegates believed that the less friction presented by international rules and regulations there was, the better. For many, such a perspective was unproblematic commonsense on the morning of 30 November. By the end of the day, however, the streets of Seattle were filled with tear gas and riot police, and images of smashed store windows and bloodied protesters dominated evening news broadcasts. What would normally be a staid meeting of bureaucrats became the subject of the most significant mass protest to hit the streets of an American city since the great anti-Vietnam War protests of the 1960s. The "Battle of Seattle," as the protest became known, had begun.

Just as the protests of the 1960s reflected tangible divisions about the place of the United States in a wider world, so too did what took place in Seattle. At issue in Seattle was arguably the most significant of contemporary geographic processes: economic globalization. Where were goods produced, under what conditions, at whose benefit and whose cost? Equally significantly, the new global economy also prompted the question of who made such decisions, and how? More broadly, then, globalization and trade were about more than simple questions of economics, but also encompassed issues of power and identity (Glassman 2002; Smith 2000). The complex – and ultimately unsolvable – problem presented on the streets of Seattle also poses serious questions for geographic study. At stake are not simply an understanding of contemporary geographic change, but also vastly different visions of the geography of the future.

21

*D. G. Janelle et al. (eds.), WorldMinds: Geographical Perspectives on 100 Problems*, 21–24.
© 2004 *Springer. Printed in the Netherlands.*

## PROTESTING GLOBALIZATION

In the decades preceding the WTO meetings, the increasing economic connections wrought by globalization transformed places throughout the world.
Beginning in the late 1960s, multinational corporations in the developed
economies of North America and Western Europe sought to cut costs by
making use of low-cost, low-skilled labor available in places like East Asia
and Mexico. In so doing, the deindustrialization of former manufacturing
powerhouses like the Northeastern United States was accompanied by the rapid
industrialization of places elsewhere. As a result, by the late 1990s, much of
the products bought by consumers in the United States or Western Europe were
made or assembled in factories in northern Mexico, or Taiwan, or China.

For proponents of globalization, these changes were largely for the good.
Consumers throughout the world benefited from a competitive global environment that yielded a wider array of goods, of higher quality at cheaper
cost. Similarly, proponents argued, the shift of low-skill manufacturing jobs
away from developed economies opened up new opportunities for displaced
American and foreign workers alike. From this perspective, economic globalization was an unmitigated success.

For critics, however, free trade simply presented multinational corporations greater freedom to take advantage of opportunities offered by various
localities as they competed in a global competitive field. Increasingly, critics
argued, the opportunities that attracted investment in distant places were the
ability to *avoid* governmental, environmental, and labor regulations, and the
unionized wage rates, of developed economies, and to exploit the low wages
and lack of regulation characteristic of many developing economies (Brecher
and Costello 1998). Globalization, as driven by new technologies and new
trade regulations, presented large corporations with global reach and greater
geographic freedom. This freedom translated into competitive pressures in
the global economy that drove down wages and environmental and labor
standards. Similarly, while global consumers benefited from better goods at
cheaper prices, the majority of the world remained too poor to be a part of
this global consumer culture.

With respect to the WTO itself – the subject of the protests in Seattle –
critics argued two things. First, as the primary organization tasked with
promoting free trade, the WTO was also a significant part of the problem.
The WTO thus presented an institutional face to the vast abstractions of
global economic change, and an easy target. Equally important, from the perspective of protesters, the political power to decide important matters that
had direct bearing on ways of life throughout the world was now being placed
in the hands of a global institution with no direct political constituency. The
WTO was thus a powerful global political force, with little public accountability.

In sum, then, protesters in general argued for greater political transparency to the WTO decision-making process, a greater voice for non-governmental organizations (NGOs) that might have important input on issues discussed in that process, and for a shift in emphasis away from purist notions of free trade, toward a greater consideration of how economic policy impacted society and environment. Such issues were crystallized in various hot-button issues. French farmers were angered about being forced to compete against U.S. producers who relied on genetically modified seeds, and European consumers were bothered by the prospects of the WTO forcing them to import hormone-injected beef from the United States. At the same time, some activists smashed the windows of Gap stores in downtown Seattle because they symbolized a corporation that, like many others, thrived because of its use of sweatshop labor in Central America or Asia.

From the perspective of the WTO, such concerns had nothing to do with trade, which was precisely the point of contention that was being fought out in Seattle. Many critics represented the concerns of the protesters as both antiquated and parochial. An editorial published in *Time* shortly after the protests entitled "Return of the Luddites" argued the following:

> The left professes concern for Third World labor. But its real objective is to keep jobs at home. That means stopping the jobs from going to the very campesinos it claims to champion – and sentencing Third World workers to the deprivation of the preindustrial life they so desperately seek to escape (Krauthammer 1999).

In another editorial in *Newsweek*, Fareed Zakaria argued that the political perspectives articulated on the streets of Seattle were – despite their superficially global and cosmopolitan character – of a decidedly parochial character; protectionism with a new, global, face: "The idea that American workers will gain from slowing down, shutting off or further regulating trade has no basis in history, economic theory or common sense. It is simply a frightened reaction to change (Zakaria 1999)." The protests in Seattle thus reflected not something fundamentally new politically, but rather an age-old conservatism in the face of important, and ultimately unstoppable, change.

## GLOBALIZING PROTEST

What was new, Zakaria argued, were the tactics. The new technologies of the Internet and satellite television allowed for the marginal perspectives of the protesters to gain much wider exposure and political reach than they would have otherwise. "What we saw in Seattle," he argued, "is the rise of a new kind of politics. Disparate groups, organized through the Internet and other easy means of communication, pursue at the supranational level what they cannot accomplish at the national level." Beyond the complex ideological contests articulated in Seattle, the protests also illustrated a more concrete transformation in the nature of protest itself. As much as the Battle of Seattle constituted a significant protest *of* globalization, it also illustrated the globalization of protest as well, in part enabled by some of the same technologies that allowed multinational corporations to coordinate the complicated global relationships of production and consumption that characterized the new global economy. For critics, however, the globalization of protest evidenced in Seattle simply allowed the intensely local interests and political perspectives of the protesters to be made into a global spectacle that masked their ultimate hollow superficiality and lack of real political substance.

Such criticisms reflect both a certain level of cynicism, as well as a rather conventional understanding of the relationships between identity, politics, and scale that may no longer be relevant to a globalized world. As another article put it, "Hitherto, it's been easy to insist that anyone opposed to 'trade' was by definition a protectionist, happy to hide behind the walls of the nation-state. That simple equation no longer holds good; one of the most important lessons of Seattle is that there are two visions of globalization on offer, one led by commerce, one by social activism" (Elliott 1999: 38). Subsequent protests in Washington, D.C., Prague, Genoa, and Québec City (Drainville 2002) suggest that perhaps there is something more to the anti-globalization movement than critics profess. It should also be noted that protests against trade liberalization and other aspects that we collectively label globalization are not limited to the developed world. In a variety of locales in the past decade, from the jungles of Southern Mexico, to the streets of Bangkok (Glassman 2002), or Seoul, or Arequipa, Peru, ordinary people have made known their deep ambivalence, and in some cases open hostility, to the nature of economic and political change under globalization. The problem posed to geographers is how to understand the complexities of globalization – and resistance to it – without taking for granted what globalization is, and what it should be. Such a more complex understanding can also help to better and more creatively envision more productive and just futures.

## REFERENCES

Brecher, J. and T. Costello, 1998. *Global Village or Global Pillage: Economic Reconstruction From the Bottom Up*, Second edition. Cambridge: South End Press.

Drainville, A. 2002. Québec City 2001 and the Making of Transnational Subjects. *Socialist Register*, 15–42.

Elliott, M. 1999. The New Radicals. *Newsweek*, 13 December: 36–39.

Glassman, J. 2002. From Seattle (and Ubon) to Bangkok: The Scales of Resistance to Corporate Globalization. *Environment and Planning D: Society and Space* 20: 513–533.

Krauthammer, C. 1999. Return of the Luddites. *Time*, 13 December: 37.

Lacayo, R. 1999. Rage Against the Machine: Despite, and Because of, Violence, Anti-WTO Protesters Were Heard. *Time*, 13 December: 35–39.

Smith, N. 2000. Global Seattle. *Environment and Planning D: Society and Space* 18(1): 1–5.

Zakaria, F. 1999. After the Storm Passes. *Newsweek* 13 December: 40.

# 5

# ALLAH'S MOUNTAINS: ESTABLISHING A NATIONAL PARK IN THE CENTRAL ASIAN PAMIR

STEPHEN F. CUNHA
*Humboldt State University*

> In this plain there are wild animals in great numbers, particularly sheep of a large size, having horns three, four, and even six palms in length . . . heaps of these horns are made at the sides of the road, for the purpose of guiding travelers. . . . For twelve days the course is along this elevated plain, which is named the Pamir. So great is the height of the mountains, that no birds are to be seen near their summits. . . . After having performed this journey of twelve days, you have still forty days to travel in the same direction, over mountains, and through valleys, in perpetual succession, passing many rivers and desert tracts, without seeing any habitations or the appearance of verdure. . . . Even amidst the highest of these mountains, there live a tribe of savage, ill disposed, and idolatrous people, who subsist upon the animals they can destroy, and clothe themselves with the skins. Marco Polo.
>
> (From Komroff 1926: 66)

Although the debate over Marco Polo's itinerary across Central Asia continues, the Venetian Merchant surely reached the Pamir Plateau and camped beside the icy waters of Lake Issikul. His 13th century narrative of this *Roof of the World*, where the Pamir and Hindu Kush converge, is very accurate. The glacial peaks, windswept plateaus, non-existent avifauna, and "heaps" of sheep horns that comprise this landscape are difficult for one uninitiated to simply imagine into being. Moreover, the nomadic Kirghiz who graze yaks here today still pursue wild ungulates for meat and clothing. This is how it appeared in July 1991, when our research team entered this highland outpost, seven centuries after Marco Polo and one year after the fall of Soviet Communism.

*D. G. Janelle et al. (eds.), WorldMinds: Geographical Perspectives on 100 Problems, 25–30.*
© 2004 *Springer. Printed in the Netherlands.*

## THE PROBLEM

This chapter explores the ongoing efforts to establish a Pamir National Park as part of an overall post-colonial development strategy in the former Soviet Republic of Tajikistan. With mountain tourism one of the world's fastest growing industries (Price 1992; Godde 1999; Gram 2002), government officials and private entrepreneurs within Tajikistan aspired to market this *terra incognita* in much the same way that neighboring China (Tibet), Pakistan (Karakoram), and India/Nepal (Himalaya) attract tourist revenues from their highlands.

Landlocked, poor, mountainous, and with no oil or nuclear weapons, Tajikistan in the eyes of world political and media foci has been subservient to conflicts in Chechnya, Afghanistan, and Iraq. Yet this fledgling republic merits global attention for several reasons. First, any capricious Tajik civil strife may easily spread into neighboring Kazakstan (and their nuclear stockpiles) and hinder Afghanistan's emergence from autocratic Taliban rule. Second, conflict here severs the already tenuous and disjointed north-south oil axis so important to industrialized economies that extends from Kazakstan to Saudi Arabia. Although the Pamirs lack oil, several proposed pipelines could pass through this Silk Road corridor. Third, the Tajik Pamirs are Central Asia's water tower. Both the Amu Darya and Syr Darya gain most of their runoff here, then carom down five states to the Aral Sea. No downstate irrigation pacts or Aral Sea mitigation will occur during Tajik upheaval. Finally, their devout Islamic citizenry will help forge, either by ballots or bullets, the brand of Islam and accompanying relationship with western nations that will carry us into the next century (Bekhrandnia 1994).

Implementing a national reserve in this struggling country required accurate geographic and biodiversity field surveys. Our project represented the first significant fieldwork here since the 1940s, when Stalin forcibly depopulated this politically sensitive multifrontier. With United Nations University funding, the Soviet and Tajik governments contracted geographers because our discipline integrates both the biophysical and human worlds. It also offers the advanced cartographic skills necessary for building a geographic information database to guide planning. In addition, our collective fieldwork dossier included stints in the Karakoram, Himalaya, Caucasus, Russian Altai, Rocky Mountains, Andes, and the mountains of Iceland and Alaska; an important consideration in the Pamirs where river crossings, glacier travel, and steep terrain were common.

From 1989 through 1992, a team of Russian, American, and Tajik geographers cooperated on summer field research in the Pamirs, and further analysis in the Tajik capitol of Dushanbe, Moscow, and the University of California at Davis. We collected and synthesized data on landforms, climate, plant and animal distribution, and assessed economic and demographic patterns. This allowed us to identify outstanding natural and cultural features, propose reserve boundaries, and assess potential tourism.

## A COVETED LOCATION

The Pamirs are a complex mountain knot where the Hindu Kush, Karakoram, Alayskiy, Tien Shan, and Kunlun Shan converge (Figure 5.1). They are antipodal (or 12 time zones) from California. An imaginary tunnel from Los Angeles through the Earth would emerge in the Pamir core, where colliding Eurasian and Indian Ocean tectonic plates uplifted an echelon grouping of ranges extending 135,000 km$^2$ (Figure 5.2). Three summits exceed 7,000 m. Peak Communism (7,495 m) was the highest mountain in the former Soviet Union. At 75 km in length, the Fedchenko Glacier is one of the largest between Alaska and Antarctica. Intercepted moisture from the Gulf of Arabia and Mediterranean Sea sustains Central Asia's irrigation lifelines. The snow leopard (*Panthera uncia*), Asiatic brown bear (*Ursus arctos isabellinus*), and Marco Polo sheep (*Ovis poli*) thrive in the upper elevations. Riparian birch forests (*Betula altaica* and *B. tianshanica*) and juniper stands (*Juniperus globosa*) provide vital habitat for numerous other species. The Pamir emerged from

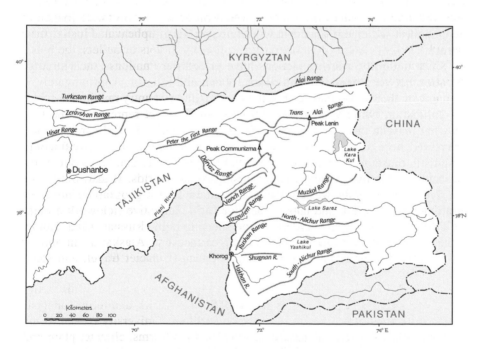

*Figure 5.1.* Map of the Pamir Mountains.

*Figure 5.2.* Peak Moskva (6,785 m) and the Sugran Glacier, High Pamir.

the Soviet era as one of the least exploited mountain environments on Earth, with modest settlement, abundant wildlife, vigorous grassland, and lush riparian forest.

Since antiquity, warring factions have prized the Pamir for their strategic location between mountain and desert, ocean and interior. Approximately 90 percent of this range falls within the Gorno-Badakshan Autonomous Oblast of Tajikistan, while Kirghizstan, Uzbekistan, Afghanistan, and China meet in what may be the world's most isolated multifrontier. Camel caravans have traversed here since the 6th century. The region remains a virtual ethno-linguistic museum. Unlike their overwhelmingly Turkic neighbors, the Tajiks descend from Iranian heritage. They worship Allah and raise potatoes and wheat on river terraces in the western Pamir. Mountain Tajiks (also called Pamirians) cultivate wheat, fruit, and nuts in the more arid canyons of the eastern Pamir (Figure 5.3). Both groups lead goats and sheep to high pastures in the summer. Still further east, nomadic Kirghiz of Mongol ancestry herd yaks in the shadow of the eastern Pamir and adjacent Pamir Plateau (Cunha 1997a).

## RESULTS

Designation of a special nature reserve in the Pamir was first suggested during the 1933 Russian and German Tajikistan-Pamir expedition. Scientific personnel of both nationalities discussed the idea on several occasions while

*Figure 5.3.* Mountain Tajik woman and daughter baking nan.

in the field, and again during the 1934 summer expedition (SCNP 1992). Two world wars, the Soviet-Afghan War, and then 40 years of Iron Curtain political constraints precluded further collaboration. While studying the Fedchenko Glacier during the 1970s, V. M. Kotlyakov proposed that both the glacier and the surrounding Academy Range should receive some form of reserve status. He promoted the idea throughout his long career with the Soviet Academy of Sciences. Russian academic journals contain numerous references to his idea. In the late 1980s, the Pamir was considered an integral component of a vast transfrontier peace park discussed by J. Ives (University of California, Davis), B. Messerli (University of Bern), Y. Badenkov (Russian Academy of Sciences), and S. Honglie (Chinese Academy of Sciences). Tajik independence in 1991 attracted foreign interest because, aside from a few climbing and research parties, this vast Asian cordillera had been closed to non-Soviets since the 1917 Bolshevik Revolution.

The Tajik Environment Ministry (Goskomproda) designed a park proposal in 1991 (Buzurukov 1991). Support and advice from numerous ex-Soviet and Western governments, our own research team, and non-governmental organizations followed. Legislation to establish the reserve passed the Tajik Parliament twice during the early 1990s, but ongoing civil war thwarted its full enactment. Persistent hurdles still delay the park and tourism development within Tajikistan. Sporadic civil strife discourages foreigners and prevents expansion of tourism infrastructure and training. War refugees from conflicts within Tajikistan and Afghanistan use designated and potential nature reserves for asylum. This sudden human influx into the mountains combined with return migrant farmers is altering vegetation, particularly in the Western Foothills. Wildlife poaching by foreign sport hunters and hungry refugees (from Tajikistan and Afghanistan) threatens long-term consequences. That the government does little to encourage private investor investment in a potentially lucrative tourism sector is another problem. In reality, it is more common for government officials to attract foreign hunting and recreation clientele, and then act as *dual professionals* to negotiate private deals (Cunha 1997b). In summary, increasing Tajikistan's exceptional tourism resource requires political stability, an organized national strategy, and significant financial investment, all of which are proving elusive (Aknazarov et al. 2002).

The recent and relatively successful Tajik peace accord has brought this initiative to the forefront again. The Swiss Agency for Development and Cooperation, the Tajik government, and the Tajik Socio-Ecological Union are pursuing a Pamir National Park. The World Conservation Union (IUCN) believes the Pamirs are an important component to a potential Central Asian transborder park linking China's Taxkorgan Nature Reserve in Xinjiang, the Khunjerab National Park in northern Pakistan, and Afghanistan's remote Wakhan Corridor. The challenge will be to overcome the military and political rivalries that characterize this *Roof of the World* so correctly described by Marco Polo.

## REFERENCES

Aknazarov, O., I. Dadabaev, and D. Meinichkov. 2002. Ecotourism in the Pamir Region: Problems and Perspectives. *Mountain Research and Development* 22: 2.

Bekhrandnia, S. 1994. The Tajik Case for a Zoroastrian Identity. *Religion, State and Society* 22: 1.

Buzurukov, A. 1991. *Pamir National Park. First Project of Tajik SSR: Collection of Scientific and Practical Papers. Conference on Problems of Protection of Animals and Special Preservation of Natural Territories.* Dushanbe: Goskomproda.

Cunha, S. 1997a. Summits, Snow Leopards, Farmers, and Fighters: Will Politics Prevent a National Park in the High Pamirs of Tajikistan? *Focus* 66: 1.

Cunha, S. 1997b. The Hunting of Rare and Endangered Fauna in the Mountains of Central Asia. *Proceedings: 8th International Snow Leopard Symposium (1995). Islamabad, Pakistan.* Seattle: International Snow Leopard Society.

Godde, P., ed. 1999. *Community-based Mountain Tourism: Practices for Linking Conservation with Enterprise. Synthesis of an Electronic Conference of the Mountain Forum. April 13–May 18, 1998.* Franklin, WV: The Mountain Institute.

Gram, S. 2002. *Community Based Conservation and Eco-tourism in Nanda Devi Biosphere Reserve.* Workshop proceedings, 29–31 October 2001, Joshimath, India.

Komroff, M., ed. 1926. *The Travels of Marco Polo (The Venetian).* New York: Garden City.

Price, M. 1992. Patterns of the Development of Tourism in Mountain Environments. *GeoJournal* 27: 87–96.

State Committee for Nature Protection of Republic of Tajikistan. 1992. *Scientific Basis of Preliminary Proposals of the Pamir National Park: February 1, 1990–January 1, 1991.* A. Buzurukov, ed. Dushanbe: Goskomproda.

# 6

# ENERGY, TERRITORY, AND CONFLICT: PLACE-BASED RESEARCH IN THE CAUCASUS

SHANNON O'LEAR
*University of Illinois*

Do energy issues play a role in conflict? Oil-rich, post-Soviet Azerbaijan is enmeshed in a territorial conflict with its neighbor, Armenia. In order to understand the complexities of the conflict, and the role that energy issues play in the resolution of that conflict, it is useful to take a place-based, geographic approach to the situation. Developing place-specific expertise can lead to insights into international implications of local situations and be very useful in advising policy-makers, in educating students and the public, and in scholarly exchanges with academic colleagues.

## ENERGY DISPARITIES BETWEEN NEIGHBORS

Since the collapse of the Soviet Union in 1991, the Caspian Sea region has been gaining attention from the United States and other western countries because it is thought to hold substantial reserves of oil and natural gas (see the U.S. Energy Information Administration website on the Caspian Sea region at http://www.eia.doe.gov/emeu/cabs/caspian.html). Azerbaijan, home to the international oil industry since the era of Tsarist Russia in the 1800s (Forsythe 1996), was among the first of the five states surrounding the Caspian to sign contracts with international oil companies following the Soviet collapse in the 1990s. International oil companies supply much-needed technology to explore and exploit these mineral resources. Since Azerbaijan is a landlocked state, pipelines are critical for the export of oil, and plans are underway to develop a pipeline that would carry oil from Azerbaijan westward through Turkey.

In contrast to Azerbaijan's oil wealth, Armenia, Azerbaijan's neighbor to the west, also faces challenges of developing its economy as an independent state (see Dudwick 1997; Bartlett 2000), but it is lacking in energy resources. Although Armenia used to export energy during Soviet times, its energy production is much lower now (O'Lear 1999). Aging infrastructure is in part to blame, and the country's economic problems also make it difficult to produce and distribute electricity, gas, and other forms of energy.

*D. G. Janelle et al. (eds.), WorldMinds: Geographical Perspectives on 100 Problems, 31–35.*
© 2004 *Springer. Printed in the Netherlands.*

*Figure 6.1.* Map of Armenia and Azerbaijan. Reprinted from *Post-Soviet Geography and Economics* 42(4): 306, with permission by V.H. Winston & Son, Inc. and Bellwether Publishing, Ltd.

When Armenia was part of the Soviet Union, a nuclear reactor was built in Armenia to generate electricity. However, shortly after the Soviet collapse, environmental activists protested the nuclear power plant and were successful in achieving the closure (or "mothballing") of the plant. The nuclear plant re-opened several years later as electricity shortages increased, but it is not a completely indigenous source of energy for Armenia since the fuel rods are imported from Russia. A few other energy options exist for Armenia, including solar and wind power, but none of them is able to meet the demands for electricity or energy in Armenia.

Just looking at a map, the solution to this uneven geography of energy resources would seem simple. On the one hand, Azerbaijan has a relative wealth of oil, but it requires export pipelines to the west. Armenia, lacking in energy resources and options for income, lies directly to Azerbaijan's west. Pipelines carrying oil from Azerbaijan could supply Armenia with a much-needed source of energy, and transit fees from oil exported through the pipelines could be a source of income for Armenia. Why not build oil export pipelines from Azerbaijan to and through Armenia? The answer is Nagorno-Karabakh.

## TERRITORIAL CONFLICT AND RIPPLE EFFECTS

During the time of the Soviet Union, the small region of Nagorno-Karabakh was recognized as a special, autonomous republic that was located within the Soviet Socialist Republic of Azerbaijan but that was populated mostly by Armenians. Shortly after the Soviet collapse in 1991, people of Nagorno-Karabakh attempted to retain a status distinct from Azerbaijan, which completely surrounds the area (Zverev 1996; van der Leeuw 2000). Armenia, in the interest of self-determination, supported this movement in Nagorno-Karabakh. As Azerbaijan tried to maintain its territorial integrity, an armed conflict erupted between the neighbors. Armenian forces now occupy not only Nagorno-Karabakh, but also a buffer zone surrounding Nagorno-Karabakh and a corridor that links the area to Armenia. This occupied area represents approximately 20 percent of Azerbaijan's territory, and the conflict is not yet resolved.

The territorial dispute in Nagorno-Karabakh would seem to be a localized problem, but there are larger scale ripple effects that effect political relations in the region and beyond (see Nuriyev 2000). Turkey, for example, is Armenia's neighbor to the east, and Turkey shares with Azerbaijan a Turkish and Muslim culture (Altstadt 1992). When Armenians in Nagorno-Karabakh initially tried to maintain their special status, Turkey protested this action by blockading its border with Armenia. This action was intended to show support for Azerbaijan, and it stopped a great deal of trade and travel across the Turkish-Armenian border, which hurt Armenia's economy significantly. Turkey is also a member of NATO, so other members of NATO became concerned with the conflict, too.

Another ripple effect of the Nagorno-Karabakh conflict was the response from the United States (O'Lear 2001). Without adequate diplomatic relations in the Caucasus when the Soviet Union collapsed, the U.S. government had only limited understanding of local situations there. An influential lobby of Armenian Diaspora in the United States convinced the U.S. government that Azerbaijan was the aggressor in the conflict, and the United States agreed to impose sanctions on Azerbaijan. These sanctions were known as Section 907 of the "Freedom for Russia and Emerging Eurasian Democracies and Open Markets Support Act" (known in shorthand as the FREEDOM Support Act). These sanctions prohibited any aid to Azerbaijan's government, such as assistance in democratic, economic, or legal reform. Although Section 907 was waived in early 2002 in recognition of Azerbaijan's support of U.S. efforts in Afghanistan, for approximately 10 years the sanctions arguably hindered several areas of development in Azerbaijan.

## TERRITORIAL SWAP AS A SOLUTION?

One of the proposals for resolving the Nagorno-Karabakh conflict, on which so many relations in the south Caucasus hinge, is a territorial swap. The first part of the proposal is that Nagorno-Karabakh could be granted a special status under the Azerbaijani government, thereby ensuring Azerbaijan's territorial integrity. The second part of proposal suggests that the Lachin corridor could be given to Armenia so as to connect the Armenian population of Nagorno-Karabakh with Armenia. That arrangement would enable Armenian self-determination. Third, in recognition that Azerbaijan would lose the territory of the Lachin corridor, it has been suggested that Armenia surrender its southern border region and allow Azerbaijan a corridor to connect the main part of Azerbaijan to Nakhichivan. This proposal would seem to satisfy all sides.

Although this solution might look reasonable on a map, it is important to understand how Armenia has dealt with its energy problems. Since independence, and given the blockade along its major borders with Turkey and Azerbaijan, Armenia has pursued closer relations both with Russia and with Iran, its neighbor to the south. In addition to the support of Armenia's nuclear reactor, Russia also engages in military support and trade with Armenia (Masih and Krikorian 1999). Relations between Armenia and Iran include several energy projects, such as gas and electricity transfer across the border and plans for a shared hydroelectric plant on the Araxes River. This narrow stretch of land on the border of Armenia and Iran, therefore, is critical to Armenia's energy and electricity supply.

Is the conflict between Azerbaijan and Armenia, then, intractable due to energy issues? Could such a territorial swap actually lead to peace in this region, or would other, unforeseen problems arise? Necessary components of a solution to this conflict are likely to include territory, energy, and ethnic or nationalistic concerns. Furthermore, it is possible that a solution to this inter-state dispute may involve non-state actors including international oil companies, populations of internally displaced persons, and other agents such as nongovernmental organizations or international organizations.

The situation between Azerbaijan and Armenia underscores the usefulness of looking at territorial conflict as more complex than a disagreement between states. If we take a place-based approach that considers aspects of physical geography, political economy, historical context, and regional and international interests in the area, it is possible to understand why simply redrawing a line on a map is unlikely to be a solution to this complex problem.

# REFERENCES

Altstadt, A. 1992. *The Azerbaijani Turks: Power and Identity Under Russian Rule.* Stanford: Hoover Institution Press.

Bartlett, D. L. 2000. Stabilization Policy in Post-Soviet Armenia. *Post-Soviet Geography and Economics* 41: 30–47.

Dudwick, N. 1997. Political Transformations in Post Communist Armenia: Images and Realities. In K. D. and B. Parrott, eds. *Conflict, Cleavage, and Change in Central Asia and the Caucasus,* 69–109. Cambridge: Cambridge University Press.

Forsythe, R. 1996. *The Politics of Oil in The Caucasus and Central Asia: Prospects for Oil Exploitation and Export in the Caspian Basin.* Adelphi Paper No. 300, International Institute for Strategic Studies. New York: Oxford University Press.

Masih, J. and R. Krikorian. 1999. *Armenia at the Crossroads.* Amsterdam: Overseas Publishers Association (Harwood Academic Press).

Nuriyev, E. E. 2000. Conflicts, Caspian Oil and NATO: Major Pieces of the Caucasus Puzzle. In G. K. Bertsch, C. Craft, S. A. Jones and M. Beck, eds. *Crossroads and Conflict: Security and Foreign Policy in the Caucasus and Central Asia,* 140–151. New York: Routledge.

O'Lear, S. 1999. Armenian Energy: Establishing or Eroding Sovereignty? *Journal of Central Asian Studies* 3(2): 2–13.

O'Lear, S. 2001. Azerbaijan: Territorial Issues and Internal Challenges in Mid-2001. *Post Soviet Geography and Economics* 42(4): 305–312.

van der Leeuw, C. 2000. *Azerbaijan: A Quest for Identity.* New York: St. Martin's Press.

Zverev, A. 1996. Ethnic Conflicts in the Caucasus 1988–1994. In B. Coppieters, ed. *Contested Borders in the Caucasus,* 13–72. Brussels: VUB University Press.

# 7

# HIDDEN KILLERS: THE PROBLEM OF LANDMINES AND UNEXPLODED ORDINANCE

CARL DAHLMAN

*University of South Carolina*

Landmines and unexploded ordinance (LM/UXO) are lethal and lasting legacies of modern conflict. Though designed to kill combatants or destroy combat equipment during conflict, LM/UXO often remain unexploded, contaminating the ground long after a conflict and posing an indiscriminant threat to civilians as well as peacekeeping and humanitarian staff. The threat of landmines is widespread as they are relatively inexpensive and easily deployed, amounting to more than 100 million mines in more than 60 countries, where they remain active for years or even decades after war (Beardsley 1997). As indiscriminate weapons, landmines kill or injure more than 15,000 people each year, mostly civilians (Strada 1996). Though mines laid and mapped by well-trained armies may be lifted after conflict, many times they are not. In many contemporary conflicts, inexperienced or untrained combatants conduct mining, with no thought of mapping the devices for post-conflict clearance. A similar threat, unexploded ordinance, includes a wide range of fired but undetonated weapons, especially grenades, artillery shells, and air-to-surface munitions. Especially pernicious are cluster bombs, designed to open over a battlefield and release many smaller bomblets, each of which explodes firing ball bearings or other body-tearing fragments across a wider area than can be reached by a single large shell. Bomblets that fail to explode during conflict, like landmines, litter the ground after major battles, posing a risk to non-combatants who may cross unaware through contaminated land or who need to quickly reestablish subsistence agricultural activities.

War zones like Bosnia, Cambodia, Mozambique, Angola, and Afghanistan are littered with millions of mines, usually produced by major foreign military powers. In some parts of Afghanistan, demining teams have found multiple layers of active mines overtop one another, laid by the different forces fighting there over the last 25 years. Even slight contamination with LM/UXO makes arable land, roads, bridges, and structures unusable, as the process of safely neutralizing these devices is specialized and expensive work requiring training and expert supervision (McGrath 2000). Reflecting the pattern of conflict today, communities affected by these devices are typically located in developing regions of the world, where conditions of poverty, social instability and weak

*D. G. Janelle et al. (eds.), WorldMinds: Geographical Perspectives on 100 Problems*, 37–41.
© 2004 *Springer. Printed in the Netherlands.*

governments are made worse still by the environmental effects of war (Finger 1991). While affected communities must live with the risk of contaminated land, they must also forego activities necessary to their survival. Meanwhile, social institutions and governments in mine-affected regions must address the political and economic challenges of post-conflict situations, made all the more fragile by LM/UXO.

Addressing the risks associated with the environmental and social effects of warfare on communities is a topic well-suited to geographical research, and can be located at the intersection of several well-established areas of expertise: geographic technologies, hazards and risk mitigation, and community development (cf. Cutter 1993; Gatrell and Senior 1999; Porter and Sheppard 1998). Research is already underway to apply geographic technologies such as geographic information systems and remote sensing to humanitarian operations in the hope of improving the efficiency and accuracy of decontamination efforts. Further, addressing LM/UXO contamination as an anthropogenic hazard and an obstacle to socio-economic development engages geography's long-standing commitment to understanding and improving human-environment relations. The remainder of this chapter examines the emerging applications of geographic technologies to short-term humanitarian demining operations and then considers the potential contributions of geographic research in identifying the long-term needs of affected communities.


## GEOGRAPHIC SOLUTIONS FOR HUMANITARIAN DEMINING OPERATIONS

As most governments in affected countries are technically, economically, or politically unprepared to address the threat of LM/UXO, humanitarian demining operations comprised of international non-governmental organizations are a common and effective response. Humanitarian organizations face the difficult task of rapidly addressing the threat on the ground and establishing national or regional mine-action programs in cooperation with government authorities. Successful demining operations must also assist humanitarian peacekeeping or socio-economic development organizations in protecting their field staff from LM/UXO risks in the short-term while, at the same time, anticipating and reducing the long-term impact of contaminated land. As in most humanitarian work, these operations require sustained donor support for a project, which means demining must compete with new crises in other countries and shifting donor priorities. The desire of all parties for a rapid neutralization of LM/UXO risks is offset, however, by the reality of demining's slow and deliberative nature. Technologies that improve response times and assist in decision-making by humanitarian demining operations serve to more fully promote the mission of neutralizing the risks posed by LM/UXO.

While the actual removal and neutralization of LM/UXO involves certain risks, it is a technically feasible task that continues to be made safer through advances in detection, neutralization, and deminer-protection technologies. Far more problematic is the sheer number of devices and their impact on local communities and regions vis-à-vis larger socio-economic development needs. Because landmines are inexpensive and easily deployed during war, but costly and slow to remove, demining operations must prioritize contaminated areas on the basis of practical needs, such as main transportation corridors, and socio-economic development objectives, such as restoring arable land for use. In addition, mine contamination almost always exceeds clearance capacity, so some areas deemed less critical will be marked, but may not be cleared for years, if at all. In most cases, moreover, national or even local development programs have yet to be planned before clearance operations commence, thus putting demining operations in the position of determining local needs with little input from governments or development specialists. These problems are compounded by the immediate confusion of post-conflict situations and the sporadic return of displaced persons whose routes and residences become immediate clearance priorities that shift resources away from other tasks. Once an area is finally selected for clearance, demining teams must carefully survey, mark, and map the area before devices can be lifted. Besides locating the contamination, surveys also include information from local residents, civilian and military authorities and humanitarian agencies on the specific type of devices, local terrain conditions, and site-specific information that will aid in the assignment of valuable demining resources (McGrath 2000).

Mine action coordinators, therefore, have before them a complex task in rapidly evaluating the human impact of LM/UXO on communities while compiling spatially coded technical data from different sources with varying standards and tolerances. In response, the humanitarian demining community has already begun to employ geography's best tools for aiding in the management and interpretation of such data: geographic information systems (GIS) and global positioning systems (GPS). Geographers have much more to contribute, however, as GIS/GPS use in the field is largely constrained to minefield mapping but can be significantly expanded to help establish socio-economic development priorities and to solve resource allocation and operational management issues. In some cases, as in the former Yugoslavia, separate national operations are seeking to improve interoperability and data-sharing, while the Kosovo Mine Action Coordination Centre is incorporating GIS as part of a larger information management system (JRC 2000; Erikson and Jean 2001). The expertise developed by GIS/GPS researchers and professionals from other applications can contribute valuably to improving demining operations (ITF 2001). Forays into the application of remote sensing technologies for humanitarian demining have opened another important research frontier for geographers (cf. Bruschini et al. 1998). The Geneva International Centre for Humanitarian Demining, for example, has recently been exploring different sources of remotely sensed data and assessing their potential in mine action programs (GICHD 2002). While there is hope that remote sensing may assist in the location of individual devices, its use in mapping minefields and establishing development priorities is already within reach. Finally, as demining organizations adapt geographic technologies to local conditions, researchers and developers must maintain flexible applications of these technologies as each new humanitarian situation involves different local data sources, environmental conditions, and response priorities.

## GEOGRAPHIC SOLUTIONS FOR ADDRESSING AFFECTED COMMUNITIES

Besides the application of geographic technology, geographers are well suited to address the problems of post-conflict societies in which LM/UXO are one of many problems impeding reconstruction and development. While humanitarian demining operations seek to clear as much land of LM-UXO as possible, many communities nevertheless continue to live with contaminated landscapes for years or decades after conflict. Reductions in arable land not only limit local carrying capacity in many mine-affected communities but also challenge the ability of cultural norms or legal practices to resolve the resulting community tensions over land-use and access (cf. Wangari et al. 1996). Contaminated land also means that some communities may face significant obstacles to their mobility, limiting access to markets, health care and neighbors. There is a need, therefore, to evaluate, *ex post facto*, the outcomes of humanitarian demining operations as well as the coping strategies developed by affected communities so as to address their continuing development needs and to improve future operations elsewhere. Furthermore, communities displaced by conflict may be unable to return to areas because of LM/UXO hazards, thereby requiring durable resettlement and potentially disrupting local populations, cultural practices or even political dynamics. Still unclear is the significance of LM/UXO hazards in contributing to internal displacement and emigration. Geographical research on the needs of affected communities would contribute importantly to an improved understanding of and response to the particular problems posed by LM/UXO. In sum, the geographer's understanding of human-environment interactions and the powerful set of analytic tools geography offers present practitioners and scholars alike with a unique opportunity to solve some of the very real problems facing communities affected by LM/UXO.

# REFERENCES

Beardsley, T. 1997. War Without End? Land Mines Strain Diplomacy as Technology Advances. *Scientific American* 276(6): 20–21.

Bruschini, C., B. Gros, F. Guerne, P. Piece, and O. Carmona. 1998. Ground Penetrating Radar and Imaging Metal Detector for Antipersonnel Mine Detection. *Journal of Applied Geophysics* 40: 59–71.

Cutter, S. 1993. *Living With Risk: The Geography of Technological Hazards.* New York: Routledge.

Erikson, D. and D. Jean. 2001. The Information Management System for Mine Action. *Landmines* (special issue): 10.

Finger, M. 1991. The Military, the Nation State and the Environment. *The Ecologist* 21(5).

Gatrell, A. and M. Senior. 1999. Health and Health Care Applications. In P. Longley, M. Goodchild, D. Maguire, and D. Rhind, eds. *Geographical Information Systems*, 925–938. New York: Wiley.

Geneva International Centre for Humanitarian Demining. 2002. Study on Remote Sensing in Mine Action Programme. http://www.gichd.ch (last accessed 27 Sept. 2002).

International Trust Fund for Demining and Landmine Victims' Assistance. 2001. Minutes of the Workshop Development/Assembly of a GIS Based 1:100,000 Scale Planning Map of S.E. Europe for Humanitarian Purposes. 20–21 June, Ljubljana, Slovenia. http://eu-mine-actions.jrc.cec.eu.int (last accessed 27 Sept. 2002).

Joint Research Centre (JRC), European Commission. 2000. Proceedings of the Workshop Towards Harmonized Information Systems for Mine Action in South Eastern Europe. 7-8 March, Ispra, Italy. http://eu-mine-actions.jrc.cec.eu.int (last accessed 27 Sept. 2002).

McGrath, R. 2000. *Landmines and Unexploded Ordinance: A Resource Book.* Sterling, VA: Pluto Press.

Porter, P. and E. Sheppard. 1998. *A World of Difference: Society, Nature, Development.* New York: Guilford Press.

Strada, G. 1996. The Horror of Land Mines. *Scientific American* 274(5): 40–45.

Wangari, E., B. Thomas-Slayter, and D. Rocheleau. 1996. Gendered Visions for Survival. In D. Rocheleau, B. Thomas-Slayter, and E. Wangari, eds. *Feminist Political Ecology: Global Issues and Local Experiences*, 127–154. New York: Routledge.

# 8

# EVALUATING THE GEOGRAPHIC COMPACTNESS OF REPRESENTATIONAL DISTRICTS

GERALD R. WEBSTER
*University of Alabama*

The federal Constitution requires that a census of the population in the United States be completed at the beginning of each decade. The results of these enumerations are used to "reapportion" or reallocate the seats, now limited to a total of 435, in the House of Representatives among the states. As a result of the 2000 Census of Population, for example, New York and Pennsylvania each had their delegations reduced by two seats while Florida and Texas each increased their presence in Congress by two Representatives.

Subsequent to the reapportionment of the House of Representatives, all states having more than a single Representative undertake "redistricting" (Webster 1997). A central purpose of redistricting is to equalize the populations in all districts in the same state to insure that citizens are represented fairly. In addition to equalizing population, those drawing new districts may also emphasize other criteria, including racial equity, partisan fairness, geographic contiguity, geographic compactness, and the preservation of communities of interest, among many others (Morrill 1981).

## GEOGRAPHIC COMPACTNESS AND REDISTRICTING

Redistricting is an innately geographic as well as political process (Morrill 1987). The redistricting criterion likely receiving the greatest attention by political geographers is the long-standing notion that representational districts should be geographically compact (e.g., Forest 2001; Webster 2002). The geographic compactness criterion focuses on a district's shape and the degree to which its area is dispersed around its core. Analyses of this criterion frequently highlight the regularity or jaggedness of a district's boundary. Supporters of the compactness criterion argue that it provides a defense against efforts to draw politically or racially gerrymandered districting plans, though debate over this issue continues (Morrill 1981; Webster 2002).

Regardless of ongoing academic debates, many state governments require that congressional or legislative districts be reasonably compact. The U.S. Supreme Court has also made clear that it views geographic compactness as

*D. G. Janelle et al. (eds.), WorldMinds: Geographical Perspectives on 100 Problems*, 43–47.
© 2004 *Springer. Printed in the Netherlands.*

a valid traditional redistricting principle. But how should "geographic shape" be measured? How compact must a district's geographic shape be to meet the approval of the legal system? Can this criterion be applied in exactly the same manner on the plains of Nebraska, the bayous of Louisiana, and the mountains of West Virginia?

These questions, among many others, were debated during the avalanche of litigation waged in the country's court system during the 1990s over the geographic compactness of representational districts. A large proportion of these court proceedings emanated from states with significant minority populations, which attempted to provide them the same probability of electing candidates of their choice as enjoyed by the white majority. These efforts were arguably limited in 1993 in *Shaw v. Reno*, a Supreme Court decision pertaining to the constitutionality of North Carolina's 12th Congressional District. This decision led some 1990s districts to be redrawn with increased levels of geographic compactness, among other alterations (e.g., Leib 1998). Today, districts whose shapes are deemed irregular may be challenged with courts making judgments about their legality. Thus, district shape is now used as a "threshold" criterion for setting judicial reviews of districting plans in motion (Pildes and Niemi 1993).

## MEASURING GEOGRAPHIC COMPACTNESS

The evaluation of the shapes of representational districts is now aided by the development of powerful geographic information systems (GIS). To date, well over two- dozen different methods of measuring geographic compactness have been proposed by geographers, political scientists, and legal scholars (Niemi et al. 1990). These different measures of geographic compactness emphasize different geometric or geographic qualities, and arguably no single method can provide a comprehensive and fair method of evaluating the compactness of all districts in all environmental circumstances. As a result, it is the norm in districting controversies for two or more compactness measures to be used simultaneously.

The two most commonly applied indicators of geographic compactness are the "geographic dispersion" and "perimeter" compactness measures (Webster 2000). Conceptually, the geographic dispersion measure evaluates the spatial concentration of the area of a district. To calculate this compactness indicator, the smallest possible circle is drawn around the district (Figure 8.1). The resulting coefficient is the proportion of the area of the circle that is also in the district. Theoretically, the scores on this measure range from 1.0 (most compact) to 0.0 (least compact). A district with a shape approximating a circle might have a coefficient approximating 0.90, while a very elongated district might produce a far lesser coefficient of 0.10. But the coefficients from this measure are not sufficiently sensitive to manipulations

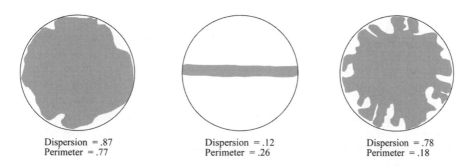

Dispersion = .87
Perimeter = .77

Dispersion = .12
Perimeter = .26

Dispersion = .78
Perimeter = .18

*Figure 8.1.* Three hypothetical districts.

that create districts with "jagged" boundaries that are areally concentrated. Thus, while this compactness measure is valuable to evaluate the geographic concentration of a district's area, it does not necessarily yield substantial information about an irregular boundary.

Due to this limitation, the geographic dispersion-compactness indicator is commonly used in conjunction with the perimeter-compactness measure. The reported coefficient for this indicator is the proportion of the area in the district relative to the area in a circle with the same perimeter (Figure 8.1). Like the geographic dispersion measure, the coefficients calculated for this indicator range from 1.0 (most compact) to 0.0 (least compact). The perimeter measure is sensitive to any "irregularity" that extends the boundary of a district without adding significant area. Thus, these two indicators should generally be used in tandem because they measure different elements of a district's compactness.

## INTERPRETING THE MEANING OF GEOGRAPHIC COMPACTNESS COEFFICIENTS

The results of geographic compactness analyses of districting plans must be evaluated cautiously. There are no absolute scales for evaluating compactness coefficients. Typically a proposed plan's mean level of compactness will be compared to the average level calculated for the past decade's map, or the "benchmark plan." If the results are essentially similar or improved, a plan is generally considered acceptable on this criterion. Most analyses will also consider the lowest level of compactness for any district in a proposed plan. If a proposed plan includes a district substantially decreasing the lowest score, inquiries about the circumstances of the district may be appropriate. It is possible that the mapmaker was attempting to provide a distinct community an opportunity for representation, such as one focused upon a meandering river. It is important to understand that coherent communities do not always concentrate themselves in geometric forms like squares or

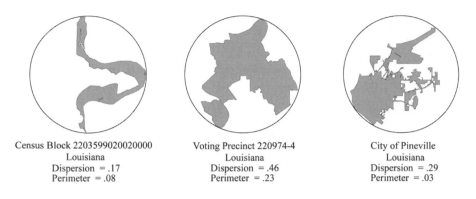

| Census Block 2203599020020000 | Voting Precinct 220974-4 | City of Pineville |
| Louisiana | Louisiana | Louisiana |
| Dispersion = .17 | Dispersion = .46 | Dispersion = .29 |
| Perimeter = .08 | Perimeter = .23 | Perimeter = .03 |

*Figure 8.2.* Real world district building blocks.

circles. Such communities should not arbitrarily be prohibited from securing effective representation.

Compactness analyses must also be interpreted in light of the geographic setting being districted. For example, representational districts are composed of other geographic units, such as census blocks, voting precincts, or counties. At times these geographic "building-blocks" have irregular shapes, making it difficult if not impossible to create highly compact districts with smooth boundaries (Figure 8.2). A visually irregular district should not be automatically judged harshly if its comparatively low level of compactness is in part the result of its inclusion of an irregularly shaped voting precinct, for example.

Geographic compactness evaluations must also be undertaken in light of other redistricting criteria. For example, the equal population criterion emanates from the Equal Protection Clause of the Fourteenth Amendment to the Constitution. It is therefore of greater legal importance to create districts that are equal in population than are highly compact. Efforts to create highly compact representational districts may also contradict the goals of other districting criteria, such as limiting the subdivision of local governments. Municipalities commonly grow in geographic size through annexation processes, which oftentimes do not consider the influence of the additional territory on the city's level of geographic compactness. Thus, if a district avoids dividing a city and follows its municipal boundary, the district's resulting shape may be jagged or irregular.

Finally, it is critical to remember that districts should not be viewed as random territorial containers of voters. The mapmaker may well have a number of goals pertaining to the representation of geographic communities, political subdivisions, supporters of different political parties, and racial or ethnic groups. A good redistricting plan is one that distributes political power as equitably as possible among all groups. It is surely better to include a moderately irregular district in a plan that provides access for a historically under-represented group than to create only highly compact districts that prevent the group from securing an equitable measure of representation. Thus, districting should be about facilitating and not thwarting the democratic process.

## REFERENCES

Forest, B. 2001. Mapping Democracy: Racial Identity and the Quandary of Political Representation. *Annals of the Association of American Geographers* 91: 143–166.

Leib, J. I. 1998. Communities of Interest and Minority Districting After Miller v. Johnson. *Political Geography* 17: 683–699.

Morrill, R. L. 1981. *Political Redistricting and Geographic Theory.* Washington, D.C.: Association of American Geographers.

Morrill, R. L. 1987. Redistricting, Region and Representation. *Political Geography Quarterly* 6: 241–260.

Niemi, R. G., B. Grofman, C. Carlucci, and H. Thomas. 1990. Measuring Compactness and the Role of a Compactness Standard in a Test for Partisan and Racial Gerrymandering. *Journal of Politics* 52: 1155–1181.

Pildes, R. H. and R. G. Niemi. 1993. Expressive Harms, 'Bizarre Districts,' and Voting Rights: Evaluating Election-District Appearances After Shaw v. Reno. *Michigan Law Review* 92: 483–587.

Webster, G. R. 1997. Geography and the Decennial Task of Redistricting. *Journal of Geography* 96: 61–68.

Webster, G. R. 2000. Playing a Game With Changing Rules: Geography, Politics and Redistricting in the 1990s. *Political Geography* 19: 141–161.

Webster, G. R. 2002. Rethinking the Role of Geographic Compactness in Redistricting. In A. Willingham, ed. *Beyond the Color Line: Race Representation and Community in the New Century*, 117–134. New York: Brennan Center for Justice.

# 9

# ENVIRONMENTAL CONFLICT, COLLABORATIVE SOLUTIONS, AND THE POLITICS OF GEOGRAPHIC SCALE

RANDALL K. WILSON
*Gettysburg College*

Since the 1990s, community-based collaborative approaches to environmental planning and conflict resolution have gained significant influence in the United States. As federal and state agencies engage more directly with local communities to manage environmental resources, collaboration has come to be seen as a means of increasing public participation, improving management decisions, and overcoming the divisive politics of environmental conflicts (Brick et al. 2001). The potential for collaborative efforts to successfully resolve conflict, however, hinge upon many variables. In this essay, I examine one of these, namely, the politics of geographic scale. Drawing upon a case study of a long-running conflict over historical land grant rights in southern Colorado, I demonstrate how the politics of geographic scale can serve to both hinder and facilitate efforts to resolve environmental disputes through collaborative means.

## WITHER A COLLABORATIVE SOLUTION?

In 1993, then-governor of Colorado Roy Romer created the Sangre de Cristo Land Grant Commission as part of a local-state collaborative effort to resolve a dispute over a mountainous parcel of land known as Taylor Ranch, or La Sierra to local residents (Wilson 1999). Consisting of approximately 77,500 acres of forest in the Sangre de Cristo range of the Rocky Mountains, including 14,000-foot Culebra Peak (Figure 9.1), the land is claimed by local Hispano residents as a commons to which they hold traditional *ejido* or usufruct rights. These rights include access for hunting, wood gathering, recreation, livestock grazing, and water use. Granted as an incentive to recruit the original Hispano settlers into the area, these rights were exercised uninterrupted for over five generations. Then, in 1960, Jack Taylor, Jr. purchased the land and successfully cleared the title of the "little cloud" guaranteeing usufruct rights to local residents in 1967. Until his death in 1988, Taylor's relationship with

49

*D. G. Janelle et al. (eds.), WorldMinds: Geographical Perspectives on 100 Problems, 49–54.*
© 2004 *Springer. Printed in the Netherlands.*

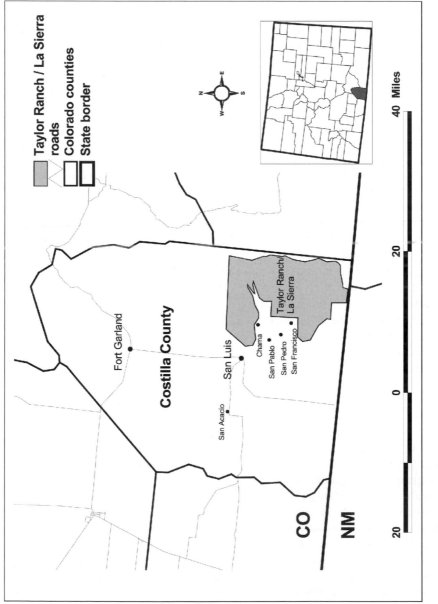

*Figure 9.1.* The location of the Taylor Ranch/La Sierra in the San Luis Valley, Colorado. Map adapted from Wilson 1999.

many local Hispano residents was strained, at times erupting into episodes of violence.

The situation changed when Taylor's son assumed control of the land and offered to sell, creating an opportunity to finally resolve the conflict. In response, the governor's Commission was formed to develop a collaborative plan to purchase the Taylor Ranch, create a state park, and restore the traditional land rights. Consisting of local residents and state officials, the Commission completed its proposal in 1993.

Four years later, however, the Commission cut off negotiations with Taylor. His refusal to accept the state's $18 million offer (he demanded $32 million), combined with his efforts to pursue logging contracts, effectively ended the collaborative effort. Though Taylor's actions were the most immediate cause of the impasse, other dynamics were also at work further complicating the implementation of the collaborative plan. Among these were the politics of scale inherent in the Commission's collaborative proposal.

## COLLABORATION AND THE POLITICS OF GEOGRAPHIC SCALE

For the purposes of this essay, the politics of scale refers to efforts to set the spatial parameters of an environmental issue or conflict. "Jumping scales" (Smith 1993; Swyngedouw 2000) refers to successful attempts to re-define the geographic extent of a conflict, say from a local concern to one of regional or national interest or vice-versa. For groups lacking political advantage at a particular geographic scale, this move provides a strategy to tip the balance of power in a conflict by mobilizing support from new actors and resources operating at other scales. Jumping scales, however, is rarely so straightforward, often rendering multiple and unforeseen effects. For instance, as the constituency for an issue expands and diversifies it can become more difficult to promote narrowly defined local interests. In short, changing scales may necessitate re-defining the issues at stake according to new political and conceptual frameworks that challenge the original rationale for mobilization (see Harvey 1996).

Both the political and conceptual implications of geographic scale are well illustrated in the Commission's effort to resolve the Taylor Ranch dispute. For many local Hispano residents, collaboration with the state was initially viewed as a positive step. The enthusiastic support of the Governor meant the mobilization of the state's vast monetary, political, and technical resources on their behalf. However, state involvement also meant that the restoration of local land rights must be re-cast as an issue of statewide interest. In order to justify the use of state funds to purchase the property, the Commission's proposal had to re-define the Taylor Ranch as land suitable for state-ownership, namely, as *public* land. This had significant implications for the ensuing regional debate over the proposal.

The state's authority over public lands and resources is premised upon a particular set of conceptual assumptions dating from the Progressive Era (Hays 1959). In contrast to private interests guided by the profit motive, state agencies are presented as disinterested actors, guided by objective scientific principles to efficiently manage resources for the greater good. In short, it is the state's presumed expertise in maximizing the utility of natural resources via rational management that justifies the retention of lands in the public domain.

In making the case for community-state stewardship of the Taylor Ranch, the Commission followed a similar logic of legitimation. As evidenced in the proposal, the land was repeatedly characterized as holding a "tremendous wealth of natural resources" (SCLGC 1993). The land's value was defined in terms of the hunting, fishing, and hiking opportunities it provided, qualities emphasizing the property's enormous potential for tourism development within the state park system. In similar fashion, local Hispano residents were also cast in utilitarian terms. The local communities were presented as potential cultural tourism attractions that might complement the recreational opportunities on Taylor Ranch.

Re-defining the conflict in this way had several implications. First, by representing the land and residents as distinct and objectified resources, the proposal tended to downplay the very socio-environmental relations it hoped to restore (i.e., traditional commons-based forms of land tenure). Second, by attempting to justify state intervention according to some objective measure of resource utility, the proposal allowed critics to undermine the plan by simply pointing to the presence of political bias. For example, during statewide debates on the issue, a *Denver Post* columnist attacked the Commission's plan as a "lousy deal" because he felt it ranked poorly as a recreational resource (Meyers 1997: 3C). However, his major objection was that the Commission appeared to be allowing *politics* to determine resource management decisions.

In sum, the effort to shift the geographic scale of the conflict from a local matter to a state concern, as necessitated by the collaborative proposal, required a concomitant reworking of the conceptual justification put forth. Rather than arguing that the restoration of local rights was the politically just course of action, the proposal relied upon a discourse of rational public land management. This muted the historical and cultural politics of the original struggle and greatly complicated efforts to cultivate statewide public support.

## TAKING ANOTHER LEAP

Community-based collaboration was not the only strategy pursued by local residents in their struggle to restore traditional land rights. Other paths included a legal challenge to Taylor's clearing of the title in 1967. In 1998, on the heels of the Commission's decision to table its proposal, members of the local Land Rights Council secured a new hearing for ten residents, citing a lapse of due process in the 1967 trial.

The residents argued that the usufruct rights documented in the title since the 1860s were legally binding. However, in both the district court trial and later, in the court of appeals, they lost their case. Then, in June 2002, the Colorado Supreme Court weighed in on the issue. This time, the ruling favored local claims, reversing the lower court decisions on three of six counts (e.g., allowing local access for grazing, firewood, and timber collection, but not for hunting, fishing, or recreation) (Supreme Court 2002).

This journey through the court system reflects another upward "jump" in geographic scale from the local to the state level. A key difference with the Commission's proposal, however, is that as a judicial body, the Colorado Supreme Court was less beholden to the current political opinions of a statewide constituency. Although bound to abide by state and federal laws, the Court could otherwise acknowledge the overtly political nature of the conflict in ways that state-based resource management institutions could not.

Meanwhile, as the court battles ensued, yet another re-scaling of the conflict was taking place, this time transforming it into a national and global scale concern. Between 1997 and 2000, Taylor sold the Ranch for approximately $23 million to Lou Pai, a former executive with the Enron Corporation. In February 2002, former Enron stockholders and the Securities and Exchange Commission began making claims on Pai's assets, including the Taylor Ranch (Frazier 2002). Allegedly, money used to purchase the Ranch included the $353.7 million he obtained via Enron's "creative" financial accounting practices. The Colorado Supreme Court ruling impacts these events insofar as it significantly lowers the appraised value of the property. Combined with the powerful new incentives to liquidate Pai's assets, these events have opened a new opportunity to revisit and implement the Commission's collaborative proposal.

## CONCLUSION

Local environmental conflicts are always constitutive of social and environmental relations and forces operating at multiple geographic scales. As this study demonstrates, this observation is especially true in community-based collaborations that intentionally reach across local and state institutional boundaries. Insofar as the full effects of jumping scales are often unanticipated, my intent has not been to promote a geographic "quick fix" to environmental conflict. Rather, by making explicit the politics of geographic scale, the case highlights how geographic concepts can lead to a better understanding of how collaborative efforts might succeed in crafting equitable and lasting solutions to complex socio-environmental problems.

## REFERENCES

Brick, P., D. Snow, and W. Van De Wetering, eds. 2001. *Across the Great Divide: Explorations in Collaborative Conservation in the American West.* Washington D.C.: Island Press.

Frazier, D. 2002. Lasso Thrown Around Taylor Ranch. *Rocky Mountain News* (16 February).

Hays, S. 1959. *Conservation and the Gospel of Efficiency.* Cambridge: Harvard University Press.

Harvey, D. 1996. *Justice, Nature and the Geography of Difference.* Cambridge: Blackwell.

Meyers, C. 1997. Taylor Ranch: Lousy Value. *Denver Post* (16 February): 3C.

Sangre de Cristo Land Grant Commission. 1993. *The Sangre de Cristo Land Grant Commission Report.* Denver: Colorado Department of Natural Resources.

Smith, N. 1993. Homeless/global: Scaling Places. In B. Curtis, T. Putnam, G. Robertson, and J. Bird, eds. *Mapping the Future: Local Cultures, Global Change*, 87–119. London: Routledge.

Supreme Court, State of Colorado. 2002. *Eugene Lobato vs. Zachary Taylor* (case number 00SC527). Certirorari to the Colorado Court of Appeals.

Swyngedouw, E. 2000. Authoritarian Governance, Power, and the Politics of Rescaling. *Environment and Planning D: Society and Space* 18: 63–76.

Wilson, R. 1999. Placing Nature: The Politics of Collaboration and Representation in the Struggle for La Sierra in San Luis, Colorado. *Ecumene* 6: 1–28.

# 10

# PUTTING MEMORY IN ITS PLACE: THE POLITICS OF COMMEMORATION IN THE AMERICAN SOUTH

DEREK ALDERMAN
*East Carolina University*

OWEN J. DWYER
*Indiana University, Indianapolis*

The past remains a passionately contested terrain in the American South. On the one hand, the memory of the Civil War is of vital importance in the region. Many white Southerners identify with romanticized images of the Confederacy (Hoelscher 2003). Alternately, a new historical vision of the region's past has emerged, one that challenges the centrality of the Confederacy. Propelled largely by African Americans, this challenge is embodied in the public commemoration of the Civil Rights Movement (Alderman 2000; Dwyer 2000). The intersection of these two competing memorial narratives has made collective memory a highly charged issue in the South, one whose emotional gravity comes from the interweaving of place and history (Figure 10.1) (Lowenthal 1975; Hayden 1995; Leib 2002). There is a need to make sense of the problematic nature of southern commemoration, particularly since these debates affect the prospects of building an inclusive culture in the region (Brundage 2000).

Geographers have made significant contributions to the study of collective memory by recognizing how public commemoration is shaped by the politically contestable and contradictory nature of space and place. Where a memorial is located is not incidental but actively shapes how social actors conceptualize and carry out memorialization (Johnson 1995). Locations may confirm, erode, contradict, or render mute the intended meanings of the memorial's producers. Our work illustrates the utility of a spatial perspective when analyzing the antagonism associated with Southern commemorations in the wake of the Civil Rights Movement. Our case studies focus on two figures who are representative of the deeply divided condition of the South's collective memory: Civil Rights leader Martin Luther King, Jr. and Nathan Bedford Forrest, a Confederate general and early leader of the Ku Klux Klan (KKK). In the case of Dr. King, we visit Danville, Virginia, and examine

*D. G. Janelle et al. (eds.), WorldMinds: Geographical Perspectives on 100 Problems*, 55–60.
© 2004 *Springer. Printed in the Netherlands.*

*Figure 10.1.* The intersection of Dr. Martin Luther King, Jr. Street and Jefferson Davis Avenue in Selma, Alabama.

the debate over renaming a street in his honor. In the case of Forrest, we visit Selma, Alabama, where controversy surrounded the installation of a Forrest monument.

## FINDING A PLACE FOR MARTIN LUTHER KING, JR. IN DANVILLE

Streets named after Martin Luther King, Jr. are increasingly common features in communities across the South as well as the nation (Figure 10.1). As of 2003, 600 cities in the United States had named a street for King. Southern states accounted for well over three-quarters of these streets. The prevalence of such streets should not divert attention away from the controversies that often accompany the naming process. In the case of Danville, African-American activists struggled – unsuccessfully – to convince local officials to rename a street. Although street naming is not confined to places with a direct connection to King's life, he delivered a speech in Danville in 1963 in support of local Civil Rights demonstrations. Despite King's presence, these protests eventually met with violence and failed because of the intransigency of city leaders.

Like street naming struggles in many cities and towns, the Danville case was not simply about determining the appropriateness of commemorating King; it was also a struggle over where his memory should be inscribed into the cultural landscape. In March 2002, African-American activist Torrey Dixon petitioned Danville's city council to rename Central Boulevard, a major commercial thoroughfare, in honor of King. He considered the thoroughfare an "appropriate street" to rename because its central location and high volume of traffic would ensure that King's name would be seen by many people. In addition, the busy downtown street crosses the Aiken Bridge, which, as Dixon suggested, memorialized a "segregationist judge." It was Judge Archibald M. Aiken who issued the injunction that led to the arrest of hundreds of Civil Rights demonstrators in the summer of 1963. For some African Americans, the commemoration of King in such close proximity to Aiken represented an opportunity to rewrite public memory, especially the memory of the Danville police attacking Civil Rights demonstrators.

The proposal to rename Central Boulevard drew considerable public resistance. A local newspaper opinion poll found 79 percent of 2,036 respondents opposed to the plan. As is the case in many communities, business owners cited – legitimately or not – the adverse effects and costs of an address change. As an alternative, government officials suggested placing King's name on High Street, a smaller road populated largely by African Americans. They pointed out that High Street Baptist Church served as a focal point for local Civil Rights activists when King visited Danville. Dixon rebuked the counter-proposal, saying: "I think Dr. King should have a major road . . . named after him. . . . I think having a road in a low-class neighborhood named after King is offensive" (quoted in Davis 2002: 3A).

Dixon interpreted the naming of a less prominent street as a degradation of memory, even when the street had a strong historical association with King. Eventually, the city of Danville named a feature after him, but it was a bridge rather than a street. City officials and some Civil Rights leaders tout the bridge as a symbol of King's legacy of reaching out to bring the races together. Dixon, however, remains convinced that a major road, because of its visibility, is the most appropriate way of educating the public about the importance of King for all Americans, past, present, and future.

## FINDING A PLACE FOR NATHAN BEDFORD FORREST IN SELMA

The themes of memory, place, and politics that formed the *leit motif* in Danville resonate with the controversy surrounding the Forrest monument (Figure 10.2). The foundations of the controversy lay with the changing context of com-

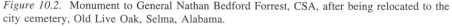

*Figure 10.2.* Monument to General Nathan Bedford Forrest, CSA, after being relocated to the city cemetery, Old Live Oak, Selma, Alabama.

memoration in Selma. The arrival of the National Voting Rights Museum (NVRM) in 1992 challenged the city's traditional memorial focus on the Civil War and whiteness. The museum commemorates the local activists whose struggle culminated in 1965 with the Voting Rights Act. In Selma, Civil Rights and Civil War commemoration demonstrate an ironic symmetry, with advances in one prompting advances in the other. The Forrest monument is a case in point.

Unveiled in October 2000, the bronze bust of Forrest sits atop a five-ton granite base inscribed with the much-debated Confederate battle flag and milestones from his military career. Erected by a coalition of local and regional neo-Confederate groups, the statue was placed at a city-owned museum in the midst of an African American neighborhood. While supporters stressed the historical connection between Forrest and the museum – it had served as a military hospital during Forrest's failed defense of the city in 1864 – area residents objected to the monument's presence on public property in their neighborhood. Additionally, monument opponents led by activists associated with the NVRM criticized the omission of any reference to Forrest's role as a slave trader, his involvement in the massacre of black Union troops, and his leadership of the KKK.

Significantly, the monument was erected a week after Selma's first black mayor, James Perkins, Jr., bested the city's previous mayor of 36 years, Joe Smitherman. Like his political mentor George Wallace, Smitherman was a political survivor. Coming into office in 1965, Smitherman opposed the Movement. In the wake of its victory, however, he adjusted with the times and managed to stay in office in a black majority city. The timing of Smitherman's defeat and the Forrest commemoration attracted national attention. The media presented the twinned events as evidence that the more things changed in Selma, the more they remained the same. A local resident echoed the media's portrayal of the situation: "You lose control of your city government and a week later you put up a statue to a Confederate general? How Southern. These heritage guys are basically saying what a lot of people around here feel: The fight goes on. The war never really ends" (quoted in *Gettleman* 2000:A-1).

Against a backdrop of protests and counter protests, four months of contentious meetings of the black-majority city council ensued. A committee recommended moving the monument inside the museum and appending a fuller account of Forrest's career. The recommendation satisfied neither party. Opponents of the monument, while eager to discuss Forrest's role in the KKK, demanded the outright removal of the monument to the city's Confederate memorial at the cemetery. Supporters claimed that it was intolerant to insist that the monument be moved and resisted discussing the KKK. In the wake of the polarizing debate, the city council narrowly approved a compromise that removed the statue to the cemetery but left its narrative content untouched. Significantly, the compromise had the backing of Selma's almost exclusively white Chamber of Commerce whose members were eager to get Selma out of the media spotlight. Predictably, the matter is now the subject of several lawsuits and awaits resolution.

## CONCLUSION

The American South's ongoing commemorative changes and challenges cannot be fully analyzed without the assistance of geographers. Memorials simultaneously draw meaning from and give meaning to their surroundings. Where a memorial is located and how it got there says a great deal about "where" the region is politically. Despite the social upheaval brought on by the Civil

Rights Movement, critics have questioned the extent to which our collective identity has become more inclusive (Baldwin 1998; Jones 2000). Landscape evidence from Danville and Selma is mixed with regards to this question. While there are a growing number of African American memorials, they are commonly located amid the remains of segregated business districts and ghettos. In contrast, Confederate memorials predominate along Main Street and cemetery row. The location of these memorials underscores the ambiguity of the changes wrought by the Movement. Despite the inclusive rhetoric that now characterizes the New South – taking its lead from Atlanta as the "city too busy to hate" – public memory remains black and white, each in its separate place.

## REFERENCES

Alderman, D. 2000. A Street Fit for a King: Naming Places and Commemoration in the American South. *The Professional Geographer* 52: 672–684.

Baldwin, J. 1998. *Collected Essays.* New York: Literary Classics of the United States.

Brundage, W. 2000. Introduction: No Deed but Memory. In W. Brundage, ed. *Where These Memories Grow: History, Memory, and Southern Identity*, 1–28. Chapel Hill, NC: University of North Carolina Press.

Davis, T. 2002. Council Rejects MLK Road Proposal. *Danville Register & Bee* March 6: 1A & 3A.

Dwyer, O. 2000. Interpreting the Civil Rights Movement: Place, Memory, and Conflict. *The Professional Geographer* 52: 660–671.

Gettleman, J. 2000. To Mayor, it's Selma's Statue of Limitation: City's Segregationist Ways Quickly Test First Black to Win Post. *Los Angeles Times* 22 October: A-1.

Hayden, D. 1995. *The Power of Place: Urban Landscapes as Public History.* Boston: MIT Press.

Hoelscher, S. 2003. Making Place, Making Race: Performances of Whiteness in the Jim Crow South. *Annals of the Association of American Geographers.*

Johnson, N. 1995. Cast in Stone: Monuments, Geography, and Nationalism. *Environment and Planning D: Society and Space* 13: 51–65.

Jones, III, J. P. 2000. The Street Politics of Jackie Smith. In G. Bridge and S. Watson, eds. *The Blackwell Companion to the City*, 448–459. Oxford: Blackwell.

Leib, J. 2002. Separate Times, Shared Spaces: Arthur Ashe, Monument Avenue, and the Politics of Richmond, Virginia's Symbolic Landscape. *Cultural Geographies* 9(3): 286–313.

Lowenthal, D. 1975. Past Time, Past Place: Landscape and Memory. *Geographical Review* 65(1): 1–36.

# 11

# BANNER HEADLINES: THE FIGHT OVER CONFEDERATE FLAGS IN THE AMERICAN SOUTH

JONATHAN LEIB
*Florida State University*

GERALD WEBSTER
*University of Alabama*

As the 21st century begins, the Confederate battle emblem, a symbol more commonly associated with the 19th century, continues to divide the population of the American South, primarily along racial lines. During the past decade, strident debates over the public display of the Confederate battle emblem by state and local governments have been waged across the region. While some have dismissed these debates as inconsequential to the daily lives of the South's population, their frequency and vitriolic character suggests otherwise. Geographers have been actively analyzing the underlying causes and ferocity of these debates using a variety of geographic concepts (e.g., Webster and Webster 1994; Leib 1995; Leib, Webster, and Webster 2000; Webster and Leib 2001; 2002; Leib and Webster 2002). In this essay, we demonstrate how the geographical concepts of public space and landscape representation can be used to better understand the dynamics of these controversies.

## BACKGROUND

The two most important events in the history of the American South were the Civil War and the Civil Rights Movement. During the past decade, the South's black and traditional white populations have been involved in a series of often harsh debates over how these events should be remembered on the region's landscape (e.g., Alderman 2000; Dwyer 2000; Leib 2002). The most contentious and visible battles have occurred over the proper display of the Confederate battle flag in the region's public spaces. While local governments, including school districts, municipalities, and counties have not been immune to these controversies, the most visible battles have been waged over the actions of state governments. Two of the states that sanctioned the most visible displays of the Confederate battle emblem also experienced the most heated debates,

*D. G. Janelle et al. (eds.), WorldMinds: Geographical Perspectives on 100 Problems*, 61–66.
© 2004 *Springer. Printed in the Netherlands.*

Alabama and South Carolina, both of which flew the flag from the tops of their state capitol buildings.


## PUBLIC SPACE, LANDSCAPE REPRESENTATION, AND THE FLAG

Geographic concepts of public space and landscape representation aid understanding of the ferocity of debates over the Confederate battle emblem. These disputes have not only focused upon *if* the battle emblem should be displayed, but also *where* it should be displayed on the Southern landscape and in the region's public spaces. Landscapes and public spaces are themselves imbedded with meanings. As Schein (1997: 660) argued, the cultural landscape is "a tangible visible entity, one that is both reflective and constitutive of society, culture, and territory." Mitchell (1996: 27) reminds us, however, that the creation and maintenance of such landscapes involves "a relation of power, an ideological rendering of spatial relations," with groups wielding power to control how people see and experience the landscape. As Mitchell (1995: 115) suggests, public spaces are

> . . . *spaces for representation.* That is, public space is a place within which a political movement can stake out the space that allows it to be seen. In public spaces, political organizations can represent themselves to a larger population.

Thus, debates over when, how, by whom, and most importantly *where* the flag should be placed on the region's landscape and in its public spaces are intrinsically political conflicts demonstrating the measure of power wielded by opposing segments of Southern society.

The controversies over if and where the Confederate battle flag should fly on the statehouse grounds in South Carolina and Alabama provide excellent examples of the importance of symbolic landscapes and public spaces. In both states, the flag flew above the state capitol dome beginning in the early 1960s. Opponents wanted the flag removed because they fervently believed that by flying it over the seat of government the states were officially sanctioning the racism that they ascribed to the flag's symbolism. Flag defenders, however, argued that moving the Confederate flag from the top of the dome was an affront to their Confederate heritage (Leib and Webster 2002).

*Alabama*

In Alabama, Governor George Wallace raised the battle flag over the state capitol dome in 1963 in a show of defiance to federal attempts to desegregate the University of Alabama. The flag (along with the U.S. and Alabama flags) were removed from the dome in 1987 during a major renovation of the capitol building. After the flags were removed, black lawmakers filed several lawsuits to prohibit the battle flag's restoration above the capitol dome upon completion of the renovations. They succeeded in a highly controversial 1992 state court decision.

The court decision led to a series of demonstrations, legislative efforts to restore the flag, and a legal attempt by the then-Republican Governor and staunch Confederate flag defender, Guy Hunt, to restore the flag to the top of the dome. Before the legal process was completed, Hunt was removed from office due to ethics violations, and his Democratic successor, James Folsom, Jr., signed an Executive Order preventing the flag's return. Folsom's Executive Order led to continuing criticism by flag supporters and a Ku Klux Klan rally on the grounds of the capitol complex to demand the flag's restoration. Folsom tried to placate flag supporters by flying the battle flag (along with three other Confederate flags) at a Confederate monument less than ten feet from the side of the capitol building. Despite his efforts at compromise, Folsom's refusal to restore the flag to the top of the dome was one factor leading to his 1994 defeat in his bid to serve another term.

Today the flag continues to fly at a Confederate monument on the grounds of the Alabama state capitol. But efforts by flag supporters continue to return the flag to the top of the capitol. Numerous bills have been filed in the legislature to either legislatively restore the flag or provide for a public referendum on the issue. In 2000, a Southern independence rally at the capitol complex drew a crowd of 2,500, with many waving the battle flag and calling for the flag to be restored to the top of the dome (Webster and Leib 2002).

*South Carolina*

In South Carolina, the battle flag was raised over the state capitol in 1961 during the Civil War Centenary and at a time when much of the state's white political establishment opposed integration. After heated debates about moving the flag from the top of the capitol throughout most of the 1990s, a compromise to do so was achieved in 2000. It satisfied neither pro- nor anti-flag supporters (Webster and Leib 2001).

In a highly charged 1997 legislative debate over the flag's location, opponents proposed removing the flag from the top of the capitol and flying it instead at the Confederate soldiers' monument in front of the capitol. The geographies of the capitol's public spaces were important in this proposal. Placing the flag at the Confederate monument would set it in a historic context, but the monument is at ground level at a main road intersection in the heart of downtown Columbia. It would therefore be more readily visible than if it remained atop the capitol building. At the same time, flag supporters argued vigorously for the flag to remain in its more symbolically important, yet more visually obscure, location on top of the capitol (Figure 11.1).

In 2000, Governor Jim Hodges helped broker a compromise that moved the flag off the capitol dome and to a Confederate monument on the capitol's grounds. But controversy ensued over *which* of two locations for the

*Figure 11.1.* Editorial cartoonist Robert Ariail's view of the dispute over the Confederate battle flag on South Carolina's State Capitol Grounds. Reprinted with kind permission from Robert Ariail. In *Ariail!!!: More Cartoons by Robert Ariail*, Columbia, SC: *The State* 2001: 159.

Confederate monuments was better. One proposal was to move the flag next to a monument honoring 19th century Civil War General and political leader Wade Hampton. Flag supporters rejected the idea because the monument to Hampton sits *behind* the capitol, and therefore the flag would be less visible from the main streets surrounding the capitol. Flag opponents also rejected the Hampton compromise because by 2000 they had concluded the flag should not fly anywhere on the capitol's grounds.

In Spring 2000, at the end of a legislative session filled with acrimonious debate on the issue, a compromise was finally agreed upon to remove the flag from atop the capitol. The compromise moved the flag to a position *in front* of the state capitol but behind the Confederate soldiers' monument (Figure 11.2). The negotiations over the legislation establishing the flag's placement on the capitol grounds included such tedious, yet highly charged, details as how tall the flag pole should be and whether, as flag supporters insisted, lights should be installed to illuminate the flag at night (Webster and Leib 2001).

As a result of its new position behind the Confederate monument, the flag is difficult to see on windless days if one is looking directly at the front of the capitol. On windy days the flag unfurls, and is far more prominent. As a result, a heated exchange occurred in December 2001 when flag supporters ordered a new lighter weight nylon flag to replace the heavier cotton one flying

*Figure 11.2.* The Confederate battle flag and new flag pole in front of the State Capitol Building and behind the Confederate Soldier's Monument, Columbia, South Carolina. Photograph by Jonathan Leib.

at the monument so that the flag would unfurl and flutter more frequently on days with lighter winds. Flag opponents, not wanting the flag to unfurl frequently, demanded the cotton flag remain. The seven-week standoff ended when an agreement was reached changing the flag's material to a cotton-silk blend (Bauerlein and Sheinin 2002).

## CONCLUSIONS

The Confederate battle flag debate in the American South will almost certainly continue for the foreseeable future, and may impact the 2004 Presidential election (e.g., Balz 2003). The issue is not exclusively over the contested meaning and symbolism of the flag, but also over *where* it should be displayed within the region's public spaces and symbolic landscapes. Questions of whether, where, and how the flag should be displayed in public spaces shared jointly by all citizens intensify the political conflict over the battle flag's meaning. Flag supporters lobby to use the public spaces of state capitols as a forum to champion their reading of the flag as a symbol of heritage. Conversely, flag opponents argue that flying the flag on the grounds of state capitols provides implicit government sanction for the region's earlier acceptance of slavery and segregation. As a result of these polarized views, there is little or no common ground for compromise over the appropriateness of displaying the Confederate battle flag on the region's public spaces.

## REFERENCES

Alderman, D. 2000. A Street Fit for a King: Naming Places and Commemoration in the American South. *The Professional Geographer* 52: 672–684.

Ariail, R. 2001. *Ariail!!!: More Cartoons by Robert Ariail.* Columbia, SC: The State.

Balz, D. 2003. Democrats Caught in Confederate Issue. *Washington Post* 9 February: A5.

Bauerlein, V. and A. Sheinin. 2002. Cotton-Silk Mix May End Flag Flap. *The State (Columbia, SC)* January 24: A1.

Dwyer, O. 2000. Interpreting the Civil Rights Movement: Place, Memory and Conflict. *The Professional Geographer* 52: 660–671.

Leib, J. 1995. Heritage Versus Hate: A Geographical Analysis of Georgia's Confederate Battle Flag Debate. *Southeastern Geographer* 35: 37–57.

Leib, J. 2002. Separate Times, Shared Spaces: Arthur Ashe, Monument Avenue, and the Politics of Richmond, Virginia's Symbolic Landscape. *Cultural Geographies* 9: 286–312.

Leib, J. and G. Webster. 2002. The Confederate Flag Debate in the American South: Theoretical and Conceptual Perspectives. In A. Willingham, ed. *Beyond the Color Line?: Race, Representation and Community in the New Century*, 221–242. New York: Brennan Center for Justice at NYU Law School.

Leib, J., G. Webster, and R. Webster. 2000. Rebel With a Cause? Iconography and Public Memory in the Southern United States. *Geojournal* 52: 303–310.

Mitchell, D. 1995. The End of Public Space? People's Park, Definitions of the Public, and Democracy. *Annals of the Association of American Geographers* 85: 108–133.

Mitchell, D. 1996. *The Lie of the Land: Migrant Workers and the California Landscape.* Minneapolis: University of Minnesota Press.

Schein, R. 1997. The Place of Landscape: A Conceptual Framework for Interpreting an American Scene. *Annals of the Association of American Geographers* 87: 660–680.

Webster, G. and J. Leib. 2001. Whose South is it Anyway? Race and the Confederate Battle Flag in South Carolina. *Political Geography* 20: 271–299.

Webster G. and J. Leib. 2002. Political Culture, Religion and the Confederate Battle Flag Debate in Alabama. *Journal of Cultural Geography* 20: 1–26.

Webster, G. and R. Webster. 1994. The Power of an Icon. *Geographical Review* 84: 131–143.

## PART II

# ADDRESSING HUMAN NEEDS:
# HEALTH AND EDUCATION

Improving the quality of life for the Earth's inhabitants must certainly be a strong motivation in the agenda of any discipline that warrants public support and recognition. Geography has responded well to this challenge. The commitments of geographical talent, tools, and thinking are illustrated in its concern for issues of human health, accessibility to basic services and education, and for the mitigation of poverty and other disparities. It is critical that science and humanism coexist as one in laying the base for a world that attends to human needs.

*Don Janelle*

# 12

# HALTING THE AIDS PANDEMIC

HAROLD D. FOSTER
*University of Victoria*

As William A. Hasteltine (1993) pointed out, "The future of AIDS is the future of humanity." Hasteltine, the then-chief retro-virologist at Harvard's Dana-Farber Cancer Institute, added that "Unless the epidemic of AIDS is controlled, there is no predictable future for our species." Later, testifying before a U.S. Senate hearing, he suggested that by 2000 there might be 50 million HIV-infected people and that by 2015 the total number dead or dying could reach one billion, one-sixth of the current global population. Hasteltine (1992) may have been a little optimistic. In *Vital Signs 2001*, Worldwatch (2001) estimated that by the end of the year 2000, 57.9 million people had been infected with HIV, 21.8 million of whom had died. The July 2002 infection estimate is 40 million (Cohen 2002), a figure that, as Hasteltine (1992) predicted, is rising exponentially. If there are no significant breakthroughs by 2015, the AIDS pandemic will have become by far the greatest catastrophe in human history, worse than the Black Death and the Second World War combined, the equivalent of eight First World Wars (Foster 2002a).

## THE GEOGRAPHY OF AIDS

Genetic evidence (Hahn et al. 2000) shows that HIV-1 originated as the chimpanzee (*Pan troglodytes*) virus $SIV_{cpz}$, while $SIV_{sm}$, a sooty mangabey (*Cercocebus atyts*), that is monkey virus, gave rise to HIV-2. Exactly where these two viruses jumped to humans is unclear, but the AIDS pandemic, driven by HIV-1, first appeared with a vengeance in East Africa in the Rakai/Kagera region (Hooper 1999) about 1980. Prior to this, a few unrecognized cases cropped up in North America, Europe, and elsewhere in Africa. In the pandemic's heartland, Tanzania, Zaire, and Uganda, the disease spread rapidly through the local population, soon reaching Kenya and elsewhere in Southern Africa (Hooper 1999). AIDS is now by far the leading cause of death in Sub-Saharan Africa, where the average life expectancy has fallen to 47 years. It would have been 62 years without the disease. The estimated number of people in Sub-Sahara Africa who are currently HIV-1 positive, as of the end of 2002, was 30,000,000, almost three quarters of the global total. Although

*D. G. Janelle et al. (eds.), WorldMinds: Geographical Perspectives on 100 Problems*, 69–73.
© 2004 *Springer. Printed in the Netherlands.*

AIDS is a significant issue elsewhere, in most countries the majority of those infected still belong to the three most susceptible groups, homosexuals, drug addicts, and hemophiliacs (UNAIDS 2002).


## DIFFUSION OF HIV-1

Despite 20 years of educational effort, HIV-1 continues to rapidly diffuse. Programs with calls for fidelity, chastity, and the use of condoms may help, but they certainly are not yet significantly slowing the pandemic. However, interesting spatial patterns are becoming apparent in the spread of this virus. Those countries, such as Zaire, Uganda, Tanzania, Kenya, and South Africa, where AIDS is now the number one cause of mortality, are all very selenium deficient (Cenac et al. 1992; James 2000; Ngo et al. 1997; Van Ryssen and Bradford 1992). Their soils contain depressed levels of this trace element, a lack of which is associated in Africa with Keshan disease (Cenac et al. 1992; Burke and Opeskin 2002), a cardiomyopathy common in the selenium deficiency belt of China (Cheng 1987). In contrast, the soils of Senegal are derived from marine sediments containing phosphorites and are selenium enriched (Keller 1992). Despite widespread unprotected promiscuous sexual activity in Senegal, HIV-1 is diffusing very slowly, if at all, amongst the Senegalese (UNAIDS/WHO 2000). Even in Dakar, HIV-1 prevalence amongst antenatal clinic attendees, in 1998, was only 0.5 percent. It is apparent that in Africa, differences in soil selenium levels are greatly influencing who becomes infected with HIV-1 and who does not. A similar relationship has been documented in the United States, where Cowgill (1997) has demonstrated an inverse relationship, especially in the Black population, between mortality from AIDS and local soil selenium levels.


## WHY SELENIUM?

There may be several reasons why the course of the AIDS pandemic is greatly affected by the selenium intake of the local population. Selenium *in vitro* blocks HIV-1 replication (Kotaro et al. 1997). Conversely, HIV-positive women who have low serum selenium levels are much more likely to infect their partners than are HIV-positive females with normal serum selenium (Baeten et al. 2001). Individuals who are infected with HIV become increasingly selenium deficient (Baum et al. 1997). This decline, which is known to undermine immune function, is not unique to HIV-infection but is seen in almost all infectious pathogens (Sammalkorpi et al. 1988). Normally, where death does not occur, selenium levels rebound soon after recovery. However, HIV-1 can effectively elude the defence mechanisms of the immune system, and can replicate indefinitely, endlessly depressing serum selenium. As a result, the immune system is compromised, allowing infection by other pathogens that continue to deplete the host of selenium so allowing HIV-1 to replicate more easily, further undermining immunity. Therefore, this relationship between selenium and the immune system is one of positive feedback in which a decline in either of these two variables causes further depression in the other. Termed the "selenium-CD4 T cell tailspin" by the author (Foster 2002b), it is the reason that serum selenium levels are a better predictor of AIDS mortality than CD4 T lymphocyte counts (Baum et al. 1997). Like other positive feedback systems, such as avalanches and forest fires, the decline in serum selenium in HIV-1 infection is extremely difficult to control and gains momentum as it progresses.

Taylor (1997) demonstrated that part of the genetic code of HIV-1 is identical to that required to produce the human selenoenzyme glutathione peroxidase. This means that as this virus is replicated, it must obtain selenium (one of the components of the enzyme) from its host. HIV-1, however, encodes the entire selenoenzyme. As it replicates, therefore, it depletes its host not only of selenium but also of the other enzyme's three components: namely, the amino-acids cysteine, glutamine, and tryptophan (Maiorino et al. 1998). AIDS, therefore, is a nutritional deficiency illness caused by a virus. Its victims suffer from extreme deficiencies of all four of these nutrients, which cause a depressed CD4T lymphocyte count, vulnerability to cancers (including Kaposi's sarcoma), depression, psoriasis, diarrhea, muscle wasting, and dementia (Foster 2002a). Associated infections produce their own unique symptoms and increased risk of death.

## HALTING THE PANDEMIC

The world is experiencing simultaneous pandemics caused by viruses that encode the selenoenzyme glutathione peroxidase. These include Hepatitis B and C and the Coxsackie B virus. The geographical and field trial evidence demonstrates that paradoxically, these viruses, including HIV-1, diffuse most easily in populations that are very selenium deficient. This, then, is the Achilles Heel of HIV-1. It cannot diffuse well in countries like Senegal where selenium intake is naturally high.

While this principle has not yet been demonstrated experimentally for HIV-1, it has been established in China for Hepatitis B and C and the Coxsackie B virus. Chinese research has shown that selenium supplementation can greatly reduce the incidence of Hepatitis B. In 1989, for example, Yu et al. (1989) published details of a three-year project conducted in Qidong County, Jiangsu Province, where the 20,847 inhabitants of one township were provided with table salt fortified with 15 ppm anhydrous sodium selenite. In contrast, residents in six surrounding townships were encouraged to use normal table salt. By the third year, a very significant drop in the incidence of hepatitis had occurred in the selenium-supplied township (4.52 per 1,000) compared with those communities using normal salt (10.48 per 1,000; 56.8% reduction, $p <$ 0.002). Similar declines in disease incidence have been caused by increasing dietary selenium levels in communities where Keshan disease (linked to the Coxsackie B virus) is endemic. A field trial, conducted in Sichuan Province from 1974 to 1976 using fortified salt showed a decrease in the average annual incidence rate of this heart disease from 3.19 to 0.195 per thousand, compared to a drop of 1.11 to 0.86 per thousand in an untreated control population ($p < 0.01$). Adding selenium to diet, therefore, greatly reduces the frequency with which the Coxsackie B virus causes cardiomyopathy (heart disease) in Chinese children (Cheng 1987).

## CONCLUSION

The Chinese have been increasing dietary selenium by adding this trace element to salt, animal fodder, and fertilizers. Agricultural supplementation programs are also already underway in New Zealand (Oldfield 1999) and Finland (Pyykkö et al. 1988) for other health reasons. This strategy, if employed on a global scale, seems the most likely to halt or at least slow the diffusion of HIV-1 and other pathogens that encode the selenoenzyme, glutathione per- oxidase. It can be applied immediately with virtually instantaneous benefits. It is relatively cheap and easy to administer, with long lasting positive effects. This strategy is also compatible with, and indeed will improve the outcome of any large scale HIV-1 vaccination program.

## REFERENCES

Baeten, J., S. Mostad, M. Hughes, J. Overbaugh, D. Bankson, K. Mandaliya, J. Ndinya-Achola, J. Bwayo, and J. Kreiss. 2001. Selenium Deficiency is Associated With Shedding of HIV-1 Infected Cells in Female Genital Tract. *Journal of Acquired Immune Deficiency Syndromes and Human Retrovirology* 26(4): 360–364.

Baum, M., G. Shor-Posner, S. Lai, G. Zhang, H. Lai, M. Fletcher, H. Sauberlich, and J. Page. 1997. High Risk of HIV-Related Mortality is Associated With Selenium Deficiency. *Journal of Acquired Immune Deficiency Syndromes and Human Retrovirology* 15(5): 370–374.

Burke, M. and K. Opeskin. 2002. Fulminant Heart Failure due to Selenium Deficiency Cardiomyopathy (Keshan disease). *Medical Science Law* 42(1): 10–13.

Cenac, A., M. Simonoff, P. Moretto, and A. Djibo. 1992. A Low Plasma Selenium is a Risk Factor for Peripartum Cardiomyopathy. A Comparative Study in Sahelian Africa. *International Journal of Cardiology* 36(1): 57–59.

Cheng, Y. 1987. Selenium and Keshan Disease in Sechuan Province, China. In G. Combs Jr., J. Spallholz, O. Levander, and J. Oldfield, eds. *Selenium in Biology and Medicine*, 877–891. New York: Van Nostrand Reinhold.

Cohen, J. 2002. Malawi: A Suitable Case for Treatment. *Science* 297(5583): 927–928.

Cowgill, V. 1997. The Distribution of Selenium and Mortality Owing to Acquired Immune Deficiency Syndrome in the Continental United States. *Biological Trace Element Research* 56: 43–61.

Foster, H. 2002a. *What Really Causes AIDS*. Victoria, B.C.: Trafford.

Foster, H. 2000b. AIDS and the "Selenium-CD4T Cell Tailspin" The Geography of a Pandemic. *Townsend Letter for Doctors and Patients* 209: 94–99.

Hahn, B., G. Shaw, K. DeCook, and P. Sharp. 2000. AIDS as a Zoonosis: Scientific and Public Health Implications. *Science* 287(5454): 607–614.

Hasteltine, W. 1992. Cited in Large AIDS Increases Predicted by Early 2000s. *The Vancouver Sun* 15 December: A12.

Haseltine, W. 1993. Cited in More Cases, Same Old Question. *The Philadelphia Inquirer* 6 June: D1.

Hooper, E. 1999. *The River: A Journey to the Source of HIV and AIDS*. Boston: Little, Brown and Company.

Hori, K., D. Hatfield, F. Maldarelli, B. Lee, and K. Clouse. 1997. Selenium Supplementation Suppresses Tumor Necrosis Factor Alpha-induced Human Immunodeficiency Virus Type 1 Replication in vitro. *AIDS Research and Human Retroviruses* 13(15): 1325–1332.

James, J. 2000. Selenium: African Studies Reported at Durban. *AIDS Treatment News* 28 July (347): 7.

Keller, E. A. 1992. *Environmental Geology*. New York: Macmillan.

Maiorino, M., K. Aumann, R. Brigelius-Flohé, D. Doria, J. van den Heuvel, J. McCarthy, A. Roveri, F. Urini, and L. Flohé. 1998. Probing the Presumed Catalytic Triad of Selenium-Containing Peroxidase by Mutational Analysis. *Zeitschrift fur Ernährungswissenschaft* 37 (Supplement 1): 118–121.

Ngo, D., L. Dikassa, W. Okitolonda, T. Kashala, C. Gervy, J. Dumont, N. Vanovervelt, B. Contempre, A. Diplock, S. Peach, and J. Vanderpas. 1997. Selenium Status in Pregnant Women of Rural Population (Zaire) in Relationship to Iodine Deficiency. *Tropical Medicine and International Health* 2(6): 572–581.

Oldfield, J. 1999. *Selenium World Atlas*. Grimbergen, Belgium: Selenium-Tellurium Development Association.

Pyykkö, K., R. Tuimala, R. Kroneld, M. Roos, and R. Huuska, 1988. Effect of Selenium Supplementation to Fertilizers on the Selenium Status of the Population in Different Parts of Finland. *Environment Journal of Clinical Nutrition* 42(7): 571–579.

Sammalkorpi, K., V. Valtonen, G. Alfthan, A. Aro, and J. Huttunen. 1988. Serum Selenium in Acute Infections. *Infection* 16(4): 222–224.

Taylor, E. 1997. Selenium and Viral Diseases: Facts and Hypotheses. *Journal of Orthomolecular Medicine* 12(4): 227–239.

UNAIDS 2002 HIV/AIDS Documentation http://www.unaids.org/hivaidsinfo/documents.html//wad.

Van Ryssen, J. and G. Bradfield. 1992. An Assessment of the Selenium, Copper and Zinc Status of Sheep on Cultivated Pastures in the Natal Midlands. *Journal of the South African Veterinary Association* 63(4): 56–61.

Worldwatch Institute. 2001. *Vital Signs 2001. The Trends That are Shaping our Future*. New York: W.W. Norton.

Yu, S., W. Li, Y. Zhu, W. Yu, and C. Hou. 1989. Chemo-Prevention Trial of Human Hepatitis With Selenium Supplementation in China. *Biological Trace Element Research* 20(1–2): 15–22.

# 13

# DETECTING SPATIAL CLUSTERS OF CANCER MORTALITY IN EAST BATON ROUGE PARISH, LOUISIANA

ESRA OZDENEROL
*University of Memphis*

NINA LAM
*Louisiana State University*

According to the Louisiana State Center for Health Statistics (1998), overall age-adjusted mortality rates for Louisiana's major causes of death over the past five years indicate an increase in cancer-related deaths. This has generated great concern from the citizens in the state. In East Baton Rouge Parish, questions were raised on whether the heavy petrochemical industry, increased air pollution, increased population growth, and increased need for transportation were related to the increase in cancer-related deaths. Before an intense epidemiological study is launched, a key question will need to be answered: do these cancer deaths tend to concentrate spatially? If so, where are the clusters and what characterize them? If significant spatial clusters are identified, then further in-depth epidemiological studies can be conducted to investigate what causes these clusters. On the contrary, if no significant clusters are found, further in-depth investigation may be useless. Therefore, spatial techniques that measure the spatial structure of cancer patterns and detect clusters can be a useful surveillance or hypothesis-generating tool for environmental and public health investigations.

In this study, we demonstrate the use of spatial correlograms in describing quantitatively the spatial patterns of cancer at multiple geographic scales (census tract and block group levels). We show that correlograms can be used to reveal how different cancers are distributed or spatially autocorrelated, thus providing useful insights into the selection of specific cancer types for further analysis, such as cluster detection. We then applied the spatial scan statistic clustering method to lung cancer, which was found from the correlogram analysis to have the highest spatial autocorrelation values among all other cancer types (Ozdenerol 2000).

*D. G. Janelle et al. (eds.), WorldMinds: Geographical Perspectives on 100 Problems, 75–79.*
© 2004 *Springer. Printed in the Netherlands.*

## METHODS

The five most common cancers in East Baton Rouge Parish: lung, female breast, colon, prostate, and pancreatic were analyzed. Age-adjusted cancer mortality rates for these five cancer types by site, race (white and black), and sex in 86 census tracts and 349 block groups from 1993–1996 were calculated and mapped. Spatial correlograms were computed for the cancer mortality rates of each cancer type by race and sex and at both census tract and block group geographic scales. Based on the results from the correlogram analysis, which shows that lung cancer had the highest spatial autocorrelation, spatial scan statistic clustering was further applied to lung cancer to detect whether spatial clusters of lung cancer mortality exist.

Spatial correlograms are diagrams depicting how spatial autocorrelation measures change with the distance scale (Lam et al. 1996). Spatial autocorrelation refers to the degree of association of a variable in relation to its location. A useful measure of spatial autocorrelation is Moran's $I$. Moran's $I$ is positive when nearby areas are similar in attributes, negative when they are dissimilar, and approximately zero when attribute values are arranged randomly and independently in space. The significance of Moran's $I$ is tested by computing $z$ scores. The resultant $z$ scores, instead of $I$ values, are plotted against spatial lags (Figure 13.1). Based on a two-tailed test with a significance level of $\alpha = 0.05$, a $z$ value outside the range of $\pm 1.96$ (shown with

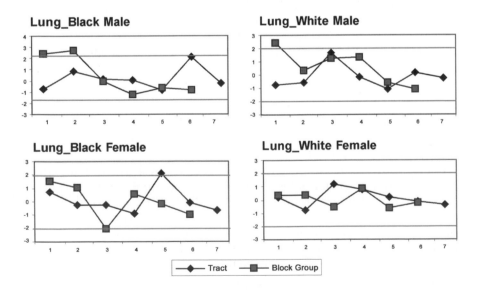

*Figure 13.1.* Correlograms for lung cancer mortality rates, 1993–1996, by race and sex at census-tract and block-group scales; the $x$ axis represents spatial lag 1, 2, to 7 tracts or block groups apart; $y$ axis shows the $z$ values at each lag; values outside the two inner horizontal lines are considered statistically significant at $\alpha = 0.05$.

two horizontal lines in Figure 13.1) is considered significantly spatially auto-correlated in either a positive or negative direction.

For cluster detection, we used the spatial scan statistic clustering technique (Kulldorff 1997; Kulldorf et al. 1997). In Kulldorff's method, the observed and expected number of cases and the relative risk are calculated for each individual census area. By repeating the same analytical exercise on multiple replications of the data set generated under the null hypothesis in a Monte Carlo stimulation, the distribution and corresponding $p$-value can be obtained to test whether the cluster is statistically significant. We then used GIS methods to extract census data for the census areas that belonged to the cluster so that we could identify the key socio-economic characteristics of the lung cancer clusters.

## RESULTS

Figure 13.1 shows the lung cancer correlograms that were found to be the most distinctive, showing high spatial autocorrelation. Other cancer correlograms revealed no significant neighborhood effects at both geographic scales. Figure 13.1 also shows that black male lung cancer mortality had the highest significant values (outside the two horizontal lines) at the block group scale in the first two spatial lags (1 and 2 block groups apart).

Since significantly high positive spatial autocorrelation for black males were found at the first two lags at the census block group scale, spatial clustering analysis was applied for black male lung cancer mortality only. Because of confidentiality concerns, maps showing the location of the clusters are not included here. In general, the clusters were found concentrating in the western part of the parish. A tabulation with the 1990 census data shows that the most likely clusters for black male lung cancer were characterized with low socio-economic status (Table 13.1). When compared with the parish's averages, these clusters had high percentages of persons below poverty, high percentage of renter units, low percentage of owner units, low percentage of high school

*Table 13.1.* Comparison of Socioeconomic Indicators (1990 Census) Between the Parish Averages and the Most Likely Clusters of Male Lung Cancer Mortality at the Census Block Group and Tract Scales.

| Most Likely Clusters | Persons below Poverty (percent) | Owner Occupation (percent) | Renter Occupation (percent) | High School Graduates (percent) | Median Household Income ($) |
|---|---|---|---|---|---|
| Census Tract | 31.83 | 17.66 | 22.78 | 14.16 | 20,093 |
| Census Block Group | 39.21 | 45.06 | 50.65 | 15.04 | 14,735 |
| East Baton Rouge Parish | 19.05 | 21.88 | 14.58 | 19.93 | 27,709 |

graduates, and low median household income. Such socioeconomic inequalities in health in other regions have also been reported (Krieger et al. 1997).

## CONCLUSION

We have demonstrated in this paper the use of correlograms as an effective means for comparing the spatial pattern of cancer mortality at multiple geographic scales and as a good indicator for further analysis. A highly spatially autocorrelated pattern for black male lung cancer mortality was found in East Baton Rouge Parish. Significant spatial clusters of black male lung cancer mortality were subsequently detected and they were further identified to have low socio-economic status, such as high percentage of persons below poverty. Our findings that poverty status appeared to be a predictor of black male lung cancer mortality are not surprising. Persons below poverty level more likely experience inadequate nutrition, poor standard of living, proximity to hazardous facilities, and limited access to health care, which could play an important role in cancer mortality. However, an in-depth study of the spatial clusters with a focus on possible causal agents is necessary to identify the causes of these clusters.

Our findings suggest that health policies targeted to reduce adverse health outcomes would be more effective if cluster location is considered. Particularly, cancer prevention and care facilities could be located close to the clusters to provide easy access for medically underserved populations. The results of this study suggest preventive measures to reduce the likelihood of similar clusters in the future.

## ACKNOWLEDGEMENT

We are grateful to Dr. Laura Fleming of the School of Public Health at the University of Miami whose comments proved very insightful and helpful.

## REFERENCES

Krieger, N., J. T. Chen, and G. Ebel. 1997. Can We Monitor Socioeconomic Inequalities in Health? A Survey of U.S. Health Departments Data Collection and Reporting Practices. *Public Health Reports* 112: 481–491.

Kulldorf, M. 1997. A Spatial Scan Statistic. *Communications in Statistics: Theory and Methods* 26: 1481–1496.

Kulldorf, M., E. J. Feuer, B. A. Miller, and L. S. Freedman. 1997. Breast Cancer in Northeastern United States: A Geographical Analysis. *American Journal of Epidemiology* 146: 161–170.

Lam, N. S.-N., M. Fan, and K.-B. Liu. 1996. Spatial-Temporal Spread of the AIDS Epidemic:

A Correlogram Analysis of Four Regions of the United States. *Geographical Analysis* 28(2): 93–107.

Louisiana Health Report Card. 1998. Five-Year Trend in Age-Adjusted Death Rates for Louisiana's Leading Causes of Death, 1993–1997. Louisiana State Center for Health Statistics. Louisiana Office of Public Health: 47 New Orleans, LA.

Ozdenerol, E. 2000. *A Spatial Inquiry of Infant Low Birth Weight and Cancer Mortality in East Baton Rouge Parish, Louisiana*. Ph.D. Dissertation, Department of Geography and Anthropology, Louisiana State University, Baton Rouge, LA.

# 14

# A GEOGRAPHIC APPROACH TO IDENTIFYING DISEASE CLUSTERS

ARTHUR GETIS

*San Diego State University*

In recent years, several interesting trends have come together to help pave the way toward a better understanding of disease diffusion. First, geography's traditional interest in map pattern analysis has been augmented by geographic information systems technology. Second, the identification of relatively new diseases, such as AIDS and ebola, and rapidly spreading diseases, such as West Nile virus and dengue, has stimulated interest in finding ways to thwart their diffusion. Third, ecologists and epidemiologists have come to recognize the usefulness of the spatial viewpoint. Finally, statisticians, ecologists, and geographers have developed new ways of identifying statistically significant disease clusters. As a result of these trends, the study of disease clusters has shed new light on the ecological characteristics of infectious diseases and on many environmentally induced health problems. In this paper, I will discuss a relatively new set of cluster statistics: local statistics. The emphasis, however, will be to show how these statistics aid in identifying the possible clustering and transmission of the dengue fever vector, the *Aedes aegypti* mosquito, in Iquitos, Peru. The female *Aedes aegypti* is the main carrier of the viruses responsible for dengue fever. The mosquito transmits the viruses by human blood feeding, which is part of the mosquito egg-laying process.

A number of statistics have been developed that measure spatial dependence in a sub-region of a larger study area (Getis and Ord 1992; Ord and Getis 1995; Anselin 1995). They identify the association between a single value at a particular site, say the incidence of a disease in a given census tract, and its neighbors up to a specified distance from the site in question. The Getis and Ord statistics, called *local statistics*, are well suited to identifying the existence of pockets or "hot spots," such as regions in which the incidence of a disease is greater than elsewhere. In addition, they can be used to identify distances beyond which no discernible spatial association obtains, that is, they may indicate the general boundaries of a disease cluster, subject to the granularity of the regions for which data are recorded. A related set of tests is available in the epidemiology literature, with the primary aim of cluster identification; for an extensive review, see Marshall (1991).

In this paper we use the local statistic, $G_i^*$, developed by Getis, a

*D. G. Janelle et al. (eds.), WorldMinds: Geographical Perspectives on 100 Problems, 81–85.*
© 2004 *Springer. Printed in the Netherlands.*

geographer, and Ord, a statistician (Getis and Ord 1992; and Ord and Getis 1995), to focus on the clustering of a disease. Our example explores the possible spatial clustering of *Aedes aegypti* in a neighborhood of Iquitos, Peru, during the period 1999 to 2002. A more thorough analysis by Getis et al. (2003) considers many of the dengue risk factors and the diffusion of the disease. These include identifying those individuals who are most susceptible to the disease (especially children), the spatial occurrence of *Aedes aegypti* larvae and pupae, and the location of rain-water-filled containers (the most culpable breeding site for the vector). These are studied over time.

## THE STRUCTURE OF THE STATISTIC

Consider an area, such as a city, subdivided into $n$ subregions, where each subregion $i$ is identified with a geo-referenced point, $i = 1, 2, \ldots, n$. Associated with each point is a value $x$ that represents an observation, say, the number of *Aedes aegypti* in subregion $i$ as one realization of a variable $X$, called number of *Aedes aegypti*. When we focus on a particular $x$ at $i$ (written $x_i$), the remaining observations are denoted as $x_j$. In the Maynas neighborhood of Iquitos, for example, there are $n = 528$ households. One of the households is designated as $i = 31$, and the surrounding households up to a distance $d$ are denoted as $j = 29, 30, 32,$ and $33$.

For the $G_i^*$ statistic we state the *clustering null hypothesis*: The sum of the numbers in $i$ and $j$ within distance $d$ of $i$ constitutes a value no higher or lower than one would expect if *Aedes aegypti* occurred at random within the city. This is a clustering hypothesis because our focus is on the sum total of mosquito activity in a region surrounding a place $i$. When we test the same hypothesis for each $i$, we can then identify those $i$ that fall within regions of statistically significant high numbers of *Aedes aegypti*. These $i$ regions are therefore members of statistically significant clusters. On a planar map, the distance is often expressed as a circle of radius $d$ from $i$, but the meaning of $d$ can be specified as something besides Euclidian distance, such as the number of houses separating $i$ and $j$.

When there is a preponderance of high values among the $i$ and $j$, the resulting $G_i^*$ will be positive. The $G_i^*(d)$ statistic is written

$$G_i^*(d) = \frac{\Sigma_j \, w_{ij} \, (d)x_j - W_i^* \, \bar{x}}{s\{[NS_{1i}^* - W_i^{*2}]/(N-1)\}^{1/2}}, \qquad \text{all } j$$

where $w_{ij}(d)$ is the $i, j$th element of a one/zero spatial weights matrix with ones if the $j$th house is within $d$ of a given $i$th house; all other elements are zero; $W_i^* = \Sigma w_{ij}(d)$, where $w_{ii}$ is included, and $S_{1i}^{**} = \Sigma w_{ij}^2$ (all $j$). The mean of the adult mosquitoes in houses is $\bar{x}$ and $s$ is the standard deviation. The value of $G_i^*(d)$ is given in normal standard deviates. Note that this statistic has as its expectation, $W_i \bar{x}$, which controls for the number of houses within $d$ of each house. Note, too, that $G_i^*(d)$ is 0 in a pattern where adult mosquitoes are randomly distributed within $d$ of house $i$.

## CLUSTER IDENTIFICATION

The local statistic, $G_i^*$, helps to identify the magnitude of a hot spot about a location $i$ by computing a value for each of a series of increasing distances from that location. An intuitively appealing definition of the boundaries of the hot spot is given by selecting the critical distance $d_c$ for which $G_i^*$ is maximized with respect to $d$. That is, for distances beyond $d_c$, the "new" $x_j$ values that contribute to the numerator of $G_i^*$ are less in sum than expected under the null hypothesis of no clustering. Usually, the $d$ that corresponds to the first local maximum satisfies the definition of $d_c$.

## A LIMITED CASE STUDY: AEDES AEGYPTI IN IQUITOS, PERU

In the Western Hemisphere, the mosquito *Aedes aegypti* is the primary vector for the transmission of dengue fever viruses from viremic individuals to susceptibles. For this limited case study (the more complete paper is Getis et al. 2003), the goal is to determine the spatial pattern of adult *Aedes aegypti* mosquitoes in the Maynas neighborhood (528 houses) of the Amazonian city of Iquitos, Peru. Spatial referencing of the mosquito survey data and the application of statistical tools like $G_i^*$ provide insights into adult mosquito dispersal behavior that can help explain patterns of human dengue infections.

We focus on the pattern of the numbers of *Aedes aegypti* in individual houses in the Maynas neighborhood. Do the mosquitoes cluster and, if so, to what degree or intensity? The full study explores the factors that may be responsible for any clustering and diffusion. Here, we identify the exact houses that can be considered as members of clusters. First, we consider the actual numbers of adult mosquitoes in each house in Maynas (see Figure 14.1). If the clustering is statistically significant within any particular household, the $G_i^*$ statistic will be above +2.575 (the 0.01 level of significance) at short distances from the household, say 10 meters. If clustering continues to near neighbors within 20 meters of a household, the value of $G_i^*$ will be higher than at 10 meters. If values of $G_i^*$ do not increase with increases in distance, then whatever clustering existed at the shorter distance ceases to exist at longer distances. The houses that are members of significant clusters at 30 meters are shown in Figure 14.2. Note that of the 528 houses in Maynas, 48 (9.1%) are members of statistically significant clusters of adult mosquitoes. The adult mosquitoes are observed to cluster strongly within houses and weakly to a distance of 30 meters beyond the household. Specific houses are identified as being members of statistically significant clusters of adult mosquitoes.

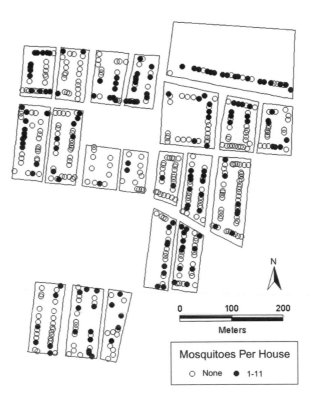

*Figure 14.1.* Adult mosquitoes per house in the Maynas neighborhood.

Pattern analysis can efficiently describe local *Aedes aegypti* populations and substantially aid in our understanding of dengue epidemiology and the development of dengue surveillance and control strategies. Development of long-term entomological risk-assessment strategies requires thorough surveys of all mosquito life stages. For purposes of investigating the dynamics of dengue transmission, our results point out the need to assess risk of human infection at the household level at frequent time intervals. Our detailed spatial approach constitutes the framework for analysis of data from ongoing longitudinal studies in Iquitos in which we will assess entomological risk at the level of the household with human dengue infection, and ultimately severity of disease. We conclude that over short periods of time the flight range and blood feeding behavior of *Aedes aegypti* are underlying factors in the clustering patterns of human dengue infections. Results indicate that any source reduction campaign, such as decreasing the number of water-holding containers, needs to be undertaken at the scale of the individual household.

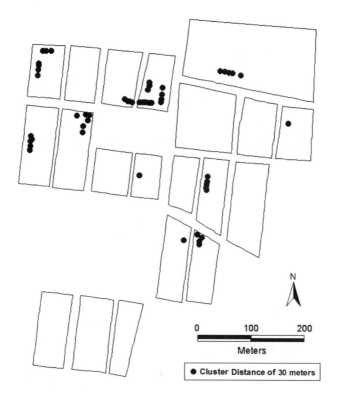

*Figure 14.2.* Clusters of adult *A. aegypti* to 30 meters in the Maynas neighborhood.

## REFERENCES

Anselin, L. 1994 Local Indicators of Spatial Association – LISA. *Geographical Analysis* 27: 93–115.

Getis, A., K. Gray, A. C. Morrison, and T. W. Scott. 2003. Characteristics of the Spatial Pattern of the Dengue Vector, Aedes aegypti, in Iquitos, Peru. Paper under review.

Getis, A. and J. K. Ord. 1992. The Analysis of Spatial Association by Use of Distance Statistics. *Geographical Analysis* 24: 189–206.

Marshall, R. J. 1991 A Review of Methods for the Statistical Analysis of Spatial Patterns of Disease, *Journal of the Royal Statistical Society, Series A* 154: 421–441.

Ord, J. K. and A. Getis. 1995. Local Spatial Autocorrelation Statistics: Distributional Issues and an Application. *Geographical Analysis* 27(4): 286–306.

# 15

# MORTALITY RATES ACROSS TIME: DOES PERSISTENCE SUGGEST "HEALTHY AND UNHEALTHY PLACES" IN THE UNITED STATES?

RONALD E. COSSMAN, JERALYNN S. COSSMAN,
TROY C. BLANCHARD, WESLEY L. JAMES and ARTHUR COSBY
*Mississippi State University*

The access to, and provision of, health care is – quite literally – a matter of life and death. Yet most health analysis is based on a-spatial (i.e., placeless) categories such as socioeconomic status, race, gender, or population density that do not facilitate the delivery of health services to the most needy populations. We propose using an explicitly spatial approach to reveal previously hidden patterns of underserved or at-risk populations. Whether this visualization technique is used at the national, state, or local level, it serves as a valuable analytical tool to reestablish the notion that "space matters."

Health and health care affect virtually everyone, either in the short or long term. We identify spatially defined "healthy and unhealthy places," in other words, places with significantly high or low death rates. We then map these rates across time. This mapping sets the stage for an important series of questions. When it comes to health, what is the role of "place" (both the physical and social environment) versus the people who inhabit those places? Also, what are the long-term implications for those who previously lived in these "healthy and unhealthy places?" Ultimately everyone may be affected to some degree by the health effects of the places in which they have lived.

## BACKGROUND

Approaching this research question entails considering the fundamental problems of transforming the data and looking for patterns (Sinton 1978; Goodchild 1992). The way in which the research question is framed and the manner in which data are manipulated will, in large part, predetermine the types of patterns that can be revealed. No analysis technique is free of these limitations. We argue that other investigators, by ignoring geography, have overlooked important dimensions of health in the United States.

Previous researchers have typically used three different a-spatial categorization schemes to examine health. First, some have used social categories

*D. G. Janelle et al. (eds.), WorldMinds: Geographical Perspectives on 100 Problems,* 87–92.

such as socioeconomic class, race, ethnicity, or gender to understand how mortality and morbidity affect different groups. The problem with this approach is that some categories are evenly distributed throughout the country. For example, socioeconomic class is ubiquitous; that is, people in different classes can be found virtually everywhere. Therefore, this type of analysis would not indicate where geographic concentrations or *places* that are healthy (or not) are located.

In a second approach, researchers have focused on specific diseases such as cancer or stroke as a means of determining what groups are most affected. The problem with this method is that illness follows many pathways and can ultimately manifest itself in a number of different outcomes (Committee on Health and Behavior 2001; House 2001; Evan et al. 1994). By focusing on the outcome, these researchers ignore the fact that people responding to the same underlying causative agent can contract different diseases that do not fall into pre-established health outcome categories. This approach also does not facilitate the delivery of health services to spatially concentrated populations.

Finally, a few researchers have looked at differences between geographic regions, such as the U.S. Bureau of the Census' designated region "South" versus the "Northeast." While this approach nominally introduces geography, it is still in the form of a category, in this case an administratively defined region.

## METHODOLOGY

We analyzed health outcomes in the United States using exploratory spatial data analysis techniques; in other words, we mapped health data in such a way as to be able to visualize spatial patterns. Implicit in this technique is the assumption of an isotropic plain; that is, a flat, featureless field that contains no physical or social barriers to movement and access to care (e.g., physical, political, and cultural – mountains, administrative boundaries, or racial discrimination). We recognize the roles that such barriers play in terms of the spatial patterns of mortality that are visually evident and do not mean to discount them.

We mapped age-adjusted death rates, to facilitate county-to-county comparisons, using five-year averages to stabilize the rates for rural counties. Each county was coded into one of three classes; more than one standard deviation above the national mean mortality rate for that five-year time period (i.e., an "unhealthy place"), within one standard deviation (health neutral), and more than one standard deviation below the national mean (a "healthy place"). By using the standard deviation as the analytical point to distinguish among counties, we were able to focus on those with statistically high or low mortality rates. We immediately noticed that both healthy and unhealthy counties tended to cluster in groups with similar characteristics.

Our next question was whether these clusters of counties were statistically significant; that is, was there a spatial correlation among the county clusters? A spatial statistic (the Local Moran's $I$) compared each county to each of its neighbors to determine if high or low mortality counties were grouped together more frequently than would be expected, given a random distribution of rates among the counties. The test confirmed that there was a statistically significant clustering of counties with high or low mortality rates.

Our final research question introduced the dimension of time. Simply put, if we mapped these county mortality rates over time, would the same spatial clusters be evident? We classified counties as having high mortality in five or six of the six time periods (a score of 5–6), four out of six (score of 4), and three out of six (score of 3). We did the same for low mortality counties. While it is possible that a county may have high and then low rates, thus canceling out abnormal scores, the emphasis here is on overall rates, as opposed to rate trends. Our visualization technique focused on the spatial persistence across time. Figures 15.1 and 15.2 show the persistence of high and low mortality rates, respectively, for the period 1968 to 1997. These maps clearly demonstrate a spatially anchored "persistence across time" of both high and low mortality counties in the United States. Some aspects of these results mirror the spatial concentration of morbidity in the United States in 1880 (Elman and Myers 1999) and the persistence of poverty in London across 100 years (Dorling et al. 2000).

## RESULTS

There are six apparent clusters of persistently high mortality counties, as shown in Figure 15.1: the Piedmont belt in the Southeast, the Mississippi Delta, part of Appalachia, the Upper Peninsula of Michigan, South Dakota, and northern Nevada. The larger (i.e., more counties) clusters tend to have their own temporal structure; that is, persistent counties (high mortality in five or six out of six time periods) form the core of some areas and seem to be surrounded by less persistent (four out of six) and least persistent (three out of six) counties in a manner resembling concentric circles, suggesting a "distance-decay function" within a temporal dimension. However, since this relationship is presented over time and the overall trend of the county rate could be increasing or decreasing, more research is necessary before such a process is hypothesized. State borders do not confine the clusters; they cross these political borders easily, creating their own regions. Another interesting aspect is that the majority of high mortality counties are in the vernacular region called "The South."

There are five low mortality clusters, as shown in Figure 15.2. The largest cluster is in the Northern Great Plains of the Midwest, followed by Colorado, Utah, Oregon, and Southwest Florida. Whereas the high mortality counties tend

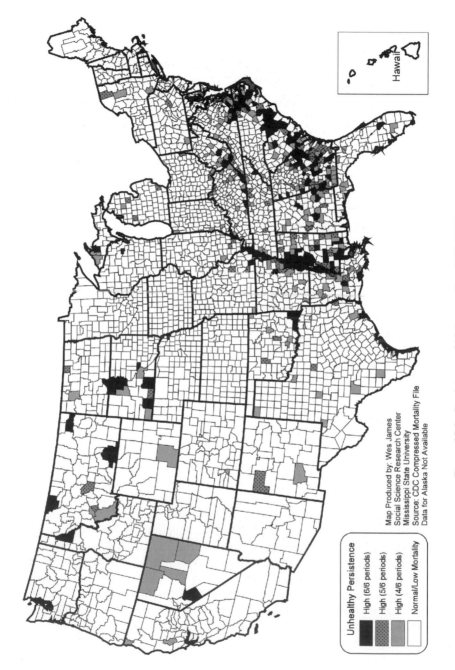

*Figure 15.1.* Persistence of high mortality, 1968–1997.

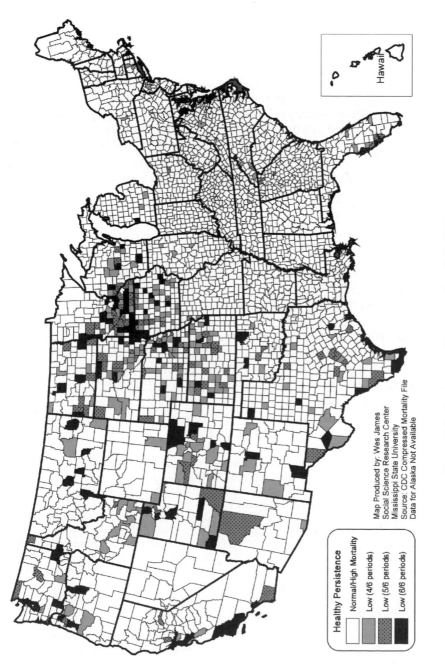

*Figure 15.2.* Persistence of low mortality, 1968–1997.

to be contiguous, the low mortality counties are more dispersed, particularly in the Great Plains.

## CONCLUSION

By mapping high and low mortality rates at the county level, we have demonstrated spatial clustering. Further, by introducing the dimension of time, we have demonstrated that high and low mortality is spatially anchored across time in these places. This project suggests that "healthy and unhealthy places" indeed exist and indicates where they are located. Future research will address the question of the relative importance of "the place" versus "the people" in these health outcomes.

## ACKNOWLEDGEMENTS

This study was made possible by grant number 4-DIA RH-00005-01-01 from the Office of Rural Health Policy of the Department of Health and Human Services through the Rural Health, Safety and Security Institute, Social Science Research Center, Mississippi State University. Its contents are solely the responsibility of the authors and do not necessarily represent the official views of the Office of Rural Health Policy. The authors thank Carol Campbell for her high quality analysis of mortality rates and Rita Belli Jackson for early drafts of the maps.

## REFERENCES

Committee on Health and Behavior: Research, Practice, and Policy Board on Neuroscience and Behavioral Health, eds. 2001. *Health and Behavior: The Interplay of Biological, Behavioral, and Societal Influences.* Washington, D.C.: Institute of Medicine, National Academy of Sciences: National Academy Press.

Dorling, D., R. Mitchell, M. Shaw, S. Orford, and D. Smith. 2000. The Health Effects of Poverty in London in 1896 and 1991. *British Medical Journal* 321: 1547–1551.

Elman, C. and G. Myers. 1999. Geographic Morbidity Differentials in the Late Nineteenth-Century United States. *Demography* 36: 429–443.

Evan, R., M. Barer, and T. Marmor, eds. 1994. *Why Are Some People Healthy and Others Not?: The Determinants of Health of Populations.* New York, NY: Aldine, De Gruyter.

Goodchild, M. 1992. Analysis: The Search for Pattern. In R. Abler, M. Marcus, and J. Olson, eds. *Geography's Inner Worlds: Pervasive Themes in Contemporary American Geography*, 138–162. New Brunswick, NJ: Rutgers University Press.

House, J. 2001. Understanding Social Factors and Inequalities in Health: 20th Century Progress and 21st Century Prospects. *Journal of Health and Social Behavior* 43: 125–142.

Sinton, D. 1978. The Inherent Structure of Information as a Constraint to Analysis: Mapped Thematic Data as a Case Study. In H. Dutton, ed. *Harvard Papers on Geographic Information Systems*, 1–17. Reading, MA: Addison-Wesley.

# 16

# LEARNING GEOGRAPHY IN THE ABSENCE OF SIGHT

REGINALD G. GOLLEDGE
*University of California, Santa Barbara*

Sight is the most comprehensive of the senses when dealing with the geospatial domain. Not only does sight reach well beyond the radii of other human senses and other body parts, but, since geography has a long tradition of representing its data and its findings in forms tailored to visual representation, it is indeed difficult to think of doing geography without sight. Usually, when one thinks of a geographic representation, one thinks of maps, images, tables, graphs, charts, sketches, photographs, videos, and, now, immersive or desktop virtual environments. Some of these representations are available in hardcopy; others are projected on a screen (including computer screens). As more and more geospatial problems have been pursued with the help of computers and electronic representational forms, such as geographic information systems (GIS), the emphasis on visualization for the representational modality has become even more obvious.

As more attention is being paid to providing equal opportunities for disabled people to learn different subjects, geography – long the preserve of the visual modality – has begun to expand its boundaries to accommodate the demands of disabled people. In particular, conventional tactile representations, originally designed to augment sight, have been expanded to determine how they can be used alone and in combination with other non-sight variables, such as sound and spatial language, to provide learning opportunities for vision impaired geographers. One of the significant problems facing the learning of geography generally is the extended range of scales at which geographic knowledge has been developed. It continues to be a challenge to the profession to find ways and means to complement the traditional world of visualization with representations in non-visual modalities that provide the primary form of contact for blind or vision disabled people. Ongoing research is indicating that alone and in combination, touch, haptics (i.e., touch and force), sound, and speech can provide interpretable and useful learning domains. As further emphasis is placed on providing equal access to information and its revealing and analytical representations, it is important for geographers to think outside the box and to extend their research interests to the non-visual domain.

What happens when sight is not available? Does this mean that it is not

*D. G. Janelle et al. (eds.), WorldMinds: Geographical Perspectives on 100 Problems*, 93–98.
© 2004 *Springer. Printed in the Netherlands.*

possible to engage in geography? Obviously, not. Technical advances have produced imaging capabilities that explore different aspects of the spectrum (e.g., infrared), have creatively used sonar to allow 3-D map representations of domains such as the ocean floor, have used satellites to explore population distribution at night by recording light intensity at different locations throughout the globe, and are building immersive virtual visual and auditory environments so that even sight disabled people can experience exotic and hard to reach places. In addition, technology has allowed us to build reasonably accurate digital elevation models (DEM) or to produce scaled table models of environmental settings. Thus, although we accept that sight provides the greatest sensory experiences and gives us the greatest volume of information about different environments at all scales, there is, nevertheless, a tradition of representing and experiencing geographic phenomena using senses such as touch and sound.

## TACTILE MAPS AND MODELS

Tactile maps and models were originally developed as three-dimensional visualizations of geospatial domains. Made from as cheap and as simple a material as papier-mâché or as complicated an industrial process as thermo-forming or vacuum packing (see Wiedel 1983 or Tatham and Dodds 1988 for more extensive treatments of building tactile maps), the tactile map or three-dimensional table model are a means of representing the diversity of terrain or land uses as a scaled-down version of their real setting. While initially concentrating on landforms, these three-dimensional maps and models have also been produced for the ocean floor, for cities, and for specialized environments (e.g., recreational areas).

As augmentations to flat cartographic maps, 3-D maps and models incorporated important concepts of slope, gradient, landform type, geographic alignment, location, adjacency, and other essential geospatial characteristics for the sighted observer. These models have become a means for those without sight to experience the geospatial domain. By touching a three-dimensional map of the world, for example, those without sight can experience the concepts of shape, location, proximity, association, adjacency, distance, direction, orientation, and other essential spatial properties of the real world that otherwise exist only in their imagination.

Much of the demand for tactile maps or models amongst the blind or vision-impaired population generally, however, emanates from their desire to interact at a local level. For use in the geography of everyday life, the conventional tactile map and model technology is usually inappropriate: it is too bulky, non-portable, and too expensive for use by individuals in local settings. An innovative and inexpensive technology evolved around the development of microcapsule paper (sometimes termed "puff paper" or "swell paper"). In essence, this paper is impregnated with chemical microcapsules that expand when heated. The expansion rate is proportional to the blackness of any image that is reproduced on the paper (e.g., by Xeroxing; see Andrews (1983) and Golledge (1991)).

## AUDITORY MAPS

Auditory maps, as opposed to auditory augmentations of standard flat maps, have not yet been explored as fully as they could be. This is because there has been little substantive work in the area of auditory localization of geo-spatial features (except in the area of stereo sound and quadric sound in music). An auditory map should include fundamental auditory information on spatial characteristics and spatial relations, such as location, shape, and pattern, and other spatial concepts that can be revealed via the medium of real or virtual sounds, or speech.

Sound lends itself to dynamic representation and also to the representation of gradients. Thus, the conventional scale bars of cartographic flat maps can be replaced by scroll bars of sound, changing pitch and tone from one place on the bar to another. Conventional maps of elevation and thematic maps (e.g., precipitation and temperature) also lend themselves to representation by graduated continuous sound changes (i.e., sound painting). Thus, geographic phenomena expressed in isoline form can be represented effectively in the auditory domain. For thematic maps generally, sound can be used to express a range of densities (as in choropleth maps) or could represent different land use types on general land use maps. One restriction, however, is that only a limited number of sound categories can be used – unless the listener has a profound ability to discriminate even slight changes in pitch or tone. In some cases, pitch can reflect distance, while tonal qualities could express like-defined regions (Krygier 1994; Weber 1998).

Perhaps the closest existing technology that can satisfy the requirement needed to be an auditory map is the auditory virtual setting. Their procedure was incorporated into the design of a personal guidance system (PGS) or wayfinding device for blind or vision-impaired travelers (Loomis et al. 1998). In this system, an individual traveling through an environment is tracked by GPS, which continually updates location in real space every second and updates location simultaneously in a digital base map carried by the traveler in a wearable computer. As the traveler moves through a real environment, a trace of movement takes place in the digitized map. As part of the electronic database, significant landmarks or other important environmental features (e.g., major obstacles) are locationally identified and related text is stored in attribute tables. As the traveler walks through the real world, a corresponding cursor moves through the electronic world, surrounded by a buffer. Designated features whose locations fall within the buffer are activated to give an identifying verbal message to the traveler. Using stereo headphones, the message is "externalized" (i.e., appears to be coming from the actual location of the feature in real space). It is possible for the traveler to turn and face an object thus identified, point to it with great accuracy, and, depending on the information given, express its distance away. In this way, both routes and layouts can be learned using the environment itself as a base map.

## SPATIAL LANGUAGE

Speech is suffused with spatial concepts. Space is often used as a metaphor for helping to interpret non-spatial phenomena (i.e., spatialization). For example, one may be given demographic data on the age structure of a population in tabular or list form. This information can be rearranged in a shape or graphic form called a population "pyramid" to enhance its interpretability.

Verbal descriptions are often laced with place and space-based nouns, adjectives, and prepositions. Spatial language is prominent in direction giving and location-based information (e.g., particularly in fieldwork situations). But spatial language, though representing a significant part of all spoken languages, is often fuzzy, incomplete, and hard to interpret. Its use is at times problematic as when giving directions in the field.

Geographic settings have long been described in literature. Even imaginative environments can be mapped from a spatial language used to describe them (e.g., Tolkien's worlds). Australian aborigines traditionally recounted the events and places experienced on a walkabout in song and dance (termed a "songline"). In this drama, experiences and information were tied to places by expressive language.

Spatial language is useful for providing general and relative spatial information, but it requires expert training to get the type of precision required to build a realistically proportional representation of a real world environment using spatial language by itself.

## EXPERIENCING GEOGRAPHY USING TACTILE-AUDIO SYSTEMS

The main concern with tactile-audio information systems is that they are static. They generally have to be set up for a single user. Each map domain has to be created as a separate tactile representation, which may mean having the ability to create one's own tactile representations. Although some tactile atlases (e.g., of Canada and the United States) have been produced, they are rare and invariably confined to an institutional setting. A recent development pioneered by Jacobson (1998) is based on the development of technology that transforms an individual and static environment into a dynamic computer screen environment. Termed "Haptic Soundscapes," this combination of hardware and software technology uses interfaces such as a haptic mouse or a vibrotactile mouse to explore onscreen-map and graphic information. This can be done in a variety of ways, each of which allows for non-visual learning:

- creating virtual boundaries around onscreen shapes or menus that help a non-sighted searcher find information or experience displayed material;
- allowing the researcher to identify specific icons by shape or transform the icon into an earcon by having its descriptive label spoken when touched; and
- using a simple query system that gives navigation instructions such as up, down, left, right to search a screen and take one to a desired location (e.g., in a menu or on a map).

## SUMMARY

Much has been achieved, particularly in the last two decades. Much still has to be achieved. Currently a number of software companies claim to have devices that navigate the Internet and access web pages using touch and sound characteristics rather than the visual/tactile movement of cursors or keyboards. But these all have shortcomings. It is highly likely that speech input will become a more viable interface than either keyboard or mouse in the future (particularly with wearable computers) and if this is the case, then geographers must expand their research interests to explore nonvisual input and output modalities, including a more precise spatial language, particularly for electronic visualizations (e.g., GIS output). They must also allocate more attention to augmenting sight with sound and touch both to increase the richness of representations and to facilitate the learning of geography without the use of sight. The demand for multimodal interfaces has already swept the video gaming environment so that in many cases, hardware exists that could be adapted for use in the geographic domain. It is a challenge for future geographers to pay attention to this interaction problem and to devise equitably accessible modes for learning geography at all scales and in real world, map-based, or onscreen environments.

## REFERENCES

Andrews, S. K. 1983. Spatial Cognition Through Tactual Maps. In J. Wiedel, ed. *Proceedings of the 1st International Symposium on Maps and Graphics for the Visually Handicapped*, 30–40. Washington, D.C.: Association of American Geographers.

Golledge, R. G. 1991. Tactual Strip Maps as Navigational Aids. *Journal of Visual Impairment & Blindness* 85(7): 296–301.

Jacobson, R. D. 1998. *Navigating Maps With Little or No Sight: A Novel Audio-Tactile Approach, Proceedings, Content, Visualization and Intermedia Representation, August 15, University of Montreal, Canada*, 95–102.

Krygier, J. B. 1994. Sound and Geographic Visualisation. In A. M. MacEachren and D. R. Fraser-Taylor, eds. *Visualisation in Modern Cartography*, 149–166. Pergamon.

Loomis, J. M., R. G. Golledge, and R. L. Klatzky. 1998. Navigation System for the Blind: Auditory Display Modes and Guidance. *Presence: Teleoperators and Virtual Environments* 7(2): 193–203.

Tatham, A. F. and A. G. Dodds. 1988. *Proceedings of the Second International Symposium on Maps and Graphics for Visually Handicapped People*. Nottingham, England: University of Nottingham.

Weber, C. R. 1998. The Representation of Spatio-Temporal Variation in GIS and Cartographic Displays: The Case for Sonification and Auditory Data Representation. In M. J. Egenhofer and R. G. Golledge eds. Spatial and Temporal Reasoning in Geographic Information Systems, 74-84. New York: Oxford University Press.

Wiedel, J. W. 1983. *Proceedings of the First International Conference on Maps and Graphics for the Visually Handicapped*. Washington, D.C.: Association of American Geographers.

# 17

# HAPTIC SOUNDSCAPES: DEVELOPING NOVEL MULTI-SENSORY TOOLS TO PROMOTE ACCESS TO GEOGRAPHIC INFORMATION

DAN JACOBSON

*University of Calgary*

This essay explores the critical need for developing new tools to promote access to geographic information that have throughout history been conventionally represented by maps. This problem is especially acute for vision-impaired individuals. The need for new tools to access map-like information is driven by the changing nature of maps, from static paper-based products to digital representations that are interactive, dynamic, and distributed across the Internet. This revolution in the content, display, and availability of geographic representations generates a significant problem and an opportunity. The problem is that for people without sight there is a wealth of information that is inaccessible due the visual nature of computer displays. At the same time the digital nature of geographic information provides an opportunity for making information accessible to non-visual users by presenting the information in different sensory modalities in computer interfaces, such as, speech, touch, sound, and haptics (computer generated devices that allow users to interact with and to feel information).

## UNDERSTANDING THE GEOGRAPHIC ENVIRONMENT

Whether blind, visually impaired, or sighted, our quality of life is highly dependent on an ability to make informed spatial choices over a wide range of locations, situations, and scales. For people with visual impairments, access to maps and graphics is more critical as maps may be the only efficient means of acquiring the structural knowledge to organize spaces and places (Jacobson 2000).

Developing an understanding and knowledge of the spatial world is fundamental to human existence. From the earliest exploration by crawling in infancy, to adulthood mobility, we operate in a world where we have to negotiate space; in the physical environment or in representations of that environment (Kitchen and Blades 2002). Our language and thinking is imbued

*D. G. Janelle et al. (eds.), WorldMinds: Geographical Perspectives on 100 Problems*, 99–103.
© 2004 *Springer. Printed in the Netherlands.*

with spatial constructs and metaphors, which shape our ways of acting and communicating. From the earliest times humans have tried to communicate and represent this spatial world to each other – cave paintings, drawings in the sand, models, maps, works of art, photographs, and into the recent era with satellite images, computer-generated worlds, and virtual environments. Throughout history these representations have been crucial in education and in the development of new ideas. With the rise of the mass media, mass communications, global cities, and the accelerating development of the Internet, the need to facilitate access and ways of understanding and interpreting this information is increasingly important.

## PROBLEMS WITH VISUALIZING THE GEOGRAPHIC ENVIRONMENT

In situations involving complex scientific information, visualization is an increasingly important method for exploring data. Information is displayed in many ways: using tables, graphs, plots, maps, and so forth. In education and in daily life similar approaches are used for navigating around structured information, such as computer software and interfaces to libraries and to the World Wide Web. In an ideal scenario, future interfaces to geographic information should be designed to be accessible to a broad spectrum of people irrespective of age, ability, or sensory impairment. Non-visual interfaces have the greatest potential impact in the daily lives and in the education of vision-impaired people, but offer benefits for all. The use of tactile, haptic, and auditory interfaces has the potential to make technology more universally accessible. Information technology transformations are affecting how we communicate, and how we store and access information. These transformations have the potential to facilitate how we receive proper medical care, how we learn, how we conduct business, how we work, how we design and build things, and how we conduct research. Developing "every-citizen interfaces to the nation's information infrastructure" (National Research Council 1997) states that there are significant benefits for all by investing in the national information infrastructure (NII) for people with disabilities in that it: removes communications and information access barriers that restrict business and social interactions between people with and without disabilities; removes age-related barriers to participation in society; reduces language and literacy-related barriers to society; reduces risk of information worker injuries and; enhances global commerce opportunities. However, the transformation of the Internet from a text-based medium to a robust multi-media environment has created a crisis – a growing digital divide in access for people with disabilities (Waddell 1999). Increasingly the Web is becoming the dominant Internet medium, with more than 60 percent of traffic. For Web traffic, almost three-fourths is in images that are inaccessible to vision impaired people. By researching the fundamental components of non-visual interfaces, the "optimal" combinations of haptic and auditory modalities, it is possible with suitable hardware to create information that is accessible to both blind and sighted people.


## DEVELOPING MULTI-SENSORY INTERFACES

The reasons for developing multi-sensory interfaces are compelling. These include: issues driven by the ethics of developing inclusionary technology; the legal context of these technologies, the Americans with Disabilities Act; and the technical benefits of having information that can be accessed by multiple senses (Powlik and Karshmer 2002). Although Geography has been an almost exclusively visual discipline, due to vision being the sense *par excellence* for viewing geographic representations, geographers are uniquely placed to develop multi-sensory access to maps. Geographers routinely integrate information from different sources, across different scales (form the local to the global); and from different geographic perspectives, from direct experience navigating through an environment, to indirect information obtained about the environment from secondary sources.

The development of multi-sensory interfaces of parallel and complementary information provides and promotes "universal access" through combinations of speech, sound maps, and haptic (force-feedback) computer-based interfaces. This (1) facilitates access for those without vision, who need ways of exploring spatial information through maps, diagrams, and graphs that are central to education; (2) promotes novel ways of augmenting visual displays; (3) enables the exploring and mining of spatial data; and (4) enhances navigation through spatial information where vision is restricted, such as within vehicle navigation systems. This multi-sensory approach can be directed to develop solutions to provide access to digital spatial based information for novice and expert users, for the young and old, for people with sensory impairments, and in situations where vision-based displays may not be optimal, such as Internet connections with low bandwidth.

With rapidly changing user interface technology and more Internet navigators requiring multi-sensory approaches to handling information via visualization procedures, the development of multi-sensory interfaces has wide relevance. Jacobson et al. (2002) describe the development of multi-sensory virtual reality for presenting geographic information involving the integration of touch and sound to navigate through map-like information (Figure 17.1).

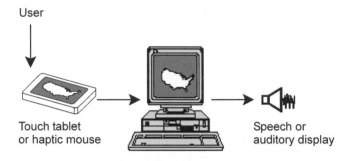

*Figure 17.1.* An example of a multi-sensory interface.

## BENEFITING FROM AN INCLUSIONARY APPROACH TO INFORMATION ACCESS

Enabling worker participation in the digital economy is perhaps the leading challenge for education in these early years of the 21st Century and is an important challenge facing information science and technology. But 54 million disabled people and 80 million elderly people in the United States (many with low vision) face an ever-widening "Digital Divide" that increasingly disassociates them from workforce participation (National Telecommunications and Information Administration 2000) Students with disabilities take fewer math and science courses and are more likely to drop out of high school, and persons with disabilities are underrepresented in science and engineering (National Science Foundation 2000).

Multi-sensory interfaces will ultimately be the best solution to the problem of Internet access by vision-impaired computer users. Due to the increased relevance of geography within society (Cutter et al. 2002), it is likely that information technology (IT) environments will include massive amounts of geographic information. Spatial representations, such as, maps, diagrams, and images are the most efficient and prevalent ways of presenting scientific data. Such representations are usually inaccessible to blind people. Geographers have argued that learning about people, places, and environments by using maps, spatial models, geographic information systems (GIS), remote sensing (satellite) imagery, geographic animations, simulations, and virtual reality provides perspectives not fostered by other learning processes (e.g., direct personal experience or indirect verbal learning) (Golledge 2002).

Although the information discussed here has focused on blind and vision-impaired people and on developing access tools for spatial information (maps, diagrams, graphs) to promote access to education and employment, there are other significant avenues for the multi-sensory presentation of geographic information. These include areas where vision is not the optimal sense for accessing spatial information. For example, in an in-car navigation system,

visual information may be too distracting. Other applications include aug-
mentation of the visual sense with information accessible by other senses,
especially when the visual sense is already saturated with information (a
common situation in some virtual environments).

Practically speaking, many potential employment opportunities are denied
to blind or vision-impaired people. Often, they are denied this because of an
inability to travel independently in their environment. Access to representations of the environment, such as maps, when presented in an accessible
multimodal form and distributed across the Internet have the potential to
provide a vehicle for learning about the environment. In a similar manner,
access to diagrams vital in education can be facilitated and improved.

## REFERENCES

Cutter, S. L., R. Golledge, and W. L. Graf. 2002. The Big Questions in Geography. *The Professional Geographer* 54(3): 305–317.

Jacobson, R. D., R. M. Kitchin, and R. G. Golledge. 2002. Multimodal Virtual Reality for Presenting Geographic Information. In P. Fisher and D. Unwin, eds. *Virtual Reality in Geography*, 382–400. London: Taylor and Francis.

National Science Foundation (NSF). 2000. *Women, Minorities, and Persons With Disabilities in Science and Engineering*. NSF 00-327. Arlington, VA: National Science Foundation.

Golledge, R. G. 2002. The Nature of Geographic Knowledge. *Annals of the Association of American Geographers* 92(1): 1–14.

Jacobson, R. D. 2000. *Exploring Geographies of Blindness: Learning Reading and Communicating Geographic Space*. Ph.D. Thesis, Queen's University of Belfast, Belfast, Northern Ireland.

Kitchin, R., and M. Blades. 2002. *The Cognition of Geographic Space*. London/New York: I. B. Tauris.

National Research Council (NRC). 1997. *More than Screen Deep: Toward Every-Citizen Interfaces to the Nation's Information Infrastructure*. Washington, D.C.: National Academy Press.

Powlik, J. J., and A. I. Karshmer. 2002. When Accessibility Meets Usability. *Universal Access in the Information Society* 1(3): 217–222.

National Telecommunications and Information Administration (NTITA). 2000. *Falling Through The Net. Toward Digital Inclusion*. Washington D.C.: U.S. Department of Commerce http://www.ntia.doc.gov/ntiahome/fttn00/contents00.html (last accessed 30 October 2002).

Waddell, C.D. 1999. *The Growing Digital Divide in Access for People With Disabilities: Overcoming Barriers to Participation*. Paper read at Understanding the Digital Economy: Data, Tools and Research, 25–26 May: Washington D.C.: U.S. Department of Commerce. http://www.aasa.dshs.wa.gov/access/waddell.htm (last accessed 30 October, 2002).

# 18

# THE FINNISH *KAAMOS*: ESCAPES FROM SHORT DAYS AND COLD WINTERS

STANLEY D. BRUNN
*University of Kentucky*

PAULI TAPANI KARJALAINEN
*University of Oulu*

RIKARD KÜLLER
*Lund University*

MIKA ROINILA
*SUNY New Paltz*

Geographer William Mead (1968: 16) wrote "Finns are first and foremost concerned with the challenge of high latitude. The principal consequence of high latitude is that winter prevails over summer." Residents of high latitudes in both hemispheres experience and adjust to daylight-nighttime regimens differently from those in the middle and lower latitudes (Tonello 2001). During the northern hemisphere winter, the prolonged periods of darkness last several months. These periods result not only in some changes in biorhythms and psychological problems, but also require adjustments in human activities and behavior (Küller and Wetterberg 1996; Küller et al. 1999). We consider how Finns cope with this period of prolonged darkness, or *kaamos*, by discussing first the physical origins of the phenomenon and then the lived worlds or coping mechanisms.

## THE PHYSICAL OR CELESTIAL *KAAMOS*

The physical *kaamos* occurs when the sun descends below the horizon. Due to the tilted axis of the earth with respect to the sun, the number of daylight hours changes from one season to the next. The true *kaamos* occurs north of the Arctic Circle where the sun never appears above the horizon during late December and early January. Finland is not only a good country to study this phenomenon because its north-south extent approaches 1,160 km (706 miles), but because the winter and consequently day/night variations

*D. G. Janelle et al. (eds.), WorldMinds: Geographical Perspectives on 100 Problems, 105–112.*
© 2004 *Springer. Printed in the Netherlands.*

are extensive. The far south has dark and cloudy days, whereas the north experiences a snow cover much of the winter. Snow may disappear in southern Finland in March, in central Finland during April, and in the extreme north in mid-May. The winters may be mild, with medium snowfall levels in the archipelagic south and around Helsinki and Turku, and bitter cold with deep snow in Lapland. The average winter temperatures in Inari (68° N) for the past three decades have been between –13 °C to –15 °C; temperatures –30 °C are not uncommon. Of the 5.2 million people in Finland, 25 percent live in Helsinki, Espoo, Tampere, Vantaa, and Turku. Joensuu, Jyväskylä, Kuopio, and Oulu are major regional centers. In sparsely populated Lapland (one quarter of Finland is north of the Arctic Circle) live approximately 192,000 people (less than 5 percent of the total population). In Kemijärvi, Rovaniemi, Kuusamo, and Kemi, the sun does not get above the horizon during late December and early January. However, there is never complete darkness as the sun immediately below the horizon reflects light. In Sodankylä, there is no direct sunlight from 20 12 to 24 12 hours and in Utsjoki there is no sun visible from 25 November to 18 January (Mielonen 1990: 40; Figure 18.1). If there is no snow cover and there are cloudy nights, the landscape is very dark. The combination of long nights and extreme conditions "have called for and continue to call for ingenious and enterprising environmental adjustments" (Mead and Smeds 1967: 16).

## TYPES OF *KAAMOS*

In addition to the physical *kaamos*, the Finns experience other types; we identify seven:

*Subjective kaamos or the kaamos of the inner self:* not considered special or a problem, because it occurs regularly. There is little or no special preparation and little variation in seasonal regimens.

*Lappish kaamos:* characterized by snow cover, no sunlight, but light reflecting from the snow or from the sun immediately below the horizon. During nighttimes there may be light from the stars and moon and the northern lights (aurora borealis).

*Eastern/continental kaamos:* continental coldness, dryness in deep forests, and silence in nature.

*Southern/seashore kaamos:* No snow, dirty snow, and a muddy/gray landscape, dark colors from coniferous forests and "islands" of woods within cities and rural areas), excessive cloudiness, a general gloomy appearance.

# NUMBER OF DAYS WITH NO SUN ABOVE THE HORIZON DURING FINNISH WINTERS

*Figure 18.1.* Number of days with no sun above the horizon during Finnish winters.

*Urban kaamos:* artificial lighting along streets, in buildings and shopping centers, and shops with light therapy lamps.

*Rural kaamos:* a very silent and dark landscape except for light coming from the moon, stars, the northern lights, and light reflected from snow.

*Tourist Kaamos:* experienced by European and Japanese tourists coming to Lapland to stay in "ice hotels" and to participate in "winter safaris."

## SYMPTOMS OF THE *KAAMOS* DEPRESSION OR INNER *KAAMOS*: THE LOSS OF LIGHT

Prolonged darkness calls for some adjustments in activities and behavior (Table 18.1). Some individuals may experience difficulties in getting and traveling to work in the dark, working much of the day in darkness, especially if engaged in outdoor occupations, and returning home also in the dark. Getting up may be tiring, because there is no natural light. Some people sleep 12 hours or more during nighttime hours.

Tiredness may lead to *kaamos* depression or *kaamosmasenns*. There are two peaks, the first in October–November and the second in late winter (March and April). October and November are often considered the darkest times marked by long periods of darkness, no snow cover, but slush (*loka*). While there are longer daylight periods in January, there is light from the stars and moon and reflected from the surface. The depression that may result is due to the metabolism of a certain hormone, melatonin, (*melatoniini* in Finnish) which can be treated with a strong light. *Kaamos* depression may diminish when there is light again; alcohol consumption may also peak during the winter in some locales. The spring tired or fatigue period (*kevätväsymys*) is a feeling of exhaustion after the dark period and is caused by low levels of vitamins C and D.

Light and lighting are very important landscape features. After the winter

*Table 18.1.* Finnish Darkness and Human Experiences.

| Causes | Natural "Cures" | Human "Cures" |
|---|---|---|
| | Very Dark Landscapes ——————— Very Light Landscapes | |
| Low Sun Angle | Bright moon | Indoor entertainment |
| No Sun | Open space | Artificially lighted places |
| Ice Cover | Twilight covers | Sauna and winter swimming |
| Snow Cover | Sunrise | Enclosed shopping centers |
| No Moonlight | Fireplaces | Consuming vitamins |
| Overcast Skies | Candles | Winter trips to the Mediterranean |
| Dark Landscapes (lakes, rivers, forests, fields) | Exercising | Music as therapy |
| | Blue moods and thoughts (alakulo) | Seasonal indoor games and hobbies |
| | Outdoor activities (skiing; walking) | Weather reports on sunrise and sunset times |
| | | Workaholism |
| | | Light therapy lamps |
| | | Seasonal rituals (Christmas and the Festival of Light) |

Compiled by authors.

solstice, Finns look forward to longer daylight periods, even if only a few minutes longer each day. The local daily and national media report sunrises and sunsets as important news items. Colors also have a special meaning during the low sun season. These include the dark colors of forests and snowless surfaces and the brightness of the snow. The colors of the evening sky are mixes of reds, blues, and yellows, the result of low sun angle, low moisture content, the reflection of snow, and heavy snow density.

One special time that some Finns await is *sininen hetki*, or the moment before darkness. They experience this time by being outside, shoveling snow, walking in parks, and preparing winter sauna. Some also watch for the bluethroat (*Cyanoslyvia svecica*) and snow bunting (*Plactrophenax nivalis*) as spring omens. Immediately before the sun sets there is a blue sky, or the blue season, which is associated with blue moods and thoughts (*siniset ajatukset*). These twilight times evoke feelings of positive sadness or serenity (*alakulo*) in which one appreciates the beauty of a multilayered landscaped. This period of twilight or gradual darkness may last for several hours depending on the latitude, and before streetlights are turned on. Still later, the northern lights become another beautiful, colorful, and therapeutic layer component of the darkened landscape.

## ESCAPING THE WORLDS OF DARKNESS

Finns cope with and adjust to the harsh coldness and kaamos in many individual ways (Table 18.1). We identify two "worlds," one living and the other escaping.

*The Lived World*

In the high latitude worlds of twilight there is a beauty in the grayness and the blue skies that comes before the sun disappears completely below the horizon (Birkeland 1999). *Kaamos* also leads one to search for the meaning of one's place in the natural environment. Feelings of escapism are noted in *marraskuu* (November, the "month of dying or death"). The dark period outside calls for turning inwards to consider what gives meaning to the self and to one's surroundings. These include more restricted modes of living, closer family life and activities, the importance of candles (for lighting and reviving the human spirit) and fireplaces (warmth and light), and meditation, yoga etc. These inner world meanings may have their roots in northern shamanism/ female existence/the north as a wilderness and search for therapeutic resources and landscapes.

Music may also be a search for the inner self. One might play the *kantele* (like a small dulcimer) or listen to *kantele* music, which is played with deep emotion. Others may enjoy the popular Finnish folk music group Värttinä, which promotes Karelian folk songs and images in the Kalevala, the Finnish national epic. This epic contains an allusion to *kaamos* in the Poem 47, which describes a period of no sun in the middle of winter and "Louhi, mistress of Pohjola, sparse-tooth matron of the far Northland, snatched the sun and caught the moon . . . and took them with her to the Dark of Pohjola" (Friberg 1990).

Rituals are an important element in societies, especially those related to seasonal holidays (Connerton 1989). In Finland there is also the sauna and jumping in the snow, a rich tradition that assumes greater significance during the winter, and *avantouinti* (winter swimming), which is becoming popular. Christmas and pre-Christian rituals, including the Festival of Light (which is held shortly after the winter solstice), are part of this lived world during the dark season.

*The Escaped World*

Finns adopt various strategies and mechanisms to cope with these dark worlds of the outer environment and the inner self. Exposure to sunlight is one remedy, which may include traveling for a weekend or more to the Mediterranean. These are trips not only for light, but also warmth. From January through March there are direct flights from major Finnish cities to Greece, Cyprus, the Canary Islands, Balearic Islands, Tunisia, Morocco, and coastal Spain. Large sections of the travel section in the *Helsingin Sanomat* (the major national newspaper) advertise package tours. Early 2003 Finnair flights from Helsinki to the Canary Islands cost about 535 Euros and to Madrid and Barcelona about 350 Euros. Most desired destinations are within four hours of Helsinki.

Living and working in "sealed spaces" represents another form of escapism or seasonal adjustment. Some Finns spend time in their second homes or cottages, which are very popular today. A related phenomenon is the changes that are taking place in indoor and outdoor living. Architects respond to these lifestyles and behaviors by designing indoor malls and constructing passage-ways between buildings. With artificial lighting, indoor work and exercise space, and enclosed shopping spaces, one can live an entire winter in Helsinki, for example, and never venture outside. The low temperatures, strong winds and air pollution keep many elderly and asthmatic residents indoors. Some city residents have little heavy winter clothing. Underground shopping areas in downtown Helsinki and ongoing plans include developing areas where one can purchase necessity and luxury goods indoors year round. Hobbies, chess clubs and bingo tournaments, bowling, swimming, walking, dancing, singing, music contests, and playing cards are additional escapes. Outdoor activities include cross-country and downhill skiing (some with lighted courses), snow-boarding, and ice hockey.

"Workaholism" is another escape, but it may bring about mental or bodily tension. Some overachieve to avoid possible burnout and getting depressed.

Since winter is considered the traditional rest period after summer months of being outside and engaged in regular or vigorous outdoor activity, the seasonal biorhythms seek to adjust to this winter period.

Light therapy (*Kirkasvalohoito*) offers another alternative, achieved by keeping the lights on all the time, facing windows where natural light enters, lighting multiple candles, or maintaining a fire in fireplaces. Light therapy lamps, *kirkasvalolamppu*, are considered by some to offer therapeutic value. These lamps, about a meter square, of soft glare fluorescent glass hang on a wall.

## SUMMARY

The adjustments societies make in their economies and livelihoods are of longstanding interest to human/environmental geographers. As this Finnish example illustrates, there are various coping strategies during *kaamos*, the prolonged period of winter darkness accompanied by very low temperatures. Adjustments to work, travel, shopping, and leisure patterns and activities as well as interpersonal relationships are observed. Additional work investigating seasonal behaviors and activities of those residing in high latitudes is merited, including how similar the Finnish experiences are to those in the high latitudes of those living in both hemispheres; SAD (seasonally affected disorders) among Finns; urban and rural variations; how *kaamos* is depicted in art, music, and literature; how *kaamos* is commodified and "sold" to foreign tourists; and also the work, leisure, and exercise activities and coping strategies adopted during the prolonged period of light, (*yötön yö* or "night without night"). Iinvestigations by geographers, environmental psychologists, and behavioral scientists will help us better understand the seasonal and daily time/space dimensions of humankind.

## ACKNOWLEDGEMENTS

We thank Harri Andersson, Richard Impola, Ari Lehtinen, Markku Löytönen, Pauliina Raento, and Pentti Yli-Jokipii for invaluable suggestions to improve the manuscript.

## REFERENCES

Birkeland, I. J. 1999. The Mytho-Poetic in Northern Travel. In D. Crouch, ed. *Leisure/Tourist Geographies: Practices and Geographical Knowledge*, 17–33. Routledge: London.
Connerton, P. 1989. *How Societies Remember.* Cambridge: Cambridge University Press.
Friberg, E. 1990. *The Kalevala, Epic of the Finnish People.* Trans. Eino Friberg. Helsinki, Finland: Finnish North American Literary Society.

Küller, R. and L. Wetterberg. 1996. The Subterranean Work Environment: Impact on Well-Being and Health. *Environment International* 22: 33–52.

Küller, R., S. Ballal, T. Laike, and B. Mikellides. 1999. *Shortness of Daylight as a Reason for Fatigue and Sadness: A Cross-Cultural Comparison.* Warsaw: Proceedings, CIE 24th Session, 1(2): 291–294.

Mead, W. 1968. *Finland.* New York: Praeger.

Mead, W. and H. Smeds. 1967. *Winter in Finland. A Study in Human Geography.* New York: Praeger.

Mielonen, M. 1990. *Astronomisen Maantieteen Perusteita.* Turku: Turun Yliopiston Maantieteen Laitoksen Opintomoniste.

Tonello, G. 2001. *Lighting, Mood, and Seasonal Fatigue in Northern Argentina. Comparison of Countries Close to and Further from the Equator.* Lund, Sweden: Lund University.

# 19

# APPLIED GEOGRAPHY IN THE PROVISION OF EDUCATIONAL SERVICES

RICHARD MORRILL
*University of Washington*

Education is a pervasive and prominent part of society. Just in the arena of the provision of educational services, opportunities for involvement abound. Schools are everywhere, arranged at defined hierarchical levels, and operating within an elaborate structure of control, for private as well as public education.

The geography of schools, of school transportation, and of school administration is incredibly decentralized, yet subject to changing state and national standards, values and goals, and responds to constant and significant change in the underlying demography of the clients – enrolled students.

I report here on a number of studies, primarily carried out as a consultant to school districts or educational administrations. I am not at all an "education geographer" either with respect to teaching methodology or curriculum. Rather the many studies in this area are primarily from an expertise in "demographics", or, as I much prefer, the interplay of population/demography and spatial analysis. Secondarily, my work derives from an interest in inequality, both economic and social; and third, from experience in political redistricting. Many of these studies involve issues that are fairly universal in occurrence, and are also recurrent, as settlement, the character of the population, and even governmental structure change. I divide the summary here into three parts: (1) standard locational analysis, (2) locational analysis with demographic analyses and forecasts and, (3) locational analysis and issues of inequality.

What may appear to be simple situations of location analysis are rarely so. The world of schools is intensely political. Schooling is expensive and looms high in household expenditures, and schools are seen by most households as vital to their and their children's ability to survive and compete in society. As a result, the geography consultant in this area must be a political geographer as well.

D. G. Janelle et al. (eds.), WorldMinds: Geographical Perspectives on 100 Problems, 113–118.

## LOCATION ANALYSIS

*School Board Director Districts*

Delimiting districts for electing school boards would appear to be the simplest application of location models; merely divide the district into the requisite number of sub-districts, using an appropriate location-allocation algorithm. As a political redistrictor knows, however, the number-one criterion is to keep incumbent directors in their districts, and the second priority is to change the system as little as possible, to keep as many voters as possible in the same districts, whatever the population change. Adjustment of attendance areas for schools, and the nesting of elementary schools within middle school area, within high school areas, is a similar problem – straightforward as to modeling, fraught with risk as to neighborhood feelings (Morrill and Svart 1975).

*School System Efficiency and School Closures*

Jumping from the frying pan into the fire, the most unpopular, but some-times necessary kind of analysis concern the closure of schools This is why administrators choose an outside consultant, such as an "objective" univer-sity scholar – to study the efficiency of schools within a district, or all the districts within a region or state, with an eye toward closing down or merging the "inefficient" or "redundant" or "underutilized" ones. Since school dis-tricts and, within districts, each school is a vital and potent definer of autonomy and of community, studies of this kind are highly threatening. In the case of closing schools within a district – here I report on a team effort for the Seattle schools, post baby-boom – the analytical problem was to define a subset of schools, which, if closed, would result in the fewest pupils moved and the smallest increase in travel time, especially of young children (Morrill and Manninen 1976). The final decisions will of course be political, and probably not as "ideal" as the consultant recommended.

I undertook a large study, with David Hodge, of the entire system of over 300 school districts for the Washington Superintendent of Public Instruction (Morrill and Hodge 1984). The analytical problem involved tradeoffs between size ("bigger is better") and remoteness. Essentially, as few as a dozen pupils more than an hour from any alternative are sufficient to retain an indepen-dent school (district). Thus, the districts vulnerable for closure were those with small numbers, under 500 students, located in suburban or exurban situa-tions and only a few miles from larger districts. We dutifully proposed a plan for consolidation, reducing the number of districts to about 200, and proposing special relationships between remote but necessary and larger districts. The resulting uproar was predictable and, in the succeeding twenty years, little has actually changed!

## LOCATION AND DEMOGRAPHIC ANALYSIS

*Location of New Schools*

More popular and satisfying than closure and consolidation is locating new schools – generally a positive good. Nevertheless, while this might appear to call for a straightforward application of a location-allocation model, perhaps complicated by consideration of how it fits into the hierarchy of schools, the "optimum" site will obviously be affected by such factors as preexisting land use, available school district property (for building or trading), micro-scale NIMBYism, and neighborhood rivalries and strengths. The end line political decision is beyond the work of the consultant, but it is still satisfying to learn of the new school opening and to find out how far from your ideal they ended up!

*Enrollment Forecasts and Locating New Schools*

School districts may need to prepare capital plans for 10 to 20 years. For those with demographic as well as spatial analysis skills, preparing geographically detailed, e.g., census tract or block group, forecasts of potential students, and analyzing the implications of the changed district to the possible location of new schools, is satisfying, integrating work. By far the most fascinating and significant project of this kind for me was work with William Beyers, for the University of Washington, involving four components (Morrill 1988b; Morrill and Beyers 1991): (1) an analysis of "unmet need" for the four-county greater Seattle-Tacoma metropolitan region (the UW "turf"), examining potential versus actual enrollment, based on the geography of persons at serious risk of attending college; (2) a demographic forecast (20 year) of this at risk of college population, geographically distributed; (3) a parallel analysis of the economy's future demand for college degrees and; and (4) use of a location-allocation model, ALLOC, to select varying numbers of new branch campuses for the university. The analysis suggested two new campuses, one to the south in the heart of Tacoma, and one at the border with the suburban county to the north. The university and subsequently the legislature accepted the precise recommendations, and both new campuses were built and are thriving.

## LOCATION ANALYSIS AND INEQUALITY

*Special Needs Students*

Another study done for the Superintendent of Public Instruction of Washington used the 1990 census to provide a detailed analysis by school district of students with special needs in education, those with particular disabilities, those in households below the poverty line, those with single mothers, those in house-holds where the primary language was not English, and those with combinations of these characteristics (Morrill 1986).

*Unequal Access to Higher Education*

As part of a 20-year enrollment forecast for all university and college presidents in the state of Washington, I did a statewide analysis of variation in educational attainment, in college and university enrollment, and of relative access to both community colleges and four-year colleges and universities (Morrill 1997). Even I was surprised by the extreme variation and by how remoteness from opportunities diminished educational achievement and the nature of the labor force. This study provided impetus for establishing several other branch campuses of the state's universities, greater cooperation with community colleges, and distance new learning programs.

*Variation Across Schools in Student Choices, Aspirations, and Success*

I'm currently in a multi-year study of graduating seniors from all Tacoma high schools. The purpose is first to evaluate the effect of Washington's Proposition 100, which repealed affirmative action in college enrollment, to find whether and where Tacoma students attend college. More generally, the study examines the tremendous variation in characteristics of students, both by area of residence and by school of attendance (there is open enrollment and much movement), the relative attractiveness of the schools to different kinds of students, the values and expectations of students, and what helps or hinders the chances of attending college. Geography, class, and race all matter.

*School Desegregation and Mandatory Bussing*

As far back as 1964, I worked in CORE (Congress of Racial Equality) on plans to desegregate the Seattle schools via bussing. Seattle implemented manda-tory bussing in 1977–1978. A decade later the Seattle schools sensed increased unhappiness and asked me to evaluate the long-term effects of bussing and to recommend changes to make the program more acceptable. I discovered that the bussing plan was not only perceived as unfair, but indeed was, and that only one-third of children born in Seattle ten years earlier remained in the public schools; that is, bussing actually increased school and residential segregation (Morrill 1988a, 1989). The school board implemented the fairer recommended revision, but took another ten years to recognize that manda-tory bussing was a failed social experiment.

*Related Work*

While schools have been the most frequent areas of location analysis and demographic work, I've done similar studies for hospitals (changing demographic character, director districts) and for fire districts (locating facilities, especially of emergency medical vehicles). These are fulfilling activities that not only help improve the functioning and performance of public institutions but also use otherwise esoteric geographic tools on real problems, demonstrating the value of geography, and, yes, supplementing one's income!

## A NOTE ON MODELS AND METHODOLOGY

The most useful techniques from the geography tool-kit have been (1) the simple transportation-assignment problem; (2) the location-allocation algorithm, in particular ALLOC (on a network); (3) the good old gravity model; (4) GIS capabilities, including address matching; and (5) geographically constrained cluster analysis.

## REFERENCES

Morrill, R. 1986. *Forecasts of Special Needs Students in Washington State.* For Superintendent of Public Instruction.

Morrill, R. 1988a. *Alternate Desegregation Plans for the Seattle Schools.* April 1987 to February 1988.

Morrill, R. 1988b. *Planning for University of Washington Branch Campuses.* Office of the Provost.

Morrill, R. 1989. School Bussing and Demographic Change. *Urban Geography* 10: 336–354.

Morrill, R. 1997. Spatial Inequality in Higher Education in Washington. *Applied Geographic Studies*: 1–20.

Morrill, R. and W. Beyers. 1991. Locating Branch Campuses for the University of Washington, *Journal of Geography in Higher Education* 15: 161–171.

Morrill, R. and D. Hodge. 1984. School District Organization and Educational Opportunities For Committee on Educational Policies, Structure and Management (Superintendent of Public Instruction).

Morrill, R. and D. Manninen. 1976. Neighborhood Impact Study. Ch. 2 31–93. *Seattle Public Schools, Schools and Neighborhood Project.*

Morrill, R. and L. Svart. 1975. King County Intermediate School District Boundary Study.

# 20

# SPATIAL ANALYSIS AND MODELING FOR THE SCHOOL DISTRICT PLANNING PROBLEM

## DAVID S. LEMBERG
*Western Michigan University*

Declining funding for school construction and maintenance, rising student enrollment, locally mandated school facilities requirements, state mandated reductions in class sizes, and natural disasters have combined to create a classroom space deficit nationwide in the United States. Other nations around the world are having similar, if not worse problems in school facilities planning and provision. The National Education Association estimated the cost of retrofitting and modernizing existing schools in the United States at $322 billion (NEA 2001). School district facility planning is a complex spatial problem that geographers are especially well equipped to contribute to understanding and to develop solutions. Geographers have been assisting school district planners in areas such as spatial allocation, multi-modal transportation planning, and enrollment forecasting on different spatial (classroom, site, attendance area, and district) and temporal (term, year, and multiyear) scales.

It has been 25 years since J. Dennis Lord published the "Spatial Perspectives on School Desegregation and Busing" (1977), the last general survey of geography and school district planning. In 1977, the focus in school district planning was on the impacts of mandated desegregation programs. In the 21st century, desegregation is just one part of a much more complex problem – the school district planning problem (SDPP). The SDPP may be defined as how to manipulate school district site capacities and enrollment assignments into feasible alternatives to best manage school district resources over the long term. This is a classical geographical problem. The SDPP revolves around tradeoffs in space and time and geographers have been exploring important issues of scale, environmental perception, spatial interaction, and temporal dynamics.

Geographers and other researchers have been developing models and techniques for school district facility planning problems for more than 40 years since Garrison (1959) first suggested modeling be applied to schools. These include districting models (Clarke and Surkis 1968; Lord 1977; Thomas 1987; Schoepfle and Church 1991; Lemberg and Church 2000), school busing models (Yeates 1963), enrollment forecasting models (Rushton et al. 1995), school

*D. G. Janelle et al. (eds.), WorldMinds: Geographical Perspectives on 100 Problems*, 119–124.
© 2004 *Springer. Printed in the Netherlands.*

equity (Talen 2001), school choice (Church and Schoepfle 1993), and class-room scheduling models (Tillet 1974). Many of these models have generated good results, but have not been widely adopted by school districts or by the commercial firms and consultants marketing solutions to school districts. After 40 years of progress in research, the major impact in the field is the use of GIS for electronic pin mapping using shortest path algorithms to allocate students to schools. The challenge for the 21st century is not only to continue to research these problems, but also to apply some of the solutions to common practice in the schools.

In planning theory, a key to successful planning is problem definition. If the decision makers cannot define the problems, they cannot create good solutions to the problems. This is often difficult as many planning problems are hard to define, hard to evaluate, and involve many parties with differing agendas.

*Table 20.1.* Constraints.

| | |
|---|---|
| Site Capacity | The ability of the district's physical sites to handle present and future enrollment – space inventory, new facilities, and portable buildings |
| Spatial Dynamics | The spatial distribution of the school sites in relation to student residences and the community transportation network – the juxtaposition of current and future students with existing and potential school sites |
| Site Resources | The spatial distribution of existing educational resources on sites and the potential for future resources and facilities – program requirements versus available and potential space |
| Enrollment Patterns | The spatial (and temporal) distribution of the student population in the district – student enrollment projections by location, by grade, and by diversity attributes |
| Travel Distance | The maximum distance that students should travel from home to school – ride-walk distances, shortest-path routing, barrier mitigation, and transportation alternatives |
| Classroom Loading | The maximum loading capacity for each classroom as determined by the district – evaluating contract-teacher ratios, technology applications, and classroom-infrastructure constraints |
| Site Scheduling | The range of scheduling options open to the district – standard, year-round, double sessions, and extended days |
| Grade Configuration | The range of grades assigned to a school site – i.e., {K-6, 7–8, 9–12}, {K-5, 6–8, 9–12}, {K-6, 7–9, 10–12}, {K-3, 4–6, 7–8, 9–12}, etc. |
| Finance | The ability of the district to finance student enrollment management alternatives – a constraint as well as an objective |
| Legal Factors | State and Federal mandated rules – including reduced class sizes, racial desegregation, access for the disabled, special education, etc. |
| Political Factors | Community expectations for district performance – spending patterns, and tolerance for reform, disruption, and change |

These "wicked" planning problems such as the SDPP can be deconstructed into a set of general objectives (see below) and constraints (see Table 20.1) that collectively are the components for school district decision-making. These objectives may be applied to various spatial and temporal scales (see Figure 20.1).

## OBJECTIVES

### Education: To Maximize Educational Performance

While this should certainly be among the prime objectives of any educational decision maker, it is difficult to operationalize as a capacity or enrollment allocation objective. To separate the spatial planning from the curriculum and teaching of the schools, this objective can be reworded as to provide the best educational environment for students and teachers. Geographers can then investigate space requirements to facilitate learning and how these space requirements change as students mature. Another fertile area of research for geographers is how new technologies will alter learning space needs.

### Safety: To Develop Solutions That Ensure Student Safety

The geographer's role is to examine safety issues on two scales. On the site scale (site layout, site repair, site maintenance, and site retrofitting), eliminating hazards that will threaten present and future populations of users. On the school attendance-area scale, the geographers are needed to analyze routing

| SDPP: Spatiotemporal Scales for Analysis | | | | | | |
|---|---|---|---|---|---|---|
| Objective | Spatial Scale | | | | Temporal Scale | |
| | Neighborhood | Site | Attendance Area | Distance | Current | Long-Term |
| Education | X | X | X | X | X | X |
| Safety | X | X | X | X | X | X |
| Stability | X | | X | | X | X |
| Contiguity | | | X | X | X | X |
| Compactness | | | X | X | X | X |
| Neighborhood Preservation | X | X | | X | | X |
| Equity | | X | X | | X | X |
| Diversity | | X | X | | X | X |
| Finance | | X | X | X | X | X |
| Feasibility | X | X | X | X | X | X |

*Figure 20.1.* SDPP: Spatiotemporal scales for analysis.

and transportation hazards so students may safely travel between home and school.

*Stability: To Minimize the Impact of Change on the Constituency (Students, Parents, Teachers)*

Implementing changes in schools is difficult. Boundary changes, school closures, school construction, grade configuration changes, etc., all create friction with the school district's community. Geographers can apply spatial analysis to minimize spatial change (the movement of students, teachers, and territory between schools) and maximize temporal stability (increase the temporal interval between changes).

*Contiguity: To Create Contiguous Boundaries*

Geographers may apply spatial analysis to create contiguous boundaries for schools or if this is not possible, to minimize the fragmentation of attendance areas.

*Compactness: To Create Compact Boundaries*

Geographers may apply spatial analysis to minimize travel distances to schools, minimizing or eliminating the need and expense for school busing.

*Neighborhood Preservation: To Avoid Splitting Neighborhoods Between Schools*

Neighborhoods are strongly tied to local schools. A change of schools in all or part of a neighborhood impacts both neighborhood identity and local property values. Neighborhood boundaries themselves are fuzzy and vary with spatial perspectives. Geographers can explore the spatial extent of neighborhoods and their relationships with schools.

*Equity: To Ensure that Each Student has Access to Equivalent Educational Resources and is Treated Equally by the District*

Geographers may apply spatial analysis to discover any variation between schools in allocation of space and resources, treatment of students, and performance of students. On the school site scale, the Americans with Disabilities Act (1990) should encourage spatial analysts to maximize access for the disabled. Spatial analysis may also be used to equalize the impacts of changes to sites, boundaries, schedules, etc., between school sites.

*Diversity: To Encourage Diversity in Site Enrollment*

While the emphasis for the past 30 years has been on racial diversity, other types of diversity among school sites are also of interest including socio-economic status, disabilities, sex, home language, student performance, etc.

*Finance: To Minimize the Financial Outlay of the District*

Spatial analysts can explore the financial impacts of alternative capacity and enrollment assignments on present and future budgets. Better methods of projecting enrollment, new construction, and tax revenue on neighborhood, school attendance area, and school district scales are required to make long term plans in a time of fiscal uncertainty.

*Feasibility: To Develop a Plan That Will Survive the Public Approval Process*

Spatial analysts can research what changes are more acceptable to the community. There are issues of political feasibility, financial feasibility, educational feasibility, administrative feasibility, structural feasibility, etc. Many of the issues of feasibility are contained in the constraints on the SDPP (Table 20.1).

## CONCLUSION

The SDPP is one of the most important problems facing the world in the 21st century. It applies directly to many of the things that humans around the world feel most strongly about – our children, our tax spending, our property values, our society, and our future. As such, decision-makers need help in developing the best feasible alternatives for our schools. Given the complexity of the problem, the concern of the people, and the billions of dollars at stake, it is surprising how little research has been done in the field. The geographer is particularly well suited in formulating and exploring these objectives and constraints because they vary through space and time and across scales.

## REFERENCES

Church, R. and O. Schoepfle. 1993. The Choice Alternative to School Assignment. *Environment and Planning B* 20(4): 447–457.

Carke, S. and J. Surkis. 1968. An Operations Research Approach to Racial Desegregation of School Systems. *Socio-Economic Planning Sciences* 1: 91–103.

Garrison, W. 1959. Spatial Structure of the Economy: II. *Annals of the Association of American Geographers* 49(4): 471–482.

Lemberg, D. and R. Church. 2000. The School Boundary Stability Problem Over Time. *Socio-Economic Planning Sciences* 34: 159–176.

Lord, J. 1977. Spatial Perspectives on School Desegregation and Busing. *Resource Papers for College Geography*, 77-3. Washington D.C.: Association of American Geographers.

National Education Association. 2000. Modernizing Our Schools: What Will It Cost? Web page. See http://www.nea.org/lac/modern/modrpt.pdf.

Rushton, G., M. Armstrong, and P. Lolonis. 1995. Small Area Student Enrollment Projections Based on a Modifiable Spatial Filter. *Socio-Economic Planning Sciences* 29(3): 169–185.

Schoepfle, O. and R. Church. 1991. A New Network Representation of a "Classic" School Districting Problem. *Socio-Economic Planning Sciences* 25(3): 189–197.

Talen, E. 2001. School, Community, and Spatial Equity: An Empirical Investigation of Access in Elementary Schools in West Virginia. *Annals of the Association of American Geographers* 91(3): 465–486.

Thomas, R. 1987. Developments in Mathematical Programming Models and Their Impact on the Spatial Allocation of Educational Resources. *Progress in Human Geography* 11(2): 207–226.

Tillet, P. 1974. An Operations Research Approach to the Assignment of Teachers to Courses. *Socio-Economic Planning Sciences* 9: 101–104.

Yeates, M. 1963. Hinterland Delimitation: A Distance Minimizing Approach. *The Professional Geographer* 15(6): 7–10.

# 21

# A MODEL FOR COLLABORATIVE RESEARCH: BUILDING A COMMUNITY-UNIVERSITY INSTITUTE FOR SOCIAL RESEARCH

JAMES E. RANDALL
*University of Northern British Columbia*

ALLISON M. WILLIAMS
*University of Saskatchewan*

BILL HOLDEN
*City of Saskatoon*

KATE WAYGOOD
*Saskatoon, Saskatchewan*

This paper addresses the social relevance of research undertaken by academic geographers to find a model that integrates the community and the university into social research issues. The challenge we faced is that many involved in the practice of social and economic change within our communities have been disillusioned by a traditional scholarly model of research that appears to be disconnected from the service delivery and policy needs of the organizations and people they are ultimately intended to serve. This chapter provides one case study to address this problem by outlining the activities of a research partnership between the University of Saskatchewan and various government, community-based, and private-sector organizations in Saskatoon, Canada, entitled the Community-University Institute for Social Research (CUISR – see www.usask.ca/cuisr).

## BACKGROUND

Geographers have long been concerned with the relevance of their work (Abler 1993; Johnston 1995) and applications to solving social problems (Briggs 1981; Taylor 1985). This latter concern is linked to the much larger issues of the perceived failure of traditional scholarly approaches to resolve community-level social and economic problems (Green and Mercer 2001), and the inability

*D. G. Janelle et al. (eds.), WorldMinds: Geographical Perspectives on 100 Problems, 125–130.*
© 2004 *Springer. Printed in the Netherlands.*

of post-secondary institutes to contribute to the betterment of society (Boyer 1994; Harkavay and Wiewel 1995). In response, different research models have emerged, including systematic research partnerships among universities, community organizations, governments and the private sector. Rubin (2000) estimates that over 200 of these have been created and funded in the past six years in the United States, and we are about to witness an explosion of work of these groups. No one has undertaken an analysis of the disciplinary origins of the people involved in these research partnerships, but the characteristics of successful geographic research suggest that we are well suited to playing a central role in these initiatives. These characteristics include an emphasis upon integrating theory, practice and policy (Lee 2002), an ability to understand and appreciate the role of an interdisciplinary approach to issues (Schoenberger 2001), and a willingness to communicate with the public (Pattison 1964).

This paper describes one of those initiatives in which geographers have taken an important role, the creation of a Community-University Institute for Social Research (CUISR). This Institute operates as a partnership between researchers at the University of Saskatchewan and practitioners in government and community-based organizations (CBOs) in the city of Saskatoon, Saskatchewan, Canada. CUISR emerged out of a series of meetings of an *ad hoc* Quality-of-Life Roundtable, comprised of academics from geography and community health and epidemiology, together with senior representatives from municipal government, community development organizations, the health board, charitable and other non-profit organizations. The overall goal of the meetings, and the appeal for geographers, was to discuss ways in which the community-at-large and social researchers at the University of Saskatchewan could collaborate in a more systematic, long-term basis. There had been concerns that previous research conducted on (rather than with) the community by researchers at the University had not addressed the issues most relevant to community needs, nor had the outcomes of the research been "community-friendly." In other words, the peer-reviewed journal articles written by academics were of little relevance to the needs of community organizations, and the link between research results and either improvements to service delivery or policy change were seen as tenuous or non-existent. What our community participants wanted was a mechanism to create a truly collaborative research environment to build capacity within the community, thereby improving the quality of life of the residents of that community.

A proposal to establish a collaborative research institute to address some of these concerns was written by the ad hoc group and submitted in 1999 to a federal government funding agency (the Social Sciences and Humanities Research Council of Canada's Community University Research alliance program), as well as the University of Saskatchewan and various community partners. Funding success led to the formal creation of CUISR with the following mission statement "To serve as a focal point for community-based research and to integrate the social research needs and knowledge of community-based organizations with the technical expertise available at the University."

In order to achieve healthy, sustainable communities, CUISR operates on a model in which an integrative approach must be employed, recognizing that economic, social, health, or environmental problems can rarely be solved in isolation (Hancock et al. 2000). The programs and projects of CUISR are oriented around three research modules: assessing and applying quality of life (QOL) indicators, community health determinants and health policy, and community economic development. It also recognizes that knowledge is produced and reproduced in universities, among other places, and that this knowledge is mediated, in the form of policy and programs, in community organizations and government (Figure 21.1). Of the eight core researchers involved in this project, four have disciplinary backgrounds as geographers, including the two Institute Co-Directors (Randall and Waygood) as well as the Community and University Co-Leads of the QOL research module (Holden and Williams). The remaining four team members were trained in epidemiology, economics, commerce, and medicine.

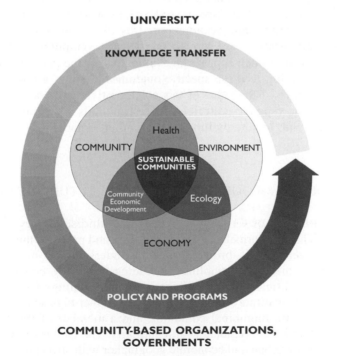

*Figure 21.1.* A conceptual model of urban sustainability used by CUISR. Source: Adapted with permission from Hancock, Labonte, and Edwards (2000).

## ACHIEVEMENTS OF CUISR

In the three years of CUISR's existence, peer-reviewed papers are appearing in journals (three published, two submitted, 12 in preparation), team members have given 10 conference presentations, and an additional 31 presentations have been made to other decision-makers or client groups. Twenty-one non-refereed research reports, on issues ranging from *Access to Food in Saskatoon's Core Neighbourhoods* to the *Experiences of Aboriginal Women in an Outreach Diabetes Education Program*, have been written and posted to the Institute website (www.usask.ca/cuisr). In addition, after several community forums to agree on research methodology and discuss preliminary results, a special 26-page supplement was disseminated to 76,000 households by one of our partners, the Saskatoon *StarPhoenix* newspaper. The research described within this insert represented the outcome of an attitudinal survey of Saskatoon residents towards their own quality of life, by neighborhood-level socio-economic status (see Figure 21.2 for an illustration of the low, medium and high socioeconomic status neighborhoods in Saskatoon). We also had the opportunity to suggest community and university "experts" to provide commentary on the storylines that emerged from the results. Training has been provided for 34 graduate students and six CBO practitioners. In the former case, students on paid internships or scholarships worked directly with CBOs on clearly defined research questions posed by the community organizations themselves. In the latter case, practitioners used community sabbaticals to undertake needs assessments, program evaluations, or the preparation of grant proposals. In addition to these specific outcomes, there is a growing awareness and acceptance of the value of this collaborative partnership. This effort is reflected in newspaper editorials, in testimonials by community-based organizations, and by the willingness of other organizations to provide resources to CUISR.

## THE ROLE OF GEOGRAPHERS AND THE FUTURE OF CUISR

One of the strengths of CUISR has been the multidisciplinary perspective that emerges. Through formal academic education and practice, the human geographers involved in this project bring an understanding of the research traditions within different disciplines, including sociology, economics, community health, political sciences, education, and planning. The Community Co-Director of the Institute (Kate Waygood) is a former high school geography teacher, a long time municipal councilor, and a member of the community development team of the regional health authority. The University Co-Director (Jim Randall) is an urban and economic geographer with strong ties to the local planning community. Bill Holden, the Community Co-Lead of the QOL module, has an undergraduate degree in Geography and is a planner with the

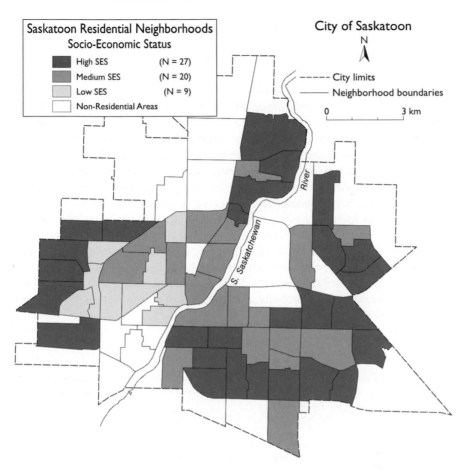

*Figure 21.2.* A grouping of Saskatoon neighborhoods by socioeconomic status.

City of Saskatoon, while the University Co-Lead of this module (Allison Williams) is a health and social geographer with a strong background in community-based research. Examination of the complex issues and perspectives that arise in research on quality of life, determinants of health, and community economic development is best undertaken in a multidisciplinary environment by individuals who are able to appreciate and integrate the contributions of team members from disparate disciplines. An openness to diverse disciplinary perspectives to research also extends both to effective management of complex projects and teams and an ability to ground concepts and theory to everyday social and economic problems or issues. We have gained credibility within the community for respecting the values and input of different stakeholders and for addressing community research concerns in a pragmatic manner. Both of the academic geographers have undertaken research in collaboration with community organizations in the past, providing a foundation for future cooperation. Although the same could be said for individual academics drawn from other disciplines, it is less common to see the same level of applied research being practiced.

The future direction of CUISR over the next several years includes the following activities: undertaking a comprehensive and critical evaluation of the impact that the Institute has had on policy change, building local research capacity and, ultimately, affecting the quality of life of Saskatoon residents, and serving as a local and national hub for enquiry on community-university partnerships. Research projects continue to unfold and the process itself constitutes an important research contribution to evaluation and assessment of processes, impacts, models, and effective practices. Strategies to achieve these objectives include, hosting an international conference on community-university research partnerships, hiring an action researcher to assist in analyzing the process of knowledge-to-policy transfer, developing stronger links with other institutions with similar goals, undertaking a second quality of life survey in order to make the link between evolving attitudes to QOL and neighborhood change, and engaging in a set of community policy forums.

## REFERENCES

Abler, R. 1993. Desiderata for Geography: An Institutional View From the United States. In R. Johnston, ed. *The Challenge for Geography. A Changing World: A Changing Discipline?* 215–238. Oxford, U.K.: Blackwell.

Boyer, E. 1994. Creating the New American College. *The Chronicle of Higher Education* 40: A48.

Briggs, D. 1981. The Principles and Practice of Applied Geography. *Applied Geography* 1(1): 1–8.

Green, L. and S. Mercer. 2001. Can Public Health Researchers and Agencies Reconcile the Push From Funding Bodies and the Pull From Communities? *American Journal of Public Health* 91: 1926–1929.

Hancock, T., R. Labonte, and R. Edwards. 2000. *Indicators that Count!: Measuring Population Health at the Community Level.* Toronto: Centre for Health Promotion/ParticipACTION Lecture Series HP-10-0207.

Harkavy, I. and W. Wiewel. 1995. University-Community Partnerships: Current State and Future Issues. *Metropolitan Universities* 6(3): 7–14.

Johnston, R. J. 1995. The Business of British Geography. In A. Cliff, P. Gould, A. Hoare, and N. Thrift, eds. *Diffusing Geography: Essays for Peter Haggett*, 317–341.Oxford, U.K.: Blackwell.

Lee, R. 2002. Geography, Policy and Geographical Agendas: A Short Intervention in a Continuing Debate. *Progress in Human Geography* 26: 627–628.

Pattison, W. 1964. The Four Traditions of Geography. *Journal of Geography* 63: 211–216.

Rubin, V. 2000. Evaluating University-Community Partnerships: An Examination of the Evolution of Questions and Approaches. *Cityscape* 5(1): 219–230.

Schoenberger, E. 2001. Interdisciplinarity and Social Power. *Progress in Human Geography* 25: 365–382.

Taylor, P. 1985. The Value of a Geographic Perspective. In R. J. Johnston, ed. *The Future of Geography*, 92–110. London, U.K.: Methuen.

# 22

# STRUGGLING AGAINST ILLITERACY IN A GLOBAL CITY: THE NEW YORK EXPERIENCE

WERNER GAMERITH
*Heidelberg University*

An eminent center of national and international migration, New York City (NYC) has had difficulty in integrating immigrants into its social fabric since it began as a colonial outpost at the southern tip of Manhattan in the early seventeenth century. As schooling began to be institutionalized as a public endeavor, questions of education and immigration immediately became intertwined. In educating immigrant children from diverse shores, the public school system has reflected both socio-political struggles within the city's elites (the clergy, politicians, and the local aristocracy) and socio-economic disparities within the city's geography. New York's role as a prime global player generates a range of problems within the city's schools. The social geography of education is a useful tool for analyzing the disparities that shape New York's numerous neighborhoods.

## CONSTANT "SCHOOL WARS"

Ever since its beginning in 1842, the nation's largest public school system has suffered from constant administrative disputes. The struggles were so severe that they came to be designated the "School Wars" in New York history (Ravitch 1974). Conflicts in the nineteenth century centered on financing religiously affiliated schools with tax money, while controversies in the late twentieth century arose from antagonistic efforts to centralize or decentralize public school administration. Behind this debate, notions of excellence versus equity focused on student achievements and the relatively poor results in the NYC public school system. Financial scarcities are yet another common feature of New York's public school system. Between 1976 and 2000, expenses for the public schools rose by 40 percent, on a constant dollar basis. Since the municipal budget has to cover for a wide range of public structures, schools are financially squeezed between other municipal concerns. That more than half the money goes into school administration instead of directly into instruction adds to the quality problem of public schooling.

*D. G. Janelle et al. (eds.), WorldMinds: Geographical Perspectives on 100 Problems*, 131–135.

## PROBLEMS IN NEW YORK CITY'S PUBLIC SCHOOLS

A highly diverse student body of nearly 1.1 million makes up a public school system comprising more than 1,200 schools. From the formative years until the present, cultural, racial, and ethnic variety has been a steady feature of New York's public schools. Roughly one-third of the students originate from immigrant families. Practically all of the 190 nations worldwide can be found in New York's public schools (DeWind 1997). A high portion of the student body lives in poverty and destitution. In 1998–1999, about 440,000 students came from families that received some form of public assistance (Hartocollis 2001). An ever-growing part of the school-age population, mainly from families with considerable financial means, chooses to bypass the public sector and to enroll in private institutions instead.

NYC's public school system is beset with a range of difficulties both in personnel and in materiel (Rivera-Batiz 1995; Ravitch and Viteritti 1997, 2000; Tanners 1997). Higher salaries in the suburbs prompt many teachers to leave inner city schools. Moreover, rising numbers of public school students cause the gap to become ever wider, so that there seems to be no alternative to the strategy of recruiting uncertified teachers or even persons with no teaching experience at all. In 1999–2000, uncertified personnel constituted 14 percent of the teaching staff, whereas the comparable figure for New York State was slightly above three percent. Nowhere in the United States is it more difficult to get highly qualified professionals for public education than in NYC. Teacher shortages hit mathematics and science courses particularly hard. Nearly 60 percent of 8,000 newly hired teachers in 1999–2000 were not certified for the courses they taught. In searching for qualified teaching staff, the city's school authorities capitalize on New York's global position to allure teachers from all over the world. In the school year 2001–2002, almost 1,000 teachers were recruited from countries as diverse as Austria, Canada, Hungary, India, Jamaica, and Spain. Teachers with native languages other than English to serve New York's bilingual students are especially welcome. Canada and India give hope of providing further well-trained personnel for science and mathematics courses. Meanwhile, New York's ambitions to cope with a rising teacher shortage have been emulated by a number of other large cities all over the nation. Media campaigns with ads in newspapers and journals reflect these attempts both at the national and international scales (Messow 2003).

Decreasing standards in teaching inevitably result in lower student achievement levels. NYC students show test scores substantially below the national and state averages, with huge gaps between different ethnic groups (Figure 22.1). In some districts, more than 80 percent of students failed the New York State Mathematics Test. About one-third of the students in public institutions never manage to get a high school diploma. This alarming dropout rate is further aggravated by the high incidence of delayed high school completers. Only half of all students graduate from high school in time (Messow

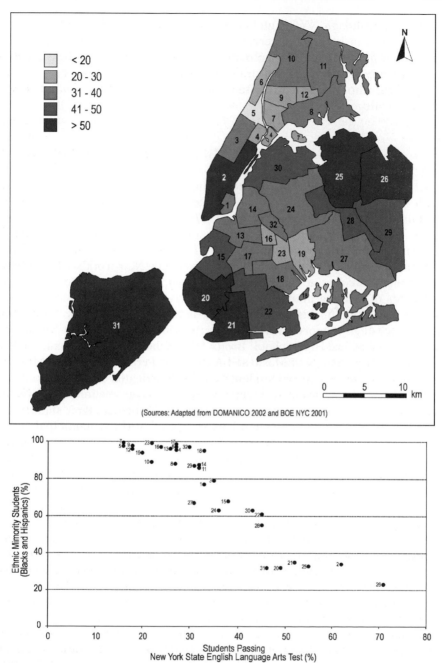

*Figure 22.1.* Spatial and ethnic disparities in New York City public education. Students passing New York State Mathematics and English Language Arts (ELA) Tests, 2001, by city school district and ethnicity (percentages). Sources: Adapted from Domanico (2002) and New York City Board of Education.

2003). Even though NYC's globalized labor market offers a variety of job opportunities in low-skilled service occupations, public school dropouts run the risk of permanent underemployment, if not unemployment. In competing with recently arrived, less well-trained immigrants, they no longer succeed. Even high school completers often fail to acquire qualifications that help them in college.

## THE IMAGE OF THE PUBLIC SCHOOL SYSTEM

Low achievement levels have eroded the New York public's contentment with its school system. Several surveys testify to the high level of dissatisfaction. In 1992, a study conducted by *The New York Times* revealed that a mere 12 percent of the interviewed persons trusted the system (Domanico 1994). In 1998, only seven percent of 450 employers evaluated the public school system as successfully preparing youth for New York's labor market. Personnel managers at Manhattan-based companies also deplored the low standards of learning in many public schools (Messow 2003).

The other side of this issue, however, is represented by New York's four specialized high schools (Stuyvesant High School in Downtown Manhattan, Brooklyn Technical High School, Bronx High School of Science, and Fiorello LaGuardia High School of Music and Art and the Performing Arts in Midtown Manhattan), which demand students pass sophisticated admission tests for entry. These schools possess outstanding reputations on a nationwide basis, are able to employ excellent teachers, and successfully prepare their students for future academic and professional careers. In addition to municipal funds, alumni clubs care for these specialized high schools and foster high levels of identification with the respective school even years and decades after graduation.

## THE SOCIAL GEOGRAPHY OF PUBLIC EDUCATION

These social differences are reproduced in spatial terms, and it is exactly that point that must be emphasized by urban social geographers if the existing problems are to be resolved. The geography of education and educational facilities can delineate both the contextuality and the spatiality of urban social phenomena at a critical point. NYC's school system exemplifies the strains of American public education as it is torn between the claims of excellence and equity – within a metropolitan framework that is both local and global. The fact that disparities can occur within one and the same NYC public school district is all the more striking since – theoretically at least – money should be spread evenly among the city's schools. The "Great Equalizer," as the system of public schools was once called, seems to be far from doing exactly that. Depending on the spatial context, which is connected with a wide range of social phenomena, students may find themselves in either disadvantaged or privileged positions when it comes to schooling and learning. This is not to say that the neighborhood and the community are exclusive predictors for students' achievement, yet the forces of social conditions are reflected spatially. Within a framework socially and ethnically as diverse as New York, geography provides outstanding tools and possibilities for analyzing the mechanisms of one of the most important systems of social reproduction.

## REFERENCES

DeWind, J. 1997. Educating the Children of Immigrants in New York's Restructured Economy. In M. Crahan and A. Vourvoulias-Bush, eds. *The City and the World: New York's Global Future.* New York: Council on Foreign Relations.

Domanico, R. 1994. Undoing the Failure of Large School Systems: Policy Options for School Autonomy. *Journal of Negro Education* 63: 19–27.

Domanico, R. 2002. *State of the New York City Public Schools 2002.* Civic Report 26. New York: Manhattan Institute for Policy Research, Center for Civic Innovation.

Hartocollis, A. 2001. Legal Portrait of System That Cheats Its Pupils. *The New York Times* 12 January 2001: A1.

Messow, E. 2003. *Schule in der Global City – multikulturelle Gesellschaften zwischen Leistung und Integration. Eine sozialgeographische Analyse des öffentlichen Schulsystems in New York City unter besonderer Berücksichtigung von Einflussfaktoren auf das Bildungsverhalten eines Schulkindes und aktuellen Reformmaßnahmen.* Heidelberg: Department of Geography, University of Heidelberg, unpublished Ph.D. dissertation.

Ravitch, D. 1974. *The Great School Wars. New York City, 1805–1973. A History of the Public Schools as Battlefields of Social Change.* New York: Basic Books.

Ravitch, D. and J. P. Viteritti. 1997. New York: The Obsolete Factory. In D. Ravitch and J. P. Viteritti, eds. *New Schools for a New Century. The Redesign of Urban Education,* 17–36. New Haven, London: Yale University Press.

Ravitch, D. and J. P. Viteritti, eds. 2000. *City Schools: Lessons From New York.* Baltimore: Johns Hopkins University Press.

Rivera-Batiz, F. L. 1995. Immigrants and Schools: The Case of the Big Apple. *Forum for Applied Research and Public Policy* 10: 84–89.

Tanners, L. 1997. Immigrant Students in New York City Schools. *Urban Education* 32: 233–255.

# 23

# REACHING NEW STANDARDS: GEOGRAPHY'S RETURN TO RELEVANCE IN AMERICAN EDUCATION

MICHAEL N. SOLEM
*Association of American Geographers*

Why do Americans need to know geography? In his collection of essays, *Me Talk Pretty One Day*, humorist David Sedaris (2000: 201) writes "When asked, any high school teacher can confidently answer that the knowledge will come in handy once the student hits middle age and starts working crossword puzzles in order to stave off the terrible loneliness." Although the thought of hitting middle age can induce anxiety in many folks, a more urgent problem is being caused by an educational system that, for many children, does not teach the practical, ethical, and intellectual values of geography. Fortunately, discernible progress has been made in reforming geography education in American schools.

This chapter tells the story of how professional geographers worked collaboratively with teachers, parents, businesses, legislators, and non-profit organizations to reform the teaching and learning of geography in K-12 education. Although geography educators have repeatedly advocated for their subject, the most profound successes have been achieved since the early 1990s in a period characterized by improvements in textbooks, multimedia, state curriculum frameworks, and teacher education programs. But the pace of reform has been uneven across America's educational landscape, and much work remains to be done before every schoolchild becomes capable of using the knowledge, skills, and perspectives of geography.

## THE PROBLEM: BEING GEOGRAPHICALLY IGNORANT

The world at the beginning of the twenty-first century is full of political, environmental, and economic change. Ethnic tensions are stirring political conflicts from the Balkans to the Pacific. Economic globalization is pressuring businesses and workers to become more competitive. Environmental pollution is altering world climates and disrupting sensitive ecosystems. Hundreds of millions of people suffer daily from miseries caused by hunger, war, and disease. As these problems intensify, world leaders struggle to formulate policies that can lead to effective and equitable solutions. Often these efforts

*D. G. Janelle et al. (eds.), WorldMinds: Geographical Perspectives on 100 Problems*, 137–141.
© 2004 *Springer. Printed in the Netherlands.*

fail to consider the geographical dimensions of problems related to international politics, development, economics, security, human rights, and the environment.

By learning geography, students can recognize patterns in human and physical systems, interpret the characteristics and interrelations of places and regions, and use maps and geographic information technologies to discover solutions to pressing problems. As leading scholars have noted (National Research Council 1997), geography has new relevance for students who will become the scientists, policymakers, government leaders, and industry workers of tomorrow. Unfortunately, most Americans seem to go about their lives without knowing the most basic facts and concepts of geography, let alone how to use that information creatively to improve conditions in their local communities and to analyze global issues critically. Consider, for instance, the results of the 1994 National Assessment of Educational Progress (NAEP) conducted by the U.S. Department of Education: only 22 percent of fourth graders, 28 percent of eighth graders, and 27 percent of twelfth graders taking the NAEP Geography test scored at or above the "proficient" level, defined as "solid academic performance that demonstrates competency over challenging subject matter" (Persky et al. 1996).

Part of the reason for such poor performance over the years has been geography's secondary status to history in American social studies education, and for many years efforts to revitalize geography were stymied by a lack of financial and political support from state and federal government. But ultimately, geographic ignorance grew too embarrassing for politicians: in 1989, the nation's governors publicly decreed that new national education standards were urgently needed in five subject areas: history, science, mathematics, English, and – in a landmark acknowledgement – geography. The political impetus had finally come to return geography to a respectable status in American schools.

## THE SOLUTION: BECOMING GEOGRAPHICALLY INFORMED

For professional geographers, the time had come to start devising clear, rigorous standards for K-12 education that emphasize more than geographic literacy (i.e., having factual knowledge about places and regions) by including skills and perspectives that characterize someone competent in the *science* of geography. Anticipating rapid changes ahead, geographers rallied by forming the Geography Education Standards Project (GESP) to navigate the momentous task of writing academically rigorous, internationally competitive geography standards. In January 1994, the U.S. Congress passed the Goals 2000 Educate America Act to promote the adoption of new education standards by the fifty states. Shortly thereafter, the nation's new national geography standards were published in the volume *Geography for Life* (GESP 1994).

*Geography for Life* remains the definitive statement of what every school-child should know, understand, and be able to do as a geographically informed person (Figure 23.1). According to *Geography for Life*, being geographically informed means having a mind inquisitive of space, place, region, and scale. It means understanding the interdependence of human and physical systems. It means being competent with the analytical power of geographic information systems, maps, satellite imagery, and other tools and technologies that help people visualize spatial patterns and relationships. And it means knowing how to use geographic knowledge not only to interpret the causes, meanings, and effects of past, present, and future events, but also to improve the habitability of Earth – "the only home that humans know or are likely to know" (GESP 1994: 23).

The 18 standards of *Geography for Life* were written as voluntary guidelines for states and school districts. By the year 2000, 45 out of 50 states had developed their own geography standards, although some states have been far more successful than others in writing academically challenging standards and in supporting their new standards with strong accountability measures (Munroe 2000).

Of course, educational change is expensive, and geography's renaissance in the nation's schools probably could not have happened without the financial support of organizations such as the National Geographic Society (NGS). By funding a national network of Geographic Alliances, the NGS supports the professional development of geography teachers, develops new curriculum materials, and promotes public awareness of geography during Geography Awareness Week every November. The Alliance network exists in all 50 states and has become a central force in placing geography in state curriculum frameworks.

## THE ASSESSMENT: BEGINNING TO SEE THE LIGHT

So what have we reaped from the efforts dedicated to K-12 geography education? A NAEP Geography report released in 2002 shows that average geography scores of the nation's fourth and eighth-graders rose significantly since the last NAEP exam in 1994, whereas scores for high school seniors were not significantly better. The improvement at grades four and eight was among lower-performing students, and although there continues to be wide disparities in test scores between whites and ethnic minorities, the scores for black fourth graders improved as well. Notably, school districts with teachers who participated in Geographic Alliance teacher training programs reported higher than average scores for their students (NCGE 2002).

The modest gains in NAEP geography scores are cause for celebration, but must be tempered with some sobering facts. Because the NAEP test organizes geographic concepts somewhat differently than *Geography for Life*,

## Essential Element 1: The World in Spatial Terms

*Standard 1:*         How to use maps and other geographic representations, tools, and technologies to acquire, process, and report information from a spatial perspective.
*Standard 2*: How to use mental maps to organize information about people, places, and environments in a spatial context.
*Standard 3*: How to analyze the spatial organization of people, places, and environments on earth's surface.

## Essential Element 2: Places and Regions

*Standard 4*: The physical and human characteristics of places.
*Standard 5*: That people create regions to interpret earth's complexity.
*Standard 6*: How culture and experience influence people's perceptions of places and regions.

## Essential Element 3: Physical Systems

*Standard 7:* The physical processes that shape the patterns of earth's surface.
*Standard 8*: The characteristics and spatial distribution of ecosystems on earth's surface.

## Essential Element 4: Human Systems

*Standard 9:* The characteristics, distribution, and migration of human populations on earth's surface.
*Standard 10:* The characteristics, distribution, and complexity of earth's cultural mosaics.
*Standard 11:* The patterns and networks of economic interdependence on earth's surface.
*Standard 12:* The processes, patterns, and functions of human settlement.
*Standard 13:* How the forces of cooperation and conflict among people influence the division and control of earth's surface.

## Essential Element 5: Environment and Society

*Standard 14:* How human actions modify the physical environment.
*Standard 15:* How physical systems affect human systems.
*Standard 16:* The changes that occur in the meaning, use, distribution, and importance of resources.

## Essential Element 6: The Uses of Geography

*Standard 17:* How to apply geography to interpret the past.
*Standard 18:* How to apply geography to interpret the present and plan for the future.

*Figure 23.1.* The 18 National Geography Standards. In *Geography for Life*, each standard is organized under one of six "essential elements" (in boldface) and introduced with the phrase, "The geographically informed person knows and understands . . ." (GESP 1994). Photos by David Rutherford (used with permission).

it is difficult to attribute the observed gains in student performance directly to the national dissemination of the geography standards. Although many states and school districts mandate that geography be taught in public schools, the quick pace of reform has left many smaller states and Alliance programs behind. And although researchers have shown the Geographic Alliances to be successful vehicles for promoting the abilities of geography teachers (Cole and Omrod 1995; Hill and Lockyear-Collop 1997), it has become evident that the Alliance network alone cannot satisfy the growing local demands for teachers with a strong background in geography.

## REFERENCES

Cole, D. and J. Omrod. 1995. Effectiveness of Teaching Pedagogical Content Knowledge Through Summer Geography Institutes. *Journal of Geography* 94: 428–433.

Geography Education Standards Project. 1994. *Geography for Life: National Geography Standards 1994.* Washington, D.C.: National Geographic Research and Exploration.

Hill, A. D. and E. Lockyear-Collop. 1998. Valuing Professional Development in the Creation of the Best Geography Teachers. *International Research in Geographical and Environmental Education* 7: 142–145.

Munroe, S. 2000. The State of State Standards in Geography. In C. Finn and M. Petrelli, eds. *The State of State Standards*, 15–18. Washington, D.C.: Fordham Foundation.

National Council for Geographic Education. 2002. NAEP Results: In Small Gains, Weaker Fourth and Eighth Graders Get Stronger. *Perspective* 30(6): 1–3.

National Research Council. 1997. *Rediscovering Geography: New Relevance for Science and Society.* Washington, D.C.: National Academy Press.

Persky, H., C. Reese, C. O'Sullivan, S. Lazer, J. Moore, and S. Shakrani. 1996. *NAEP 1994 Geography Report Card: Findings From the National Assessment of Educational Progress NCES 96087.* Washington, D.C.: National Center for Education Statistics, U.S. Department of Education.

Sedaris, D. 2000. *Me Talk Pretty One Day.* Boston, MA: Back Bay Books.

## PART III

# ENHANCING LIFE IN CITIES

Half the human race today lives in cities, and the majority will do so in the near future. The complexity of cities and the rapidity of their change add to the severity and intensity of the problems they face – congestion, conflicts over the juxtaposition of activities, and the struggles of people from different backgrounds to live together. Yet, as dense nodes of interaction cities have also traditionally harbored and harnessed the creative energies for humanity's most productive enterprises. Geographers have shown a long-standing interest in the structure and functioning of cities. The geographical perspective adds an essential dimension for capturing the benefits of high densities and for reducing the costs of incompatibilities in the patterns and process of land use and urban social change.

*Barney Warf*

# 24

# POVERTY AND GEOGRAPHICAL ACCESS TO EMPLOYMENT: MINORITY WOMEN IN AMERICA'S INNER CITIES

SARA McLAFFERTY
*University of Illinois*

VALERIE PRESTON
*York University*

The feminization of poverty and the growing concentration of poor people in inner-city neighborhoods are two of the most enduring trends in American cities. More poor people live in cities, and an increasing fraction of the urban poor are women and children. Minority women, often the main bread-winners for their families, are especially vulnerable to poverty. This essay explores how geographical barriers contribute to poverty and joblessness for minority women in inner-city areas and how social policies, including welfare reform, are affecting these women's access to stable, well-paid jobs.

## THE GEOGRAPHICAL BASES OF DEPRIVATION

Decades of research show that poverty and unemployment reflect the interlocking effects of social and geographical deprivation. For minority women in inner-city communities, deprivation is partly rooted in *place* – in the communities in which women live and how those communities are situated within larger urban, regional, and global systems. The localized availability of jobs, transportation, housing, and social, retail, and educational services critically affects women's workforce participation.

The *spatial mismatch* between jobs and minority workers is a crucial factor in urban poverty. Spatial mismatch refers to the combined effects of residential segregation and the suburbanization of jobs on access to employment. Living in segregated inner-city neighborhoods, distant from and poorly connected to areas of employment concentration and growth, minority residents have few local job opportunities available and difficulty traveling to more distant opportunities (Preston and McLafferty 1999). In many cities, the propensity to work in one's local neighborhood is lower in inner-city, African-

*D. G. Janelle et al. (eds.), WorldMinds: Geographical Perspectives on 100 Problems*, 145–149.
© 2004 *Springer. Printed in the Netherlands.*

American neighborhoods than in other communities (Immergluck 1998). Long commutes are not compensated by higher wages for many inner-city residents (Johnston-Anumonwo 1997). These spatial mismatch constraints are especially important for minority women, who earn less than their male counterparts and must juggle household and childcare responsibilities along with paid employment.

Transportation also poses a significant barrier to employment for inner-city women. American cities are built around the automobile – the most flexible form of transportation; yet inner-city minority women have the lowest rates of car ownership of any group, making them dependent on slow public transit in traveling to and from work. Furthermore, the hub-and-spoke design of most mass transit systems is poorly structured for reverse commuting to suburban job centers. Trips from inner-city areas to suburban job hubs can take one to two hours each way – a significant time burden for women with responsibilities at home and in the paid workforce.

Geographical barriers to employment are compounded by the limited availability and poor quality of services in inner-city neighborhoods (Wilson 1996). Lack of childcare and after school programs, the poor quality of local schools, and high rates of crime limit the place-based support networks women need in order to find and keep good jobs. Although women actively create such networks to provide for themselves and their families (Gilbert 1998), the process is much more difficult in the service-poor environment of inner city neighborhoods.

Geographical barriers cannot be isolated from the constellation of forces that contribute to poverty and unemployment for inner city women. Other important factors include the lack of skills and education needed to obtain jobs in the "new economy," employer discrimination in hiring, and health problems that limit the ability to work. These intersect with geographical barriers, leading to complex patterns of disadvantage. Improving geographical access to job opportunities and improving the place-based support systems that make it possible to work offer important tools for reducing poverty among inner-city women and enhancing social and economic well-being.

## ANTI-POVERTY POLICIES

Congressional passage of the Personal Responsibility and Work Opportunity Reconciliation Act (PRWORA) in 1996 signaled a fundamental shift in federal policies towards poor women – a shift from welfare to workfare, a change associated with the rise of neo-liberal politics in many western industrialized countries (Peck 2001). Workfare describes a concerted effort to remake the lives of many inner-city women by encouraging and coercing them to move from welfare into paid employment, no matter how insecure or poorly paid. Workfare failed to acknowledge the social and spatial constraints on women's ability to find well-paid employment. Recognizing the limitations of workfare, new policies are being developed to reduce poverty among inner-city women. These policies concentrate on three different strategies; improving geographical access to existing job opportunities, enhancing women's place-based support systems and improving the job opportunities in inner-city neighborhoods.

Improvements to geographical access take several forms. Transportation subsidies that reduce the cost of work trips have been an important tool for enabling and enhancing workforce participation. Subsidies have been provided through direct financial assistance to unemployed families and through programs that pay employers' costs of providing transportation for workers who were formerly welfare-recipients. Moving beyond subsidies, the Job Access program, part of the Transportation Equity Act for the 21st Century, funds the development of transportation services that link isolated workers with employment centers and other employment-related services. Similarly, the Reverse Commute program funds transportation services to suburban employment centers for all populations. These programs require a collaborative planning process among nongovernmental and governmental agencies at state and municipal levels and employers. Providing geographical access to employment has emerged as a logistical problem that may be solved readily with adequate funding and careful planning (Palubinsky and Watson 1997).

Despite their success in improving geographical access to employment, transportation policies do not benefit all women equally. Minority women with higher levels of education and work experience are more likely to take advantage of transportation subsidies than are other women (Chapple 2001). For women with high school education or less and little work experience, building links to employers, particularly by increasing the density of economic and social opportunities in women's neighborhoods, is likely to be a more effective way to ameliorate poverty. Transportation policies also do not help women who encounter place-based social barriers to employment. Fear of crime, and the lack of quality childcare and education services inhibit women's labor force participation, and these barriers are not addressed in transportation programs.

Policies that address social barriers to employment include the Moving to Opportunity program that provides housing subsidies to public housing families to assist them in moving out of extremely poor neighborhoods. Several hypotheses concerning neighborhood effects on employment guided the program including the expectation that living in low-poverty areas would offer a broader range of employment opportunities, reduce job search and commuting costs, and expose welfare recipients to community norms that are more supportive of work. Relocation was also expected to improve physical health and reduce exposure to stress and danger in the local environment, perhaps contributing to increased employment and earnings. Demonstration programs were implemented in Baltimore, Boston, Chicago, Los Angeles, and New York between 1994 and 1999. Preliminary evaluations indicate that these demonstration programs have had mixed results. By 2001, there was no significant difference between the employment rates for households that had moved and those that had not moved (Popkin et al. 2002). The impacts of relocation on employment were limited by psychological factors such as motivation and social barriers to employment such as medical problems, limited education, a criminal record, and inadequate transportation. The findings confirm that poor women are a diverse group whose needs will not be met by programs that adopt a "one-size-fits-all" strategy (Chapple 2001).

An alternative approach is to emphasize economic development and job creation in inner-city areas. These place-based policies involve the provision of subsidies, training programs and infrastructure to attract new industries to inner-city communities. The policies go by various names – the current incarnation is enterprise zones – and, in general, they have had limited success (Greenbaum and Engberg 2000). Many have failed to generate sustained positive impacts on employment or income, because the policies did not alter the profitability of inner-city locations and benefits leaked out from inner-city neighborhoods to other areas.

## A GEOGRAPHICAL ASSESSMENT

Many welfare-to-work programs now recognize the isolated geographical context in which many poor women live in American cities. The programs are gender-sensitive to the extent that they target female-headed single-parent households, the majority of the urban poor. Programs now take some account of women's child-care responsibilities through child care subsidies and transportation services to child care centers. Even though experts are still debating the adequacy of current subsidies, their existence represents an important step forward.

Nevertheless, current policies pay little attention to the occupational segregation that contributes to women's poverty (Hanson and Pratt 1995). Women work in a relatively small number of female-dominated occupations that pay less than occupations dominated by men. Although federal money is available to promote the hiring of low-income persons in sectors such as transportation, policy statements rarely mention the occupations in which women will be hired. The omission is serious in light of growing evidence that poorly paid jobs discourage successful transitions from welfare to work (Popkin et al. 2002).

Reducing poverty among minority women in inner city communities requires multi-pronged approaches that are sensitive to the particularities of place, race, and gender. The most effective mix of policies is likely to vary among cities depending on the density and patterning of urban development, transportation infrastructure, and social-educational services. Policies also need to recognize the strong interconnections between geographical disadvantages and disadvantages based on race, gender, and class. By addressing these issues, geographical research will contribute to the elimination of racial and gender discrimination in job and housing markets, and improvements in health, education, and transportation services in inner city neighborhoods – critical steps in ameliorating the life chances for poor urban women and their families.

## REFERENCES

Chapple, K. 2001. Time to Work: Job Search Strategies and Commute Time for Women on Welfare in San Francisco. *Journal of Urban Affairs* 23: 155–173.

Gilbert, M. 1998. "Race," Space and Power: The Survival Strategies of Working Poor Women. *Annals of the Association of American Geographers* 88: 595–621.

Greenbaum, R. and J. Engberg. 2000. An Evaluation of State Enterprise Zone Policies. *Policy Studies Review* 17: 29–46.

Hanson, S. and G. Pratt. 1995. *Gender, Work, and Space.* London and New York: Routledge.

Immergluck, D. 1998. Job Proximity and the Urban Employment Problem: Do Suitable Nearby Jobs Improve Neighborhood Employment Rates? *Urban Studies* 35: 7–23.

Johnston-Anumonwo, I. 1997. Race, Gender and Constrained Work Trips in Buffalo. *The Professional Geographer* 49: 306–317.

Palubinsky, B. Z. and B. H. Watson. 1997. *Getting From Here to There, The Bridges to Work Demonstration First Report to the Field.* Philadelphia, PA: Field Report Series, Public/Private Ventures.

Peck, J. 2001. *Workfare States.* New York and London: Guilford Press.

Popkin, S., L. Harris, and M. Cunningham. 2002. *Families in Transition: A Qualitative Analysis of the MTO Experiment, Final Report.* Washington, D.C.: Office of Policy Development and Research, U.S. Department of Housing and Urban Development.

Preston, V. and S. McLafferty. 1999. Spatial Mismatch Research in the 1990s: Progress and Potential. *Papers in Regional Science* 78: 387–402.

Wilson, W. 1996. *When Work Disappears: The World of the New Urban Poor.* New York: Knopf.

# 25

# ETHNIC SEGREGATION: MEASUREMENT, CAUSES, AND CONSEQUENCES

DAVID H. KAPLAN
*Kent State University*

The tendency of people to separate themselves into groups is well known and well studied. Social scientists of all persuasions analyze the manner by which groups are formed and the ramifications of these formations. Geographers examine these groups over the landscape, attempting to understand the spatial distribution of groups, the processes by which distributions occur, and why such group formation is significant.

Ethnic segregation refers to the geographical separation of people on the basis of ethnicity. The term "race" is often used as well, but both race and ethnicity are constructions based on the categorization of particular groups of people by self-identified members of the groups themselves and by people outside the group. Ethnic segregation, most often considered within an urban setting with the appearance of ethnic neighborhoods, quarters, and ghettoes, became an object of academic concern as cities grew in size, area, and scope.

## SEGREGATION AS A PROBLEM

Why is segregation a problem? It certainly does not have to be. After all, people may choose to live with others for whom they feel a strong affiliation. We note concentrations of Amish, for example, with affection and not alarm. Yet, in many cases, ethnic segregation – especially when characterized as "racial" segregation – is viewed with concern. Because the patterns of people on the ground reflect broader social relations, segregation may be recognized as a problem of group incorporation and urban fragmentation.

More specifically, ethnic segregation may foster the spatial isolation of a particular group. In this regard, ethnic segregation has the capacity to influence the members of a group in myriad ways. Often, the segregation of a group corresponds with substandard housing and urban services as groups are channeled into neighborhoods with limited quality housing stock, fewer retail outlets, poor schools, and inadequate services. Segregation can create a gap between the power of residents to affect policy and the extent to which public policy affects them. Choices about where to locate highways, for instance, may

*D. G. Janelle et al. (eds.), WorldMinds: Geographical Perspectives on 100 Problems*, 151–156.
© 2004 *Springer. Printed in the Netherlands.*

be decided without feedback from residents, even though these residents will clearly be impacted. Segregation fosters a perceptual connection between group and place. Ghettoization becomes another method to stigmatize a population by stigmatizing the neighborhoods they inhabit.


## ANALYZING SEGREGATION

While segregation as a phenomenon has been with us since the earliest cities, systematic study began with the Chicago School of urban sociologists. These scholars looked at ethnic communities within growing cities as akin to biotic communities. Over time, new ethnic groups would enter into the neighborhood (invasion) and then replace the existing ethnic population with a new ethnic population (succession).

The examination of segregation was developed further by sociologists like Karl Taeuber and Stanley Lieberson. They used the Index of Segregation to measure the relative evenness of the population. This index calculated what percentage of a particular group would have to move in order for their spatial distribution to be the same as the overall population. A Segregation Index of 0 meant that the group was perfectly integrated, while a Segregation Index of 1 (or 100) indicated that the group was completely segregated. Other measures were also developed. The importance of these techniques is that the geographical distribution of groups could now be quantified. Geographers made segregation come alive with the use of maps. For example, Figure 25.1 shows how Chicago's "black belt" expanded and deepened over time. With their knowledge of spatial techniques, geographers have been able to contribute to further refining these measures, especially with the widespread adoption of geographic information systems (GIS). Wong (2002), for instance, provided a means of extending ArcView (a popular GIS package) with spatial segregation measures that incorporate the boundaries between different spatial units.


## EXPLAINING SEGREGATION

Given the facts of segregation, there has been the question of why. If segregation is viewed as a problem – and for the most part it has been – then an understanding of the causes can point the way to a solution. Clearly, in the case of certain groups, such as the African Americans, segregation was a historically imposed condition. The location of African Americans was dictated by Jim Crow type laws and reinforced by discrimination. An understanding of these causes led the way to clear solutions. Since the establishment of fair housing guidelines, however, segregation has continued, and geographers have been involved in ascertaining the reasons. In the 1960s, Morrill (1965) demonstrated the importance of "tipping" in generating black segregation even

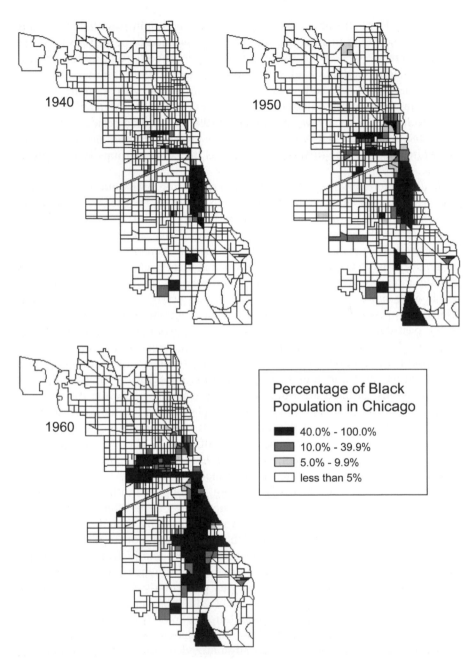

*Figure 25.1.* This map shows the concentrated growth of the African American population in Chicago between 1940 and 1960. Reproduced with permission from the University of Chicago Press from Arnold R. Hirsch, 1998. Making the Second Ghetto. *Race and Housing in Chicago, 1940–1960.* Second Edition: 6–8.

without any legal strictures, while at the same time implicating the real estate, financial, and governmental institutions that continued to perpetuate it. White homeowners were likely to move out of neighborhoods with a high proportion of black residents and few white residents would move into such neighborhoods. The result was that neighborhoods with a small African American population could tip into a predominantly African American neighborhood in just a few years. Clark (1991) examined residential preferences and concluded that African Americans prefer integrated neighborhoods, but whites do not. The differing preferences of the two groups lead to segregation.

While preferences may play a role in perpetuating segregation in American cities, the geographic community has been involved in researching some of the remaining ways in which it can be perpetuated and deepened. One mechanism is through the home mortgage loan sector. Early research by Harvey and Chatterjee (1974) exposed the inequities of mortgage credit in Baltimore. Later, geographers used special data to determine that there was less mortgage activity in predominantly African American neighborhoods and that these neighborhoods were more likely to suffer much higher mortgage denial rates (Shlay 1989). This research opened the door to more regulations of lending activity. In the past decade, another segregation related problem – that of predatory lending – has come to the fore. In this instance, lenders may actively seek out certain populations and neighborhoods and sell residents a mortgage loan with interest rates and fees that are onerous. Foreclosure may result. The neighborhoods targeted are often segregated and poor, and the impact is to extract a tremendous quantity of assets from places with limited assets to begin with. Existing research on this issue is still new but, among geographers, this author has analyzed the activities of one predatory lender in the Washington, D.C. metropolitan area and noted the coincidence of loans with blocks that were highly segregated and black.

The role of segregation in stigmatizing a group is another area for geographic inquiry. The stereotype that members of an ethnic group were unhygienic and squalid, for instance, was bolstered by their forced concentration into neighborhoods filled with substandard housing. Turn-of-the-20th-century immigrant communities suffered especially from this linkage, fostering lingering perceptions that went on to influence public policy. The Chinese population on the west coast was particularly victim to the conflation of group and place. Anderson's (1987) study of Vancouver's Chinatown details how the growing Chinese population was cordoned into an undesirable section of Vancouver, given few options about where and how to live within this neighborhood, and then stigmatized by the association between the view of "Chinatown" and the Chinese group. Here, an understanding of the connections between group and place helps to engender a greater awareness of the origins of prejudice.

One aspect of segregation rests in its broader impact on the economic and social opportunities of people who live within segregated neighborhoods. Geographers have contributed to an investigation of segregation's varied impacts, from the perpetuation of an urban "underclass" to the promotion of ethnic economies. In 1971, Harold Rose looked at some of the political consequences of residence within the African American ghetto, focusing upon its role in fostering conditions for collective violence or riots. While well-known sociologists like William Julius Wilson have described the economic desperation of America's segregated inner cities, geographers have helped to more clearly define the geographic boundaries of these blighted zones, pointing the way to place-based solutions. Some geographers have examined the so-called spatial mismatch theory that posits that inner-city jobs have become scarcer, and residents of these zones – often segregated – have suffered as a consequence. Holloway (1996) found that the impacts of spatial inaccessibility for African-American teenagers had diminished between 1980 and 1990. Kaplan's (1999) study suggested, however, that segregated African American communities in Cleveland had only about one-half as many nearby job opportunities in 1990 as predominantly white neighborhoods.

Finally, geographers have begun to examine the correspondence between segregation and ethnic economic activity. McEvoy (Aldrich et al. 1985), in his study of British cities, noted the importance of protected markets. Li (1998) focused her attention on suburban ethnic enclaves – what she terms "ethnoburbs" – especially among the Chinese population of Los Angeles's San Gabriel Valley that have become more economically complete. While such suburban zones are a far cry from the traditional inner-city ghettoes, they are still segregated ethnic neighborhoods. And, they have become more economically complete over time, with financial and other producer services added to the stores and restaurants often associated with ethnic economies.

## CONCLUSIONS

Ethnic segregation is a phenomenon that covers the circumstances of many different groups in many different contexts. As a result, each situation demands to be treated on its own terms. Yet ethnic segregation has played a dominant role in shaping our cities and it has brought in its wake problems that may require some intervention. When segregation is the consequence of enforced discrimination (whether legal or not), the resolution is clear. When segregation's cause is more ambiguous, then the possible styles of intervention may vary. Geographers have studied segregation in concert with a number of other social science scholars, but they have helped guide the debate through clearly defining segregation's spatial dimensions, causes, and consequences.

# REFERENCES

Aldrich, H., J. Cater, T. Jones, D. McEvoy, and P. Velleman. 1985. Ethnic Residential Concentration and the Protected Market Hypothesis. *Social Forces* 63: 996–1009.

Anderson, K. 1987. The Idea of Chinatown: The Power of Place and Institutional Practice in the Making of a Racial Category. *Annals of the Association of American Geographers* 77: 580–594.

Clark, W. 1991. Residential Preferences and Neighborhood Racial Segregation: A Test of the Schelling Segregation Model. *Demography* 28: 1–19.

Harvey, D. and L. Chatterjee. 1974. Absolute Rent and the Structuring of Space by Governmental and Financial Institutions. *Antipode* 6(1): 22–36.

Hirsch, A. 1983 [1998]. *Making The Second Ghetto: Race and Housing in Chicago, 1940–1960.* Cambridge, U.K.: Cambridge University Press/Chicago: University of Chicago press.

Holloway, S. 1996. Job Accessibility and Male Teenage Employment, 1980–1990: The Declining Significance of Space? *The Professional Geographer* 48: 445–458.

Kaplan, D. 1999. The Uneven Distribution of Employment Opportunities: Neighborhood and Race in Cleveland, Ohio. *Journal of Urban Affairs* 21: 189–212.

Li, W. 1998. Anatomy of a New Ethnic Settlement: The Chinese Ethnoburb in Los Angeles. *Urban Studies* 35: 479–501.

Morrill, R. 1965. The Negro Ghetto: Problems and Alternatives. *Geographical Review* 55: 339–361.

Rose, H. 1971. *The Black Ghetto: A Spatial Behavioral Perspective.* New York: McGraw-Hill.

Shlay, A. 1989. Financing Community: Methods for Assessing Residential Credit Disparities in the Metropolis. *Journal of Urban Affairs* 11: 201–223.

Wong D. 2002. Spatial Measures of Segregation and GIS. *Urban Geography* 23: 85–92.

# 26

# FLAT BREAK-UPS: THE BRITISH CONDOMINIUM CONVERSION EXPERIENCE

CHRIS HAMNETT
*Kings College London*

This chapter details my involvement in both basic research and a govern-ment committee of inquiry, where I was appointed research director into the problems experienced in the apartment sector in England and Wales in the 1970s and 1980s. This involvement led to a series of proposals produced by the Committee that fed directly into the Landlord and Tenant Act of 1987, which received multi-party support in the Houses of Parliament and was passed into law. The work made a significant improvement to the conditions and rights of tenants and long leaseholders in Britain. Although the legislation did not go far enough, it provided the basis for subsequent legislation, which has further strengthened the rights of these groups and reduced the power of preda-tory landlords. This chapter sets out the background to the problem, the nature of the problems, and the work of the Committee.

## BACKGROUND

Central and inner London is generally seen as a city made up of terraced houses built in the eighteenth and nineteenth centuries, while suburban London consists of semi-detached houses built in the 1920s and 1930s (Jackson 1974). In the post-war period, large parts of inner London and other major cities were also redeveloped in the form of high-density social housing estates (Dunleavy 1981; Garside 1983). But London also has a large number of pri-vately rented apartment blocks, mostly built either in the late nineteenth century or in the 1930s. These were owned, for the most part, either by private resi-dential property companies or by large institutional landlords, including insurance companies. The apartment blocks were concentrated in central and inner London and provided a high standard of market accommodation, primarily for upper middle class tenants (Hamnett and Randolph 1985, 1988). They are akin to some of the luxury apartments on the upper east and west sides of New York or Lakeshore Drive in Chicago.

Three things happened in the late 1960s onwards. First, there was increasing criticism of the dubious practices of some "cowboy" landlords that gave private

*D. G. Janelle et al. (eds.), WorldMinds: Geographical Perspectives on 100 Problems, 157–161.*
© 2004 *Springer. Printed in the Netherlands.*

landlordism a bad name. Second, a combination of tough rent control and security of tenure legislation reduced the attractiveness of private landlordism to many institutional owners. They were looking for an escape route. This search was enhanced in the early 1970s by the first of the post-war house price booms, which doubled residential property prices in Britain between 1971 and 1973. This process made selling privately rented apartments a very attractive proposition for landlords hit by rent control (Hamnett and Randolph 1984, 1986, 1988).

The changing economic conditions provided the context for the emergence of a new type of owner: the speculative property company, who bought blocks of apartments from their traditional owners with a view to selling off individual rented flats for owner occupation. Their objective was simple: short-term profit maximization. To achieve this, they needed to sell as many flats as possible in as short a time. These could either be sold to existing "sitting tenants" who had security of tenure, or vacant flats could be sold on the market. Vacancies could be achieved either by offering existing tenants a cash sum to vacate the unit, by sales on the death of elderly tenants, or by making life difficult for tenants who refused to buy or vacate. The latter could be done by raising service charges, which landlords are allowed to levy for maintenance and upkeep of the common parts, or by other means.

Unlike the United States, where a rented building has to be converted outright into a condominium in which individual buyers own a share of the common parts of the building once converted, English property law distinguishes between freehold ownership, where an owner owns both the building and the ground on which it stands, and leasehold ownership, where the individual owns a long lease on an individual apartment. This would generally be 99 years, but can be longer (up to 999 years). In these cases, the freeholder retains the ownership of the land and the common parts of the building and the long leasehold buys a long lease on an individual apartment. As a result, it is possible for the freeholder to sell individual long leaseholds on apartments in a building where they retain freehold. The freeholder is responsible for the overall maintenance of the building and for ensuring that building works are done, and insurance is purchased, etc. At the expiration of the long leasehold, the apartment reverts to the freehold owner.

## THE TENANTS' PROBLEMS

This division of ownership rights and responsibilities offers the possibility for a variety of conflicts between freeholder and leaseholder and for ruthless freeholders to exploit the situation of the leaseholder. In the early mid-1970s, a growing number of complaints were recorded in the press and raised with central London Members of Parliament. The problems concerned rapidly rising service charges, poor services, high charges for building work, and failure of the landlord to respond to tenants' complaints. This problem was exacerbated by the fact that most landlords operated via management companies, and many of them were holding companies based overseas for tax reasons. Complaints were simply fielded by the management companies and tenants felt that they were unable to contact the landlords or often to even know who they were. They were on the receiving end of a process of tenure change over which they had little or no control.

## THE RESEARCH

I was intrigued by this new phenomenon on which no academic research had been done, and I obtained a small ESRC grant in 1979 to examine it. Bill Randolph was my researcher and we set out to document the scale and pace of the transformation that was taking place, the identity of the owners, and the causes of the process. The empirical work lasted two years, and we gained a good picture of the importance of this sector of the housing market in central and inner London. In theorizing the economics of tenure change, we developed the idea of a "value gap" between the tenanted capital value of flats (measured in terms of a multiple of annual rental income) and their potential capital value if sold for owner occupation. The objective of the new breed of speculative landlords was to capture this value by the sale of rented flats aided by generating additional income from increases in service charges and the like.

## THE COMMITTEE OF INQUIRY AND ITS FINDINGS

The significance of the research only emerged in 1984, when, in response to growing volume of complaints, the Minister of Housing established a Committee of Inquiry on the Management of Privately Owned Blocks of Flats, chaired by Edward Nugee (Department of the Environment 1985). Its terms of reference were simple: "To collect and examine evidence on the nature, scale and incidence of problems for landlords and tenants arising from the management of privately owned blocks of flats; to assess the difficulties caused by the these management problems and to make recommendations on how they might be resolved." As the only academic who had undertaken research on this topic, I was invited to become research director of the committee. The composition of the committee was important as it contained both landlord and tenant representatives as well lawyers. Problems arose immediately as one section of the committee felt that the group already knew the problems and did not need to undertake any systematic new research. We argued against this view, and a large program of research was initiated, including a large-scale write-in survey of problems experienced by landlords and tenants. It was crucial to include both, as it was politically important that the committee examined all sides of the issue. It quickly became clear, however, that a significant proportion of tenants were experiencing a wide variety of problems. The results of the tenant surveys were queried as to their representativeness and accuracy and it was argued that a write-in survey would have probably overstated the extent and scale of the problems, as it would contain a disproportionate number of tenants with problems. As a result, a third, random national survey was initiated which resolved these issues.

The politics of the committee were fascinating. The landlords' representatives were slow to acknowledge the severity of the problems and, when accepted, they and the lawyers were extremely reluctant to seriously infringe the property rights of landlords or to move towards any right of tenant right of collective purchase of badly managed blocks of flats or to reform leasehold property law. The proposals for change that were accepted were therefore marginal and incremental, though nonetheless significant. They included, for example, a provision that all demands for payment of service charge and building work charges had to have a British address to which tenants could serve a legal notice. If there was no such address the service charge demand was deemed invalid. Landlords also had to provide proof to tenants that they had taken out valid building insurance before they could claim payment for insurance. After publication of the report and acceptance of its main findings by the minister, it was then passed to the lawyers who attempted to put several of its key proposals into government legislation. As the bill was going through Parliament in 1987, Margaret Thatcher announced a general election. This generally causes all unapproved legislation to fall, but quite remarkably all parties unanimously accepted the provisions of the bill and it was rushed through Parliament and into law. We were very fortunate, as were the hundreds of thousands of affected tenants and long leaseholders, whose interests have been further protected by subsequent legislation on leasehold reform which has extended the rights of leaseholders to buy the freehold of their home even if their landlords are opposed. To this extent, the legislation opened the door to a raft of later reforms that have modernized antiquated real property law in Britain.

## IMPLICATIONS

I learned several lessons from the research. First, policy relevance does not come just from policy-relevant research. Our research was fundamentally not policy driven, and the policy implications only emerged subsequently. Second, when doing survey research with policy implications, do not cut corners to save time. Opponents will always manage to pick some holes in the methodology. Third, do not back off from difficult findings or policy implications: once must hold his or her ground because opponents will hold theirs. Finally, getting involved in policy research and formulation can be immensely rewarding. If one is fortunate, it can enable research to make a positive difference to people's lives rather than simply being read by a small number of specialists in academic journals.

## REFERENCES

Department of the Environment. 1985. *Report of the Nugee Committee of Inquiry on the Management of Privately Owned Blocks of Flats*, volumes I and II. London: HMSO.

Dunleavy, P. 1981. *The Politics of Mass Housing in Britain, 1945–75*. Oxford: Oxford University Press.

Garside, P. 1983. Intergovernmental Relations and Housing Policy in London 1919–1970: The Density and Location of Council Housing. *The London Journal* 9(1): 39–57.

Hamnett, C. 1999. *Winners and Losers: The Home Ownership Market in Modern Britain*. London: Taylor and Francis.

Hamnett, C. and W. Randolph. 1984. The Role of Landlord Disinvestments in Housing Market Transformation: An Analysis of the Flat Break-up Market in Central London. *Transactions of the Institute of British Geographers* 9: 259–279.

Hamnett, C. and W. Randolph. 1985. The Rise and Fall of London's Purpose Built Blocks of Privately Rented Flats: 1853–1983. *The London Journal* 11(2): 160–175.

Hamnett, C. and W. Randolph. 1986. Tenurial Transformation and the Flat Break-up Market in London: The British Condo Experience. In N. Smith and P. Williams, eds. *Gentrification of the City*. New York: Allen and Unwin

Hamnett, C. and W. Randolph. 1988. *Cities, Housing and Profits: Flat Break-up and the Decline of Private Renting*. London: Hutchinson.

Jackson, A. 1974. *Semi-Detached London*. London: Allen and Unwin.

# 27

# THE GEOGRAPHY OF ENVIRONMENTAL INJUSTICE IN THE BRONX, NEW YORK CITY

JULIANA A. MAANTAY
*Lehman College/City University of New York*

Although the mainstream environmental movement of the 1950s and 1960s alerted the public to the dangers posed by pollution and environmental degradation, these impacts on people's health and the environment were not generally acknowledged (or thought) to be spatially or socially differentiated: everyone was presumed to be affected just about equally. The understanding that environmental problems may impact certain locations and people more than others (and in a predictable pattern based on race and income) is a relatively new concept called environmental injustice.

Environmental injustice can be defined as the disproportionate exposure of communities of color and the poor to pollution, and its concomitant adverse effects on health and the environment. It also encompasses the unequal environmental protection and environmental quality provided by laws, regulations, governmental programs, enforcement, and policies (Bryant 1995), the limited participation of people of color in mainstream environmental organizations, and the general loss of control over the environment experienced by marginalized groups (Pulido 1996). The definition is often extended to include all vulnerable populations, such as the very young, the elderly, infirm, pregnant, immune-compromised, and future generations (Greenberg 1993).

## GEOGRAPHICAL ANALYSES OF ENVIRONMENTAL INJUSTICE

Within the past decade, it has become increasingly prevalent to "map" instances of environmental injustice. This has usually been done with geographic information systems (GIS) to plot facilities or land uses suspected of posing an environmental and human health hazard and then to determine the racial, ethnic, and economic characteristics of the potentially affected populations in comparison to a reference population. This approach often results in dramatic maps showing toxic facilities concentrated in areas with high proportions of African-Americans, Latinos, Native Americans, and/or poor people (cf. Maantay 2002 for a review of such studies). The geographic analysis of

*D. G. Janelle et al. (eds.), WorldMinds: Geographical Perspectives on 100 Problems*, 163–169.
© 2004 *Springer. Printed in the Netherlands.*

environmental injustice has led to wide spread public awareness and political acknowledgement of its existence, which in turn has helped hasten the implementation of new laws and policies to address the problem. It has also fostered, in many cases, community empowerment and increased self-determination.

The groundbreaking environmental justice study *Toxic Wastes and Race in the United States: A National Report on the Racial and Socio-Economic Characteristics of Communities With Hazardous Waste Sites* was produced in 1987 under the auspices of the United Church of Christ's Commission for Racial Justice (United Church of Christ 1987). This report presented maps of the locations of the country's hazardous waste facilities in conjunction with the characteristics of the nearest populations (by zip code), using indicators such as race, ethnicity, and income. As compared to the areas that were not hosts to a hazardous waste facility, the host areas showed an unmistakable statistical and spatial correspondence to "minority" populations. The term "minority" is used in many of these studies and refers generally to those who are not Non-Hispanic White.

## ENVIRONMENTAL INJUSTICE MAPPING AND POLITICAL ACTION

If *Toxic Wastes and Race* was the seminal study that helped propel the issue of environmental justice to the forefront of the public's consciousness in the late 1980s and early 1990s, it was certainly not the first environmental justice analysis. These issues have been researched extensively since at least the late 1960s, and study after study throughout three decades has shown the existence of disproportionate environmental impacts based on race and/or income (Goldman 1993).

The Commission for Racial Justice's report is an example of how geographical analysis has been instrumental in promoting awareness of environmental injustice and in focusing attention on possible public policy and regulatory solutions. The report galvanized the environmental justice movement, and was a major impetus to the organization of a national environmental justice conference in 1991, the First National People of Color Environmental Leadership Summit, held in Washington, D.C. (Bullard 1994). Shortly thereafter, in 1994, President Clinton issued Executive Order 12898, which directed each federal agency to identify and address "disproportionately high and adverse human health or environmental effects of its programs, policies, and activities on minority populations and low-income populations" (Clinton 1994).

Although these actions have not necessarily resulted in the end of environmental injustice or a definitive improvement in environmental conditions in communities of color or poorer neighborhoods, federal government acknowledgement of the existence of environmental injustice has lent a degree of legitimacy to the grassroots environmental justice advocacy movement, and has increased the amount of monies (and the political will) available to combat and remedy environmental burdens, and to support community environmental education and "Right-to-Know" programs. In some communities in New York City (NYC), it has resulted in the establishment of community environmental "watch person" offices and in the use of community-led GIS to monitor environmental conditions and develop pro-active planning strategies to improve the community and help achieve environmental justice.

## ENVIRONMENTAL INJUSTICE IN NEW YORK CITY

New York City, like many densely settled urban areas, often has industrial districts in close proximity to residential areas. Approximately 22 percent of New York's population lives in census tracts within an industrially zoned district (Maantay 2001b). This situation stems from historic settlement patterns, the imposition of zoning regulations and zoning changes that perpetuate the status quo land use distribution, and urban renewal and transportation policies that over time had the effect of siting many public housing projects and major highway infrastructure in or near industrial areas.

Thus, many poor and minority neighborhoods are in or adjacent to industrial districts, and are burdened by the resulting environmental hazards. Industrial land uses typically generate greater environmental burdens than do purely residential uses. In recent decades, many industrial areas in NYC have become the repositories of waste-related facilities, as manufacturing firms have left the city. Facilities commonly found in many industrial zones include solid waste transfer stations, medical waste treatment and disposal facilities, marine transfer stations, recycled materials handling facilities, auto salvage yards, scrap metal processing plants, construction and demolition debris processing plants, combined sewer overflow outfalls, and junkyards, in addition to sewage treatment plants and sewage sludge pelletization plants in certain parts of the city, such as the South Bronx.

Potential environmental impacts of these noxious uses include: emissions of toxic substances to air, soil, and water; hazardous materials use and storage; adverse impacts from truck traffic (safety, air pollution, noise, vibration, traffic congestion); illegal dumping of hazardous and toxic materials; and visual blight. Individual waste transfer stations can generate up to 1,000 truck trips per day each, and these trucks travel through residential neighborhoods to access these facilities. Most of these environmental impacts potentially have adverse health consequences.

A number of community-based organizations and environmental justice advocacy groups in NYC have undertaken geographic analysis of local conditions and environmental injustice (Ahearn and Osleeb 1993; Chan and Lowe 2001; Maantay et al. 1997). Beginning in 1995 at The Center for a

Sustainable Urban Environment at Hostos Community College and at Lehman College, we have been exploring environmental justice concerns in the Bronx, the City's poorest borough, and the one with the highest proportion of minority residents. We examined the locations of Toxic Release Inventory (TRI) facilities and polluting waste-related facilities in relationship to demographic and socio-economic indicators of the population nearest to the noxious facilities in the Bronx. TRI facilities are those that manufacture, import, or process more than 25,000 pounds per year, or use 10,000 pounds per year, of each of one or more of about 650 listed toxic chemicals, and TRI facilities must report annually to the federal Environmental Protection Agency. The maps, spatial analyses, and tabular data resulting from our study confirmed what many people have long believed to be true based on anecdotal information: the people living in close proximity (within 1/2 mile) to TRI and waste-

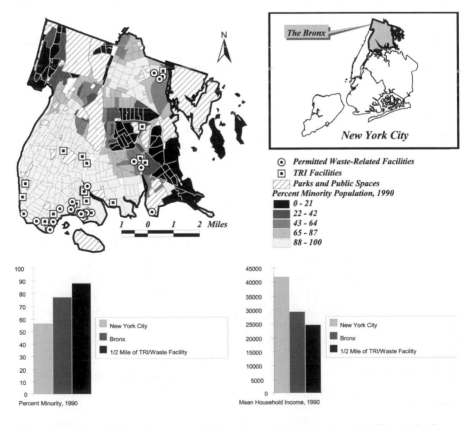

*Figure 27.1.* Toxic Release Inventory (TRI) and permitted waste related facilities in the Bronx, New York, with percent minority population by census tract. Data sources: New York State Department of Environmental Conservation 2000; New York City Department of Sanitation 2000; U.S. Environmental Protection Agency 2000; U.S. Census Bureau 1990.

related facilities tend to be poorer than the average Bronx resident, and more likely to be a minority (Maantay et al. 1997; Maantay 2001a). As seen in Figures 27.1 and 27.2, the noxious facilities are concentrated in areas with high proportions of minority and poor people, while the more affluent and less minority neighborhoods are relatively free from hazardous land uses.

The geographic analyses and maps were useful in making the community's voice heard at public hearings pertaining to planning decisions and the future of their neighborhoods. Spatial data are effective in lending credibility to the community's concerns, so that these concerns can no longer be easily overlooked by decision-makers as uninformed opinions or emotional responses. With the help of community-based organizations, such as the South Bronx Clean Air Coalition, and major medical institutions in the Bronx, such as Montefiore Medical Center and Albert Einstein College of Medicine, this on-going study is now exploring the possible spatial correspondence among concentrations of polluting land uses (such as TRI facilities and major highways), areas of high air pollution, and childhood asthma hospitalization rates in certain communities in the Bronx that are 250 percent higher than

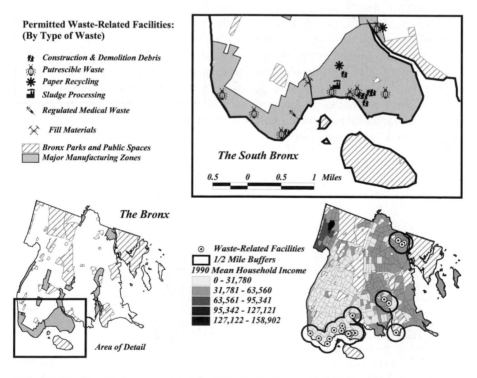

*Figure 27.2.* Permitted waste-related facilities in the Bronx, New York with average household income by census tract. Data sources: New York State Department of Environmental Conservation 2000; New York City Department of Sanitation 2000; New York City Department of City Planning 1993; U.S. Census Bureau 1990.

that of NYC as a whole, and 1,000 percent higher than New York State (New York City Department of Health 1999).

## USING GEOGRAPHIC ANALYSIS TO MAKE THE CONNECTIONS BETWEEN ENVIRONMENTAL JUSTICE AND ENVIRONMENTAL HEALTH

Geographic analysis and mapping are not panaceas for addressing environmental injustice. However, mapping has proved to be unusually effective in bringing attention to the problem and in crafting possible solutions. Although most environmental justice studies have demonstrated a clear link between disproportionate environmental burdens and race/ethnicity and/or income, the real work in terms of geographic analyses is just beginning. Showing environmental inequity regarding the distribution of polluting facilities is important, but it is probably more critical at this point to demonstrate linkages between environmental burdens and adverse health outcomes.

When the spatial correspondence is clear, public health and environmental protection officials, the medical research community, health care providers, and pollution prevention scientists can better develop solutions to existing environmental injustices and resulting health effects. It will become increasingly important to be able to show the disproportionate effects of pollution, rather than just the fact that disproportionate distribution of pollution sources exists. With advances in geographic information systems, better spatial statistical capabilities, more complete and accurate data collection and accessibility, and increased community participation, this will undoubtedly be possible.

## REFERENCES

Ahearn, S. and J. Osleeb. 1993. The Greenpoint/Williamsburg Environmental Benefits Program: Development of a Pilot Geographic Information System. In *Proceedings of the GIS/LIS*, 10–18. Minneapolis, MN: GIS/LIS.

Bryant, B., ed. 1995. *Environmental Justice: Issues, Policies, and Solutions.* Washington: Island Press.

Bullard, R., ed. 1994. *Unequal Protection: Environmental Justice and Communities of Color.* San Francisco: Sierra Club Books.

Chan, E. and L. Lowe. 2001. New York City Environmental Justice Alliance. In C. Convis, ed. *Conservation Geography: Case Studies in GIS, Computer Mapping, and Activism.* Redlands: Environmental Systems Research Institute Press.

Clinton, W. 1994. Federal Action to Address Environmental Justice in Minority Populations and Low Income Populations. In D. Camacho, ed. *Environmental Injustices, Political Struggles: Race, Class, and the Environment*, 43. Durham: Duke University Press 1998.

Goldman, B. 1993. *Not Just Prosperity: Achieving Sustainability With Environmental Justice.* Washington: National Wildlife Foundation.

Greenberg, M. 1993. Proving Environmental Inequity in Siting Locally Unwanted Land Uses. *Journal of Risk – Issues in Health and Safety* 4: 235–252.

Maantay, J., L. Timander, G. Graziosi, and L. Meyers. 1997. *The Bronx Toxic Release Inventory Report*. New York: Center for a Sustainable Urban Environment/United States Environmental Protection Agency, Region 2.

Maantay, J. 2001a. Race and Waste: Options for Equity Planning in New York City. *Planners Network* 145(1): 6–10.

Maantay, J. 2001b. Zoning, Equity, and Public Health. *American Journal of Public Health* 91: 1033–1041.

Maantay, J. 2002. Mapping Environmental Injustices: Pitfalls and Potential of Geographic Information Systems (GIS) in Assessing Environmental Health and Equity. *Environmental Health Perspectives* 110 (suppl 2): 161–171.

New York City Department of Health. 1999. *Asthma Facts*. New York, NY: New York City Childhood Asthma Initiative.

Pulido, L. 1996. A Critical Review of the Methodology of Environmental Racism Research. *Antipode* 28: 142–159.

United Church of Christ's Commission for Racial Justice. 1987. *Toxic Wastes and Race in the United States: A National Report on the Racial and Socio-Economic Characteristics of Communities With Hazardous Waste Sites*. New York: United Church of Christ.

# 28

# COMMUTING, CONGESTION, AND URBAN TRANSPORT SUSTAINABILITY

## MARK W. HORNER
*Texas State University – San Marcos*

Dispersed urban growth following from rising affluence and mobility presents serious challenges to the future operation of urban transport systems. Indeed, recent demographic and social trends in the United States and other industrialized countries seem counter to the sustainability of urban transport. First, already large metropolitan populations grew substantially during past decades, particularly in the United States, where they expanded by 13.9 percent from 1990 to 2000. As of 2000, metropolitan persons accounted for roughly 80 percent of the total U.S. population (Perry and Mackun 2001). Second, quarterly U.S. gross domestic product (GDP) grew at robust rates (2 to 8 percent annualized) during the boom period of the last decade (1996 to 2000) (Bureau of Economic Analysis 2001). This prosperity has facilitated an ability to obtain material goods, particularly automobiles, thereby adding to levels of personal mobility. Today, increased numbers of automobiles are appearing on highways; evidence from the 2000 U.S. census shows that automobile ownership is approaching almost 2 cars (1.7) per household (Pisarski 2002). Third, homeownership has reached unprecedented levels, with about two-thirds of households owning their own home (Pisarski 2002; U.S. Census 2002). New housing development has occurred at the expense of previously undeveloped land on the outskirts of urban areas. Personal mobility enhancements have allowed the building of homes in suburban and exurban locations, which are more distant from traditional urban centers (Krizek and Power 1996; Pisarski 2002).

Despite the positive impacts that improvements in material conditions have had on near-term quality of life, increased mobility coupled with dispersed metropolitan growth raises questions about longer-term future social, economic, and environmental well-being. The convergence of these sociodemographic factors has already begun impacting the operating characteristics of urban transportation systems.

*D. G. Janelle et al. (eds.), WorldMinds: Geographical Perspectives on 100 Problems*, 171–175.

## THE PROBLEM

There are several problems that threaten the sustainability of contemporary urban transportation systems, though the impacts of congestion are perhaps at the forefront. Illustrating this, a study by the Texas Transportation Institute (2002) shows that the average commuter in Los Angeles, California lost 136 hours of time in 2000 due to congestion. Congested highways also contribute to air and water pollution, while finite fuel resources are expended (Black 1997). Emissions increase during congested periods because vehicles are less efficient at slower speeds, and because being on the road longer simply allows vehicles more time to expend emissions. In short, the processes driving urban congestion must be better understood if their effects are to be lessened.

Congestion is quite complex, involving the interplay of demographics, personal choice, local policies, roadway infrastructure, land use, and other factors. However, geographers' emphasis on spatial relationships allows a basic insight into proposing solutions to this problem: some congestion is attributable to *work travel*, and work trip lengths follow from the degree of spatial separation between people's residences and workplaces. Certainly many urban roadway facilities are running near capacity in the peak hours (i.e., the morning and evening periods when most people travel to and from work). During these times roads are congested, mainly because people are commuting (Redmond and Mokhtarian 2001).

Geographers recognize that commute lengths are influenced by the spatial separation of peoples' homes and workplaces because intuitively, a person living *far* from their place of work must commute *farther* than someone living closer to their work. At the urban scale, if cities are oriented *spatially* such that employment and residences are far away from one another on average, then commuting will likely be greater than it could be (Horner 2002). Conversely, urban areas with more spatially integrated residential and workplace locations should experience less commuting (Cervero 1989).

Turning to congestion, when too many vehicles pile on to a stretch of roadway, speeds slow and it is considered "congested." Vehicles entering a roadway with relatively light traffic volumes have a smaller effect on speeds (i.e., little slowing of vehicles) whereas additional vehicles added at higher traffic volumes have a greater effect on speeds (i.e., a substantial slowing of vehicles). Therefore, slight changes in the number of cars on roadways during peak hours could have significant impacts on congestion.

Bringing these ideas together, decreasing the commute distances of people would potentially reduce traffic volumes and in turn, help to alleviate congestion. One way that commute reductions could be achieved is through implementing targeted urban development policies. Yet, before such policies can be pursued, basic research is required. A spatial perspective is needed to explore the interplay of home/workplace location and commute behavior.

Urban geographers would tackle this problem by collecting georeferenced data on peoples' residential and travel choices, and likely make use of tools such as geographical information systems (GIS) to build models of commuting and congestion. This approach differs substantially from say an economic perspective, where tolling roads might be considered as a means of reducing congestion, or an engineering viewpoint stressing new highway construction to satisfy travel demand. The next section describes geographic approaches focused on commuting and urban structure.

## RESEARCH DIRECTIONS

The concept of *excess commuting* is one of many possible starting points for researching urban spatial structure. Excess commuting is used to *benchmark* and compare urban travel behavior. When applying this benchmark, the *observed level of commuting* in a city is compared to an estimated *theoretical minimum commute* for the same city. Observed commuting is given directly by data on choices people make about where they live and work. On the other hand, the theoretical minimum commute is estimated from data. It is a special commuting pattern in which we fix home and work locations and assume all commuters collectively choose workplaces to minimize total region-wide travel.

To illustrate the theoretical minimum benchmark, imagine all of a city's workers congregating in a large meeting hall to collectively decide each individual's commute. The group is free to make individual members change their commutes and work at other locations, but the physical locations of jobs and housing cannot change. Their final answer is the lowest possible (optimal) citywide costs of commuting.

Clearly the minimum commute is a theoretical construct because people's jobs are generally not interchangeable (e.g., doctors do not work at law firms), and logistically, such a meeting is impossible! In practice, a computer solves this problem given the appropriate data. The minimum commute reveals a baseline level of commuting that a city's observed commuting can never fall below. Geographically speaking, it is the intrinsic amount of commuting required due to the spatial layout of jobs and housing.

Excess commuting is determined at the urban scale using data from a specific place and time. The most recently available data source capable of supporting excess-commuting analysis is from the 1990 census. Example estimates for two cities are illustrated in Figure 28.1. Data on home and work locations are input into a spatial model (see Horner 2002). The minimum theoretical average commute found for Columbus, OH is 3.31 miles, whereas the corresponding statistic for San Antonio is 2.81 miles (Horner 2002).

In Figure 28.1, the difference between the observed commute and the theoretical minimum benchmark is 4.04 miles (7.35 − 3.31) in Columbus and 4.66 miles (7.47 − 2.81) in San Antonio. On a relative basis, about 55

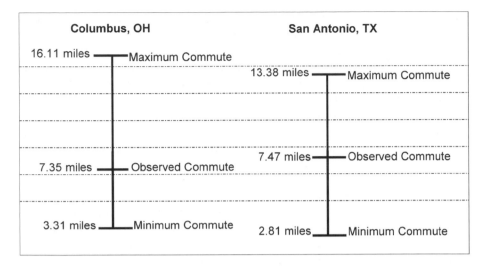

*Figure 28.1.* Comparison of cities' commuting statistics. Estimates based on 1990 Census Transportation Planning Package (CTTP).

percent of the observed commute in Columbus would be considered excess (4.04 / 7.35), while about 62 percent of San Antonio's observed commute is excess (2.81 / 7.47). Thus, after accounting for the commuting mandated by the spatial structures for the two cities, commuting in Columbus is more efficient.

When presented with the minimum and observed commute, a natural question to ask is, how much higher could the observed commuting go? To answer this, we calculate a *maximum theoretical commute* that provides an upper bound. Under Columbus' present urban structure it is impossible for its regional average commute to be greater than 16.11 miles (see Figure 28.1). The range between the minimum and the maximum commute can be used to create a continuum for evaluating observed commuting. Essentially, this continuum expresses the urban region's *commuting carrying capacity* because observed commuting must fall within these extremes. In the case of Columbus, the range would be about 12.80 miles (16.11 − 3.31 miles).

## NEXT STEPS

This essay has provided an overview of ways that geographers approach commuting and congestion. The described approach emphasizes basic research on the spatial relationships between home and workplace. Understanding more about these issues is important to our cities' social, environmental, and economic well-being.

One avenue for future exploration involves developing meaningful transport sustainability indicators to be tracked over time. The standard excess commuting statistic offers some help in this regard. For instance, as a city becomes progressively more decentralized due to new suburban development, its theoretical maximum commute will grow larger. Similarly, personal mobility improvements may be reflected in larger observed commutes. Such metrics would interest governing bodies desiring new ways of assessing their growth and development patterns.

Second, fundamental questions involving commuting and congestion remain to be explored. For instance, do places with less efficient travel, as measured by "excess commuting," experience more congestion? Conceptually, the smaller the magnitude of the theoretical minimum-commute estimate, the more proximal jobs are relative to housing in the region. And, research has shown that the theoretical minimum commute and the observed commute are positively correlated with one another (Horner 2002). So, by improving the proximity of jobs and housing to one another, what specific reductions in commuting/congestion might occur? All of these issues are open for further geographic inquiry.

## REFERENCES

Black, W. 1997. North American Transportation: Perspectives on Research Needs and Sustainable Transportation. *Journal of Transport Geography* 5(1): 12–19.

Bureau of Economic Analysis. 2001. Quarterly Gross Domestic Product, (GDP) Growth. See: http://www.bea.doc.gov/briefrm/gdp.htm.

Cervero, R. 1989. The Jobs-Housing Balance and Regional Mobility. *Journal of the American Planning Association* 55: 136–150.

Horner, M. 2002. Extensions to the Concept of Excess Commuting. *Environment and Planning A* 34(3): 543–566.

Krizek, K., and J. Power. 1996. *A Planner's Guide to Sustainable Development.* American Planning Association, Report Number 467.

Perry, M. and P. Mackun. 2001. *Population Change and Distribution: Census 2000 Brief.* Washington D.C.: U.S. Census Bureau. See: http://www.census.gov/prod/2001pubs/c2kbr01-2.pdf.

Pisarski, A. 2002. *Mobility Congestion and Intermodalism.* Testimony Before the United States Senate, Committee on Environment and Public Works, Washington D.C., March 2002. See: http://www.senate.gov/~epw/Pisarski_031902.htm.

Redmond, L. and P. Mokhtarian. 2001. The Positive Utility of the Commute: Modeling Ideal Commute Time and Relative Desired Commute Amount. *Transportation* 28: 179–205.

Texas Transportation Institute (TTI). 2002. *2002 Urban Mobility Study.* Texas A&M University. See summary tables: http://mobility.tamu.edu/ums/.

U.S. Census Bureau. 2002. *Housing Vacancies and Homeownership Historical Tables: Table 14. Homeownership Rates for the U.S. and Regions: 1965 to Present.* See: http://www.census.gov/hhes/www/housing/hvs/historic/histt14.html.

# 29

# CHARTING URBAN TRAVELERS 24-7 FOR DISASTER EVACUATION AND HOMELAND SECURITY

FREDERICK P. STUTZ
*San Diego State University*

Major shifts in urban regions' population distribution occur throughout the 24-hour day and the seven-day week. On weekday mornings, the population of residential zones declines and the population of educational, commercial, and employment zones increases. Weekday evening hours see a reverse of this pattern. The differences of population distribution throughout the urban region, especially the differences between night and day distributions, are extreme in most cities, and, therefore, of great importance to a variety of public officials, including transportation planners, highway patrol, and public service officials. The location of resident (night) population is documented through the U.S. census of population. Local metropolitan planning organizations (MPOs) usually update such statistics for their own purposes at least every five years. On a typical weekday between the hours of 7 a.m. and 7 p.m., most people are not at home, but have relocated themselves elsewhere in the urban region for a variety of activities. Zonal counts are required for 24-7 disaster preparedness. No such methods exist to target 24-7 population concentrations until now. The approach employed here provides ninety-five percent accurate zonal population counts across the urban region for each hour of the day and for each day of the week based on dynamic trip generation and trip distribution models and a GIS.

Public safety officials are influenced by the location of daytime populations within the region. Areas with high numbers of employees but few residential sites still require urban infrastructure, including electricity, water and sewer capacity, street access, and parking, in addition to police protection. Hazardous material routing must occur through a minimum of populated areas that change by hour of day and day of week. Disaster preparedness and terrorist evacuation models must consider variations in daytime and nighttime populations in order to dispatch emergency vehicles and determine evacuation route plans for city subareas.

*D. G. Janelle et al. (eds.), WorldMinds: Geographical Perspectives on 100 Problems*, 177–182.
© 2004 *Springer. Printed in the Netherlands.*

## SPACE-TIME LITERATURE

Space-time patterns within the field of geography have identified out-of-home, non-employment activities as a crucial component of the dynamic dependencies between home and work. GIS-based three-dimensional visualization techniques now compare space-time patterns of both employment and non-employment activities for various population subgroups (Kwan 1998, 1999). The significance of the temporal dimension and its interaction with the spatial dimension in structuring the daily space-time trajectories of individuals has been clearly revealed. To illustrate, Kwan (1998, 1999) and Kwan and Hong (1998) assigned travel times based on assumptions about driving speeds over different street types. Brainard et al. (1997, 1999) used a similar approach to create isochrone surfaces (equal travel time) for predicting the number of visitors of recreational facilities. Miller (1999) developed a method for interpolating travel times between nodes within a transportation network. A desktop GIS application has been developed that can automatically generate isochrones for travel by auto or by public transit in response to disaster evacuation (O'Sullivan et al. 2000). Aided by new spatial data capture technologies, such as very high resolution remote sensing satellites and global positioning system (GPS), relatively accurate and comprehensive digital data sets of metropolitan areas are now collected and maintained by MPOs, such as the San Diego Association of Governments (SANDAG). The data are becoming widely and often freely available (Longley 1998; Kwan 2000).

## AROUND-THE-CLOCK POPULATION ESTIMATES

Geographers (working for SANDAG) prepared daytime-to-nighttime population estimates. The task involves producing around-the-clock population estimates for the 29,000 census blocks located within the 4,100 square mile western one-third of San Diego County, encompassing three million people. The so-called 24-7 model combines trip generation data with information from the 1995 Regional Travel Behavior Survey. This survey gathers the daily travel and activity patterns of a representative, stratified random sample of San Diego county households. Household demographic data are included in the survey as well as temporal and spatial aspects of all trips made during a 24-hour period by each member of the household. Variables include: trip start-time, trip arrival-time, trip-origin address, trip-destination address, mode of travel, and purpose of trip. The regional travel behavior survey was then expanded mathematically to represent all 1.3 million households in the San Diego region. Geographical data in the census block GIS were aggregated to census tracts, to major statistical subregions, and to the 18 incorporated jurisdictions within the county of San Diego.

The 24-7 model begins with each census block's U.S. census (nighttime) figure. This basic figure for each census block includes residents who live in that zone plus people staying in hotels, motels, and government military barracks. Factors of multiplication were then developed from the regional travel behavior survey variables for each census block and for each hour of the week 24-7. The factors encompass the type of trip made, time of day the trip was made, trip purpose, and traveler socioeconomic characteristics. The factors become trip-generation coefficients when multiplied by the resident (nighttime) population. The 24-7 model subtracts the people leaving each census block for each hour of the day and, likewise, adds those people entering the census block for the same time period. Through standard trip-generation modeling techniques and an ArcInfo GIS format, the number of people present in each census block on an average weekday is estimated within 95 percent accuracy. SANDAG now applies the same methodology to 2020 projected population forecasts to produce future 24-7 population concentrations. The region's 2020 projected population model, discussed elsewhere, includes regional projected population growth, land zoning, the general plan for each subregion, land accessibility, land-traveler accessibility, and land attractiveness, to produce population and land use maps for 2020.

## JOBS/HOUSING BALANCE

The ratio of jobs to residential units in any subregion is called the jobs/housing balance. It is a well-accepted figure (expressed as a ratio) that reflects daily population shifts throughout the region. A relatively equivalent balance between jobs to homes in a subregion suggests that there is an opportunity for workers to work in close proximity to their residences. In 2000, the region as a whole had 1.19 jobs for each housing unit. Only 11 of the 41 subregional areas in San Diego County had a ratio higher than the region's average of 1.19. This implies that employment in San Diego County is concentrated in just a few areas. Ten of San Diego County's 18 incorporated jurisdictions lose population during daytime hours. By the year 2020, that number is expected to drop to only eight. The largest numeric drop between 2000 and 2020 is projected to occur in the unincorporated regions of the county because the majority of the region's jobs and schools are located in incorporated jurisdictions.

## TEMPORAL POPULATION SHIFTS

Populations during daytime hours are, in general, more concentrated than at nighttime. Maps of San Diego County displaying daytime population densities yield eleven fewer census tracts in the densest category (6,000+/square mile) than the nighttime maps. In other words, the daytime concentrations are much denser than the nighttime concentrations. During the nighttime the densest census tract in the region displays 28,000 persons per square mile. Eight others display densities of 20,000 or more people per square mile. During the weekday, however, at 11 a.m. only five census tracts have densities of greater than 20,000 people per square mile. Only four registered more than 30,000 people per square mile. The downtown San Diego tract has a daytime

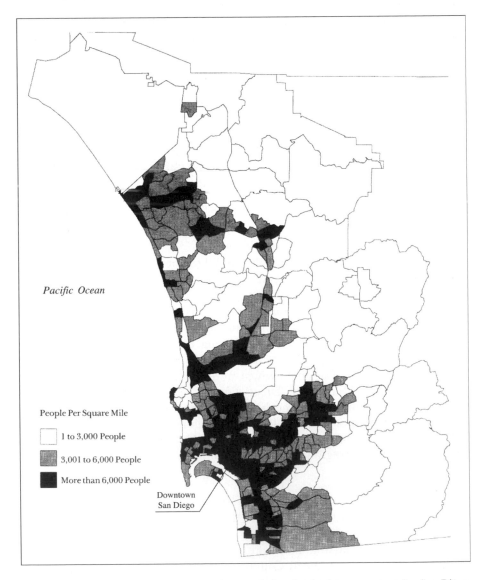

*Figure 29.1.* Estimated 11:00 a.m. weekday population density by census tract for San Diego in 2020.

population of over 90,000 people per square mile for the year 2000. Other areas showing high concentrations for the year 2000 during daytime hours include Palomar airport in Carlsbad (commercial), Sorrento Valley (commercial), and several U.S. military bases located in Coronado, National City, and Camp Pendleton.

The nighttime-to-daytime population shift within San Diego County continues through 2020. Again, fewer census tracts have high densities in the daytime as opposed to the nighttime. During daytime hours for 2020, 11 tracts have densities above 30,000 people per square mile, and five have more than 50,000 per square mile. The San Diego downtown census tract will house 140,000 per square mile by 2020 according to SANDAG's population-projection model.

Figure 29.1 shows the 11 a.m. population density by census tract for 2020 in San Diego County. By this time, the County will have added an extra one million residents and five-hundred-thousand jobs. These totals are represented in Figure 29.1. Between 2000 and 2020, 36 census tracts move from the lowest category of 3,000 people per square mile to a higher density category. Once again, the difference between 2000 and 2020 suggests that even fewer tracts have high densities in the daytime, but those that do have markedly higher densities. The Figure 29.1 totals for both 2000 and 2020 can be produced for any year in between, by census tract for each day of the week and hour of the day. This information, when applied with an evacuation model, will give adequate planning assistance for disaster preparedness in the event that the region must be cleared.

## ACKNOWLEDGEMENTS

The author acknowledges contributions to this research by former San Diego State University geography students now at the San Diego Association of Governments (SANDAG): Bob Parrott – Deputy Director of SANDAG, Steve Kunkel – GIS/Cartography, Mike Hix – Transportation, Sue Carnevale – Environmental, John Hofmockel – Research Division, Mike Calandra – Research Division, and Tom King – Research Division. Paul Kavanaugh, the Senior Planner, directed the research effort at SANDAG.

## REFERENCES

Brainard, J. S., A. Lovett, and I. Bateman. 1997. Using Isochrone Surfaces in Travel-Cost Models. *Journal of Tranport Geography* 5(2): 117–126.

Brainard, J., A. Lovett, and I. Bateman. 1999. Integrating Geographical Information Systems Into Travel Cost Analysis and Benefit Transfer. *Geographical Information Science* 13(3): 227–246.

Longley, P. 1998. GIS and the Development of Digital Urban Infrastructure. *Environment and Planning B* 53: 56.

Kwan, M.-P. 1998. Space-Time and Integral Measures of Individual Accessibility: A Comparative Analysis Using a Point-Based Framework. *Geographical Analysis* 30: 191–217.

Kwan, M.-P. 2000. Interactive Geovisualization of Activity-Travel Patterns Using Three-Dimensional Geographical Information Systems: A Methodological Exploration With a Large Data Set. *Transportation Research C* 8: 185–203.

Kwan, M.-P. and X. Hong. 1998. Network-Based Constraints-Oriented Choice Set Formation Using GIS. *Geographical Systems* 5: 139–162.

Miller, H. J. 1991. Modeling Accessibility Using Space-Time Prism Concepts Within GIS. *International Journal of Geographic Information Systems* 5: 287–301.

O'Sullivan, D., A. Morrison, and J. Shearer. 2000. Using Desktop GIS for the Investigation of Accessibility by Public Transport: An Isochrone Approach. *Geographical Information Science* 14(1): 85–114.

Silva, N. 2002. Designing a Spatial Decision Support System for Evacuation Planning. *International Journal of Mass Emergencies and Disasters* 20(1): 51–68.

Silva, N. and R. W. Eglese. 1996. A Simulation Model of Emergency Evacuation. *European Journal of Operational Research* 90(3): 413–419.

Wolshon, B. 2002. Planning for Evacuation of New Orleans. *Institute of Traffic Engineers Journal* 8123: 44–88.

# 30

# THE WALKING SECURITY INDEX AND PEDESTRIANS' SECURITY IN URBAN AREAS

BARRY WELLAR
*University of Ottawa*

The Walking Security Index (WSI) was proposed to the Region of Ottawa-Carleton in 1995 for inclusion in its Transportation Environment Action Plan (TEAP). Four related problems were behind the idea of developing an index (Wellar 1996, 1998). First, the Region attached "high priority" to the walking mode in its Official Plan and Transportation Master Plan. However, it had no means of methodologically evaluating how well any of its 875 signalized intersections met the needs of pedestrians. Second, due to the apparent premise of the engineering field and of the automotive industry that vehicle operators have an "entitlement" to convenience, comfort, and safety, transportation research in North America has focused overwhelmingly on moving cars, trucks, vans, SUVs, and buses (Highway Research Board 1965; Transportation Research Board 1994; Wellar 1996, 1998, 2000, 2002a, b). An index could be a means to articulate the concerns of pedestrians, and to identify pedestrian-sensitive solutions to urban transportation problems. Third, by training and tradition, the Region's transportation planning and traffic engineering staff concentrated its efforts on moving vehicles. Very little in-house talent and resources were dedicated to serving and promoting pedestrians' safety, comfort, and convenience. The development of an index could reduce the technical imbalance, and provide a basis for pedestrian-oriented initiatives in Ottawa-Carleton's transportation, public safety, and planning departments. Fourth, the lack of quantitative measures – such as indexes – worked against making factual, evidence-based arguments for pedestrians' interests at meetings of decision-making bodies. That situation would change with a Walking Security Index at the disposal of walking-mode advocates.

Impelled by community associations, several municipal and regional councillors, and the TEAP-Citizen Advisory Group, the Region of Ottawa-Carleton agreed in 1996 to fund the WSI project. The mission was to design an index to measure the levels of safety, comfort, and convenience that pedestrians expect and experience at signalized intersections, and to guide remedial actions by the regional government.

*D. G. Janelle et al. (eds.), WorldMinds: Geographical Perspectives on 100 Problems*, 183–189.
© 2004 *Springer. Printed in the Netherlands.*

## KEYWORD-BASED SEARCHES

Course assignments for more than 100 students included the use of keywords to search the literature and to "surf" the Internet for researchers engaged in index-related projects. The keyword-based searches covered the learned and popular literature (newspapers and magazines), as well as the literature of professional associations, technical associations, government agencies, and vested and public interest groups (Wellar 1996, 1998, 2000).

Thousands of pertinent documents were identified by the searches. However, no index formulation (equation, expression, spreadsheet, etc.) was located that could be used "as is" to relate WSI variables. Further, the formulations for other transport modes provided no insights into how to design an index to measure the safety, comfort, and convenience expectations or experiences of pedestrians.

The Internet surfing activity was similarly unsuccessful: no researcher or firm designing indexes pertinent to our mission was identified. However, the searches did identify academics, consultants, and members of pedestrian advocacy groups and professional/technical organizations in North America and abroad wishing to receive publications.

## INVOLVING EXPERTS

The vast majority of "experts" involved in transportation and traffic research have been engineers. A smattering of planners, public safety officials, and other professionals play support roles. The WSI project recognized two other types of experts: (1) elected officials, since they decide who or what goes where, and (2) citizens, since they know how they feel about using intersections, and they know what kinds of changes they want made to intersection designs, traffic volumes, maintenance programs, and driver behavior.

The WSI experts proposed variables, rated variables (essential, high, medium, low priority), and provided feedback on the criteria used to evaluate variables (pertinence, degree of difficulty, level of support, enforceability, and data availability). And, they continue to participate in WSI – based challenges to transportation practices or decisions that are driven by the opinions of traffic engineers.

## SPECIFYING THE INDEXES

To represent the combined opinions of pedestrians, professional staff (engineering and works, planning, police, etc.), and elected officials, three types of index were identified. The reader is referred to the project reports for details about the specification process (Wellar 1998).

*Intersection Volume and Design Index (IVDI)*

Findings from the literature and interviews with experts revealed major gaps in engineering-type studies, and especially their failure to account adequately for differences in vehicle types (cars, trucks, buses) or intersection design concerns (number of lanes, slope, geometry, etc.). To address this gap, a series of rudimentary indexes were designed, and then combined to form the Intersection Volume and Design Index (IVDI):

$$IVDI = V_1 \cdot V_2 \cdot V_3 \cdot V_4 \cdot V_5 \cdot V_6 \cdot V_7 \cdot V_8 \qquad [1]$$

where,
$V_1$ = number of passenger car equivalents[2]/hour
$V_2$ = number of pedestrians/hour
$V_3$ = number of lanes rating
$V_4$ = number of turn lanes by type rating
$V_5$ = intersection geometry rating
$V_6$ = intersection slope rating
$V_7$ = direction(s) of traffic flow rating
$V_8$ = number of channels adjacent to intersection rating.

*Quality of Intersection Condition Index (QICI)*

A pedestrian safety conference (Wellar 1996) and interviews with pedestrians were instrumental in identifying the need for an index that incorporates variables on intersection construction and maintenance features. An audit-type formulation was designed, and is shown as Figure 30.1.

*Driver Behavior Index (DBI)*

Pedestrians' safety, comfort, and convenience when traversing intersections is directly impacted by the behavior of drivers. That cause-effect relationship occurs because drivers who run the lights, or commit fail-to-yield infractions, cause death or injury to pedestrians. And, aggressive driver behavior causes fear, unease, frustration, and sometimes rage among pedestrians who feel that they are treated as second-class citizens by drivers who act as though they "own the road" (Wellar 1996, 1998, 2000).

The initial, incident-specific aggressive driving indexes were combined into a general formulation:

$$DBI = \frac{ALI}{P} + \frac{RLI}{P} + \frac{FTYI}{P} \qquad [2]$$

where,

$$\frac{ALI}{P} = \text{amber-light incidents per phase} \qquad [3]$$

| ID | Variable Names for Intersection Design and Maintenance Features | Condition Met ? | | | | |
|----|---------------------------------------------------------------|-----------------|------|------|------|------|
| | | Yes | No (Quadrant) | | | |
| | | | NW | NE | SE | SW |
| 1 | Sidewalk corner capacity | | | | | |
| 2 | Height of curbing | | | | | |
| 3 | Condition of curbing | | | | | |
| 4 | Sidewalk width capacity | | | | | |
| 5 | Sidewalk condition | | | | | |
| 6 | Crosswalk surface condition | | | | | |
| 7 | Median (refuge) capacity | | | | | |
| 8 | Median (refuge) condition | | | | | |
| 9 | Traffic calmer(s) | | | | | |
| 10 | Channel island (refuge) capacity | | | | | |
| 11 | Crosswalk capacity | | | | | |
| 12 | Crosswalk signed and painted | | | | | |
| 13 | Stop bar signed and painted | | | | | |
| 14 | Pedestrian signage | | | | | |
| 15 | No sight line obstruction | | | | | |
| 16 | Street furniture proximal to corner | | | | | |
| 17 | Ice/snow/slush removal | | | | | |
| 18 | Water drainage | | | | | |
| | Totals | | | | | |
| | Overall Score (Yes-No=) | | | | | |

*Figure 30.1.* Quality of Intersection Condition Index (QICI).

$$\frac{RLI}{P} = \text{red-light incidents per phase} \qquad [4]$$

$$\frac{FYTI}{P} = \text{fail-to-yield incidents per phase} \qquad [5]$$

## COMPLETING THE DESIGN PHASE

The final report in the design phase, Walking Security Index, was submitted in 1998. In addition to creating ten indexes, the report also contained 17 recommendations on intersection modifications to serve and promote pedestrians' safety, comfort, and convenience (Table 30.1).

*Table 30.1.* Summary of Proposed Intersection Modifications That Serve and Promote Basic Needs of Pedestrians.

- Install photo (red-light) cameras
- Install camera radar and strictly enforce the 60 kph maximum
- Increase separation of stop lines/stop bars from crosswalks
- Adjust light cycle duration on the green phases
- Remove pedestrian walk signals: Pilot study
- Increase enforcement of crosswalk and stop line/stop bar by-laws
- Restrict right turns on red: Pilot program
- Modify light cycles: eliminate delays from red to green
- Petition the Government of Ontario for a change to Section 140 of the *Highway Traffic Act* to properly recognize the risks to pedestrians in channel crossovers
- Change yield to pedestrian signs to stop signs: Pilot program
- Modify posted and painted roadway signage: yield to pedestrians
- Modify roadway marking materials: paint
- Provide proper maintenance
- Ensure adequate lighting from the pedestrians' perspective
- Ensure adequate sight lines from the pedestrians' perspective
- Modify and standardize intersection features so as to eliminate obstacles and nasty surprises that make intersection usage difficult and even dangerous for pedestrians with disabilities

Source: Wellar (1998).

## PILOT STUDY TEST OF OPERATIONALITY – TERMS OF REFERENCE

In 1999 the Region of Ottawa-Carleton funded a pilot study to test the indexes for operationality. The terms of reference included taking field observations at 33 intersections in different parts of the region to test for spatial variations, and a study period of 16 months to ensure fieldwork coverage for all seasons (Region of Ottawa-Carleton 1999; Wellar 1999, 2001).

## KEY PILOT STUDY FINDINGS AND LESSONS LEARNED

Pilot study findings and lessons learned are contained in more than a dozen publications (Wellar 2002b), and have also been presented at conferences, public meetings, and workshops over the past four years. The following observations indicate the kinds of findings and lessons learned to date (Wellar 2002a, b).

1.  All indexes tested positive for operationality, that is, the *IVDI, QICI,* and *DBI* formulations satisfied the evaluation criteria (pertinence, support, difficulty, enforceability, data availability), and all can be implemented.

2.  Involving elected officials, professional staff, and citizens as experts helped define the research problem, and enabled us to rigorously specify and test the indexes for operationality.

3. The "intersection" was found to be an overly aggregate concept for evaluating pedestrians' security. The more appropriate spatial scale for fieldwork, as well as for diagnostic and prescriptive purposes, is the quadrant. A focus on quadrants (or quarters of an intersection) yields more informative data on index variables, leads to significantly more robust index scores and rankings, and explicitly points to where changes are needed to improve pedestrians' security.
4. There are significant seasonal variations in index scores and rankings, which discount the traffic engineering practices of limiting traffic count programs to the summer months, and using annual averages.
5. There are significant peak hour variations (a.m., noon, p.m.) in index scores and rankings, which discount the traffic engineering practice of using daily averages.
6. There are significant variations in index scores and rankings as a function of geographic location (downtown zone, inner suburban zone, outer suburban zone). Consequently, the spatial aspect or "geo-dimension" must be explicitly respected in order to make rational decisions about roadway or traffic modifications, enforcement programs, and development projects that affect pedestrians' security.
7. The indexes proved to be user-friendly for ease of understanding and conducting the fieldwork program. Third- and fourth-year students in Geography, University of Ottawa, performed tests of the *QIC* and *DB* indexes successfully, and field experiments by several community associations helped make refinements to the *QICI* form and procedure.
8. Feedback indicates a high level of transferability and acceptability among citizens, pedestrian advocates, injury prevention, and other groups regarding WSI concepts, formulations, field instruments, field procedures, and index scoring and ranking procedures (Wellar 2002a, b).

## CONCLUSION

The Walking Security Index is a product of applied geographic research, and pedestrian-oriented agencies, groups, and individuals in many municipalities in Canada, the United States, and abroad now use materials from the project.

The value of bringing a geographer's perspective to a field that has been largely within the purview of traffic engineering cannot be over-emphasized. There is still much to be done by geographers, however, in terms of refining the indexes, building and maintaining the indexes through the application of geographic information systems (GIS), and ensuring the implementation of public policies, programs, and plans that serve and promote pedestrian travel, the most sustainable, benefit-efficient mode of urban transportation.

# REFERENCES

Highway Research Board. 1965. *Highway Capacity Manual.* Washington, D.C.: National Academy of Sciences – National Research Council.

Region of Ottawa-Carleton. 1999. *Departmental Recommendations on Walking Security Index.* File No. 50 20-99-0101. Ottawa: Region of Ottawa-Carleton, Environment and Transportation Department.

Transportation Research Board. 1994. *Highway Capacity Manual.* Washington, D.C.: National Academy of Sciences – National Research Council.

Wellar, B. 1998. *The Walking Security Index.* Ottawa: Region of Ottawa-Carleton and the University of Ottawa.

Wellar, B. 1999. *Use of the Walking Security Index (WSI) to Evaluate Regional Intersections: Pilot Study Proposal.* Ottawa: Region of Ottawa-Carleton and the University of Ottawa.

Wellar, B. 2000. *Newspapers as a Source of Fact and Opinion on Pedestrians' Safety, Comfort, Convenience: A Keyword-Based Literature Search and Review.* Ottawa: Region of Ottawa-Carleton and University of Ottawa.

Wellar, B. 2001. The Pilot Study as a Step in the Process of Implementing Transportation Innovations: Findings From the Walking Security Index (WSI) Project. In G. Tobin, B. Montz, and F. Schoolmaster, eds. *Papers and Proceedings of the Applied Geography Conferences,* Vol. 24, 243–252. Denton, TX: University of North Texas.

Wellar, B. 2002a. *Walking Security Index Pilot Study.* Ottawa: Region of Ottawa-Carleton and University of Ottawa. (http://aix1.uottawa.ca/~wellarb/walking_security_index_pilot_stu.htm)

Wellar, B. 2002b. Lessons Learned from the Walking Security Index (WSI) Project on How to Achieve Street-Smart Urban Transportation Improvements. *Proceedings, 2002 Annual Conference, Canadian Institute of Planners.* (http://www.cip-icu.ca/English/conference/proceedings/02proc15.pdf)

Wellar, B., ed. 1996. *Perspectives on Pedestrian Safety.* Ottawa: Pedestrian Safety Conference Committee, Region of Ottawa-Carleton.

# 31

# CLEAN STREETS – CLEAN WATERWAYS: STREET SWEEPING, STORM-WATER RUNOFF, AND POLLUTION REDUCTION

GRAHAM A. TOBIN and ROBERT BRINKMANN
*University of South Florida*

> If a man is called to be a street sweeper, he should sweep streets even as Michelangelo painted, or Beethoven composed music, or Shakespeare wrote poetry. He should sweep streets so well that all the hosts of heaven and earth will pause to say, here lived a great street sweeper that did his job well.
>
> Dr. Martin Luther King, Jr.

The sentiments expressed by Dr. King in reference to street sweepers are pertinent to the geographer and geographical research in several ways. Like Michelangelo, Beethoven, and Shakespeare, the street sweeper plays an important role in our urban setting. Human health, environmental quality, and pollution control, for example, are all inextricably linked to street sweeping. In essence, the street sweeper holds a critical key to the sustainability of urban society. Here then is the science of street sweeping, for the cleaning of the urban environment is directly related to the quality of life. If litter and wastes are not removed from the roads, then there is a danger that they will enter the storm water system where they will eventually co-mingle with surface water bodies. Without such street cleaning, natural water bodies are susceptible to contamination from many pollutants generated by human activities. Such issues do not end there, for the street sweeping process itself spawns yet another concern, namely how to manage the tons of sediment that are collected daily. Two broad issues, therefore, need to be addressed, storm water quality and waste management.

## STREET-SWEEPING FRAMEWORK

The importance of street sweeping cannot be determined without careful consideration of the whole range of processes taking place within cities. Some of these processes are firmly based in the natural sciences, including

*D. G. Janelle et al. (eds.), WorldMinds: Geographical Perspectives on 100 Problems, 191–196.*
© 2004 *Springer. Printed in the Netherlands.*

aspects of climatology, geology, pedology, topography, biology, and hydrology. The significance of street sweeping to pollution reduction, for instance, can only be addressed through rigorous research that incorporates virtually all these physical variables. Other processes are human in scope, such as land use changes, transportation routes and traffic density, industrial and commercial activities, and residential developments. Thus, it is only through a multi-disciplinary, geographic approach that a true picture of street sweeping can be assessed. A framework for looking at street-sweeping practices and their impacts is outlined in Figure 31.1 (Brinkmann and Tobin 2001).

## PHYSICAL AND HUMAN CONTEXTS

Figure 31.1 depicts two contextual foci – the physical and human environ-ments. In the physical context, local characteristics will both facilitate and

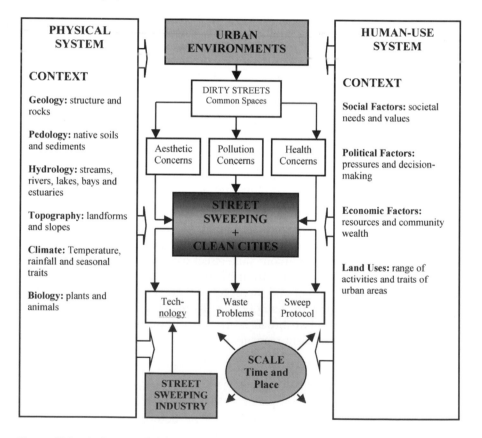

*Figure 31.1.* A framework for an examination of street-sweeping practices and protocols recognizing the context of the physical environment and the human-use system.

impede street sweeping operations. It is reasonable to assume, for instance, that sediment and waste production in Duluth, Minnesota will be significantly different to those in Tampa, Florida, and will lead to different management problems. The composition of street-sweepings generally consists of two broad components: physical attributes, comprising sediments, organic matter, litter, and urban dust, and chemical constituents incorporating different metals, fertilizers, pesticides, and organic constituents such as oil and grease. Similarly, from a human contextual perspective, different conditions will prevail in different communities incorporating a whole range of social, economic, and political forces. For example, larger and wealthier communities may well have greater access to resources and hence adopt different sweeping strategies than smaller, less wealthy communities.

## URBAN ENVIRONMENTS: DIRTY STREETS – COMMON SPACES

While street sweeping has traditionally been conducted for aesthetic reasons and public-health concerns, recently there has been a growing interest in using street sweeping to mitigate storm water contamination. Inevitably, though, common spaces, such as roads, become depositories for many wastes. Indeed, the tragedy of the commons, as described by Hardin (1968) is pertinent here, since roads are treated in much the same way as the commons. The resource or road is exploited and few are concerned with overall upkeep and maintenance, so dirt, garbage, litter, and fecal matter continue to accumulate. In this respect, the consequences of non-action can be profound as recognized over 160 years ago by the Poor Law Commissioners in their report "Inquiry into Sanitary Conditions of the Laboring Population of Great Britain," which stated:

> . . . effective town cleansing fosters habits of the most abject degradation and tends to the demoralization of large numbers of human beings, who subsist by means of what they find amidst the noxious filth accumulated in neglected streets and bye-places. Chadwick 1842: 423.

Related to this, the Clean Water Acts, passed at the Federal level, have certainly raised the quality of water bodies throughout the United States, and have had an impact on local resources. However, recent efforts to improve storm water quality and reduce pollution in streams and bays are difficult to evaluate. Federal regulations do not prescribe how to achieve lower levels of contamination, only that it must be accomplished. One way that has been recommended to accomplish this is to sweep streets. While there is little information available on the real effectiveness of street sweepers in reducing storm water pollution, some locally-based studies have been completed, which suggest that street sweeping holds promise in reducing storm-water pollution (Brinkmann et al. 2001).

## STREET-SWEEPING INDUSTRY

The street-sweeping industry (Figure 31.1) faces a variety of issues including technological difficulties, sweeping protocols, and waste management, that are associated with the implementation of street-sweeping strategies.

### Street Sweeping Machines/Technology

Street sweepers primarily come in three main types: mechanical, regenerative air (vacuum) and combination. Each street sweeper type has strengths and weaknesses (Tobin and Brinkmann 2002) and thus selecting a sweeper may be a critical management decision. Mechanical rotary brush sweepers represent the workhorses of the industry and are typically used to remove the bulk of sediment from roads. Vacuum sweepers are commonly used to collect litter and fine material, while combination sweepers are very good at removing both bulk debris and fine particles. See Figure 31.2.

### Street-Sweeping Protocols

City managers must decide on an acceptable sweeping schedule to maintain their streets. Unfortunately, there is a dearth of information on what frequency of sweeping is most effective, and hence managers typically make decisions on sweeping schedules based on other criteria. An investigation in Tampa, Florida, tried to address this and demonstrated that sweeping at least once every week may be the most effective strategy for reducing storm-water contamination in sub-tropical, sandy environments (Brinkmann and Tobin 2001).

*Figure 31.2.* Street sweeping.

*Waste Management*

Attitudes towards management of street sweeping waste vary considerably. On one hand, many consider the waste to be relatively benign material, and hence suggest that precious space in landfills should not be taken up with it. On the other hand, others have expressed reservations about the quality of the waste material collected, and advocate that it should be disposed of in a more controlled way. However, due in large part to fears of litigation related to future clean-up of hazardous sites, many sanitary landfills operators are now rejecting this material for disposal because of the possibility of contamination by heavy metals and toxic organic compounds (Kidwell-Ross 1993).

Unfortunately, little information exists on the precise physical and chemical composition of street sweepings, which makes such management decisions difficult (Pitt 1979; Rogge et al. 1993). A few communities have tested a small number of waste samples for their physical and chemical traits, but no comprehensive, city-wide study has been undertaken to characterize these wastes from a temporal or geographic perspective (Bannerman et al. 1983; Brinkmann and Ryan 1997; Billus 1999; Hepp 1997; Pitt et al. 1982; Sartor and Boyd 1972). Furthermore, there has been no detailed examination of spatial patterns in sediment generation that might enhance street-sweeping protocols.

## SPECIFIC GEOGRAPHIC CONCERNS

Given the context outlined above, several themes of geographic research are worth developing:

(1) Macro-scale geographic variability of street sediment. Comparative research on the variability of street sediments and sweepings between urban areas is limited. Additional research, then, is necessary to determine how and why street sediment chemistry and quantity vary from one urban area to another.

(2) Micro-scale geographic variability of street sediment. Little evidence exists to conclude how street sediment and sweepings chemistry and content vary within cities. It would be useful to know how and why street sediment and sweepings chemistry and quantity vary with land use. What roles, for instance, do traffic density, population density, and industrial base play in contributing to sediment accumulation and storm-water contamination?

(3) Impact of weather and climate on the temporal variability of street sediment and sweepings. Researchers should determine how different precipitation events contribute to street sediment and sweepings chemistry and quantity. Similarly, other climatic and meteorological factors must be assessed, such as antecedent wetness conditions, extreme temperatures, or wind strength.

(4) Impact of urban street-sweeping protocols on the geographic variability of street sediment. It is unclear how effective sweepers are in reducing sediment quantities and minimizing different chemical contaminants from surfaces in different environments. A scientifically based, nationwide study of different sweeping protocols and frequencies, therefore, is necessary to determine sweeping effectiveness to reduce sediment levels in different environments.

(5) Public policy and street sweeping. Decision-making regarding street sweeping, such as sweeping schedules and waste deposition, needs to be addressed to determine public health outcomes and environmental consequences.

## REFERENCES

Bannerman, R., K. Baun, and V. F. Boh. 1983. *Evaluation of Urban Non-Point Source Pollution Management in Milwaukee County, Wisconsin. Volume I. Urban Storm Water Characteristics, Sources and Pollutant Management by Street Sweeping.* Chicago, IL: EPA.

Billus, M. 1999. *Investigation into the Application of Street Sweeping Waste as a Soil Amendment.* University of South Florida, Masters Thesis.

Brinkmann, R. and J. Ryan. 1997. *Chemical and Physical Composition of Street Sediments.* Gainesville, Florida: FCSHWM.

Brinkmann, R. and G. A. Tobin. 2001. *Urban Sediment Removal: The Science, Policy, and Management of Street Sweeping.* Boston: Kluwer Academic Publishers.

Brinkmann, R., G. A. Tobin, and J. Ryan. 2001. *Street Sweeping and Storm Water Runoff. Phase II: Testing the Effectiveness of Different Street Sweeping Strategies.* Technical Report prepared for Florida Department of Transportation, District 7.

Chadwick, E. 1842 [M. W. Flinn, ed. 1965]. *An Inquiry into the Sanitary Condition of the Labouring Population of Great Britain.* A Report to Her Majesty's Principle Secretary of State for the Home Office from the Poor Law Commissioners. Edinburgh: Edinburgh University Press.

Hardin, G. 1968. The Tragedy of the Commons. *Science* 62: 1243–1248.

Hepp, M. 1997. *Best Management Practices (BMPs) for Management and Disposal of Street Wastes.* www.schwarze.com/secure/topics/waseco795/.

Kidwell-Ross, R. 1993. www.schwarze.com/secure/v2n2/v2n2trash.html.

Pitt, R. 1979. *Derivation of Non-point Pollution Through Improved Street Cleaning Practices.* EPA Report 600/2-79-161.

Pitt, R., CH2MHill, Cooper and Associates, Inc. and Consulting Engineering Services, Inc. 1982. *Washoe County Urban Storm Water Management Program.* Washoe Council of Governments.

Rogge, W. F., L. M. Hildmann., M. A. Mazurek., G. R. Cass, and P. R. T. Somneit. 1993. Sources of Fine Organic Aerosol. 3. Road Dust, Tire Debris, and Organometallic Brake Lining Dust: Roads as Sources and Sinks. *Environmental Science and Technology* 27: 1892–1904.

Sartor, J. D. and G. B. Boyd. 1972. *Water Pollution Aspects of Street Surface Contaminants.* Project Report prepared for Office of Research and Monitoring. Washington D.C.: U.S. Environmental protection Agency (EPA-R2-72-081).

Tobin, G. A. and R. Brinkmann. 2002. The Effectiveness of Street Sweepers in Removing Pollutants from Road Surfaces in Florida. *Journal of Environmental Science and Health Part A: Toxic Hazardous Substances and Environmental Engineering* 37: 1687–2000.

# 32

# RECLAIMING BROWNFIELD SITES: FROM TOXIC LEGACIES TO SUSTAINABLE COMMUNITIES

MARK D. BJELLAND
*Gustavus Adolphus College*

The economic landscape is continually changing under the influence of global competition and technological innovation. While capital moves with relative ease, toxic chemical residues in soils and groundwater may persist long after industrial, commercial, or waste disposal activity has ceased. Thus, brownfield sites – idle, abandoned, or underused parcels of land with real or potential environmental contamination – are one of the most serious challenges facing communities across the industrialized world. If communities are unable to find ways to clean up and reclaim brownfield sites for new uses, these sites will pose risks to human health and the environment and act as a blight and economic drain on the surrounding area. Further, if effective means of reusing brownfield sites are not found, urban development will be pushed ever outward onto undeveloped greenfield sites at the metropolitan fringe. Thinking geographically is critical to addressing the toxic legacy of the past. Geographic research methodologies are standard practice for identifying past land uses or off-site pollution that may contribute to site contamination. Geographers have helped policymakers understand the underlying causes, magnitude, spatial extent, and consequences of brownfield sites. Further, geographers are involved in formulating effective policy solutions by creating site inventories, consulting with local residents, and evaluating alternative plans for reclaiming brownfield sites and creating sustainable communities.

## BROWNFIELDS: A CRITICAL URBAN ENVIRONMENTAL CHALLENGE

Brownfield sites emerged in the 1990s as one of the major economic and environmental challenges for communities, particularly those with a strong legacy of industrial activity. Brownfields are defined by the co-presence of both environmental degradation and economic decline. Environmental degradation is the result of past land uses involving waste and chemical handling while the economic component typically involves the mothballing or closure of a facility. Brownfields range from former steel mills to closed gas stations

*D. G. Janelle et al. (eds.), WorldMinds: Geographical Perspectives on 100 Problems, 197–202.*
© 2004 *Springer. Printed in the Netherlands.*

with leaking underground storage tanks. Some brownfields are merely under-used sites while others may go through tax-forfeiture and place a severe burden on local governments (Figure 32.1).

Real or even potential environmental contamination can be a major concern for owners or prospective purchasers of land who may find themselves liable for past contamination. Within the United States, the Comprehensive Environmental Restoration, Compensation, and Liability Act (CERCLA) of 1980 established the legal framework for contaminated sites. CERCLA imposes retroactive, strict, joint and several liabilities for environmental contamination. This means that any party associated with a site can be held responsible for all cleanup costs regardless of fault and regardless of whether the activities were legal at the time. The liability problems under CERCLA and the sub-stantial financial costs and uncertainties in cleaning up and reusing brownfield sites have allowed them to proliferate. The U.S. General Accounting Office estimates there are 130,000 to 425,000 contaminated sites across the country of which only 27,000 have been identified by the U.S. Environmental Protection Agency (U.S. General Accounting Office 1987). In 1996, the U.S. Conference of Mayors declared the situation an emergency and in 1998 declared brownfields their highest legislative priority (Platt 1998). In response,

*Figure 32.1.* Abandoned tax-forfeited brownfields may endanger the environment; act as blight on surrounding neighborhoods; and place severe burdens on local governments. This aban-doned waste oil refinery is located in a floodplain, adjacent to a low-income, predominantly minority neighborhood in Minneapolis, Minnesota. A U.S. EPA brownfields pilot grant is being used to explore cleanup and reuse of the site for either housing or park space.

• Location of U.S. EPA Brownfields Pilots

*Figure 32.2.* Location of 437 U.S. Environmental Protection Agency brownfields pilot projects, as of January 2003. Brownfields pilot projects in Alaska, Hawaii, and Puerto Rico are not shown.

wide arrays of brownfields reclamation efforts have emerged at the federal, state, and local levels (Figure 32.2).

## GEOGRAPHY AND BROWNFIELD SITES

Chemicals in the environment pose risks because spatial patterns of human behavior intersect with spatial patterns of potentially harmful chemicals. Geography is critical to identifying brownfield sites because past land uses are the key to understanding potential site contamination, and future land uses are key to determining possible future exposures to contaminants. As a spatially aware discipline that embraces both environmental and social science, geographers are well suited to contribute to both the management of individual brownfield sites through the site assessment process and to the development of public policy aimed at comprehensively addressing brownfields. Formulating and sustaining effective public policy solutions for brownfields requires understanding the magnitude of the problem, its underlying causes, spatial patterns, effects on communities, and likely prospects for redevelopment. Geographers have made important contributions on each of these fronts.

## GEOGRAPHIC RESEARCH TOOLS FOR ENVIRONMENTAL SITE ASSESSMENTS

Given concerns about environmental liability, Phase I environmental site assessments are performed as part of nearly all transactions involving commercial or industrial properties. Standard practices for environmental site assessments rely heavily on geographic research tools and define appropriate inquiry for purposes of using CERCLA's innocent landowner defense should contamination be subsequently discovered (ASTM 2000). Environmental consultants examine topographic maps, fire insurance maps, aerial photographs, city street directories, and public records. The consultant seeks to understand the geologic, hydrologic and topographic setting of the site and reconstructs past land uses looking for evidence of chemical storage, handling, or waste disposal facilities. Because contaminants may migrate across property boundaries, standard practice calls for performing a geographic information system (GIS) based radial search around the target property looking for features of environmental concern such as hazardous waste sites or leaking underground storage tanks. If potential concerns are found, a Phase II assessment involving chemical testing would be recommended.

## TOWARDS A BROADER GEOGRAPHIC UNDERSTANDING OF BROWNFIELD SITES

Many brownfield sites are not listed in environmental databases but are the result of long-forgotten industrial activities. Craig Colten, a geographer at Louisiana State University, has documented historic land uses associated with potential environmental contamination and created a predictive model for the location of unidentified waste disposal sites based on typical patterns of urban growth, industrial development, and waste disposal in U.S. cities (1990). Leigh and Coffin adopted Colten's approach to identifying the potential for unknown sites and created a GIS-based industrial legacy database for Atlanta, Georgia. Their GIS database identified potential locations of yet-to-be-identified brownfield sites and serves as a model for other cities seeking to determine the extent and magnitude of brownfield site problems in their jurisdictions (2000). Geographical analysis of brownfields in the Minneapolis-St. Paul metropolitan area demonstrated the connections between the region's historical industrial development and the geography of brownfield sites (Bjelland 2000).

Michael Greenberg, a geographer at Rutgers University, directs the National Center for Neighborhood and Brownfields Redevelopment. Greenberg's work has focused on the effect of brownfield sites on communities and has connected issues of environmental contamination to issues of public health and environmental justice. Greenberg's research has documented how brown-fields contribute to the creation of environmentally devastated communities as they interact with other social and technological hazards and he has argued for brownfields programs that address neighborhood quality and public health concerns alongside toxics cleanup (Greenberg and Schneider 1996).

The most effective solutions to brownfield sites have come through state programs that offer liability protections to parties who participate in a vol-untary investigation and cleanup process and through grants to communities to fund the investigation, cleanup, and redevelopment of brownfield sites. The staggering costs of cleaning up all brownfield sites to pristine condi-tions means that it is necessary to base cleanup plans on a realistic site reuse plan. Thinking geographically is necessary to devise reuse plans appropriate to the site and situational characteristics of individual brownfields. While most brownfields programs initially tried to restore brownfields to industrial use, De Sousa, a geographer at the University of Wisconsin-Milwaukee, noted that conversion of many centrally located brownfields to housing could be economically profitable while conversion of other brownfields to green space could improve the livability of cities (De Sousa 2000). Bjelland found that grant programs were effective in encouraging the cleanup and redevelopment of those brownfields with favorable site and situational characteristics, while small brownfield sites located in distressed neighborhoods were often over-looked. Yet, it was the brownfields not being addressed by existing programs that were most likely to create public health and environmental justice concerns (Bjelland 2000). Greenberg studied a primarily Latino community in Perth Amboy, New Jersey and found that community preferences for brownfields redevelopment were typically for parks and community facilities rather than the high intensity commercial and industrial projects favored by public offi-cials (Greenberg and Lewis 2000). Greenberg has also provided justification for brownfields programs by documenting the economic and environmental benefits of brownfields redevelopment versus greenfields development (Greenberg et al. 2001). Continued geographic research is critical to improving brownfields policies and to devising brownfield cleanup and reuse plans that are both cost-effective and tailored to community needs.

## CONCLUSIONS

Brownfields, when allowed to proliferate, are a critical example of unsus-tainable human-environment interactions. Geographic research tools play an integral role in the identification of individual brownfield sites through the site assessment process. Further, geographers have contributed critical policy-relevant research that places brownfield sites within their broader neighbor-hood, regional and environmental systems context. Thinking geographically has assisted many communities in their efforts to identify and reclaim brownfield sites and thereby create more environmentally sustainable places to live and work.

## REFERENCES

American Society for Testing of Materials. 2000. *Standard Practice for Environmental Site Assessments: Phase I Environmental Site Assessment Process, E 1527-00*. Philadelphia, PA: ASTM.

Bjelland, M. 2000. Brownfield Sites: Causes, Effects, and Solutions. *Center for Urban and Regional Affairs Reporter* 30(1): 1–10.

Colten, C. 1990. Historical Hazards: The Geography of Relict Industrial Wastes. *The Professional Geographer* 42: 143–156.

De Sousa, C. 2000. Brownfield Redevelopment Versus Greenfield Development: A Private Sector Perspective on the Costs and Risks Associated With Brownfield Redevelopment in the Greater Toronto Area. *Journal of Environmental Planning and Management* 43: 831–853.

Greenberg, M., K. T. Miller, K. Lowrie, and H. Mayer. 2001. Surveying the Land: Brownfields in Medium-Sized and Small Communities. *Public Management* 83(1): 18–23.

Greenberg, M. and M. Lewis. 2000. Brownfields Redevelopment, Preferences and Public Involvement: A Case Study of an Ethnically Mixed Neighbourhood. *Urban Studies* 37: 2501–2514.

Greenberg, M. and D. Schneider. 1996. *Environmentally Devastated Neighborhoods: Perceptions, Policies, and Realities*. New Brunswick, NJ: Rutgers University Press.

Leigh, N. G. and S. Coffin. 2000. How Many Brownfields Are There? Building an Industrial Legacy Database. *Journal of Urban Technology* 7(3): 1–18.

Platt, R. 1998. Recycling Brownfields. *Urban Land* 57(6): 30–35, 96.

U.S. General Accounting Office. 1987. *Superfund: Extent of Nation's Potential Hazardous Waste Problem Still Unknown*. Washington, D.C.: GAO/RCED-88-4.

# 33

# LAND COVER CHANGE IN THE SEATTLE REGION: LINKING PATTERNS OF GROWTH TO PLANNING EFFORTS THROUGH LAND COVER CHANGE ASSESSMENTS

MELISSA WYATT

*University of Minnesota*

Sprawl issues and growth management have become a larger part of American cultural consciousness in recent years. A variety of responses, including political and grass-roots movements, have attempted to address uncontrolled growth and its economic, ecological and social consequences. These efforts have generally met with limited success. In many cases, political solutions as well as proposed land use plans can often be poorly designed, despite careful study. My case study of land-cover change in the Seattle metropolitan area during 1999 illustrated that the 1991 Washington State Growth Management Act did not successfully curb growth outside the established urban growth boundary. The language of the legislation, implementation and enforcement of the law failed to curb further development within the designated rural use areas. Land-cover change in the rural area actually was more rapid than within the designated urban area. Case studies such as this one, observing the effectiveness of land use legislation, are too rare.

To better address this issue, we need to develop a mechanism to observe the larger picture, that of a regional landscape changing one site at a time. Assessments of public policy and its intended and unintended effects need to be studied more often, in a rigorous manner, and at regular intervals. In addition, we need to be able to adapt legislation to achieve long-range goals.

## SMART GROWTH POLICIES

"Smart growth" has become part of the planning vocabulary of the general public. Most regions have several examples of this approach to control sprawl, direct growth, and provide infrastructure and affordable housing, among other goals (DeGrove 1992; Lewis 1996; Porter 1996). The over-arching hope is to preserve or to improve quality of life. Many state and regional agencies have adopted policies encouraging counties, municipalities, and smaller regions to

*D. G. Janelle et al. (eds.), WorldMinds: Geographical Perspectives on 100 Problems*, 203–207.
© 2004 *Springer. Printed in the Netherlands.*

develop and adopt comprehensive plans with smart-growth concepts in mind.

Planning often encounters obstacles at jurisdictional boundaries, as coordination between cities and counties complicates the jurisdictional-centered planning process. American planning bodies, most often based in a city or county planning board, generally do not facilitate cross-region planning discussions. The notable exception would be transportation planning, due to federal and state funding for transportation infrastructure. Regional coordination of housing, commercial expansion, and provision of goods and services remain primarily city planning issues. Growth occurs independently in the constellations of smaller communities within a metropolitan region. The unfortunate truth is that often city councils do not have the expertise to coordinate all aspects of growth in a community. The result has been the array of bedroom communities in many metropolitan areas, fragmenting and separating schools from homes, homes from jobs and basic needs, and forcing reliance on private automobiles for acceptable transportation solutions.

The review of land-cover change currently occurs sporadically in many regions. For example, the State of Minnesota, through the Legislative Commission on Minnesota Resources (LCMR), has funded several land-cover inventories since 1980. These assessments have occurred at intervals dependent on the availability of funding. A group can procure a sum of money and conduct an inventory in the way it sees fit, so long as it fulfills the obligation of producing a product for the state. A myriad of different methodologies and products can result from this process, and any given product may not correspond to a previously produced inventory. These discrepancies make the rigorous assessment of land-cover change hazy; in the simplest sense, one assessment cannot be measured against another.

## THE KING COUNTY EXAMPLE

In 1999, I conducted an analysis of Washington State's Growth Management Act (GMA) and its effects on land-cover change in King County, WA (see Washington State 1992, 1997). King County includes the Seattle metropolitan area. Growth has been rapid since 1980, driven by a booming local economy. Growth patterns are constrained by a varied topography, including a number of large inland waterways, Puget Sound, and the Cascade Mountains to the east.

The Washington Growth Management Act required each municipality in the state to develop a comprehensive plan that would inventory existing infrastructure and denote an Urban Growth Boundary (UGB) for the next 20 years. Growth would be focused in the urban-designated area. Outside the UGB, land would be designated "rural" (King County Department of Natural Resources 1996). Growth in rural areas would be restricted through several factors, including lot size. In addition, rural growth would not be allowed if it required improvement of urban services such as sewer line expansion, road improvements, and the provision of fire or police protection.

A study area in eastern King County was selected (Figure 33.1). The study area included land on both sides of the UGB spanning from Interstate 90 in the south to the King/Snohomish County line in the north. Satellite imagery (Landsat) data from 1991, the year the GMA was passed, and 1998, the most recent data available, was broadly classed into eight land-cover classes. The eight broad land cover classes included complete forest cover along the gradient to complete impervious cover. Spectral analysis allowed a general, yet coarse, assessment of land-cover. Land use could be inferred loosely from land cover.

Datasets for 1991 and 1998 were classified in the same manner and compared. General conclusions showed that land cover change, despite the edict of the law, continued to occur at a more rapid rate outside the growth boundary (the rural area) than within. The eight classes were collapsed to general categories for a broad analysis, one class representing undeveloped resource land, one showing development, and one category depicted water (Table 33.1). The three classes are illustrated by two images, with darker areas representing undeveloped land, lighter areas showing developed land, and black illustrating water. Fragmentation and loss of resource lands over the period of seven years is clearly illustrated by the loss of large contiguous areas of green.

## SOLUTIONS

The initial conclusion from the case study was that growth was occurring outside the UGB despite the intentions of the growth management policy and goals. Several factors could contribute to this phenomenon, including

*Figure 33.1.* Study area in 1991 (left) and in 1998 (right). Landsat imagery for Eastern King County, Washington.

*Table 33.1.* Developed/Undeveloped Land-use Change, 1991–1998 (number of 30 × 30 m pixels within study area).

| Class of Landcover | Urban Area 1991 | Rural Area 1991 | Urban Area 1998 | Rural Area 1998 | Urban Change (percent to other landcover) | Rural Change (percent to other landcover) |
|---|---|---|---|---|---|---|
| Developed | 170,681 | 81,537 | 216,215 | 283,121 | 26.7 | 247.2 |
| Undeveloped | 166,448 | 352,766 | 121,334 | 152,653 | –27.1 | –56.7 |
| Water | 24,971 | 4,302 | 24,551 | 2,831 | –1.7 | –34.2 |
| Total 30 × 30 Pixels | 362,100 | 438,605 | 362,100 | 438,605 | | |
| Percentage of Developed Landcover | 47.14 | 18.59 | 59.71 | 64.55 | | |

the incentive of the county to building tax-base revenue in the rural portions of the county, the availability of septic system permits for new housing, and inadequate assessment of the need for improved infrastructure (specifically roads) to accommodate increasing populations. It is evident that the intentions of the policy were not being met. Case studies such as this one are often relegated to the motivations of interest groups to test policy against effective gains or losses. The process needs to change, implementing a feedback process into policy legislation.

To avoid the circumvention of a growth policy, a region should adopt a rigorous, regularly scheduled plan to accurately assess land cover change. This process could be modeled with four steps. First, a system of land-cover analysis via satellite imagery should be established. This process must be replicable and documented. A rigorous system of assessment would allow for tracking over time. Second, following the land cover change analysis, the intent and language of growth policy should be documented. This review would be compared with results of the land cover change analysis, observing trends since implementation. A third step would be review and revision of existing policy to assess the legislation's shortcomings. Appropriate revisions would be made and implemented. The fourth and final step would require that such a review should occur at a regular interval, ideally no longer than five years. Many current policies imply that a ten or twenty year interval for planning is adequate. My contention is that it is not; the King County case study illustrates how rapid the rates of change can be. Change and learning occur more rapidly than we often realize. Planning policy, for the most part, is a reactive endeavor at this time. Legislation and policy need to be flexible and adaptable, and employ a proactive, rather than reactive, stance.

# REFERENCES

DeGrove, M. 1992. Planning and Growth Management in the States. Cambridge, MA: Lincoln Institute of Land Policy.

King County Department of Natural Resources. 1996. Farm and Forest: A Strategy for Preserving the Working Landscapes of Rural King County. Seattle.

King County Office of Budget and Strategic Planning. 1997 (update) The 1994 King County Comprehensive Plan. Seattle.

Lewis, P. 1996. Tomorrow by Design: A Regional Design Process for Sustainability. New York: John Wiley.

Porter, D., ed. 1996. Profiles in Growth Management. Washington, D.C.: Urban Land Institute.

Washington State, Department of Community Development, Growth Management Division. 1992. The Art and Science of Designating Urban Growth Areas: Part I and Part II. Olympia, WA: Washington State Government.

Washington State, Department of Community, Trade and Economic Development, Growth Management Division. 1997. State of Washington's Growth Management Act and Related Laws. Olympia, WA: Washington State Government.

34

# HOW SMART IS *SMART GROWTH*?
# THE CASE OF AUSTIN, TEXAS

DANIEL Z. SUI, WEI TU, and JOSE GAVINHA
*Texas A&M University*

Since the publication of the *Costs of Sprawl* by the Real Estate Research Corporation in 1974, scholars and policy makers have been debating the pros and cons of suburban urban sprawl (Kunstler 1993, 1996; Ewing 1997; Gordon and Richardson 1997; Burchell et al. 1998; Katz and Bradley 1999; Duany et al. 2000). Although a wide spectrum of opinions regarding the economic, social, and environmental impacts of sprawl have been reported in the literature, a general consensus seems to be emerging around three kernel points on the urban-sprawl debate: (1) the environmental, social, and fiscal costs outweigh the benefits of the urban sprawl; (2) ultimately, urban sprawl is not a sustainable form of development; and (3) the nation can't cease to grow and new growth will require more urban development. To combat the problems of urban sprawl and, at the same time, allow new urban development to accommodate the demands of a growing urban population, a variety of planned-growth initiatives have been tried across the nation during the past two decades. They include, but are not limited to, growth management, new urbanism, regionalism, urban-growth boundaries, and the promotion of more compact city forms (Gillham 2002).

## WHAT IS SMART GROWTH?

Smart growth is a relatively recent nation-wide movement that seeks to address problems caused by urban sprawl, the latest addition to the list of growth-management initiatives that emerged in the second half of the last century (Weitz 1999). It quickly became an umbrella concept endorsed by a range of diverse groups seeking ways to alleviate problems of urban sprawl. Smart growth does not oppose growth *per se*. It emphasizes the concept of developing more livable cities and towns to ease the social, fiscal, and environmental problems caused by a specific type of growth – urban sprawl. Despite diverse interpretations, smart growth advocates generally prefer infill development, concentrated development, and redevelopment (Krieger 1999; Chen 2000; Downs 2001). The primary goals of smart growth are promoting fiscal equity,

*D. G. Janelle et al. (eds.), WorldMinds: Geographical Perspectives on 100 Problems, 209–214.*
© 2004 *Springer. Printed in the Netherlands.*

environment sustainability, and a better sense of community. For newly developing areas, smart growth emphasizes alternative patterns of development to create a balanced mix of land uses, and a transportation system that provides easier and safer mobility for pedestrians, bicyclists, public-transit users, and automobiles. The multiple dimensions of smart-growth measurement and policy initiatives are summarized in Table 34.1.

## THE CASE OF AUSTIN, TEXAS

Using the city of Austin as a case study, the goal of this chapter is to provide a preliminary assessment on how smart is *smart growth*. Popularly known as the "Silicon Hill," Austin was ranked recently by *Fortune Magazine* as one of the 25 most promising cities in the United States. As an emerging Sunbelt high-tech metropolis, the city enjoyed unprecedented growth during the 1990s, when its population grew from 465,000 in 1990 to about 650,000 in 2000. Between 2000 and 2010, the city is projected to grow by another 200,000. The Austin metropolitan region (including Travis, Hays, and

*Table 34.1.* Smart Growth Measures and Policies.

| Smart Growth Measure | Sample Policies |
|---|---|
| Open-space Conservation | • Regulatory controls (environmental restrictions, zoning controls, transfer of development rights, etc.)<br>• Easements and deed restrictions<br>• Tax incentives<br>• Land acquisition |
| Growth Boundaries | • Local urban growth boundaries<br>• Regional urban growth boundaries |
| Compact Developments | • Traditional neighborhood developments<br>• Transit-oriented developments<br>• Transit villages |
| Revitalization of Older Areas | • Downtown and main street redevelopment programs<br>• Brownfield redevelopment<br>• Greyfield redevelopment |
| Public Transit | • Local transit programs<br>• Regional transit programs |
| Regional Planning Coordination | • Regional governments<br>• Regional authorities<br>• Regional infrastructure service districts<br>• State planning initiatives |
| Equitable Sharing of Resources/Burdens | • Regional revenue sharing<br>• Regional affordable housing programs |

Source: Adapted from Downs (2001) and Gillhan (2002).

Williamson County) is expected to grow from 1 million in 2000 to 1.5 million by 2010. Concomitantly, typical symptoms of urban sprawl, such as extensive suburbs, increasing traffic jams, deteriorating environmental conditions, and erosion of the central-city tax base, began to surface in Austin. Both citizens and government officials are aware of these problems – one of the most popular bumper stickers in the city recently has been "Don't Houstonize Austin!"

To alleviate the problems brought by urban sprawl, the city of Austin initiated its Smart Growth Initiative (SGI) to manage rapid growth in 1998. Austin's SGI consists of a number of programs designed to contain urban growth within designated areas, enhance existing neighborhoods in the urban core, and preserve environmentally sensitive areas. The SGI also aims to establish new development patterns and to implement transportation improvements that maximize opportunities for multimodal transportation (City of Austin 1999).

More specifically, Austin's SGI aims to accomplish three main objectives, (1) determine and manage where and how Austin is and should be growing; (2) improve the quality of life for all citizens; and (3) enhance the tax base. To start implementing the SGI, two primary zones were identified in a smart-growth map – the desired development zone (DDZ) and the drinking-water protection zone (DWPZ) (Figure 34.1). The DDZ encompasses roughly the eastern two-thirds of Austin, including most of the traditional downtown areas; it is to this area that the city wants to direct most of its future growth. The DWPZ comprises watersheds that supply Austin with drinking water, including a portion of the Edwards Aquifer that feeds Barton Springs. Also included, as part of the DWPZ is an area of steep slopes and shallow soils of the Texas Hill country – the habitat of endangered species that are not well suited for intensive development. The SGI tries to discourage additional development in the DWPZ due to the environmental constraints found in the zone.

In implementing the SGI, the city has relied on a variety of planning goals and incentive policies, including: (1) restoring community life and vitality to the urban core by investing in the city center; (2) protecting the character of existing residential areas through neighborhood planning; (3) protecting the quality of the environment by preserving sensitive environmental features and encouraging more efficient development patterns; (4) making efficient use of public investments in facilities such as streets and sidewalks, water and sewer lines, fire and emergency services, parks, open space, and schools; (5) creating development that is pedestrian and transit friendly by permitting a mix of land uses and by increasing densities where appropriate; (6) decreasing automobile congestion by providing alternative modes of transportation, such as bus lines, light rail, bicycle lanes, and improved pedestrian facilities, such as side walks and neighborhood parks; and (7) rewarding developers with reduced fees and more flexible and faster application processing for development projects that meet smart-growth goals.

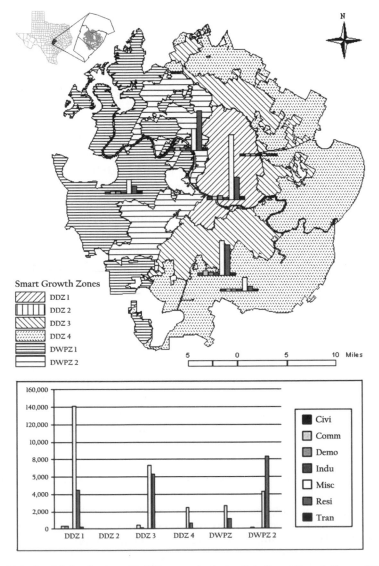

*Figure 34.1.* Spatial distribution of building permits in Austin's Smart Growth Zones: 1997–2000 (Source: City of Austin).

The SGI has been in existence for more than 4 years. Have the above policies and incentives achieved the goals and objectives defined in the SGI? Or, in other words, how smart has the SGI in Austin been? In the following section, we describe a preliminary assessment on the effectiveness of SGI in Austin based on the first objective of Austin's SGI: to verify whether recent growth has occurred in the desired development zones or not.

## METHODS

Despite the new challenges that geographic information systems poses for urban studies, GIS has become an indispensable tool for mapping communities and monitoring growth management in recent years. To address the research question, we used ArcView 3.3 GIS to tabulate the number and kinds of building permits issued in each DDZ and DWPZ. The data were obtained from the Austin's planning division, which updates information on building permits on a quarterly basis. During the study period (1997–2000), a total of 56,724 new building permits of different kinds were issued. These building permits were geo-referenced via street address matching, and a geo-referenced point-data layer was created. Using ArcView's "select-feature-by-theme" operation along with several Avenue scripts specifically developed for this project, the number and type of building permits were tabulated by the different zones defined in the smart-growth map. Currently there are 4 DDZs and 2 DWPZs, which cover both the city of Austin and extraterritorial areas (Table 34.2). At the same time, the 71 different codes used by the City of Austin's planning division to classify building permits were grouped in seven major categories, shown in Table 34.3 and Figure 34.1.

## FINDINGS AND CONCLUSIONS

Out of the 56,724 building permits issued from 1990 to 2000, about 97 percent of the cases were within the jurisdiction of the city. About 30 percent were located in drinking water protection areas (DWPZ1 and DWPZ2), and the remaining 67 percent were located in desired growth zones (DDZ 1 to 4). The share of new construction in the DWPZ has remained unabated during the study period. On average, by 2000 it was even slightly higher than in 1997 – the only year without the smart-growth program during our study period. As far as the first objective of SMI is concerned, there were no

Table 34.2. Building Permit Distribution in Smart Growth Zones, 1997–2000.

| Smart Growth Zone | Building Permit Distribution (%) | | | | |
|---|---|---|---|---|---|
| | 1997 | 1998 | 1999 | 2000 | 1997–2000 |
| DDZ 1 | 40.28 | 34.09 | 34.04 | 32.53 | 34.78 |
| DDZ 2 | 0.02 | 0 | 0 | 0.01 | 0.01 |
| DDZ 3 | 26.61 | 25.01 | 25.45 | 26.27 | 25.86 |
| DDZ 4 | 3.51 | 4.10 | 5.27 | 8.33 | 5.71 |
| DWPZ 1 | 21.60 | 27.75 | 7.98 | 20.59 | 23.27 |
| DWPZ 2 | 5.41 | 6.77 | 24.37 | 7.58 | 7.07 |
| Outside | 2.56 | 2.28 | 2.89 | 4.69 | 3.30 |

Source: Calculated by authors.

*Table 34.3.* Smart Growth Zone Categories.

| Sub-zone | Description |
| --- | --- |
| DDZ1 | Urban areas within full jurisdiction |
| DDZ2 | Non-Urban areas within full jurisdiction |
| DDZ3 | Urban areas within extraterritorial jurisdiction |
| DDZ4 | CBD+CURE+UT* |
| DWPZ1 | Extraterritorial jurisdiction |
| DWPZ2 | Full jurisdiction |

* CBD: Central Business District; CURE: Central Urban Redevelopment; UT: University of Texas.

discernable improvements in preventing further development in DWPZ. For the four DDZs, the proportion of building permits has been steady, even declining slightly – which is consistent with the slight increase in the DWPZ.

Although far from being conclusive, this preliminary assessment on the effectiveness of the smart-growth program in Austin does cast doubts on the IQ of current smart growth in Austin. Based on the data available, we conclude that Austin's SGI has neither discouraged growth in the two DWPZs nor stimulated new development in four DDZs. Perhaps, in most cases, mere incentive programs won't work because most developers view profit as their business bottom line. Will a more draconian regulatory framework be more effective in curtailing urban sprawl? That is a question for further debate.

## REFERENCES

Burchell, R. W. (nine others). 1998. *The Costs of Sprawl – Revisited.* Washington, D.C.: National Research Council – Transportation Research Board.

Chen, D. 2000. The Science of Smart Growth. *Scientific American* 283(6): 84–91.

City of Austin. 1999. *Smart Growth Initiative: Matrix Application Packet.* Planning, Environmental & Conservation Services Department, Austin, Texas.

Downs, A. 2001. What Does Smart Growth Really Mean? *Planning* 67(4): 20–25.

Ewing, R. 1997. Is Los Angeles-style Sprawl Desirable? *Journal of the American Planning Association* 63(1): 107–126.

Gillham, O. 2002. *The Limitless City: A Primer on the Urban Sprawl Debate.* Washington, D.C.: The Island Press.

Gordon, P. and H. W. Richardson. 1997. Are Compact Cities a Desirable Planning Goal? *Journal of the American Planning Association* 63(1): 95–106.

Katz, B. and J. Bradley. 1999. Divided We Sprawl. *Atlantic Monthly* 284(6): 26–42.

Krieger, A. 1999. Beyond the Rhetoric of Smart Growth. *Architecture* 88(6): 53–57.

Kunstler, J. H. 1993. *The Geography of Nowhere.* New York: Simon and Schuster.

Real Estate Research Corporation. 1974. *The Costs of Sprawl: Environment and Economic Costs of Alternative Residential Development Patterns At The Urban Fringe* (vol. I: Detailed Cost Analysis; vol. II: Literature Review and Bibliography). Washington, D.C.: The U.S. Government Printing Office.

Weitz, J. 1999. From Quiet Revolution to Smart Growth: State Growth Management Programs, 1960 to 1999. *Journal of Planning Literature* 14(2): 267–338.

## PART IV

# INTEGRATING LOCAL AND GLOBAL ECONOMIES

The integration of local places into national and international divisions of labor has been a persistent theme running throughout geographic analyses of the human occupancy and use of the earth and its resources, local patterns of work and consumption, and the uneven distributions in the quality of human life. Mindful of the need to remain engaged with the unique specifics of localities, geographers have also stressed the growing significance of global systems of commerce and politics. Understanding the linkages across geographical scales and the interactions among regions is a fundamental necessity to minding the unfolding world order of development, underdevelopment, and exchange.

*Barney Warf and Don Janelle*

# 35

# CULTURAL EXPLORATION AND UNDERSTANDING: A FRAMEWORK FOR GLOBAL BUSINESS

NANDA R. SHRESTHA, WILBUR I. SMITH, and KENNETH R. GRAY
*Florida A&M University*

> Culture is a sort of theater where various political and ideological causes engage one another. Far from being a placid realm of Apollonian gentility, culture can even be a battleground on which causes expose themselves to the light of day and contend with one another (Said 1993: xiii).

Growing business competition demands that companies demonstrate cultural responsiveness to establish competitive advantages in the global market where a heightened sense of national/ethnic identity is on the rise (Holden 2002). This is true whether a company is "going global" to extend its *market* (consumer focus) or to relocate *production* (to reduce operating costs), and whether the company adopts a *multicountry* or *multidomestic* strategy (i.e., operating relatively autonomous business units in multiple countries) or a *global* strategy (operating business units in multiple countries using standardized products and business practices) to expand globally (Yip 2003). Yet the treatment of culture in the business literature thus far has largely ignored the complexities of culture, opting for broad generalizations that amount to sophisticated stereotypes and misconceptions, often leading to poor business performance or even failure. To address this tendency, we propose a multitiered framework of cultural exploration and understanding. Although it is proposed in the context of global business, the framework can be applied in other contexts, in that it portrays a more complete picture of cultural landscapes.

## PROPOSED FRAMEWORK

"Culture" is one of the most commonly – and loosely – used words in the English language. Invariably evoked to highlight distinctiveness, it is generally interpreted as a way of life. Take, for example, agriculture (agri + culture), which means a cultural way of life practiced by a segment of the population

217

*D. G. Janelle et al. (eds.), WorldMinds: Geographical Perspectives on 100 Problems*, 217–224.
© 2004 *Springer. Printed in the Netherlands.*

whose economic livelihood revolves around farming. As it molds the common way of life and value system, culture acts as a foundation of shared group identity whether rooted in geography (e.g., the Middle East), religion (e.g., Buddhism), ethnic/racial/tribal affiliations or nationality (e.g., African-Americans), or something else. However, culture goes beyond "the way of life" and group identity formation to encompass virtually every aspect of life. Culture is at once *value-forming* and *value-contingent*. As such, it shapes and is shaped by a people's social views and values and political/ideological orientations and outlooks. For some, it is a ruling dogma or shared mindset such as the imperial culture that some European nations practiced for some four hundred years to rally their national populations around the claim of European superiority, drawing a racialized cultural line between Whites ("the West") and non-Whites ("the Rest").

In the global business context, groups of people and their "cultures" are often reduced to food, fiesta, or some other "cultural" traditions and *dos* and *don'ts* or are aggregated into vastly generalized categories such as Western vs. non-Western. Such hyper generalization is reflected in two frequently used global business management models: Hall and Hall's (1990) concept of *low-context* (Anglo-Germanic) and *high-context* (e.g., African, Asian, and Latin American) cultures and Hofstede's (1991) limited measures of power distance, uncertainty avoidance, individualism, and masculinity. The following state-ment typifies how culture is conceptualized in business texts:

> With respect to assumptions and attitudes, U.S. and Canadian management systems have been greatly influenced by their societies' belief in self-determination. In many other countries, however, the culture is much more fatalistic – a dominant belief is that human beings cannot really control their own future. Problems may arise in other areas as well – for example, with respect to time: being "on time" is very important in most Western societies but is of little importance in China. In fact, being late is more common there than being on time. With respect to personal beliefs, aspirations, and motivations, the need to achieve is usually much less significant in non-Western countries than it is in Western societies (Higgins 1994: 95).

This type of cultural perspective is troublesome because it overlooks internal cultural dynamics and nuances. There is rarely any attention paid to the sociopolitical dimensions of non-Western cultures and their impact on people's orientations and outlooks toward the United States and the West. Where polit-ical variables are considered in global business, they are invariably discussed in relation to the host government (the ruling faction) or the political system in general and the risk and reward associated with it rather than from a cultural perspective, i.e., in relation to the general populace (Punnett and Ricks 1997). Yet it is not uncommon to experience a substantial difference between gov-ernmental receptivity and public reactions to foreign businesses (*The Economist* 2002). Going global requires foreign companies to deal with the host gov-ernment as it entails operating within a different national boundary and legal system. But that is not necessarily acting locally, which demands reading the pulse of the public, be it political, economic, or social. With this in mind, we propose a preliminary model to explore and understand culture, specifically focusing on cultural identity and cultural orientation and outlook.

## CULTURAL IDENTITY

Culture is a fundamental source of identity (Said 1993). As a basis for group formation, it engenders a sense of belonging, security, and pride. However, cultural identity is not uniform across space, but is a contested terrain where its internal (insiders') image comes into sharp contrast with the external perceptions of outsiders. Global business requires a spatially multi-tiered framework for exploring and understanding various levels of cultural identity in a given society. The level could certainly vary, along with the criteria used to define it. For instance, in a large country with a diverse population, a firm needs to capture great geographical and ethnic diversities. The appropriate level of cultural analysis depends on the geographic size and demographic makeup of the country and on the company's objective (production, marketing, or both) and minimum market requirements (see Figure 35.1 for an example of India).

Global business has generally relied on the nation-state as the fundamental unit of cultural analysis (Wild et al. 2003). Consequently, culture has normally served as a synonym for nation without any further conceptual grounding. While this tendency can be partly explained by certain national/legal requirements and the need of businesses to assure a minimum market threshold, it is a function of two underlying assumptions. First, nationality is seen as an overriding cultural indicator of the entire population. Second, modernization is believed to have produced a relatively uniform landscape of consumer behavior and preferences: globalization of Western values. The focus on the national culture is convenient as it is readily definable in terms of political and legal demarcations. However, the national level not only obscures the terrain of cultural identities, it also fails to provide insights into the socio-political dimensions of culture in relation to global business. As a result, an exclusive focus on national culture may prove to be counterproductive to companies as they expand globally.

## CULTURAL ORIENTATION AND OUTLOOK

Contrary to the assumption that globalization fosters worldwide cultural convergence, Huntington (1993: 22) sees cultural conflicts: "[T]he fundamental source of conflict in this new world," he says "will not be primarily ideological or primarily economic . . . [but] cultural. . . . The clash of civilizations will dominate global politics." What dominates global politics tends to determine the global economy and business trends. His contention is that as people define

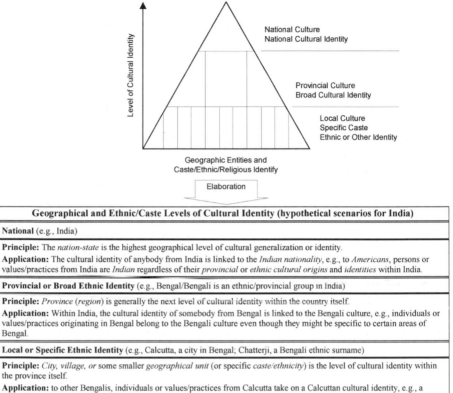

*Figure 35.1.*  Cultural identity.

their identity in ethnic and religious terms, they are likely to see an "us" versus "them" relation between themselves and other religious or ethnic groups, creating "differences over policy issues, ranging from human rights to trade and commerce" (Huntington 1993: 29). But beyond broadly linking cultural conflicts with cultural differences, where do cultural fault lines actually lie? What triggers their eruption into open hostilities? Immersed in Cold War thinking, Huntington instinctively (but wrongly) locates fault lines at the national level. To him, the nation-state represents both the cultural identity and the sovereign government as an all-encompassing force. Omitted from this conceptualization is any consideration of differences between government policies and public perceptions, orientations, and outlooks.

For example, in Indonesia the government is supportive of the U.S. position on terrorism. Yet the public reaction is just the opposite, as middle class youth, along with other demographic segments, are openly hostile to the American global policy. In their eyes, America has little respect for them and their cultural values, and the Bush administration's war on terrorism is merely a pretext to demonize Islam and dominate the world. This sentiment is widespread throughout the Muslim world and beyond. As Sevareid (2002: 14A) reports, a teacher in Qatar says the Arab leaders should not welcome U.S. troops, and a prayer leader in Saudi Arabia believes that "U.S. troops in the kingdom represent America's greediness and effort to monopolize the wealth of the Muslim world" while it supports "Israel with modern weapons to kill Palestinian children."

So the cultural fault line between the West and "the Rest" resides not with national governments, but in the public domain, at the street level. The more their rulers and cultures are perceived to be emasculated by the United States, the deeper the general public's cultural animosities toward the United States tend to be. Their resentment deepens further as globalization tends to widen the socioeconomic gap between the rich and poor (*Business Week* 2000: 74–75). Frustrated and resentful, people fight back, mobilizing opposition by appealing to cultural identity (e.g., religion). Where institutionalized economics and politics fail people, popular religion or culture fill the gap, a situation that often leads to adversarial public reaction to global (usually American) business. For example, "[A]cross the Middle East, America's war on terror and its threats to Iraq have inspired consumers to boycott American brands from Pampers nappies to Heinz ketchup," while many of their governments remain in the American fold (*The Economist* 2002: 65). It is precisely these street-level cultural perceptions and outlooks that global business needs to understand to gain competitive advantages.

To perform a more informative cultural analysis for global business, we propose a matrix format in Figure 35.2. Such an analysis is not only equipped to generate comparative profiles of different cultural identity groups within a country with respect to their cultural orientations and outlooks toward foreign businesses, but also within each group based on certain economic, demographic, or sociopolitical dimensions such as education, generation (age), and social class. In the case of the Indian example in Figure 35.2, we have used religion as a source of cultural (group) identity and a set of variables, segmented into subgroups (e.g., upper, middle, and lower classes), to decipher their varying cultural perceptions, outlooks, and reactions to the West so that global business decisions can be made on an informed basis.

However, in conducting an actual cultural analysis for a global company, an analyst has the flexibility of selecting different criteria for both the cultural identity and sociopolitical dimensions (including demographic and economic variables), depending on the specific requirements and focus of the company. Furthermore, even though the discussion and examples provided in this paper are geared toward Western companies expanding to non-Western countries, the present model is reversible.

| Cultural Identity (e.g., religion) | Orientation and Outlook | Location | Education | Social Class | Generation | Economic/Globalization | Ideology/Belief System | Cultural Relations |
|---|---|---|---|---|---|---|---|---|
| *Hindus* | Pro-West | Large cities | Technical and professional | Upper class | Youth | Business sectors benefiting from globalization | Adopters of Western economic values | *Cultural convergence* |
| | Anti-West | Large/small cities | Social sciences and humanities | Middle class; the deprived | Univ. students; older generation | Business sectors and laborers hurt by globalization | Nationalists; fundamentalists; socialists | *Cultural conflicts* |
| *Muslims* | Pro-West | Large cities | Technical and professional | Upper class | Youth | Business sectors benefiting from globalization | Adopters of Western economic values | *Cultural convergence* |
| | Anti-West | Large/small cities | Social sciences and humanities | Middle class; the deprived | Univ. students; older generation | Business sectors and laborers hurt by globalization | Nationalists; fundamentalists; socialists | *Cultural conflicts* |
| *Christians* | Pro-West | Almost everywhere | Almost all | Almost all | Almost all ages | Minimal influence | Minimal influence | *Cultural assimilation* |
| | Anti-West | Almost nowhere | Minimal | Minimal | Minimal | Minimal influence | "Liberation Theologists" | *Minimal cultural conflicts* |

*Figure 35.2.* Cultural orientation and outlook (hypothetical scenarios for India).

## CONCLUSION

We propose to address the shortcomings of existing models by going beyond the nation-state as the unit of cultural analysis. A mulitiered framework provides a better assessment of "cultural fault lines" because it generates the type of fine-grained information needed to assess the culturally based political and economic risks of global business. It can help build comprehensive risk-reward profiles for expanding operations across cultural boundaries. The flexibility of our approach allows levels of cultural identity and cultural orientation and outlook to be varied to suit a company's specific requirements and focus. The company planning to expand overseas can apply the present model in designing its globalization strategies – that is, to choose between the multicountry and global approaches to geographic expansion. Where cultural analysis uncovers large variations in cultural identity and outlook, the company would be better off adopting a multicountry strategy (Thompson and Strickland 2001). This approach, by sacrificing corporate uniformity and some productivity, allows the company to be responsive to local cultural, economic, political, and competitive conditions. Conversely, a global strategy, which requires the company to operate in the same manner in all its locations with standardized products, business practices, and strategic foci, is a better fit where cultural analysis reveals little country-to-country cultural variations (see Hitt et al. 1999). Cultural knowledge, where developed as a competency, is an organizational resource that provides a competitive advantage for global business.

## REFERENCES

*Business Week.* 2000. Global Capitalism: Can It Be Made to Work Better? 6 November: 72-100.

*Economist.* 2002. Brand Wars in the Middle East: Regime Change. 2 November: 65.

Hall, E. and M. Hall 1990. *Understanding Cultural Differences.* Yarmouth, ME: Intercultural Press.

Higgins, J. 1994. *The Management Challenge: An Introduction to Management.* New York: Macmillan.

Hitt, M., R. Ireland, and R. Hoskisson 1999. *Strategic Management: Competitiveness and Globalization.* Third Edition. Cincinnati: South-Western College Publishing.

Huntington, S. 1993. The Clash of Civilizations? *Foreign Affairs* 72(3): 22–49.

Hofstede, G. 1991. *Culture and Organizations: Software of the Mind.* London: McGraw-Hill.

Holden, N. J. 2002. *Cross-Cultural Management: A Knowledge Management Perspective.* London: Prentice Hall.

Punnett, B. and D. Ricks 1997. *International Business* (Second Edition). Cambridge, MA: Blackwell Publishers.

Said, E. 1993. *Culture and Imperialism.* New York: Alfred Knoff.

Sevareid, S. 2002. U.S. Troops Face Anger as War Talk Heats Up. *The Tallahassee Democrat* 3 November: 14A.

Thompson, A., Jr. and A. Strickland III. 2001. *Crafting and Executing Strategy: Text and Readings* (Twelfth Edition). Boston: McGraw-Hill/Irwin.

Wild, J., K. Wild, and J. Han 2003. *International Business*, Second edition. Upper Saddle River, NJ: Prentice-Hall.

Yip, G. 2003. *Total Global Strategy II.* Upper Saddle River, NJ: Prentice Hall.

# 36

# GEOGRAPHY IN A CRISIS: PERSPECTIVES ON THE ASIAN ECONOMIC CRISIS OF THE 1990s

PHILIP KELLY
*York University*

JESSIE P. H. POON
*University at Buffalo – SUNY*

Since 1997, more than 500 journal articles have been published on the financial and economic crisis that developed in East and Southeast Asia in the late 1990s. This intensity of academic interest reflected the profound impact of these events on regional economies, governments, and livelihoods. It also highlighted the challenge that the crisis presented to mainstream explanations of the developmental success stories of Pacific Asia, and the nature and consequences of globalization in a region closely integrated with the world system. Geographers have authored a small part of the scholarly literature precipitated by the crisis. We argue that a geographic sensibility has nonetheless proven essential to understanding the processes that occurred, their impacts, their amelioration, and their future prevention.

## GEOGRAPHIES OF PLACE AS WELL AS SPACE

Economists, political economists and sociologists dominated initial discourses on the causes of the crisis. On the one hand, economists tended to blame Asia's problems on "morally hazardous" behavior, in which their financial institutions engaged in risky lending practices on the assumption that governments would ultimately bail them out; allocating credit had, after all, been a key strategy of the region's developmental states. These banking practices occurred in a context of exchange rates pegged to the U.S. dollar, declining export competitiveness, widening current account deficits, and unhedged short-term foreign loans. This confluence of circumstances led to precipitous declines in exchange rates, stock markets, and foreign loans in the second half of 1997. Political economists, on the other hand, have emphasized the effectiveness of corporate and financial arrangements in Asian contexts until external pressures forced many countries to deregulate their capital flows (and thereby creating the potential for crisis). When trouble started in mid-

*D. G. Janelle et al. (eds.), WorldMinds: Geographical Perspectives on 100 Problems, 225–229.*
© 2004 *Springer. Printed in the Netherlands.*

1997, capital markets exhibited "herd" behavior, pulling funds out of Asia regardless of national economic circumstances – a panicked response exacerbated by the "one-size fits all" remedies of the International Monetary Fund.

In both cases, however, arguments were constructed at highly generalized geographical scales, i.e., global economic and regulatory characteristics were assumed to be "regional" in applicability. What was missing was a geographical sensibility to the multiple scales at which explanations need to be situated. Thus, an early geographical contribution to the crisis debate emphasized that the crisis "contagion" spread across very diverse national circumstances in East and Southeast Asia (Poon and Perry 1999). In comparison with South Korea, for example, Indonesia has weaker regulatory institutions, a much larger traditional urban as well as agricultural sector, and a more entrenched history of rent-seeking behavior between businesses and government. Demands for greater capital liberalization across the board appeared oblivious to these distinctions.

For this reason, geographers tended to prefer context-based explanations and solutions, drawing upon local and regional knowledge of political-economic systems. A good example is Dixon's (2001) insights on Thailand, where the crisis originated. Dixon observed that Thailand's allegedly strong macroeconomic indicators (e.g., high savings rate) did not necessarily reflect sound economic fundamentals as was the case with Taiwan or Singapore. Thailand did not fit the corporatist model that had allowed the governments in Taiwan and South Korea to remain relatively insulated from business interests in the 1970s and 1980s. By integrating political, economic and historical contexts, Dixon demonstrated that potential solutions to Thailand's economic crisis lie somewhere between the two competing models; that is, internal reforms that could help decrease rent-seeking behavior between businesses and government, and, more genuine development that creates a domestic capitalist class, and which speeds up agrarian reforms and human capital accumulation.

## COMMITMENT TO FIELDWORK

If geographers have exhibited a strong attachment to the examination of how "place" fits into "space," then a corollary has been the commitment to empirical data and fieldwork. When done well, this involves the generation or testing of conceptual ideas in "the field," whatever that place happens to be. Geographers were among the first to generate empirical material on the impacts of the crisis, and by extension the appropriate policies for its amelioration or prevention.

A number of studies used firm-level interviews or surveys to assess the impact of the crisis on the spatial strategies of corporate capital. In 1998, for example, Poon and Thompson (2001) surveyed over 300 transnational corporations in Hong Kong and Singapore and found an understandable short-term pessimism about market potential in the region, but also a longer term optimism concerning regulatory reforms and reduced corruption. Using case study interviews with major Japanese electronics manufacturers in Southeast Asia, Edgington and Hayter (2001) similarly concluded that foreign direct investment in the region was less footloose than might be imagined. These findings highlight the importance of focusing upon the faction of capital being discussed in relation to the crisis and its impacts. The economic land-scape perceived by foreign direct investors in this case contrasted markedly with, for example, the pessimism of the London financial community examined by Clark and Wójcik (2001). The implication is that the longer term prosperity of Asian economies might best be served by openness to foreign direct investment, but a much more cautious approach to global portfolio and loan capital.

Moving from the firm to other scales of analysis, geographers demonstrated the ramifications of the crisis and the complexity of responses. Several studies in Indonesia brought to light scarce material about "everyday" reactions to the crisis. McGee and Firman (2000) demonstrated that the crisis was quite differently experienced in rural versus urban settings, with the implication that a strong rural and informal economy can create a safety net of sorts against the worst impacts of crisis in the urban formal sector of the labor market. Furthermore, households shifted their livelihood strategies between these sectors as the crisis developed. Pursuing a parallel argument, Turner (2000) used field data from Sulawesi to highlight the subnational and sectoral dif-ferentiation of crisis impacts. She showed that small scale entrepreneurs in the urban economy of Makassar were relative unaffected in any direct manner by the economic crisis, especially in comparison with the larger scale indus-trial enterprises in the Javanese core. Also in Sulawesi, Silvey (2001) demonstrated that the uneven experiences of economic crisis could be observed even within households, as young women returning to their villages from declining industrial zones shouldered a disproportionate burden in assisting the household through difficult times relative to their older and/or male family members. At a micro-scale, the impacts of the crisis were thus both gender and generationally differentiated. These studies highlight the activities and sectors that can most effectively provide a buffer against future crises and the social groups that are most vulnerable to the effects of economic downturns.

## REPRESENTATIONS OF SPACE

Geographers have increasingly paid attention not just to spatial patterns and processes, and their manifestations in particular places, but also to the ways that spaces are socially constructed and imbued with significant meanings.

A particularly important issue is the way in which spaces and spatial processes are represented in metaphorical terms. The Asian economic crisis was rife with metaphorical representations – analogies that served to explain the complexity of events in simplified terms, but which also sought to advance particular interpretations of those events. Epidemiological readings had the crisis spreading across Asia as a contagion (e.g., the "Asian Flu," "Bahtulism") infecting the sick economies of the region. Meteorological metaphors spoke to the need to build solid national defences against future storms. Military analogies suggested that outside aggressors were deliberately undermining the success of Asian economies. Few commentators noted the significance of these representations, but geographers have started to explore their implications. Smith (1998) highlighted a number of these metaphors and noted their importance in shaping our understanding of economic crisis. Kelly (2001) developed this argument by linking the distinctive metaphors used by political leaders in Singapore and Malaysia to render the crisis intelligible, to the quite different bases for political legitimacy in each context, and, in turn, the policy responses that each government pursued. The value in deconstructing these metaphors lies in showing how they constrain thinking about the causes of crisis (and about spatial relationships in the economy more broadly), and thus limit the conceivable solutions.

## CONCLUSIONS

Geographers have not been major players in the debate over the Asian economic crisis, but in the application of geographical sensibilities (to place as well as space, to fieldwork at a micro scale, and to representations of space) provided important and distinctive insights. They demonstrated a healthy skepticism to sweeping generalizations at a regional or global scale that imply catch-all explanations and solutions. They highlighted a readiness to seek answers to big questions at small scales. And they questioned the ways in which popular representations of spatial processes obscure alternative interpretations and solutions.

## REFERENCES

Clark, G. and D. Wójcik. 2001. The City of London in the Asian Crisis. *Journal of Economic Geography* 1: 107–130.

Dixon, C. 2001. The Causes of Thai Economic Crisis: The Internal Perspective. *Geoforum* 32: 47–60.

Edgington, D. and R. Hayter. 2001. Japanese Direct Foreign Investment and the Asian Financial Crisis. *Geoforum* 32: 103–120.

Kelly, P. 2001. Metaphors of Meltdown: Political Representations of Economic Space in the Asian Financial Crisis. *Environment and Planning D: Society and Space* 19: 719–742.

McGee, T. and T. Firman. 2000. Labour Market Adjustment in the Time of Krismon: Changes in Employment Structure in Indonesia, 1997–98. *Singapore Journal of Tropical Geography* 21: 316–335.

Poon, J. and M. Perry. 1999. The Asian Economic Flu: A Geography of Crisis. *Professional Geographer* 51: 184–196.

Poon, J. and E. Thompson. 2001. Effects of the Asian Financial Crisis on Transnational Capital. *Geoforum* 32: 121–131.

Silvey, R. 2001. Migration Under Crisis: Household Safety Nets in Indonesia's Economic Collapse. *Geoforum* 32: 33–45.

Smith, N. 1998. El Nino Capitalism. *Progress in Human Geography* 22: 159–163.

Turner, S. 2000. Globalisation, the Economic Crisis, and Small Scale Enterprises in Makassar Indonesia: Focusing on the Local Dimensions. *Singapore Journal of Tropical Geography* 21: 336–354.

# 37

# MODELING SPACE FOR REGIONAL REGENERATION: HIGH-TECH DISTRICTS IN CHINA

SUSAN M. WALCOTT
*Georgia State University*

China seeks to vault into the 21st century by taking its place in the global economy. This research examines challenges faced in implementing China's strategy of encouraging technology-intensive business clusters by targeting spatially distinct zones as privileged spaces. China has proved to be a popular site for multinational firms seeking low wage, routine labor since its "Opening and Reform" movement began in 1978. As a national strategy, the intent is for these firms to generate a higher skill level of jobs with greater economic returns in selected cities.

Questions arise as to whether China's "science and technology industry parks" (STIPs) fit model practices in more advanced countries, or whether they constitute a unique form "with Chinese characteristics." If the latter, would regional variations occur in this large and varied country the size of the continental U.S.? What might be the role of proximity among companies in a zone for promoting corporate relationships within STIPs among multinational companies and/or interspersed Chinese companies? This chapter reviews four leading high-tech district sites to assess different responses to the strategy of utilizing designated high technology business spaces to accelerate modernization. The cities examined include central coastal Shanghai at the mouth of the Yangtze River delta, southern coastal Shenzhen in the Pearl River delta, Xi'an in the interior western region, and northern Beijing, the capital city (Figure 37.1).

## APPROACHING THE PROBLEM

Geographic treatment of the economic development consequences flowing from clusters of companies engaged in high technology activities focuses on outcomes attributable to the use of space, and differences in the nature of various places. Locations gain a reputation for the companies they keep, as well as the sectoral type of companies and their stage in the production process. Technopoles (science park-based successful planned developments) flourish in locations around the world (Castells and Hall 1994). Globalized production

*D. G. Janelle et al. (eds.), WorldMinds: Geographical Perspectives on 100 Problems*, 231–235.
© 2004 *Kluwer Academic Publishers. Printed in the Netherlands.*

*Figure 37.1.* Science and Technology Parks examined in this study.

creates a spatial scale favoring different locations at various stages in a new international division of labor (Dicken 1998). High technology companies require supportive locations (or "milieu") where innovations can be developed, tested, and turned into marketable productions. This process uses the gamut of labor-skill sets, supplier-producer-market links, infrastructure connections, global and local networks to embed clusters, and the prevalent cultural framework of both place and players.

Additional geographically related issues particular to the Chinese setting include the role of central government directives, new policies on land marketization, the location of new worker residences, new transportation infrastructure affecting the commute to work, an increasingly mobile population, and selective regional development incentives. The success of foreign and native businesses depended on constructing an efficient learning and production environment in designated places that were alien to both (Malecki et al. 1999). Chinese problems include how to manage innovation, entrepreneurship, and businesses; for capitalist companies, concerns involve locating suppliers, protecting intellectual property, and penetrating the new market. Methodological requirements for exploring STIP settings necessitate building personal contacts, understanding local culture, conducting targeted interviews, and accessing foreign data sources.

## CHINESE CASE STUDIES

Statistics compiled from national, municipal, and science and technology year-books to assess the comparative situation in each park reveal clear distinctions (Table 37.1). While Beijing has a huge lead in numbers of workers, and tech-related revenue, its companies are smaller and produce for a domestic market (small computer-related firms in university-heavy setting). Shanghai predominates in exports overall, but Shenzhen exports more per worker (via Hong Kong). Xi'an eclipses Shenzhen in workers and revenue, but produces for a very different, domestic market due to their vastly different locations.

Interviews in Shanghai, the nation's business heart, were conducted with park officials and company managers in the three major high-tech parks of Caohejing, Zhangjiang (in fast developing Pudong), and Minhang. STIP managers often deliberately located Chinese companies around Western ones, in hopes that propinquity would lead to new relationships fed by such close-ness. However, the paths of business heads seldom crossed due to the lack of shared settings such as eating facilities or socializing situations for building personal relationships that might lead to learning opportunities. As in the West, attraction to STIPs came from prestige derived from locational affiliation with a park, and closeness to a top research university (Walcott and Xiao 2000). The fit between a particular industry and its operational needs also varied by location, depending on the competitive advantage available in each area (Walcott 2001).

Trips and interviews in southern Shenzhen, western Xi'an, and northern Beijing demonstrated that clusters varied in each place reflecting local endow-ments in leadership and research. Western Xi'an benefited from its historic strategic military stock of top engineering universities, whose applied research translated into new high tech product applications from students with roots in that region of Inner China. Beijing also used knowledge transfers from top national universities for an early edge in computer innovations, but companies are susceptible to being co-opted by large multinational corpora-tions in the capital city. Shenzhen prospers from its proximity to Hong Kong

*Table 37.1.* Worker Technology Productivity in Selected Cities, 2000.

| City | Average Number of Tech Workers per Company | Tech Revenue Produced per Worker | Exports Produced per Worker (US$1,000) |
|---|---|---|---|
| Beijing | 47.2 | 55.9 | 6.2 |
| Shanghai | 208.7 | 20.3 | 25.7 |
| Shenzhen | 521 | 1.2 | 32.8 |
| Xi'an | 61.5 | 12.3 | 1.1 |

Source: Adapted from *China Statistical Yearbook on Science and Technology*, 2001.

and transplanted Beijing university graduates encouraged to relocate there. Chinese firms in STIP clusters function on two tracks: one based on learning from foreign examples, and the other based on Chinese university technology spillovers (Walcott 2002).

## MODEL DISTRICTS

High-technology parks in China fit one of three basic forms, which are themselves modifications of industrial districts found in earlier developing areas of Asia and extensions of Western models (Park and Markusen 1995; Markusen 1996):
1. In the *Multinational Satellite Platform* district, foreign headquartered firms heavily predominate. The district is located on the outskirts of a city. No significant connections exist with Chinese research institutions, and local suppliers are minimal. Low-cost labor and access to the Chinese market provide the major attractions. Suzhou, 80 kilometers west of Shanghai, is one example of this setting.
2. For the *District Hub and Spoke*, the STIP as a spatial entity functions as the hub; the surrounding region provides the spoke setting, with mixed local and multinational corporate activity, as in Shanghai and Shenzhen. Some ties are constructed with research institutions, but not contiguous interaction. An on-site incubator nurtures some local start up activity. Foreign firms may co-locate with their own imported suppliers. Technology parks are in or near a major urban area.
3. The *Local Innovation District* characterizes Beijing and Xi'an. The park closely locates in the proximity of major associated research institutions, harbors a large incubator, and is also close to a retail row. The domestic market is a major production target, compared to the more export orientation of the other types.

The latter two types of districts are particularly attractive for returned students (with at least one year in a foreign university), bringing new intellectual and fiscal capital as a technology transfer. For China, the local innovation district holds the greatest possibility for functioning in regional centers of strength, rectifying in part the present geographic inequality of development within the country. Several years ago, a delegation from the capital city of even more remote Sichuan province visited Xi'an to learn from the experience of universities setting up their own STIPs.

The macroeconomic Chinese institutional features that need to be addressed in each location include: the absence of effective intellectual property protection, the desire for more open capital markets and incentivized risk capital investments, financially overburdened local governments, insufficient corporate-university internships, and inadequate infrastructure links. In addition, each city examined carried its own culturally and locationally embedded features that inevitably affected the outcome of its STIP policies. In China (as elsewhere) place matters. Science and technology parks constitute spaces for experimentation and creation of new processes linking this linchpin of Asia to the global economy as it "crosses the river by feeling for stones."

## REFERENCES

Castells, M. and P. Hall. 1994. *Technopoles of the World: The Making of Twenty-First Century Industrial Complexes.* London: Routledge.

Dicken, P. 1998. *Global Shift: Transforming the World Economy,* 3rd Edition. London: Paul Chapman Publishing.

Malecki, E., P. Oinas, and S. Park. 1999. On Technology and Development. In E. Malecki and P. Oinas, eds. *Making Connections: Technological Learning and Regional Economic Change,* 261–275. Aldershot, U.K.: Ashgate.

Markusen, A. 1996. Sticky Places in Slippery Space: A Typology of Industrial Districts. *Economic Geography* 72: 293–313.

Park, S. and A. Markusen. 1995. Generalizing New Industrial Districts: A Theoretical Agenda and an Application From a Non-Western Economy. *Environment and Planning A* 27: 81–104.

Walcott, S. 2001. Growing Global: Learning Locations in the Life Sciences. *Growth and Change* 32: 511–532.

Walcott, S. 2002. Chinese Industrial and Science Parks: Bridging the Gap. *The Professional Geographer* 54: 349–364.

Walcott, S. and W. Xiao. 2000. High-technology Parks and Development Zones in Metropolitan Shanghai: From the Industrial to the Information Age. *Asian Geography* 19: 157–179.

# 38

# WHY OFFSHORE? EXPLORING THE GEOGRAPHIES OF OFFSHORE FINANCIAL CENTERS

SHARON C. COBB

*University of North Florida*

The global economy of the 21st century is characterized by the increasing differentiation between the haves and the have-nots, the educated and illiterate, the skilled and unskilled. This inequitable trend is well documented in the arena of technology and the concomitant analysis of global geographies of production, but less attention has been paid to factors associated with the hypermobility of capital and the services provided by those functioning in offshore finance centers (OFCs).

## THE PROBLEM

Much tension exists between the onshore and offshore worlds, with developed nations blaming on OFCs as facilitators of money laundering activities, and unfair tax practices and asset protection strategies because of the perceived lack of rigid financial regulation and supervision acceptable to the onshore world. Concern exists that countries with relatively high tax/GDP ratios when compared to OFCs will see increasing corporate and individual flight to lower-tax jurisdictions, resulting in even higher taxation for the ones that are left behind and a potential downward development spiral. Supranational institutions such as the Organization for Economic Cooperation and Development (OECD) and the European Union (EU) are concerned with promoting equity with respect to jurisdictional taxation issues and have established policies aimed at extinguishing some of the more spurious offshore services and products.

## THE OFFSHORE PHENOMENON

Offshore financial centers – often labeled tax havens – are jurisdictions that "host financial activities that are separated from major regulating units (states) by geography and/or by legislation. This may be a physical separation, as in an island territory, or within a city such as London or the New York International Banking Facilities" (Hampton 1994: 237). OFCs grew in

*D. G. Janelle et al. (eds.), WorldMinds: Geographical Perspectives on 100 Problems*, 237–241.
© 2004 *Springer. Printed in the Netherlands.*

importance in the aftermath of the creation of the Eurodollar market in the post-Bretton Woods years of the late 20th century. Current estimates show that as much as half of the world's stock of money either resides in or flows through these places (Laulajainen 1998). Today, OFCs promote themselves as low-cost (low-tax), confidential, stable, and reputable places able to offer appropriate financial expertise to manage business and individual assets while simultaneously being challenged by the onshore world to create a more level jurisdictional-tax playing field.

Why do small island economies at the literal and figurative margins of the modern global economy choose to deliberately create an artificial economic sector of offshore financial services? Although the offshore phenomenon is not new, contributions by geographers over the past decade have helped to provide rigorous alternative analyses of a social process previously constrained solely within the neoclassical business world framework (McCarthy 1979; Johns 1983; Roberts 1994). Geographers have, in the words of Hudson (1999: 139), "addressed one of the latest stages in the historical geography of money, the way in which the development of new monies relates to the emergence of new spatialities of power and social relations, and the development of regulatory spaces."

Discourses of financial geography by economic geographers were largely neglected until the late 1980s, when a growing body of literature grappled with seemingly new issues of social and cultural characteristics of money and the evolving global geographies of money (e.g., Leyshon and Thrift 1997). Theorizing spatial features of money and finance drew upon critical discourses of culture, power, and social relations, rather than revenue, profit, and loss statements so typical of traditional neoclassical analyses. Human interactions, actor network linkages and the importance of power relations replaced study of costs and benefits, target sales numbers and level of market penetration in the quest to forge a new understanding of the spatial patterns of money and finance in the modern global economy.

Critical analysis of the socio-spatial processes defining "offshore" began with a seminal paper by Roberts (1994) using a robust postmodern theoretical framework defined by Harvey (1989). Roberts (1994: 92) asserts that

> a series of little places – islands and microstates – have been transformed by exploiting niches in the circuits of fictitious capital. These places have set themselves up as offshore financial centers; as places where the circuits of fictitious capital meet the circuits of "furtive money" in a murky concoction of risk and opportunity.

These small places number as many as 40 scattered throughout the planet and use accommodating (labeled by many as lax) legislative and regulatory regimes to attract capital flows with the aim of tax minimization and confidentiality of service and product.

Roberts' theoretical exposition was furthered by Hudson (1998), who argues that money is a social relation and, as such, all actions relating to offshore streams of capital necessitate the establishment of trust – both of people and place. Hudson asserts that offshore financial centers are socially constructed entities shaped by local and extra-local actors that exhibit fluidity of interaction between local government officials and multi-national financial services providers in an attempt to resolve issues of trust and risk.

From a perspective of economic development, these small places on the margin – for example, Cayman Islands, Jersey, Guernsey, Bahamas, Bermuda, and the Isle of Man (Figure 38.1) – compete with each other to create reputations of trustworthiness at individual, corporate, and jurisdictional levels. The

**EURASIAN OFCS**
Andorra, Cyprus, Guernsey, Jersey, Monaco, Gibraltar, Isle of Man, Liechtenstein, Dublin, Madeira, Luxembourg, Switzerland, Bahrain, Dubai, Lebanon

**PACIFIC OFCS**
Cook Islands, Hong Kong, Samoa, Labuan, Macao, Marianas, Naura, Singapore, Vanatu, Marshall Islands

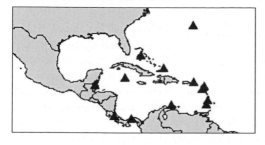

**CARIBBEAN OFCS**
Antigua, Aruba, Bahamas, Belize, Bermuda, BVI, Cayman Islands, Costa Rica, Panama, St. Lucia, St Kitts & Nevis, Turks & Caicos, St Vincent & Grenadines

*Figure 38.1.* Offshore financial centers. Source: adapted from United Nations Office for Drug Control and Crime Prevention, 1998, www.globalpolicy.org/nations/finhav99.htm.

exploration of the role of actor networks functioning to craft appropriate legislative and regulatory regimes for the creation of trust and concomitant successful and legitimate offshore financial services provision found that public-private sector interaction served as an important mechanism to create competitive advantage (Cobb 1999).

Hampton (1994: 237) explores the potential developmental trajectories for small island economies by arguing that "the increasing global integration of financial capital with the development of transnational banks and an international network of financial centers, has opened up a limited development option." He cautiously concludes that hosting offshore financial services could prove to be a useful developmental strategy if that jurisdiction has the appropriate attributes of proximity to a large developed country, basic communications infrastructure, and political stability. Warf (1999) recognizes the critical enabling feature of technological infrastructure – particularly related to advances in telecommunications – as a necessary condition for the globalization of finance and the resulting explosion of "offshore."

As Roberts (1994: 111) concludes, "new patterns of offshore financial flows overlay and interact with older geographies of tax haven activities and present-day geographies of hot money flows. The offshore financial centers are nexuses where various intertwined circuits of global capital come into focus." In recent years, criticism by supranational institutions regarding the tax-discriminatory and opaque nature of many offshore financial centers and services and has led to increasing transactional transparency (OECD 1998). Furthermore, other supranational bodies (particularly the EU) are concerned with promoting fair-practice with respect to jurisdictional taxation issues although creating a level playing field for taxation is possibly one of the most difficult goals to achieve.

## OFCS AND MONEY LAUNDERING

An associated problem also directly related to the differential regulatory and legislative topography of finance, and currently receiving much attention by researchers and policy makers, is that of money laundering. In 1989, in response to mounting concern over money laundering, the Financial Action Task Force on money laundering (FATF) was established by the G-7 economies. Money laundering is recognized by the FATF as the processing of illegal profits from a variety of activities including: illegal arms sales, smuggling, drug trafficking, prostitution and other activities of organized crime as well as such white collar crimes as embezzlement, insider trading, bribery, and computer fraud schemes.

According to FATF, criminals seek to launder illegal profits by disguising the sources or moving the funds to a place where they are less likely to attract attention. There is no doubt that the spatial trajectories of money laundering activities and offshore financial services provision have the propensity to collide, as those monies being "washed clean" have to be processed without attracting attention to the underlying illegal activity or person(s) involved and one of the primary OFC marketing strategies to provide confidentiality. OFCs, therefore, have come under serious scrutiny by FATF and many centers strive to establish anti-money laundering systems in compliance with FATF guidelines. The baseline for offshore anti-money laundering policies emanates from the now-accepted "know you customer" requirements that many OFCs currently use.

Analysis of the geography of money laundering, theoretical discourses, and empirical evidence documenting attempts to arrest this phenomenon have not yet been fully explored, although much greater attention has been paid to this issue in the light of recent global terrorist actions. Supranational pressures, whether addressing money laundering, unfair tax practices, or the opaque nature of service provision are likely to create pressures on offshore jurisdictions as they strive to maintain credible reputations in the onshore world.

## REFERENCES

Cobb, S. 1999. The Role of Corporate, Professional and Personal Networks in the Provision of Offshore Financial Services. *Environment and Planning A* 31: 1877–1892.

Johns, R. 1983. *Tax Havens and Offshore Finance – A Study of Transnational Economic Development*. New York: St. Martins' Press.

Harvey, D. 1989. *The Condition of Postmodernity*. Oxford: Blackwell.

Hampton, M. 1994. Treasure Islands or Fool's Gold: Can and Should Small Island Economies Copy Jersey? *World Development* 22: 237–250.

Hudson, A. 1998. Placing Trust, Trusting Place: On the Social Construction of Offshore Financial Centers. *Political Geography* 17: 915–937.

Laulajainen, R. 1998. *Financial Geography: A Bankers View*. Gothenburg, Sweden: Gothenburg School of Economics and Law.

Leyshon, A. and N. Thrift. 1997. *Money Space: Geographies of Monetary Transformations*. London: Routledge.

McCarthy, I. 1979. Offshore Banking Centers: Benefits and Costs. *Finance and Development*. December: 45–48.

OECD. 1998. Harmful Tax Competition: An Emerging Global Issue. http://www.oecd.org/pdf/M00004000/M00004517.pdf.

Roberts, S. 1994. Fictitious Capital, Fictitious Places: The Geography of Offshore Financial Flows. In S. Corbridge, R. Martin, and N. Thrift, eds. *Money, Power and Space*, 91–115. Oxford: Blackwell.

Warf, B. 1999. The Hypermobility of Capital and the Collapse of the Keynesian State. In R. Martin, ed. *Money and the Space Economy*, 227–240. London: Wiley.

# 39

# U.S. CALL CENTERS:
# THE UNDISCOVERED COUNTRY

DAVID L. BUTLER
*University of Southern Mississippi*

The phrase "the death of distance" provokes in the minds of many the idea that electronically mediated commerce has overcome the bounds and limits of geography that have traditionally constrained business. This essay argues that these notions are incorrect, and that, in fact, geography is as relevant as ever. It demonstrates the importance of geographical concepts by examining an information technology-dependent industry, the U.S. call center industry.

## CALL CENTERS

Call centers, defined by Bodin and Dawson (2002: 39), are "place[s] where calls are placed, or received, in high volume for the purpose of sales, marketing, customer service, telemarketing, technical support or other specialized business activity." Though Bodin and Dawson offer a clear technical definition, Belt, Richardson, and Weber (2000) point out that there is only limited agreement in the literature as to what types of activities occur within a call center, and even less agreement on how to identify and label call centers. Size variations are one reason why defining a call center is a difficult task; some micro centers have as few as five representatives on the phone, while other mega centers can house several thousand employees simultaneously. Two additional reasons for disagreement on a call center definition are: (1) lack of a synthesized literature focused on call centers, and (2) lack of recognition of call centers collectively as an industry by many national governments and thus no attempt to centralize pertinent data (Richardson 1998). Lacking these two components, call centers have remained in the shadow of other well-known information-dependent industries such as financial services.

Call center research in Britain has continued at a steady pace, with researchers such as Richardson (2002) examining the economic development aspects of these teleservice industries. However, the same cannot be said regarding research on U.S. call centers, even though the United States houses the largest concentration of call centers in the world. In an attempt to fill

*D. G. Janelle et al. (eds.), WorldMinds: Geographical Perspectives on 100 Problems, 243–248.*
© 2004 *Springer. Printed in the Netherlands.*

this vacuum, this essay examines the death of distance and footloose industries to determine how these concepts are related to U.S. call centers.

The concept of footloose industries is encapsulated in the book *The Death
of Distance*, where Cairncross (1997: xi) suggests that a main principle of
new information technologies is that "no longer will location be key to most
business decisions. Companies will locate any screen based activity anywhere
on earth." Cairncross suggests that as long as an industry has information as
its main product, businesses within this industry have almost infinite flexibility
in location. To challenge this statement, first consider a few facts. Call centers
(1) are a screen-based activity; (2) do not require proximity to raw materials; and (3) do not require a storefront within a given market area to attract
flows of customers.

Given these characteristics, it would appear that call centers are the epitome
of a potential footloose industry, defined by the *Dictionary of Human
Geography* as "an industry which can locate virtually anywhere because it
has no strong material or market orientation in its locational requirement"
(2000: 274). If, however, it can be demonstrated that call centers, the quin-
tessential footloose industry, are indeed grounded in geographical space or
place, then it would suggest that in other telecommunications-dependent
industries location criteria will have a strong geographical component as
well.


## THE GEOGRAPHY OF U.S. CALL CENTERS

To understand the locational criteria of call centers, one must first find out
where they are located, then determine what criteria are influencing their
location. Given that call centers are not considered an industry within the
United States, no federal or state government-collected data on this issue.
Instead, one must rely on poorly researched consulting reports and trade
publications. Given these limits, the method of choice for obtaining data on
this industry is for the researcher to collect the primary data themselves.

It is estimated that there are between 80,000–100,000 call centers within
the United States, which employ between three and five percent the total
national labor force (Uchitelle 2002). Centers can range in size from a dozen
people within an organization to massive freestanding buildings with several
thousand customer service representatives. The author compiled a database
of 20,017 call centers from primary data sources, creating a large representative sample of U.S. call centers. The majority is located in urban or suburban
locations (Figure 39.1). Why are centers concentrated in urban location, where
costs are higher, if they possess the spatial freedom that Cairncross (1997)
suggests? Interviews with corporate executives and call center directors, and
a survey of call center employees illuminate one reason for this urban pattern:
labor costs and availability. Before the advent of fiber optics lines within the

*Figure 39.1.* Map of 20,013 U.S. call centers, 2001. Source: David L. Butler.

United States and the simultaneous deregulation of the long distance phone system, call centers were fewer and the locational pattern much different. However, since the 1990s, American companies can tap into voice and data fiber lines, process 800 number phone calls relatively cheaply, and have calls digitally routed with relative ease. The major recurring cost for most call centers is now labor.

## DATA

Besides collecting primary data on call center locations, this ongoing research project interviewed corporate executives from companies who control a large number of call centers. The corporate executives were also asked what are the most important variables in locating their call centers (Butler 2001). Call center directors were asked the same set of questions and there was surprising similarity in their answers (Table 39.1). Of the top four most frequently mentioned answers, labor costs account for two. The other two most commonly cited variables were tax incentives and the availability of pre-existing sites.

*Labor*

Interviews and surveys yield the following observations regarding the relation of call center labor to location: (1) labor availability is critical; (2) skilled labor

*Table 39.1.* U.S. Call Center Location Criteria.

| Sector | Labor Costs | Tax Incentives | Labor Skills | Pre-existing Site | Infra-structure | Neutral Accent | Politics | Quick Move-in | Quality of Life | Other |
|---|---|---|---|---|---|---|---|---|---|---|
| Airline | | * | * | | | | * | * | * | * |
| Insurance | * | * | | | * | | | | | |
| Pharmaceuticals | * | * | * | * | | * | | | | |
| Healthcare | * | | * | * | * | | | | | |
| Government | * | | | * | | | * | | | |
| Travel | * | * | | * | | * | | * | | |

is ideal; (3) regions with little or no labor union activity are desirable; (4) women, ideally married, possibly with school-age children, and some college education, are the typical call center employee (Nelson 1986; England 1993); and (5) most call center employees who drive to work will commute on average 20 minutes (Butler 2002).

Because the average annual turnover rate in call centers is between 35–40 percent, it is necessary to site call centers in locations with as many potential employees as possible within a given labor shed (Uchitelle 2002). At the high rate of turnover, most call centers will lose 100 percent of their workforce in three to four years. This means that these centers need to locate where there is an abundance of labor that they can continuously draw upon. If not, they will run out of potential employees quickly. Because of the labor demands, most call centers will seek out areas with large concentrations of employees (i.e., cities) and locations with high concentrations of ideal non-unionized labor (particularly suburbs). Furthermore, given the highly mobile nature of capital and the relatively immobile nature of workers, call centers must eventually land in locations where there are enough employees to fill their insatiable labor demand due to the high levels of employee turnover.

*Incentives and Sites*

Tax incentives, like high concentrations of labor, are not universal. More wealthy states and communities can and do offer greater incentives to companies to (re)locate to their communities. Given this reality, it is not surprising that wealthy locations can generate the tax revenues necessary to attract new industries with incentives. This relationship becomes a positive spiral for specific wealthy cities and communities and a negative spiral for the more poor locations.

The same logic holds for pre-existing sites. Many call centers are turnkey operations, so that all they have to do is be handed the keys to their building when they arrive. To build multiple turnkey sites for call centers, or other industries, a community must have enough capital in place or the ability to borrow the capital with assurance of a return on investment (i.e., new businesses). Once again, larger communities with past success in attracting industries usually have the capital to reinvest in pre-constructed turnkey businesses.

## CONCLUSION: GEOGRAPHY MATTERS!

Concepts such as the death of distance and footloose industries, although sexy and provocative as headlines, in reality do not pertain to the U.S. call center industry. Call centers do have a greater flexibility than traditional manufacturing industries since they do not have any raw material orientation.

Because call centers have a great appetite for labor, they must locate in an area where a substantial potential labor pool exists. Likewise, locations that can offer the best incentive packages are more attractive to call centers, as are locations with pre-existing sites. Even though call centers do not have a strong material or market orientation, this does not mean that they have universal locational flexibility. In fact, many information technologies simultaneously enable some locational flexibility in regard to some inputs (e.g., raw materials) while at the same time creating rigidity in another (i.e., labor). Therefore, challenges should be issued to proclamations suggesting that information technology mediated industries (screen-based) are completely footloose. By using geographical tools, researchers can examine and challenge hyperbolic statements such as the death of distance, reinforcing the importance of geography as a lens to examine social and economic issues.

## REFERENCES

Belt, V., R. Richardson, and J. Webster. 2000. Women's Work in the Information Economy: The Case of Telephone Call Centers. Information. *Communication, Society* 3: 3.

Bodin, M. and K. Dawson. 2002. *The Call Center Dictionary: The Complete Guide to Call Center and Customer Support Technology Solutions.* New York: CMP Books.

Butler, D. 2002. Culture Matters! Retaining Employees and Increasing Profitability. In E. Phillips, ed. *Retaining Your Best Employees.* ASTD Publishing.

Butler, D. 2001. *Deregulation, Information Technology, and the Changing Locational Dynamics of the U.S. Airline Industry.* Unpublished Ph.D. dissertation, University of Cincinnati.

Cairncross, F. 1997. *The Death of Distance.* London: Orion Publishing.

England, K. 1993. Suburban Pink Collar Ghettos: The Spatial Entrapment of Women? *Annals of the Association of American Geographers* 83: 225–242.

Johnston, R., P. Haggett, D. Smith, and D. Stoddart, eds. 2000. *The Dictionary of Human Geography*, 4th edition, Oxford: Blackwell.

Nelson, K. 1986. Labor Demand, Labor Supply, and the Suburbanization of Low-Wage Office Work. In A. Scott and M. Storper, eds. *Production, Work, Territory.* Boston: Allen Unwin.

Richardson, R. 2002. *Information and Communications Technologies and Rural Inclusion.* Centre for Urban and Regional Development Studies, University of Newcastle.

Uchitelle, L. 2002. Answering '800' Calls Offers Extra Income but No Security. *New York Times* 27 March: 1.

# 40

# GENDER AND GLOBALIZATION:
# MAQUILA GEOGRAPHIES

ALTHA J. CRAVEY
*University of North Carolina, Chapel Hill*

Gender is a useful analytical tool for understanding contemporary globalization. In this regard, geographers have examined gendered representations of globalization, gender and global production patterns, and gendered migration flows. This body of research has been crucial for developing a critical understanding of globalization. Notions of the complexity of contemporary globalization geographies allow us to imagine and construct alternative models of globalization that might be oriented toward internationalist and/or social justice concerns rather than neo-liberal principles and policies promoted by international financial institutions and their allies (Enloe 1989). In this essay, I focus on the gendered geographies of global production, an aspect of globalization that has profound consequences for those directly involved in assembly operations in the "global factory," or *maquila*. Now used in many global contexts, this term derives from *maquiladora*, the popular name for Mexico's export-oriented assembly plants.

Geographers have much to offer on this issue. For workers involved in shop-floor organizing, a geographical perspective is useful for bringing local experiences to bear in improving workplace democracy and working conditions. A geographical perspective also gives workers key tools for analyzing the competition between places (for factory investment) and thus for strategizing to keep their jobs "in place." For international solidarity activists, who advocate and support workers' rights around the world, a geographical perspective is essential. Without such a perspective, solidarity actions can sometimes undermine worker goals. For example, anti-sweatshop activists in the United States have sometimes targeted certain brand names or companies without realizing that one possible outcome is the loss of employment for the workers they intended to support. In this essay, I use the example of the apparel industry to make wider arguments about gender and global production.

*D. G. Janelle et al. (eds.), WorldMinds: Geographical Perspectives on 100 Problems, 249–253.*
© 2004 *Springer. Printed in the Netherlands.*

## MAQUILAS, SWEATSHOPS, AND SPATIALITY

One of the more visible manifestations of contemporary globalization is the spatial reorganization of the apparel industry. The global assembly line – or alternatively, the global commodity chain – continues to reshape and reconfigure geographical patterns of production in a volatile search for higher profits and increased corporate flexibility. Export Processing Zones (EPZs) have played a role in the process of creating new and profitable *maquila* locations worldwide. EPZs are growth zones demarcated by individual nation states to control and simultaneously de-regulate or re-regulate labor. Geographical insularity facilitates a high degree of labor control that, in turn, attracts transnational investment. The success of the zones themselves is a dynamic process, and they may become less important over time, as is the case of Mexico. More recently, nation states such as Guatemala have defined such zones in diffuse non-contiguous patterns (Traub-Werner and Cravey 2002). The point to be emphasized is that the spatial concentration of investment and the isolation of transnational export enclaves (however they may be configured locally) have been crucial to the formation of competitive factory regimes in contemporary global patterns of production.

The use of spatially isolated production zones facilitates labor control systems that require transformed gender norms. A transnational factory regime is emerging in these enclaves where the costs of social reproduction (and wages) can be held artificially low to increase profitability (Burawoy 1985; Cravey 1998a, b). Thus, state-led differentiation of the national space-economy, including the creation and state-enforced regulation of Export Processing Zones (EPZs), facilitates a broad reformulation of gender and household dynamics. Discontinuous global growth spaces within newly industrializing countries, however fragile and variously shaped by locally specific struggles, thus emerge as a kind of transnational growth space at the global scale. Furthermore, a new global labor force and a new profoundly gendered transnational factory regime have been created within the disparate export processing zones (EPZs) worldwide.

*Maquila* workers in the apparel industry have responded to global reorganization with creative strategies of their own. Shop floor struggles in factories throughout the world have joined forces with *mixed coalitions* of diverse social groups from wide-ranging locations. For instance, coalitions based in the western hemisphere create and adapt communications networks that allow North American consumers and former politically oriented solidarity committees to keep abreast of worker struggles in Central America and elsewhere. Campus-based student movements in the United States, such as United Students Against Sweatshops, with affiliates on more than 200 campuses, have targeted lucrative university athletic clothing contracts, and have been especially effective at providing energy, momentum, and creativity in these mixed coalitions. Because retail ownership is highly concentrated in the apparel industry (in stark contrast to factory ownership), retailers and brand merchandisers have sometimes been effective targets of international solidarity campaigns as well. Other allies in these efforts include consumer groups, labor unions, worker federations, religious organizations, and socially responsible investing groups. In this way, *mixed coalitions* of diverse and dispersed social actors and groups have influenced global production geographies.

Such contemporary international solidarity activities need critical geographical insights to effectively confront fast-moving uneven geographies of global production (Waterman and Wills 2001; Herod 2001). Furthermore, volatility is exacerbated in industries like apparel production because the investment threshold is low (Hale and Shaw 2001; Gereffi 1999; Figueroa 1996). The reliance on a predominantly female workforce – and the complex gender dynamics involved – presents additional challenges to effective solidarity strategies (Hale and Shaw 2001). Across global production landscapes, advocates who support the rights of *maquila* workers encounter vastly different contexts, distinct regulatory regimes, embedded social histories, dynamics, and expectations; and contradictory personal (and collective) impulses. Unions, trade union confederations, and other non-governmental organizations create, use, and extend vast networks of communication that link supporters to local place-based struggles. In this way, workers create and recreate local, regional, and global geographies (Herod 1998). In this way, labor struggles and their outcomes become visible in the landscape at specific geographical scales, and these geographical outcomes become the terrain for future struggles (Herod 1998). For this reason, it is important to consider the way in which global (or potentially global) efforts to broaden support for place-based worker struggles, may impact upon socio-spatial outcomes. The mixed coalitions described above are aligned with these worker-led efforts. It is important to note that the collaboration of a wide range of advocates in dispersed locations (such as consumer-oriented groups in the global North) introduces even more complexity and contradictory impulses to worker geographies (Traub-Werner and Cravey 2002).

## PRODUCTION AND SOCIAL REPRODUCTION

To use gender effectively as an analytical geographical category, it must be considered from two perspectives: spatial divisions of labor and spatial divisions of social reproduction. Attention to social reproduction, specifically the tasks necessary for the daily and generational restoration of each worker's capacity to produce, illuminates and expands the terrain of analysis. In this way, we can explore the ways in which changing geographies of production influence and are influenced by, changing geographies of social reproduction. This understanding in turn, highlights the way in which contemporary globalization processes are profoundly gendered and racialized. While commodity production has been increasingly organized at larger geographical scales in the past three decades, social reproduction has moved in the opposite direction – toward increasingly smaller geographical scales. From the vantage point of nation-states, this process can be seen in the nearly universal retrenchment of social provision in these years. Welfare states and social programs are dismantled under the guise of neoliberal strategies for economic growth. Thus many social reproductive tasks have been privatized or linked directly to a particular employment relation (Burawoy 1985; Cravey 1998a). In other words, the large-scale territorial arrangements known as welfare states and as "hegemonic regimes" have given way to a prioritization of market principles, or "hegemonic despotism" (Burawoy 1985). Meanwhile, "living wage" campaigns in the United States have tried to reconnect social reproduction and production logics at smaller geographical scales. "Living wage" ordinances in some 30 U.S. cities and towns seek to build new political momentum and new place-based points of leverage for social justice.

Workers and ordinary people therefore, are caught in a bind. To increase profit rates, *maquila* owners strive to increase productivity. When structural conditions make this difficult or impossible, capitalists take measures to lower the costs of social reproduction. Either strategy can increase profitability for the capitalist owner, yet the later strategy gives workers less access to means of subsistence (by lowering the "socially necessary labor time"). In earlier research, I found this to be the case for Mexican *maquiladora* workers. In stark contrast, industrial workers in central Mexico saw wages and standards of living rise during four long decades of high growth rates from the 1940s through the 1970s (Cravey 1998a).

While production geographies exist and are being dramatically reworked by globalization processes, there is also a definite geography to social reproduction. That is, in particular places and regions, the costs of producing (and maintaining) workers are considerably higher than in other places. These costs fluctuate in response to the average needs and expectations of workers in a particular country or a particular place (i.e., socially necessary labor time). Of course, these needs in turn are the result of place-based historical struggles. During the height of Fordist production arrangements, these contradictions were largely resolved at the geographical scale of the nation state, producing welfare states (and states that moved in the direction of welfare states by providing more public goods). In recent decades, these place-based arrangements have unraveled and have been under attack such that individuals, households, and communities, and social networks, take up the slack and assume the burdens of social reproductive activities that were previously provided by the state. Workers then find themselves squeezed between policies that lower the costs of social reproduction and that increase the supply of workers through such means as opening more places to capitalist production, as in the abrupt commodification of Mexican land and agriculture in the early 1990s.

Geographers have contributed in quite specific ways to our understanding of the complexities of globalization and its impact on ordinary people. Geographical knowledge is beginning to shape strategies that are increasingly employed by *maquila* workers, their supporters, labor union confederations, mixed coalitions, and non-governmental organizations. By taking account of gender relationships and other relationships of unequal power, a feminist perspective on the geography of globalization provides insights on the dynamic nature of this contemporary phenomenon.

## REFERENCES

Burawoy, M. 1985. *The Politics of Production: Factory Regimes Under Capitalism and Socialism.* London: Verso.

Cravey, A. 1998a. *Women and Work in Mexico's Maquiladoras.* Boulder: Rowman and Littlefield.

Cravey, A. 1998b. *Toothless Tigers and Mouldered Miracles: Geography and a Global Gender Contract in the NICs.* Unpublished manuscript.

Enloe, C. 1989. *Bananas, Beaches, and Bases: Making Feminist Sense of International Politics.* Berkeley: University of California Press.

Figueroa, H. 1996. In the Name of Fashion: Exploitation in the Garment Industry. *NACLA: Report on the Americas* January: 34–40.

Gereffi, G. 1999. International Trade and Industrial Upgrading in the Apparel Commodity Chain. *Journal of International Economics* 48: 37–70.

Hale, A., and L. Shaw. 2001. Women Workers and the Promise of Ethical Trade in the Globalised Garment Industry: A Serious Beginning? *Antipode* 33: 510–530.

Herod, A. 1998. The Spatiality of Labor Unionism: A Review Essay. In A. Herod, ed. *Organizing the Landscape: Geographical Perspectives on Labor Unionism.* Minneapolis: University of Minnesota Press.

Herod, A. 2001. *Labor Geographies: Workers and the Landscapes of Capitalism.* New York: Guilford.

Traub-Werner, M. and A. Cravey. 2002. Spatiality, Sweatshops, and Solidarity in Guatemala. *Social and Cultural Geography* 3: 383–400.

Waterman, P. and J. Willis. 2001. Introduction: Space, Place and the New Labour Internationalisms: Beyond the Fragments? *Antipode* 33: 305–311.

# 41

# PROMOTING DEVELOPMENT IN
# THE CANADIAN NORTH

JAMES C. SAKU
*Frostburg State University*

More than two decades ago, Orvik (1976) argued that the Canadian north
was confronted with complex problems and that a general theory of Northern
Canadian development was required to adequately address them. While Orvik
may have made a remarkable appeal for a general theory to deal with northern
Canadian problems, the region remains confronted with insurmountable human
and physical issues associated with living in Artic and sub-Artic environments.
While several top-down development initiatives, including housing (Chislett
et al. 1987), the Eskimo Loan Fund, and the Northwest Territories Small
Business Loan Fund, were implemented to solve these problems, the settlement of modern treaties in the Canadian north within the last three decades
represents an example of the bottom-up approach to development. This chapter
examines modern treaties within the context of the bottom-up approach to
solving economic and social problems in the Canadian north.

## THE CANADIAN NORTH AND ITS PROBLEMS

The geographic extent of the Canadian north is not easily definable. There
are as many "Norths" as there are people using the word (Wonders 1987).
Several approaches were used in the past to define the Canadian north. Bone
(1992), for example, considers the Canadian north as a political region comprising of the Yukon, the Northwest Territories, and the northern areas of
the seven provinces. From a physical perspective, the Canadian north is viewed
as two natural zones, the Arctic and the sub-Arctic (Figure 41.1). Overall,
the Canadian north accounts for about 77 percent of the total land area of
Canada (Table 41.1).

Notwithstanding the complexity of delineating the Canadian north, the
uniqueness of the region is reflected in the cultural, economic, and environmental characteristics. Culturally, the Canadian north is dominated by Native
Canadian culture that includes hunting, fishing, trapping, and gathering. With
the intrusion of a modern economic system and the migration of non-aborig-
inal Canadians into the region, the situation has changed tremendously. One

255

D. G. Janelle et al. (eds.), WorldMinds: Geographical Perspectives on 100 Problems, 255–260.
© 2004 Springer. Printed in the Netherlands.

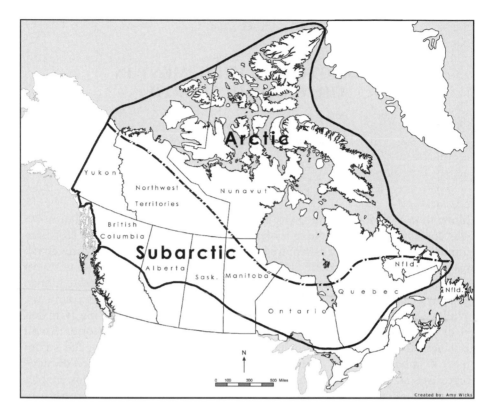

*Figure 41.1.*

*Table 41.1.* Approximate Land and Freshwater Areas (square miles).

| Province or Territory | Total Area | Northern Area | Percentages | |
|---|---|---|---|---|
| | | | North | Canada |
| Canada | 3,850 | 2,954 | 100 | 76.7 |
| Alberta | 255 | 120 | 47 | 3.1 |
| British Columbia | 366 | 145 | 40 | 3.8 |
| Manitoba | 251 | 185 | 74 | 4.6 |
| Newfoundland | 157 | 116 | 74 | 3.0 |
| Northwest Territories/Nunavut | 1,323 | 1,323 | 100 | 34.4 |
| Ontario | 413 | 270 | 65 | 7.0 |
| Quebec | 595 | 483 | 81 | 12.6 |
| Saskatchewan | 252 | 125 | 50 | 3.3 |
| Yukon | 187 | 187 | 100 | 4.9 |

Source: Adapted from Hamelin (1988).

major problem of the region, therefore, is changes in the cultural heritage of Native Canadians.

In addition to changes in the cultural landscape of the region, the Canadian north is confronted with several other problems (Table 41.2). One of these is its remoteness from southern Canada. The great distance between the Canadian north and the south has prevented the full economic integration of the two regions. As a result, the economic prosperity that southern Canada experienced in the past decade did not trickle down to the north. The region is still characterized by low personal income, high transfer payments, economic deprivation, and dependency.

Apart from its remoteness, the population dynamics of the region inhibit economic development. The Canadian north is characterised by small population scattered over very large geographic area. In Nunavut, a population of 26,745 is dispersed over an area of about 743,420.18 sq. miles. Small population limits the capacity for economies of scale. For example, the cost of service delivery is very high in the region because of the dispersion of the population.

Furthermore, the Canadian north is highly dependent on resource development for its economic prosperity. Bone (1992) observed that the development of these resources depend primarily on global demand for them. While a boom in the global economy creates a significant activity in resource development in the Canadian north, economic downturn promotes negative effects. The cycle of economic boom and bust is a common phenomenon in the region.

Additionally, the region experiences high economic leakage because a significant proportion of investment is redirected back to the south through wages, salaries, and equipment purchase. For example, about 85 percent of mining jobs in Nunavut are held by non-residents (www.nunavut.com 2003). Moreover, since multinational corporations are the main agents of resource development in the region, the profits generated by these companies are repatriated out to other parts of the world.

*Table 41.2.* Selected Problems of the Canadian North.

1. Remoteness
2. Resource base economy
3. High economic leakage
4. Dependence on transfer payment
6. External ownership of businesses
7. Sparse population
8. High population growth
9. Environmental destruction
10. Dependence on store food

## THE BOTTOM-UP APPROACH AND ECONOMIC DEVELOPMENT IN THE CANADIAN NORTH

Since the 1960s, the federal, provincial, and territorial governments in Canada have initiated top-down development policies to improve the social and economic conditions of Native northerners. A number of development analysts have questioned the effectiveness of the top-down approaches to development in peripheral regions such as the Canadian north. An alternative approach to economic development in the remote regions was therefore suggested. In the 1980s, the bottom-up approach was initiated to address these issues. This type of development advocates for an indigenous and locally controlled economic system. The approach involves the promotion of local development with greater emphasis on local empowerment.

The bottom-up approach advocates economic growth and structural changes within communities. Improving the conditions of people living in these localities and regions is often the priority of such initiatives. Residents of such communities share common political, social, and economic identity and aspirations. They also operate similar institutions within a common cultural heritage. Through empowerment, the communities identify a common objective and focus on specific issues related to their basic needs. Ironside (1990) observed that the fundamental objective of the bottom-up approach to development is one of organizing the collective resources, ingenuity, and spirit of neighboring small communities that are not viable individually.

Saku and Bone (2000) applied the bottom-up approach to the Canadian north when they examined the role of modern treaties on the economic and social development of the region. Modern treaties are used to define a variety of rights and privileges for Native Canadians living in areas of the Canadian north where historic treaties were not signed (Saku 2002). These benefits include ownership of land in the settlement region, cash compensation, right to wildlife harvesting, and the participation in environmental management within the settlement region. Other benefits are sharing of revenues from resources, measures to stimulate economic development, and the right to self-government (Saku 2002). Overall, these agreements are designed to promote better economic and social conditions for aboriginal northerners. Several agreements, including the 1975 James Bay and Northern Quebec Agreement and the Western Arctic Inuvialuit Final Agreement (1984), have been achieved (Saku 2002).

Saku and Bone (2000) claim that modern treaties share similar characteristics of the bottom-up approach to development. This is because these agreements are geared towards local control of both human and natural resources. Modern treaties also create a strong opportunity in bringing small northern aboriginal communities together to achieve economic development. Modern treaties are designed to help stimulate and promote regional economic development. More importantly, the role of modern treaties is to improve the living standards of beneficiaries through the provision of basic community needs.

While the bottom-up approach is highly suitable to the Canadian north, the main problem of the approach is that it is based on the concept of a closed economy. As the world's economy becomes more globalized, and regional economic integration is becoming highly desirable, it is very difficult to justify isolated economies. Moreover, in northern Canada, the approach is inhibited by several structural problems (Saku and Bone 2000). For example, regional population is very small and scattered over a large geographic area.

## CONCLUSION

The Canadian north is a complex and dynamic region. Its complexity is reflected in its economic, physical, and human problems. As the knowledge and the geopolitical system of the region change, new solutions to its problems are being explored. For example, the settlement of modern treaties in the Canadian north is a form of bottom-up approach to achieving development in the region that emerged in the 1970s. These agreements require the active participation of northern aboriginal Canadians in the economic development process. The assumption is that this approach changes the dynamics of regional economic development. Whether this is attainable or not is yet to be determined (Saku 2002).

## REFERENCES

Bone, R. M. 1992. *The Geography of the Canadian North: Issues and Challenges.* Toronto: Oxford University Press.

Chislett, K. L., M. B. Green, and R. M. Bone. 1987. Housing Mismatch for Metis in Northern Saskatchewan. *The Canadian Geographer* 31: 341–346.

Government of Nunavut. 2003. Basic Facts: The Economy. Available online at: www.nunavut.com/basicfacts/english/basicfacts/3economy.html.

Hamelin, L.-E. 1988. 'Nordicity'. In J. H. Marsh, ed. *The Canadian Encyclopedia* Vol. 3: 1504–1506. Edmonton: Hurtig Publishers .

Ironside, R. G. 1990. Regional Development Aid in the Peripheral Region of Northern Alberta. In P. J. Smith and E. L Jackson, eds. *A World of Real Places.* Edmonton: Department of Geography.

Orvik, N. 1976. Toward a Theory of Northern Development. *Queen's Quarterly* 83: 93–102.

Saku J. C. 2002. Modern Land Claim Agreements and Northern Canadian Aboriginal Communities. *World Development* 30: 141–151.

Saku, J. C. and R. M. Bone. 2000. Looking for Solutions in the Canadian North: Modern Treaties as a New Strategy. *The Canadian Geographer* 44: 259–270.

Wonders W. C. 1987. The Changing Role and Significance of Native People's in Canada's Northwest Territories. *Polar Record* 23: 661–671.

# 42

# U.S. LAND COVER AND LAND USE CHANGE:
# 1973–2000

DARRELL NAPTON
*South Dakota State University*

THOMAS LOVELAND
*USGS EROS Data Center*

People change the land to improve their quality of life, but these changes may have adverse consequences that affect other places, water, air, people, or species. Issues resulting from land alteration affect all and include changes to climate, ecosystems, human health, water quality, and vulnerability to hazards. To manage and balance the consequences of land-use and land-cover changes, we must first better understand them.

The U.S. Geological Survey (USGS), with support from the Environmental Protection Agency (EPA) and the National Aeronautics and Space Administration (NASA), is addressing three questions regarding changes to the Earth's land surface: What kinds of changes are occurring and why? What are the impacts of these changes? And how do these impacts, in turn, affect the land surface? The USGS Land Cover Trends project is documenting the regional rates, causes, and consequences of land-use and land-cover change across the conterminous United States between 1973 and 2000 (Loveland et al. 2002). The study focuses on geographic assessments of change in each of the nation's 84 ecoregions (Omernik 1987). Ecoregions are areas with similar natural resources and land-use (Omernik and Bailey 1997). Remote sensing, sampling statistics, field observation, regional analysis, and other geographic tools are being used to explain the geography of change in each ecoregion. The results of the research will provide answers to the following questions: (1) What are the spatial and temporal dimensions of land-use and land-cover change? (2) What are the driving forces of change? (3) What are the consequences of land-use and land-cover change, locally, regionally, and globally?

## WAYS IN WHICH GEOGRAPHY CONTRIBUTES TO A SOLUTION

Currently information about U.S. land change is sparse, and sometimes data sources disagree with each other (Hart 2001). There are a number of space

*D. G. Janelle et al. (eds.), WorldMinds: Geographical Perspectives on 100 Problems*, 261–266.
© 2004 *Springer. Printed in the Netherlands.*

age tools and data sets, including satellite imagery, which provide new ways to evaluate these changes. Satellites are not restricted by dense forests, mountains, or other barriers and therefore are excellent sources for mapping and measuring land-cover change. Satellite images, however, do not tell the entire story of geographical change. When the results of interpreting remotely sensed images are combined with socioeconomic data, field investigation, and the results of published studies, we get a more complete picture of ecoregion land-use dynamics.

## METHODS

We estimated land-use and land-cover change on an ecoregion-by-ecoregion basis using a probability sample of square blocks 10- and 20-km$^2$ in size randomly selected within each ecoregion. This approach provides estimates of change that are as localized as possible. For each block, Landsat imagery from 1973, 1980, 1986, 1992, and 2000 was used to map land-cover. To ensure accurate and consistent land-cover data, manual interpretation of satellite imagery was supplemented with larger-scale aerial photographs. Image interpreters examine the colors, patterns, textures, associations, and other image characteristics to map the different types of land-cover. The imagery had 60 m resolution (approximately one acre), which allows detection of local land changes. The sample-block interpretations were compared by mapping land-cover for each date, then overlaying the interpretations for each date, and using a geographic information system to produce maps and tables that show the land-cover transformations. Then the sample results were statistically extrapolated to produce change estimates for the entire ecoregion. Our goal was to detect within one percent of the actual amount of change within each ecoregion at an 85 percent confidence level.

We also documented the driving forces of change. Driving forces are the human and natural processes that shape how the land is changed or maintained (Turner and Meyer 1994). Typical driving forces include political, economic, social, cultural, demographic, and technological changes. Driving forces encourage or support human land modification activities. For example, in migration might result in forest clearing and subsequent house, school, and road construction. We use traditional geographic field observation to identify signals that illuminate the forces of change, and we study the geographic literature and socio-economic trends for each ecoregion so that we can identify the driving forces of change in each ecoregion.

## RESULTS

The overall amount of land that changed from one land-cover type to another for the four ecoregions area (627,619 km$^2$) – consisting of the Mid-Atlantic Coastal Plain, Southeastern Plains, Piedmont, and Blue Ridge – from 1973 to 2000 was 17.4 percent (Figure 42.1). This sequence of ecoregions spanned a topographic sequence from the coastal plains, across the interior plains and piedmont, to the mountains. The amount of land that changed per period increased steadily from 1973 to 2000 (Figure 42.2). Urban lands increased by 2.3 percent while forests decreased by 2.3 percent. Cropland decreased by 1.9 percent while mechanized disturbed lands (i.e., lands cleared by mechanical means, such as clear cuts, that have not yet changed to a higher level land-cover class) increased by 1.9 percent. These numbers indicate that the total change level of 17.4 percent is too high. However, statistical summaries may mask land where multiple changes have taken place.

The geographic variations in change characteristics are lost when all four ecoregions were considered together. The large Southeastern Plains (335,482 km$^2$) had an overall areal change of 20.8 percent. The largest change in land-use types was the loss of agricultural land (3.0 percent), followed by a 2.7 percent increase in mechanically disturbed land. Forest cover lost approximately 1.0 percent of the total area over the 27-year period. These numbers unfortunately mask the real land-use and land-cover situation occur-

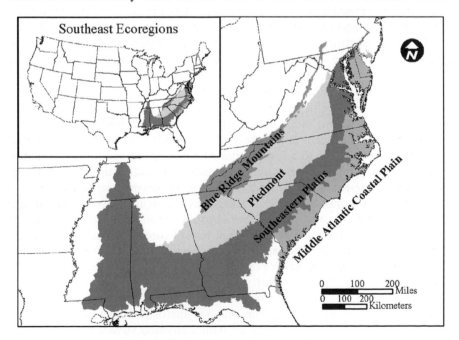

*Figure 42.1.* Selected eastern ecoregions.

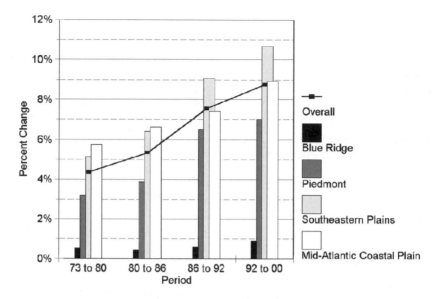

*Figure 42.2.* Overall ecoregion rates of land-use/land-cover change, 1973–2000.

ring in the Southeastern Plains. Our analysis showed that the ecoregion lost agricultural land and natural woodlands and in their place, pine plantations occupied an increasingly larger percentage of the area. The pine plantations are being harvested on very short 20–25 year cycles, meaning that the same land may support up to four uses in 27 years following a sequence from agriculture to forestland to mechanically disturbed land, and back to forest (Figure 42.3a). As a result, there can be major changes in land-use and land-cover while the statistical composition of the overall ecoregion appears to be stable. This ecoregion is the core of the world's most productive industrial forest. The warm, moist climate encourages rapid tree growth, and the flat to gently rolling land is easy to harvest with industrial forest equipment (Figure 42.3b). Within this forest, market forces have encouraged the conversion of farmland to forests and large-scale, industrial timber production. Trees were clear-cut and new trees planted to begin another cycle.

At the other extreme, the Blue Ridge ecoregion land area changed only 2.0 percent over 27 years. The major changes in this relatively small ecoregion (46,980 km$^2$) included a 1.3 percent loss of forestland, and a 1.1 percent increase in urban lands. The Blue Ridge's scenic vistas, abundant parks, and the East's southernmost ski slopes led to tourism developments and second home construction (Figure 42.3c).

Change in the Mid-Atlantic Coastal Plain was estimated to be 17.6 percent of the 81,058 km$^2$ ecoregion. The major transformation was the forest to mechanical disturbed cycle associated with plantation silviculture. While this result is similar to the Southeastern Plains, it also differed because of the overall

*Figure 42.3.* (a) Harvested wheat in foreground with slash from recently cut forest in background. (b) Small-scale logging operation. (c). Condominium and ski slope in Watauga County, North Carolina. (d) New subdivision west of Atlanta.

larger rate of forest loss (3.3 percent) and a substantially larger increase in urbanization (2.5 percent). Also noteworthy was the loss of 1.3 percent of the wetlands in this coastal region. During the 1970s, farms and industrial forest companies drained wooded wetlands to take advantage of access to the European market, high prices for crops and wood (Healy 1985), and federal incentives that encouraged wetland drainage (Hartmann and Goldstein 1994). Forest company land entered a cycle of planting, growing, and harvesting trees. Coastal tourism and retirement was accompanied by new subdivisions, roads, golf courses, and related developments (Kovacik and Winberry 1987).

Approximately 14.9 percent of the Piedmont (164,099 km$^2$) changed from 1973 to 2000. The unique change characteristics of this ecoregion were the result of forest conversions associated with both forest plantation land-uses and rapid urban growth (4.5 percent). The overall loss of forest (4.8 percent decrease) was the highest of the three ecoregions. Driving forces included dramatic population increases associated with the migration of people and businesses to the Sunbelt (Figure 42.3d). Businesses were attracted by the region's low wages, low taxes, and near absence of unions. Much of the growth and consequent land change was along Interstate Highways and at the edge of cities.

## CONCLUSION

Land use and land-cover change is a global issue. The accumulation of local changes has global consequences, and local changes are pervasive across the globe. The technical strategy used in this project is extendible to provide continental and global coverage. Stratified sampling using ecoregions provides a cost-effective, scalable approach to studying larger geographic areas. Because ecoregions can be defined hierarchically at multiple scales, they can be extended at appropriate scales to the continent or globe. Through the use of probability sampling, we are able to draw statistically sound estimates of land-use and land-cover change while only analyzing satellite data covering a small portion of each ecoregion. That use of advanced technology, when combined with old-fashioned geographic observation and assessment provide a way to understand the geography of change.

## ACKNOWLEDGEMENTS

We gratefully acknowledge the support of the U.S. Geological Survey's Geographic Analysis and Monitoring Program, the U.S. Environmental Protection Agency's Landscape Ecology Branch in Las Vegas, NV (Interagency Agreement DW14938108-01-0), and the Land Cover Land Use Change Program of NASA's Earth Science Enterprise.

## REFERENCES

Hart, J. 2001. Half a Century of Cropland Change. *The Geographical Review* 91: 525–543.
Hartmann, J. and J. Goldstein. 1994. North Carolina – the Pocosins and Other Freshwater Wetlands. In *The Impact of Federal Programs on Wetlands*, 281–300. Washington D.C. U.S. Department of the Interior.
Healy, R. 1985. *Competition for Land in the American South: Agriculture, Human Settlement, the Environment*. Washington, D.C.: The Conservation Foundation.
Kovacik, C. and J. Winberry. 1987. *South Carolina: The Making of a Landscape*. Columbia, South Carolina: University of South Carolina Press.
Loveland, T., T. Sohl, S. Stehman, A. Gallant, K. Sayler, and D. Napton. 2002. A Strategy for Estimating the Rates of Recent United States Land Cover Changes. *Photogrammetric Engineering and Remote Sensing* 68: 1091–1099.
Omernik, J. 1987. Ecoregions of the Conterminous United States. *Annals of the Association of American Geographers* 77: 118–125.
Omernik, J. and R. Bailey. 1997. Distinguishing Between Watersheds and Ecoregions. *Journal of the American Water Resources Association* 33: 935–949.
Turner, B. and W. Meyer. 1994. Global Land-Use-Cover Change: An Overview. In W. Meyer and B. Turner, eds. *Changes in Land Use and Land Cover: A Global Perspective*, 3–10. Cambridge University Press.

# 43

# PERIODIC MARKETS NOW AND THEN

ALLISON BROWN

*Pennsylvania State University*

Is there anyone who is not fascinated by the swirling colors and pungent smells of the market place? Whether it is the fish markets of the Caribbean or the cloth markets of India, the covered souks of Fez, or the farmers markets of New York City, travelers' tales have long included descriptions of the wonders found in the local markets.

Geographers, too, have been seduced. For two centuries or longer, using a range of techniques, researchers have diligently cataloged the social and economic characteristics of vendors and customers and picked apart market-place management complexities. This painstaking work has yielded a rich literature spread across a handful of disciplines. The topical areas of study range from the physical layout of the market site to the social laws of market participants to the origins and trade paths of the goods on offer. The literature on periodic markets interprets the effects of market places on war and conflict, on social status, on marriage, gift giving, farming systems, gender roles, and the economics of the household, farm, firm, region, and nation (Smith 1979, 1980), all to answer the haunting question of why there is a market at this place and at this time. But, in the past half-century, the study of periodic markets has waned. Geographers now generally shop in more fashionable districts leaving the dusty bazaars to non-geographers. Newcomers sometimes unwittingly recapitulate earlier work.

Things may be changing, however. The development of geographic information systems (GIS), coupled with new business management research techniques, has given geographers powerful new tools and brought them, once again, to the forefront of the study of business location. GIS technology facilitates a more comprehensive analysis of the spatial elements of commercial behavior and has become an important retail business-planning tool. Still, discussion of modern periodic markets and farm location remains rare, which is unfortunate because this work has important implications for preserving farmland, limiting sprawl, and improving farm livelihoods around the world.

*D. G. Janelle et al. (eds.), WorldMinds: Geographical Perspectives on 100 Problems*, 267–270.
© 2004 *Springer. Printed in the Netherlands.*

## THEN

The earliest investigator of the relationship between markets and farm location was J. H. von Thünen, whose seminal book was published in 1826. Thünen looked at the effects of farmer access on the structure of agriculture. Using a combination of thought experiment and data gathered from his own region, he derived a model demonstrating that clusters of farms producing similar classes of crops will develop in response to the characteristics of the marketplace. The shape and composition of these clusters, Thünen showed, evolve in response to improvements in communication and transportation, and to changes in the structure of the marketing system. His work outlined the dimensions of a basic economic principle: if there is no accessible market, only subsistence farming is possible. Most critical to today's discussions is Thünen's observation that the region closest to the urban center that he called the Free Crop Zone will be tenanted by producers of highly perishable products operating in an economic system shaped not so much by cost of production as by the whims of customers – what we call product life-cycle effects. We observe this effect today in modern farmers markets, a kind of periodic market that has grown in popularity over the past 30 years.

Farmers markets in the United States declined rapidly immediately following World War II to a low of about 350 in 1970. Yet, by the end of the century, the number of farmers markets had risen almost ten-fold (Brown 2001). These farmers markets are the primary market channels for many thousands of farmers and, while estimates vary, certainly add billions of dollars each year to regional economies (Brown 2002; Payne 2002).

Although the impact of this rapid and very unexpected growth is an ideal research topic, there is scant published work to date and almost no scholarly discussion of farm location in relation to modern markets (Brown 2002). This lack of attention is surprising because the resilience of the Free Crop Zone flies in the face of earlier predictions that agriculture would eventually migrate from expensive peri-urban areas to regions of comparative production advantage (cf. Chisholm 1962). Farmers markets and other local marketing structures were supposed to wither, supplanted by a global production and transport system. Yet a close look reveals that farm distribution looks surprisingly like Thünen's nineteenth-century prediction: market gardens flourish near population centers, commodity farms are located where costs are minimized, subsistence farms persist where conditions do not permit commercialization (Bowler 1992; Fugita et al. 1999).

Earlier theorists went astray at the intersection between economics and business management. Economists generally hold that agriculture is an immobile business; a farm cannot be moved to a better location and, indeed, commodity farming is largely immobile. Modern business theory, on the other hand, views market gardening as an agile, mobile business capable of producing many hundreds of distinct products in a given season and able to change production plans on a moment's notice. The farm itself can, in some cases, be movable. Market gardens behave like non-agricultural small businesses, not like traditional farms.

In business management terms, market gardening occupies a different strategic niche from commodity agriculture, one that is highly dependent on physical and political location. Although there are little hard data on the subject, the number of market gardens seems to be growing in response to the demand created by a proliferation of regional market channels, including farmers markets. Physical and cultural conditions near cities support market gardening and obstruct commodity farming; under these conditions, Thünen's pattern of agriculture location emerges, distinct and measurable.

Why is this important? Farmers markets and direct marketing are credited with an astonishing array of positive effects, the most environmentally important of which is that by boosting farmer income, farmers markets preserve farm land and slow sprawl. If these claims are true, then it makes excellent public policy to promote farmers markets in a range of contexts around the world. Unfortunately, extant research does not clarify the impact of farmers markets and other direct marketing activities on the regional or household economy (Brown 2002). My colleagues and I are trying to remedy this situation by undertaking some classic periodic market research.

## NOW

The most critical area of our analysis is the investigation of the persistence of peri-urban agriculture and reports that patterns of agricultural production change over time in response to the success of a cluster of farmers markets. Our research program is intended to define the geographic limits of the Free Crop Zone and the relationship between local and direct marketing systems and the rate of farmland conversion. We approach the question by using GIS to analyze the flow of goods and services between farmers and consumers. This analysis will, in turn, be compared with land use records.

Several modern theories of commercial behavior predict that recognizable patterns of business location develop. It is well known in modern retailing that site selection (including virtual location) is an important component of commercial success. We are looking at business patterns in successful and unsuccessful farmers markets. Some of the research questions include:

- Market rings and radius of trade: What are the factors that define the distance that farmers and consumers will travel to attend farmers markets?
- Market distribution: Market managers know that a region must have enough markets so that the best farmers can develop sustainable market rings. How many markets is enough?
- Market periodicity: Is the optimum number of market days the same for consumers and growers?

- Synergy: Newly revitalized periodic farmers markets seem to be nurturing new customers and new farmers who in turn support new farmers markets in what is called, in the business literature, a "virtuous circle." What are the dynamics of this interaction?
- Business life cycle: Is the rate that market gardens cycle in and out of business affected by changes in access to local market channels including farmers markets?
- Industry life cycle: Is it important that local agriculture marketing structures are mature in some locations and newly forming in others?

We have begun to use GIS site analysis techniques to develop tools to help market managers choose the best location for new or expanded farmers markets. We do this as geographers have always done it: by looking at where and how farmers and consumers travel to markets and where markets are in relationship to each other. GIS makes this work easier than ever before.

Research funding is limited and the antipathy of some geographers to the study of agriculture location and periodic markets remains a problem. But as GIS becomes more widespread, applied investigations of periodic markets as a facet of regional agriculture economies are becoming more common. Periodic markets are a venerable human institution that has transformed itself just in time to meet a technological revolution in geography.

## REFERENCES

Bowler, I. 1992. *The Geography of Agriculture in Developed Market Economies.* Essex: Longman.

Brown, A. 2001. Counting Farmers Markets. *Geographical Review* 91: 655–674.

Brown, A. 2002. Farmers Market Research 1940–2000, An Inventory and Review. *American Journal of Alternative Agriculture* 17: 167–176.

Chisholm, M. 1962. *Rural Settlement and Land Use.* London: Hutchinson.

Fugita, M., P. Krugman, and A. Venables. 1999. *The Spatial Economy: Cities, Regions and International Trade.* Cambridge, MA: MIT Press.

Payne, T. 2002. *U.S. Farmers Markets – 2000. A Study of Emerging Trends.* Washington, D.C.: USDA.

Smith, R. 1979. Periodic Market-Places and Periodic Marketing: Review and Prospect – I. *Progress in Human Geography* 3: 471–505.

Smith, R. 1980. Periodic Market-Places and Periodic Marketing: Review and Prospect – II. *Progress in Human Geography* 4: 1–31.

Thünen, J. 1826. *Der isolierte Staat in Beziehung auf Landwirtschaft und Nationalökonomie.* Published by the author, Rostock, Germany. An abridged edition translated by C. Wartenberg and edited by P. Hall was published in 1966 as *Von Thünen's Isolated State.* Oxford: Pergamon.

## PART V

# MAINTAINING HEALTHY ECOSYSTEMS

Geography provides important perspectives on the interrelationships between human activities and the functioning of natural systems. The documentation and interpretation of changes in land cover, habitat, and biodiversity give clues to the origins and possible remediation of problems that stem from the contamination of the life support systems of water, air, soil, and vegetation. If survival of life is valued, then maintenance of a healthy earth support system is an essential priority for attention by the custodians of science and policy.

*Kathy Hansen*

# 44

# THE BIODIVERSITY CRISIS

JOHN A. KUPFER
*University of Arizona*

GEORGE P. MALANSON
*University of Iowa*

We are now in the midst of a biological extinction event that rivals the great events of prehistory. Unlike previous events, however, the current threats to biodiversity are related to human activities, including habitat loss, degradation, fragmentation, and the introduction of invasive species. Along with scientists from other disciplines, geographers have contributed to a variety of studies on the geography of nature reserves and natural areas in fragmented landscapes. These contributions range from early descriptions of habitat destruction effects to subsequent theoretical advances such as island biogeography theory to recent technological developments in remote sensing, modeling, and geographic information systems (GIS). Geographers continue to be central in clarifying the effects of habitat change on species patterns and persistence and in helping to develop ways to mitigate impacts associated with habitat loss and fragmentation.

## BACKGROUND

Nineteenth century works by scholars such as George Perkins Marsh and Russian geographer Alexander Woeikof addressed the effects of habitat destruction on soil erosion, stream hydrology, climate, and (to a lesser degree) plant and animal communities. Insights provided by such works also served as the basis for later studies in the early- and mid-twentieth century. Convened in 1956 and chaired by geographer Carl Sauer, the influential international symposium "Man's Role in Changing the Face of the Earth" included contributions from more than a dozen geographers, including H. C. Darby's analysis of deforestation in Europe, Andrew Clark's treatise on the alteration of mid-latitude grasslands, and Arthur Strahler's paper on the effects of habitat change on fluvial geomorphic processes. Papers from this symposium also raised some of the earliest concerns about how human actions were altering patterns of biodiversity and leading to species extirpations and extinctions in both natural and semi-natural habitats.

*D. G. Janelle et al. (eds.), WorldMinds: Geographical Perspectives on 100 Problems, 273–277.*
© 2004 *Springer. Printed in the Netherlands.*

Ecological concerns about the effects of deforestation and other forms of habitat alteration grew during the mid-to late-twentieth century and, coupled with the quantitative revolutions in ecology and geography, led to efforts to develop prescriptive models for maximizing species diversity and protecting critical habitat for endangered species. Early quantitative links between habitat loss and species extinctions were developed out of concepts formalized within the *Theory of Island Biogeography* (MacArthur and Wilson 1967), which stated that the number of species on oceanic islands is a function of island size (linked to extinction rates and habitat heterogeneity) and isolation (influencing the arrival of potential colonists). Ecologists and geographers drew parallels between oceanic islands and natural habitats in a "sea of hostile human land uses" and suggested that biodiversity in reserves would be highest in large reserves and in reserves that are less isolated from one another.

Although the validity of the analogy between oceanic islands and remnants of natural habitat in human-dominated landscapes has been largely dismissed, some of the key ideas from island biogeography theory were fundamentally correct (the relationship between reserve size and species richness, for example) and have been incorporated into principles within the field of landscape ecology (Kupfer 1995). In the case of nature reserves, a landscape ecological approach broadens the focus beyond the boundaries of protected areas to include the interactions with other reserves and the surrounding landscape elements. At present, work by geographers is illustrating that a view of landscapes composed solely of "suitable" vs. "unsuitable" habitat is outdated and needs to be replaced with a view representing landscapes as complex systems in which even modified habitats may serve some role in species' persistence (e.g., Malanson 2002) (Figure 44.1).

In recent years, geographers working in the area of landscape ecology have made significant contributions to our current understanding of reserves and have advanced our knowledge of the factors influencing the functioning of natural areas. Research on biodiversity in remnant natural areas has addressed the importance of disturbance regimes within nature reserves (Baker 1989) and examined the effects of invasive species on ecosystem composition and function in protected areas (Allen and Kupfer 2001). Geographers have addressed how spatial patterns (e.g., fragmentation) and processes (e.g., dispersal) affect vegetation dynamics in response to human-induced changes such as climatic change and altered disturbance regimes and placed these findings directly in the context of nature reserve design (e.g., Malanson and Armstrong 1996). The implementation of conservation corridors, areas of protected habitat that link larger areas to facilitate the spread of organisms across the landscape, is based on fundamental geographic concepts.

*Figure 44.1.* The complexity of human-altered landscapes is clear in the countryside around Indian Church, Belize, which contains a range of forest stages (primary, mature, successional) and agricultural uses (shifting cultivation, pasture, garden, intensive row cropping). On-going geographical research in the area is looking at the effects of such spatial heterogeneity on plant succession and biological conservation.

## GISCIENCE

As our understanding of deforestation and forest fragmentation effects have grown so have the tools available for quantifying and analyzing patterns of landscape pattern and change. The developments of GIS and remote sensing, along with the increased accessibility of spatial data, have helped to revolutionize conservation planning and land management. For example, the USGS Gap Analysis Program (GAP), one of the most well-developed conservation programs in the world, used GIS to identify biological "hotspots" and areas of high conservation value in unprotected areas by linking biogeography and land use (Scott et al. 2001). Such an approach is also at the heart of the Nature Conservancy's ecoregional planning methods, which involve identifying a "portfolio" of sites that fully represent the native species and natural communities characteristic of an ecoregion.

While advances in GIS and other geospatial tools have greatly improved our ability to document and analyze spatial patterns of species, communities, and ecosystems, the utility of such work is ultimately rooted in our ability to link changes in such patterns to the effects on fundamental ecological processes like succession, energy flow through trophic webs, pollinator movement, and species migration. Given the tremendous complexity and difficulty in identifying these linkages, while also making such information accessible and useful to land managers and conservationists, geographers have been involved with attempts to develop ecological indicators to facilitate management and monitoring of biodiversity in protected and unprotected natural areas (Lindenmayer et al. 2002). One important finding that has emerged from much of the work on species preservation in fragmented landscapes is the importance of scale. Bailey et al. (2002), for example, used a national database to examine the effects of fragmentation on a range of species (including plants, birds, insects and mammals) and documented the different ways in which species respond to variables at a range of spatial and ecological scales.

## GEOGRAPHIC APPROACHES

While geographers are only one of the many groups of scientists examining issues associated with biodiversity conservation, they often possess a more interdisciplinary perspective than ecologists or other natural scientists. This viewpoint is evident in the increased focus on research that unites ecological, social, and remote sensing sciences to provide an understanding of the dynamics of habitat alteration and fragmentation. Such projects frequently work towards spatially-explicit assessments and models that can be used to monitor and project habitat change under different assumptions (Turner et al. 2002). Geographers are involved in recent advances that link changing landscape patterns quantified with remote sensing to field surveys of demographic characteristics and behavior (Walsh et al. 1999), as well as efforts to formulate multivariate, spatial models of land-cover change trajectories that incorporate a wide range of anthropogenic, environmental, and socio-economic variables (Mertens and Lambin 2000). An important insight is that we are unable to manage reserves taken not only out of their spatial context but even more so out of their human context. Geographers, thus, raise critiques of conservation as a political process that ecologists usually ignore (e.g., Bryant 2002).

As we stand at the brink of the twenty-first century, the preservation of global biodiversity in the face of increasing human populations and resource demands is one of most serious environmental challenges that we face. Geographers have helped clarify the processes that create and maintain diversity within natural areas in a range of different geographic regions and have called for conservation needs to focus explicitly on species preservation in the small habitat fragments that comprise many remnants in current and future landscapes (Kellman 1996). They continue to be involved in developing the technical tools that are often used to assess and monitor changing landscape patterns and the models that are used to project the attendant effects on biodiversity. The biodiversity crisis has been recognized as being as much geographical as it is biological, and geographers are at the forefront of the endeavor.

## ACKNOWLEDGEMENTS

This work was supported by a grant from the National Commission for Science on Sustainable Forestry.

## REFERENCES

Allen, T. R. and J. A. Kupfer. 2001. Spectral Response and Spatial Pattern of Fraser Fir Mortality and Regeneration, Great Smoky Mountains. *Plant Ecology* 156: 59–74.

Bailey, S. A., R. H. Haines-Young, and C. Watkins. 2002. Species Presence in Fragmented Landscapes: Modelling of Species Requirements at the National Level. *Biological Conservation* 108: 307–316.

Baker, W. L. 1989. Landscape Ecology and Nature Reserve Design in the Boundary Waters Canoe Area, Minnesota. *Ecology* 70: 23–35.

Bryant, R. L. 2002. Non-Governmental Organizations and Governmentality: 'Consuming' Biodiversity and Indigenous People in the Philippines. *Political Studies* 50: 268–292.

Kellman, M. 1996. Redefining Roles: Plant Community Reorganization and Species Preservation in Fragmented Systems. *Global Ecology and Biogeography Letters* 5: 111–116.

Kupfer, J. A. 1995. Landscape Ecology and Biogeography. *Progress in Physical Geography* 19: 18–34.

Lindenmayer, D. B., R. B. Cunningham, C. F. Donnelly, and R. Lesslie. 2002. On the Use of Landscape Surrogates as Ecological Indicators in Fragmented Forests. *Forest Ecology and Management* 159: 203–216.

MacArthur, R. H. and E. O. Wilson. 1967. *The Theory of Island Biogeography*. Princeton, NJ: Princeton University Press.

Malanson, G. P. 2002. Effects of Spatial Representation of Habitat in Competition-Colonization Models. *Geographical Analysis* 34: 141–154.

Malanson, G. P. and M. P. Armstrong. 1996. Dispersal Probability and Forest Diversity in a Fragmented Landscape. *Ecological Modelling* 87: 91–102.

Mertens, B. and E. F. Lambin. 2000. Land-Cover-Change Trajectories in Southern Cameroon. *Annals of the Association of American Geographers* 90: 467–494.

Scott, J. M., F. W. Davis, R. G. McGhie, R. G. Wright, C. Groves, and J. Estes. 2001. Nature Reserves: Do They Capture the Full Range of America's Biological Diversity? *Ecological Applications* 11: 999–1007.

Turner, B. L., S. C. Villar, D. Foster, J. Geoghegan, E. Keys, P. Klepeis, D. Lawrence, P. M. Mendoza, S. Manson, Y. Ogneva-Himmelberger, A. B. Plotkin, D. P. Salicrup, R. R. Chowdhury, B. Savitsky, L. Schneider, B. Schmook, and C. Vance. 2001. Deforestation in the Southern Yucatan Peninsular Region: An Integrative Approach. *Forest Ecology and Management* 154: 353–370.

Walsh, S. J., T. P. Evans, W. F. Welsh, B. Entwisle, and R. Rindfuss. 1999. Scale-Dependent Relationships Between Population and Environment in Northeastern Thailand. *Photogrammetric Engineering and Remote Sensing* 65: 97–105.

45

# BIO-DIVERSITY INVENTORY:
# KAMIALI WILDLIFE MANAGEMENT AREA

F. L. (RICK) BEIN
*Indiana University Purdue University at Indianapolis*

In 1997–1998, a team of scientists lead by the author conducted a biodiversity Inventory of the Kamiali Wild Life Management Area (KWMA), 60 km southeast from Lae, Papua New Guinea along the Huon Gulf coast. The 434-square-kilometer area was registered with the Papua New Guinea (PNG) government as a wildlife management area in 1996 (Figure 45.1). The village of 500 people chose a committee to work with The Village Development Trust, a nationally based non-government organization to help manage the area that extends 12 km out to sea and 17 km inland. The high relief environment ranges from 1,080 m below sea level up to 2,012 m above mean sea level. A transect method was used to sample biomes beginning with open sea, coral reef, mangrove, beach, village, sago swamp, garden, riparian environments, tropical forest, and cloud forest. This unique high biomass environment serves as a habitat for many rare species but faces potential threats of: (1) commercial clear-cut logging and (2) population-pressure induced hillside gardening and over fishing of reefs. This biodiversity inventory establishes a baseline from which to employ conservation efforts to maintain its uniqueness in rural Papua New Guinea.

## CONSERVATION OF BIODIVERSITY

Biological diversity is the richness of our planetary ecosystem and includes genetic diversity, species diversity, and ecological diversity (Miller 1996). The conservation of the biodiversity is important not only to ensure survival of millions of species but also to support the living systems of the world's people who depend on natural resources and the ecosystem. Leading ecological thinking considers high bio-diversity to be desirable for the planetary ecosystem (McNeely 1990). Following the ecological premise that every living thing has a "niche" or functional purpose in serving the greater ecosystem, it would hold that by maximizing the number of living species, there would be more opportunities to maintain ecological balances. As a part of the global ecosystem, humans would benefit the most from high biodiversity. Benefits

279

*D. G. Janelle et al. (eds.), WorldMinds: Geographical Perspectives on 100 Problems*, 279–285.
© 2004 *Springer. Printed in the Netherlands.*

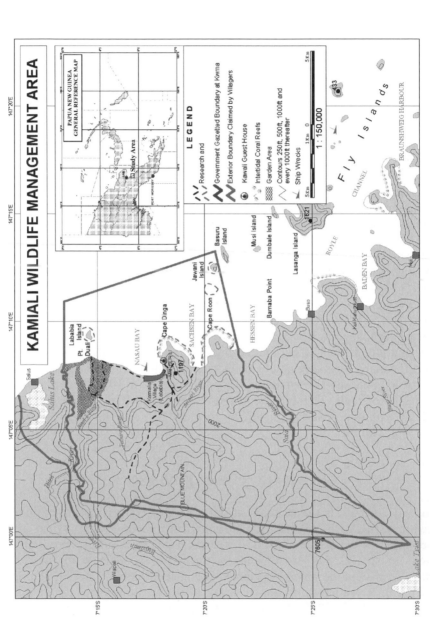

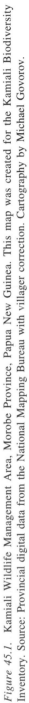

*Figure 45.1.* Kamiali Wildlife Management Area, Morobe Province, Papua New Guinea. This map was created for the Kamiali Biodiversity Inventory. Source: Provincial digital data from the National Mapping Bureau with villager correction. Cartography by Michael Govorov.

found in nature take the form of foods, medicine, building materials, clothing materials, and chemical substances. Many discoveries have benefited humankind and have over the millennia fostered the expansion of the earth's human population. Maintaining bio-diversity provides continuous human benefit.

It further happens that tropical biomes, such as those in Papua New Guinea, contain some of the highest bio-diversity on the planet. Hundreds of species, many yet to be named, thrive in the Papua New Guinea rain forests and coral reefs. Papua New Guinea is still 60 percent forested, much is pristine and contains some of the richest biodiversity in the world (Filer and Sekhran 1998). There is great fear that Papua New Guinea will lose this resource if current trends of deforestation continue.

Papua New Guinea, on the other hand, is striving to develop economically and, coupled with a 30-year population-doubling rate, it must exploit its basic resources. Traditional paradigms and lifestyles of the village are changing. Attitude toward the environment is also changing. Western paradigms promoting the monetary economy begin to erode the old lifestyles, and the environment becomes expendable for the purpose of seeking immediate wealth. During this paradigm shift, the natural environment becomes vulnerable and easily exploited to support the new way of thinking. Environmental exploitation occurs during the transition when old thinking gives way to, or merges with, new thinking.

## THE ROLE OF CONSERVATION AND THE WILDLIFE MANAGEMENT AREA

Wildlife Management Areas (WMAs) have been established for years in a number of countries (including Papua New Guinea) to slow down the process of exploitation of natural resources. In recent years, the "Integrated Conservation and Development Area" (ICAD) has been added to introduce the concept of "conservation." The ICAD concept takes a more central position and promotes sustainable use of resources, which better describes the working conditions of the KWMA. In Papua New Guinea, WMAs have been more commonly established because there are government regulations to follow. Regulations for the implementation of ICADs did not exist in Papua New Guinea until recently. In KWMA and as in many situations in Papua New Guinea, the WMAs now include the parameters and philosophy of the ICADs by incorporating the concept of "sustainable use of resources."

## THE KAMIALI WILDLIFE MANAGEMENT AREA

Kamiali became a WMA because of several unique circumstances (Figure 45.2). (1) It is the only coastal village along the Huon Gulf between Lae and the border with Oro Province, which chose not to sell their trees to the Malaysian logging companies during the early 1990s. (2) The logging companies passed over the area in favor of more productive areas further along the coast. (3) The high elevation areas of the interior portion of the KWMA are protected from settlement by neighboring tribes by an up-thrust fault that has created a huge escarpment along its western boundary (Leffler 1977). (4) The subsistence food-garden soils in the fertile Bitoi River flood plain provide adequate food for the villagers and there has been no need to clear the steep hillsides for farming. As a result of these four factors, much of the forest remains pristine, and the coral reefs fringing the coast of Kamiali are equally pristine because they are free from the smothering effect of sediments eroding from the land. Since Kamiali is mainly a fishing village, there is very little hunting pressure and the original wildlife is intact. The seven village hunters have had little problem finding wild game within short distances of the village and, as a result, rarely venture up into the more remote areas above 300 m elevation.

*Figure 45.2.* Kamiali Wildlife Management Area view of multiple biomes. Training center and dormitories in foreground.

## METHODOLOGY AND LOGISTICS

The biodiversity inventory was conducted between July 1997 and July 1998. When the Lababia community established the Kamiali Wildlife Management Area, the Village Development Trust in Lae, Papua New Guinea obtained a grant from the New Zealand High Commission to finance this study. The grant provided for the listing of species found in the various biomes throughout the Wildlife Management Area for recommendations regarding further study and better conservation of the wildlife management area. The sampling technique comprised a transect from the open sea through the coral reef, mangrove swamp, beach, freshwater mangrove, sago swamp, lowland rainforest, lowland hill forest, and upland and montane cloud and moss forest. Also sampled were sites deemed to be notably different, such as Lababia Island and the Bitoi food gardens. No attempt was made to numerically quantify any part of the biodiversity except in the case of the fish study. Nor was there any attempt to sample the genetic diversity.

A signed "scientist, good faith, research agreement" has been established to provide an incentive for the villagers to host scientific researchers and to protect their intellectual property rights. This includes an understanding that participating scientists: (1) build into their research proposals and grants a cash contribution to the community; (2) employ villagers whenever possible in the data collection to financially reward and educate them, and to build enthusiasm and expertise to support future research; and, (3) commit to share the proceeds, should the scientist or representative institution make a profit from their discoveries at Kamiali.

Facilities to accommodate field research scientists include the comforts of a guest house with prepared meals, mountain-fed water, clean showers, toilets, storage capability, easy access to most biome areas, transect trails to the more remote environments, and boat transportation from Lae. Up to 16 people can be accommodated in four-person screened bunkhouses and another 20 people in an open-air dormitory bunkroom.

## SUMMARY OF RESULTS

Thirteen participating scientists focusing on different parts of the biota have found many unique aspects in the biodiversity of the Kamiali area (Bein et al. 1998). Briefly, the findings show that:
- The fringing coral reef is one of the highest biomass reefs in the world (33% more diverse that the entire Caribbean Sea; Figure 45.3).
- The undisturbed lowland hill forests next to the ocean shoreline are unique because they have not been periodically cleared for food gardening like most Papua New Guinea coastal communities.

*Figure 45.3.* Local Kamiali woman collecting shellfish on reef for food and ornaments. Photo by John Kupfer.

- Cloud forest vegetation typical of 2,000 m elevation is found as low as 600 m on ridge slopes facing toward the sea.
- A vast area of land appears not to have been untouched by humans for at least the last 50 years.
- Lababia Island serves a major roosting site for frigate birds and pied imperial pigeons.
- The broad sand beaches of Kamiali are major nesting sites for leather back turtles.
- Several previously undescribed insect species and plant species have been recorded.

The complete inventory contains the appended reports of each scientist, including all species found and those species, not seen, but which are probably there because of their association with the known species. Also included are recommendations made by the scientists for maintaining the biodiversity and the base for future research.

## A BASE FOR FURTHER RESEARCH

The biodiversity inventory at Kamiali, with its logistical facilities in place, provides an excellent basis for further research. The potential for future research in such a pristine environment is unending. Remoteness from the rest of the world is a major obstacle.

This inventory is not all-inclusive, as the sampling technique probes only a small area. Possibilities include: (1) Further sampling of the various biomes; of note, the cloud forest biome has been barely examined for plants and insects. (2) The coral reefs, mangroves, and beech areas offer many research opportunities. (3) Over 100 leather back turtles nested in Kamiali in the 1997–1998 season. (4) Ethno-botanical studies are beginning to uncover a variety of local medicinal plants and herbs.

## ACKNOWLEDGEMENTS

This project was made possible by: (1) The Village Development Trust, Non Government Organization, Lae 411, Papua New Guinea; (2) the New Zealand High Commission; (3) the presence of the author at the Papua New Guinea University of Technology, as director of the Environmental Research and Management Centre from 1996-1999, funded by the Asian Development Bank in conjunction with the Minnesota-based Midwestern University Consortium of International Associations (MUCIA); and (4) Lababia Village. All photos were taken by the author.

## REFERENCES

Bein, F. L., J. Goodwin, K. Powell, A. Jenkins, P. Led, J. Sumaga, J. Mukiu, F. Bonaccorso, B. Iova, J. Genolagani, P. Kulmoi, J. Meru, and C. Unkau. 1998. *Kamiali Wildlife Management Area Bio-Diversity Inventory*. Auckland New Zealand: New Zealand High Commission.

Filer, C. and N. Sekhran. 1998. *Loggers, Donors and Resource Owners*. Port Moresby: National Research Institute.

Loeffler, E. 1977. *Geomorphology of Papua New Guinea*. Canberra: Australian National University Press.

McNeeley, J. A., K. R. Miller, W. V. Reid, R. A. Mittermeier, and T. B. Werner. 1990. *Conserving the World's Biological Diversity*. Washington, D.C.: World Resources Institute.

Miller, G. T., Jr. 2002. *Living in the Environment* (12th ed). Belmont, CA: Wadsworth/Thompson Learning.

# 46

# ANTHROPOGENIC SOILS AND
# SUSTAINABILITY IN AMAZONIA

WILLIAM I. WOODS
*Southern Illinois University Edwardsville*

Despite its great biotic concentration and diversity, the Amazon Basin (Figure 46.1) has been seen as a region capable of supporting only a low population density based upon non-intensive subsistence activities and that all attempts at intensifying the area's food production and increasing its sustainable populations are doomed to failure (e.g., Meggers 1996). However, questions have been raised about this traditional viewpoint and it has been proposed that complex societies with large, sedentary populations were present for at least a millennium before European contact (Heckenberger et al. 1999). It appears that early aboriginal cultivators had developed field management mecha-

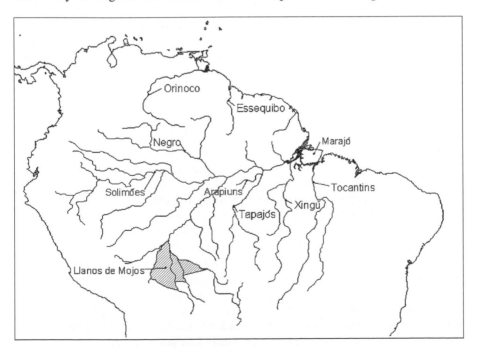

*Figure 46.1.* The Amazon Basin.

287

D. G. Janelle et al. (eds.), WorldMinds: Geographical Perspectives on 100 Problems, 287–292.
© 2004 Springer. Printed in the Netherlands.

nisms that mimicked the processes long operant within habitation areas where basic soil properties were radically changed through the addition of ash and other organic materials (Hecht and Posey 1989). As a result, long after the decimation of these peoples during the early historic period, their enriched soils still are a major resource for agriculture in Amazonia (Glaser et al. 2001) and a proper understanding of these soils' properties may provide a major resource for future populations (Mann 2002). Geographers, including William Denevan, Susanna Hecht, Joseph McCann, Nigel Smith, and William Woods, have played an integral part in the interdisciplinary research investigation of the anthropogenic soils. Contrary to expectations, they have found enormous pre-Columbian modifications of the soil landscape and unexpected answers to questions of the long-term past and future sustainability of food production in the Amazonian environment (Denevan 2001).

## SETTING

Soils, along with energy from the sun, water, and the cultivated plants themselves, constitute the major environmental resources for agricultural societies and are, therefore, critical to discussions of productivity and sustainability. In the Amazon region, often depicted as a "Counterfeit Paradise," the highly weathered, strongly acidic soils of the uplands are thought of as extremely forbidding. With few available nutrients and sometimes having aluminum in concentrations high enough to mine, one couldn't imagine a worse regime for agriculture, particularly when associated with nucleation of population. Indeed, even in the flood plain with somewhat better soils, crop production has been seen as a risky endeavor because of the unpredictability of the high-water regime.

However, it's a matter of scale when dealing with humans and their interaction with the land; and, for soils this is particularly true. Regional depictions simply are not appropriate here. Anthropogenic soils tend to form a continuum of expression within any given microhabitat and enormous complications come into play due to the extreme heterogeneity of the types, amounts, and distribution of inputs and withdrawals through time. The reality of our pre-European and current smallholder farmers, their basic decisions, and the majority of outcomes were and are at the level of the individual and the household. The point of articulation with the environment at this scale is narrow, mainly including only the limited zone of exploitation surrounding the place of habitation. The result is an extremely heterogeneous mix of adaptive modifications whose resulting soils vary in the extreme. When examined at the microscale, one finds both great variety in and enormous pre-Columbian modifications to the soil landscape. Indeed, estimates of the total area of these discrete patches of enriched soil within lowland portions of the Amazon Basin provide an enormous potential productive capacity.

## TERRA PRETA

Early explorers in Amazonia described dense villages extending for kilo-meters along river bluff edge settings with roadways linking these to settle-ments in the interior. Often associated with the presumed village locations are soils termed "Indian black earth," or *terra preta do Índio*. The naturalist Herbert Smith worked in the region during the 1870s and was well aware of these soils: "This is the rich *terra preta*, 'black land,' the best on the Amazons. It is a fine, dark loam, a foot, and often two feet thick. Strewn over it every-where we find fragments of Indian pottery, so abundant in some places that they almost cover the ground." (Smith 1879: 144–145).

The heightened fertility status of these soils has long been recognized by the indigenous and *caboclo* inhabitants of the region (Smith 1980). The most nutrient-demanding crops, including maize, beans, squash, sweet potatoes, melons, and tobacco, are planted on them (Figures 46.2 and 46.3). When not cleared for agriculture, a distinctive vegetation structure and species com-position is recognized, and the unique and abundant assemblage of useful wild and semi-domesticated plant species occurring on them is exploited (Woods and McCann 1999).

Throughout Amazonia, *terra preta* occurs in a wide variety of landscape contexts, in circumscribed patches ranging in size from less than a hectare

*Figure 46.2.* The author in newly sprouted maize field on terra preta site, Belterra, Pará, Brazil.

*Figure 46.3.* Close-up of field surface, showing healthy maize plant and numerous potsherds.

to many square kilometers. The evidence strongly supports the idea that dark earths are cultural deposits created through the accretion of waste around habitation areas and the manipulation of organic additions to associated cultivated zones (Woods 1995). Fire is a crucial component in their formation as it contributes charcoal and ash, which increase soil pH and thereby suppress aluminum activity toxic to plant roots and soil microbiota. The consequent heightened microbiological activity adds minute organic decomposition products to the soil matrix. These, along with the byproducts of incomplete combustion, provide charged surfaces largely absent in the local soils and increase nutrient retention capacity, thus setting up a synergistic cycle of continued fertility (Woods and McCann 1999).

## PROSPECT

Many questions are still not fully answered with respect to the origin, distribution, variation, and past and potential use of the Amazonian dark earths and further exploration of these soils is highly significant. Not only do they contain invaluable information about settlement and environmental management strategies during the pre-European period, but also current use of these soils by colonists provides an important resource for food production within Amazonia. Discussions of the questions and hypotheses revolving around the dark earths have led to the recent formation of a multi-disciplinary and multi-institutional research group under the overall rubric of the *Terra Preta Nova* Project; with participation of the geographers mentioned above, in association with soil scientists, archaeologists, biologists, and anthropologists from South and North America and Europe (Sombroek et al. 2002). The primary goal of this group is to recreate the highly sustainable soil enhancement practices of the past and to enable smallholder farmers to improve their standard of living through more productive and environmentally friendly crop production; not only in Amazonia, but throughout the humid tropics.

## REFERENCES

Denevan, W. M. 2001. *Cultivated Landscapes of Native Amazonia and the Andes.* New York: Oxford University Press.

Glaser, B., L. Haumaier, G. Guggenberger, and W. Zech. 2001. The 'Terra Preta' Phenomenon: A Model for Sustainable Agriculture in the Humid Tropics. *Naturwissenschaften* 88: 37–41.

Hecht, S. B., and D. A. Posey. 1989. Preliminary Results on Soil Management Techniques of the Kayapó Indians. *Advances in Economic Botany* 7: 174–188.

Heckenberger, M. J., J. B. Petersen, and E. G. Neves. 1999. Village Size and Permanence in Amazonia: Two Archaeological Examples from Brazil. *Latin American Antiquity* 10: 353–376.

Mann, C. C. 2002. The Real Dirt on Rainforest Fertility. *Science* 297: 920–923.

Meggers, B. J. 1996. *Amazonia: Man and Culture in a Counterfeit Paradise*, Revised edition. Washington, D.C.: Smithsonian Institution Press.

Smith, H. H. 1879. *Brazil: The Amazons and the Coast.* New York: Charles Scribner's Sons.

Smith, N. J. H. 1980. Anthrosols and Human Carrying Capacity in Amazonia. *Annals of the Association of American Geographers* 70: 553–566.

Sombroek, W. G., D. C. Kern, T. Rodrigues, M. S. Cravo, T. Jarbas, W. I. Woods, and B. Glaser. 2002. Terra Preta and Terra Mulata: Pre-Columbian Amazon Kitchen Middens and Agricultural Fields, Their Sustainability, and Their Replication. Paper presented at *The 17th World Congress of Soil Science*, Bangkok, Thailand.

Woods, W. I. and J. M. McCann. 1999. The Anthropogenic Origin and Persistence of Amazonian Dark Earths. *The Yearbook of the Conference of Latin Americanist Geographers* 25: 7–14.

Woods, W. I. 1995. Comments on the Black Earths of Amazonia. In F. Andrew Schoolmaster, ed. *Papers and Proceedings of the Applied Geography Conferences*, 18: 159–165.

# 47

# TROPICAL DEFORESTATION

## DAVID L. CARR
*University of North Carolina, Chapel Hill*

> As the global crisis deepens, deforestation . . . cries
> out for the geographer's attention. Human manip-
> ulation of the plant cover, especially through
> agricultural clearing . . . is the most evident of all
> human relationships with the physical earth and is
> thus central to cultural geography (Parsons 1994).

The most extensive footprint of human habitation of the Earth's surface is inscribed on the landscape through forest conversion for agriculture, logging, and urbanization. The planet's intact old-growth forests have dwindled to approximately one-fifth of their original cover.

During recent decades, deforestation has accelerated and is now almost totally concentrated in the tropics. While not one developed nation had a positive rate of deforestation during the 1990s, only India and Vietnam (with reforestation rates of 0.1 percent and 0.5 percent respectively), experienced net afforestation among tropical nations with substantial endowments of humid forests (FAO 2001).

Estimates of land-use change over the last few millennia are crude, but research suggests that approximately 50 percent of the area of tropical forests has been removed by human influence in Africa and 40 percent in Latin America and Asia (CSAEHT 1993). As forests have dwindled worldwide, deforestation rates have accelerated (approximately 40 percent from the mid-1980s to the mid-1990s). A slightly greater percentage of forests were cleared in Asia and Africa (34–38 percent) than in Latin America (approximately 28 percent) during the last century, though a far greater absolute amount was cleared in Latin America due to the vastness of the Amazon basin (Carr and Bilsborrow 2001). The lion's share of this deforestation occurred during the twentieth century's last four decades. If tropical deforestation rates continue as during the last decade, the most biologically rich forests on the planet will be erased within 50 years.

Understanding human-environment dynamics has increasingly been recognized as a priority of global environmental change research. Yet what we know about tropical deforestation is limited by the paucity of existing data

*D. G. Janelle et al. (eds.), WorldMinds: Geographical Perspectives on 100 Problems*, 293–298.
© 2004 *Springer. Printed in the Netherlands.*

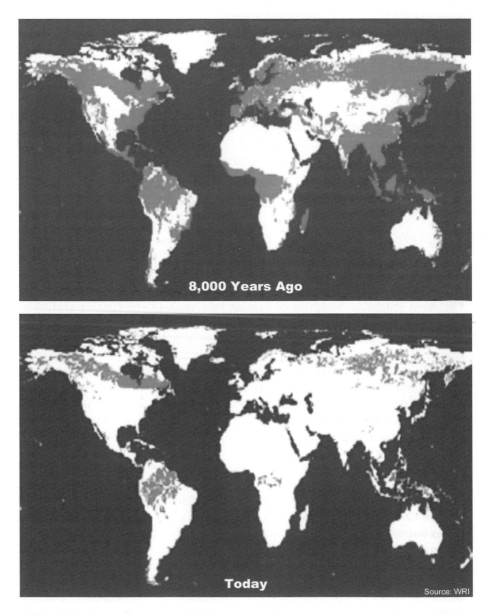

*Figure 47.1.* Frontier forests

at the household level and by poor estimates for forest cover at the macro scales. Geographers have played a key role in redressing some of these inadequacies. But research by geographers represents only about a fifth of all research on the topic (Geist and Lambdin 2001). Thus, much work remains for geographers to apply our comparative advantage linking human-environment interactions at diverse spatial scales.


## FRAMEWORKS

The great diversity of human systems of forest change carves widely varied patterns across regions and nations. Our understanding has moved beyond simple Malthusian assumptions to dynamic and complex multi-scaled causal mechanisms. Recent frameworks consider a host of proximate and underlying causes to land-use and land-cover change (e.g., Lambdin et al. 2001; Geist and Lambdin 2001). From the research on tropical deforestation explicitly categorizing proximate causes, three essential types of land use emerge: agricultural expansion, timber extraction, and infrastructure development. Literature on the underlying causes of deforestation typically identifies the following broad types of factors: demographic, socioeconomic, technological, political-economic, and environmental. To date, macro-scale analyses are based on rough estimates, and local and household studies remain largely de-coupled from regional patterns. A principal challenge ahead is linking these processes across different spatial scales and over time.


## CAUSES

A burgeoning literature has identified a host of causes thought to be driving tropical deforestation (e.g., Geist and Lambdin 2001). As mentioned above, it is increasingly evident that a concatenation of variables interacts across spatial and temporal scales (Mather et al. 1999; Turner et al. 2001). Small farmer agricultural expansion along forest frontiers is probably the primary proximate cause of forest clearing on the planet – followed by in-situ agriculture and pasture expansion, timber felling for fuel and construction, and infrastructure expansion. The last two processes often antecede frontier expansion; the first two often follow it. Underlying these proximate causes are demographic, political, economic, and environmental processes.

Some scholars have estimated that population explains half or more of the variation in worldwide deforestation patterns and indeed the great majority of deforestation case studies involve demographic dynamics at some level. However, population is never the sole cause but interacts with other proximate and underlying factors. Political economic factors also play key roles. For Example, Mather and Needle (1999) have observed that forest impacts are low in the early phase of development, accelerate during development, and again are reduced at later stages when primary resource extraction is moved to a new developing region. Politically, the wealthiest and most democratic countries usually enjoy stable or expanding forests, while poor and despotic countries tend to experience rapid forest loss.

Much of the research on tropical deforestation has been conducted in Latin America. The region harbors the greatest area of closed tropical forests in the world and over half of all freshwater on Earth. Small farmer agricultural expansion along forest frontiers has been the primary proximate cause of forest clearing in Latin America. Examples of rapid forest conversion following colonization are abundant in the literature and satellite imagery has illustrated particularly high rates of clearing adjacent to new roads. Thus, a prerequisite to frontier deforestation is road building resulting from corporate and state policies favoring certain regions for economic or geopolitical reasons.

In these remote rural environments of abundant (but often insecure) resource access and scarce labor, most forms of agricultural intensification represent an unnecessary labor burden and are uneconomical, inefficient, or too risky for small, semi-subsistence producers, leading to an extensive swidden land-use pattern. Soil nutrients are depleted in oxidized tropical soils in a matter of two to four years, encouraging farm abandonment once the swidden cycle is complete. Over time, farms are consolidated in the hands of rural elites, who take advantage of abandoned farms for raising cattle, spurring exodus among the poor, often to a subsequent forest frontier, where the deforestation cycle begins anew (e.g., Walker and Moran 2002). An important point that is neglected in the literature is that demographic, ecological, and political-economic pressures elsewhere foment migration to the frontier, and this migration process is a necessary antecedent to frontier forest clearing (Carr 2002).

Contrary to Latin America, in Africa a greater proportion of deforestation has come from the expansion of sedentary, intensive (non-frontier) agricultural expansion and fuel wood harvesting. International trade has brought indebtedness, maintaining pressures to produce for export, and has promoted increased agricultural expansion and timber extraction. Among peasant farmers, high rural population densities depending on scarce woodland for fuel have also been a major driver.

In Asia, despite a rapid fertility transition in many countries, notable deforestation continued through the 1990s due to increasing demands on timber resources and the continued migration of swidden rice farmers into formerly remote areas opened by logging roads. Today, in some of the smaller Southeast Asian nations, such as Malaysia, deforestation has claimed virtually all but a handful of forest reserves.

## IMPACTS

Deforestation may be largely confined to the tropics, but its impacts are global. Tropical forest conversion influences global biogeochemical cycles, hydrological flows, and soil degradation. Tropical forest conversion also threatens to exacerbate climate change at local and global scales. The geographic literature has highlighted spatial variation in the environmental impacts of each of these processes. For example, forest clearing is likely to warm the Earth's climate disproportionately in the tropics and subtropics (Defries and Belward 2000). Similarly, geographic research has elucidated the importance of not merely the total amount of forest change but also of how patterns of forest clearing distinctly impact physical landscapes (Walsh et al. 2001). For example, forest fragmentation can inhibit forest re-growth, decrease biodiversity, and threaten the integrity of ecological systems (Roth 1999). Temporal scales are also important. Some scholars have shown, for example, how interannual land-cover changes can be considerably more notable than long-term change in some regions (Lambin and Ehrlich 1997).

Geographers have also investigated how forest conversion threatens the rich biological integrity of tropical ecosystems (e.g., Phillips 1997). In recent years, virtually all species extinctions have occurred in this biome covering only 7 percent of the Earth's terrestrial surface. Species elimination irreparably damages the planet's biological gene pool, invaluable for the advancements of science, medicine, and food production.

A disproportionate number of global species extinctions are concentrated in those places set up to protect them, such as national parks and ecological reserves. As governments expand the area of wild lands under protection and as little unclaimed forest remains outside of these areas, protected areas represent an increasingly large proportion of unoccupied land available to migrant farm households.

## CONCLUSION

Our understanding of the processes of deforestation remains inchoate; indeed, even estimates of current tropical forest cover remain notoriously unreliable. We are learning more every day and geographers have contributed a formidable corpus of research on tropical deforestation. But given the enormity of the phenomenon to human and environmental systems, we can do much more. Practitioners in our field conduct a fraction of the research on the topic. Yet geographers are strategically positioned to pioneer future research endeavors. The multidisciplinary strengths of geographers are featured in the widely diverse research methods on the topic, including remote sensing, GIS, ecosystem processes modeling, surveys and interviews, participant observation, and stakeholder analyses. Tropical deforestation is quintessentially geographical. Forest clearing represents the most salient mark of the human ecological footprint on the Earth's surface and is inherently linked to place and space. When it comes to research on the causes and consequences of tropical deforestation – to echo UN Secretary General Kofi Annan – "the great adventure of geographic exploration is far from over" (Annan 2001).

# REFERENCES

Annan, K. 2001. Address to the 97th Annual Meeting of the Association of American Geographers. New York City.

Carr, D. L. 2002. The Role of Population Change in Land Use and Land Cover Change in Rural Latin America: Uncovering Local Processes Concealed by Macro-Level Data, In M. H. Y. Himiyama and T. Ichinose, eds. *Land Use Changes in Comparative Perspective.* Enfield, NH, and Plymouth, U.K.: Science Publishers.

Carr, D. L. and R. E. Bilsborrow. 2001. Population and Land Use/Cover Change: A Regional Comparison Between Central America and South America. *Journal of Geography Education* 43: 7–16.

Committee on Sustainable Agriculture and the Environment in the Humid Tropics (CSAEHT). 1993. *Sustainable Agriculture and the Environment in the Humid Tropics.* Washington, D.C.: National Academy Press.

Defries, R. S. and A. S. Belward. 2000. Global and Regional Land Cover Characterization From Satellite Data: An Introduction to the Special Issue. *International Journal of Remote Sensing* 21(6–7): 1083–1092.

FAO. 2001. The Global Forest Resources Assessment. 2000. *Summary Report. Report No. COFO-2001/INF.5.* Rome: Committee on Forestry, Food and Agriculture Organziation (FAO) of the United Nations.

Geist, H. J. and E. F. Lambdin. 2001. *What Drives Tropical Deforestation? A Meta-Analysis of Proximate and Underlying Causes of Deforestation Based on Sub-National Case Study Evidence.* Louvain-la-Neuve, Belgium: LUCC International Project Office.

Lambdin, E. F. and D. Ehrlich. 1997. Land-Cover Changes in Sub-Saharan Africa (1982–1991): Application of a Change Index Based on Remotely Sensed Surface Temperature and Vegetation Indices at a Continental Scale. *Remote Sensing of Environment* 61(2): 181–200.

Lambdin, E. B. et al. 2001. The Causes of Land-Use and Land-Cover Change: Moving Beyond the Myths. *Global Environmental Change Human and Policy Dimensions* 4.

Mather, A. C. et al. 1999. Environmental Kuznets Curves and Forest Trends. *Geography* 84(362): 55–65.

Parsons, J. A. 1994. Cultural Geography at Work, In K. E. Foote, P. J. Hugill, K. Mathewson, and J. M. Smith, eds. *Re-Reading Cultural Geography*, 281–288. Austin, TX: University of Texas Press.

Phillips, O. L. 1997. The Changing Ecology of Tropical Forests. *Biodiversity and Conservation* 6(2): 291–311.

Roth, L. C. 1999. Anthropogenic Change in Subtropical Dry Forest During a Century of Settlement in Jaiqui Picado, Santiago Province, Dominican Republic. *Journal of Biogeography* 26(4): 739–759.

Turner, II. B. L. et al. 2001. Deforestation in the Southern Yucatán Peninsular Region: An Integrative Approach. *Forest Ecology and Management* 154(3): 353–370.

Walker, R. E. et al. 2000. Deforestation and Cattle Ranching in the Brazilian Amazon: External Capital and Household Processes. *World Development* 28(4): 683–699.

Walsh, S. J., K. A. Crews-Meyer, T. W. Crawford, and W. F. Welsh. 2001. A Multiscale Analysis of LULC and NDVI Variation in Nang Rong District, Northeast Thailand. *Agriculture, Ecosystems, and Environment* 85(1–3): 47–64.

# 48

# DEFORESTATION OF THE ECUADORIAN AMAZON: CHARACTERIZING PATTERNS AND ASSOCIATED DRIVERS OF CHANGE

STEPHEN J. WALSH and LEO ZONN
*University of North Carolina, Chapel Hill*

JOSEPH P. MESSINA
*Michigan State University*

Beginning in the mid-1970s, spontaneous colonization occurred in the Oriente region of the northeastern Ecuadorian Amazon (Figure 48.1), made possible on petroleum industry roads that increased the geographic accessibility of land to migrants. Settlers seeking access to land began transforming primary and secondary forest to agricultural crops and pasture on rectangular household farms, or *fincas*, measuring approximately 50 ha in size and organized in a "piano-key" pattern. In addition to individual farm dwellings, clusters of dwellings also evolved around major road intersections and petroleum encampments, and sometimes grew into market towns. Lago Agrio is the largest city in the region, but still with only about 30,000 people in 2000.

Population dynamics, migration patterns, socio-economic forces, geographic accessibility, and resource endowments are among the central drivers of deforestation and agricultural extensification in the *Oriente*, a region in which household decision-making is viewed as the primary direct agent of land-use and land-cover change (LULC) (Walsh et al. 2002). But other factors function at intersecting scales as well. For example, global factors associated with commodity prices, primarily timber, and petroleum; regional issues such as the political instability in neighboring Colombia and Peru; and local factors associated with socio-economic, demographic, and biophysical characteristics of farms linked to households, as well as communities within the region, all interact in complex and scale-dependent ways to affect LULC dynamics. The usual sequence of land clearing is that settlers first convert small areas to food crops for subsistence, then cash crops, and finally pasture as soil fertility declines. Eventually, settlers may abandon land and move on to start new farms, as is common in other parts of the Amazon. Farmers in the *Oriente* deforest new patches within their farms as soil fertility declines, other economic opportunities occur, or land is sub-divided through kinship ties

*D. G. Janelle et al. (eds.), WorldMinds: Geographical Perspectives on 100 Problems*, 299–304.
© 2004 *Springer. Printed in the Netherlands.*

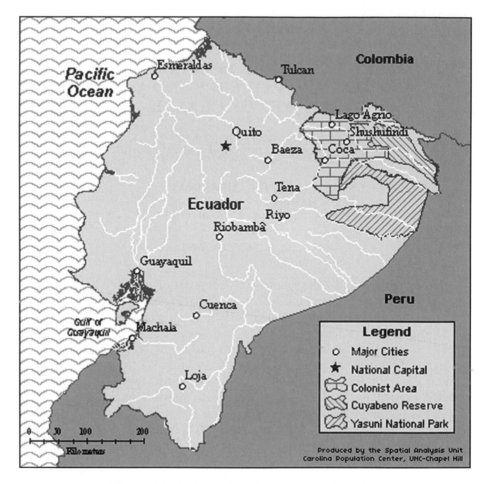

*Figure 48.1.* Study area location, northeastern Ecuador.

and formal land sales. While the Ecuadorian government facilitated the provision of permanent or provisional land titles to in-migrant families to
encourage settlement in the 1970s and 1980s, most new arrivals and children
inheriting plots do not have legal land titles, because of the change in the
authority of the now-responsible government agency. The resulting increased
insecurity of land titles may be influencing subsequent land-use practices in
new ways, as well as the permanence of migrant settlement.

## RATIONALE

Concern for tropical forests is well justified because of the global anxiety
for the fate of forests, including their biological diversity, carbon assimila-

tion rates, and sequestration patterns, and the important global ecological services provided by these ecosystems. In the *Oriente*, several areas of extraordinary biodiversity are represented, making it a priority conservation zone. Indeed, the region is considered one of the eleven ecological "hot spots" in the world (Meyers et al. 2000), so the ongoing deforestation has important implications for biodiversity, resilience, and sustainability. In light of this ecological sensitivity and the in-migration and deforestation patterns that are occurring, research is underway to examine: (1) the rates, patterns, and mechanisms of forest conversion to agricultural and urban uses; (2) the relative importance of exogenous and endogenous variables on these land uses; (3) the associated scale-dependent drivers of LULC dynamics and patterns operating across socio-economic, demographic, biophysical, and geographical domains; and (4) plausible scenarios of future land-cover change and their policy implications, as assessed through multi-level models and spatial simulations of LULC dynamics through cellular-automata approaches.

A pervasive challenge facing the population-environment community is to link people to the land so that social behavior can be studied in relation to changes in LULC, landscape conditions, and associated ecosystem processes (Fox et al. 2002). When the role of human agency is considered relative to multi-scale decision-making and LULC patterns and dynamics, we can consider the likely future fate of forests and the plausible feedbacks between people, place, and environment. This research examines the scale, pattern, and process relationships affecting LULC patterns, the fluxes in LULC, as well as the associated drivers of change in an important frontier environment, the headwaters of the Amazon Basin. The problem is one of assessing the pattern of LULC dynamics and in considering deforestation as an outgrowth of multidimensional and scale-dependent factors. Mapping and modeling of deforestation, agricultural extensification and intensification, and urbanization are critical to LULC sustainability, particularly so, in an ecologically sensitive environment such as the *Oriente*.

## THEORETICAL CONTEXT

Drawing upon a growing body of population-environment literature (Bilsborrow 1987), the effects of the growing frontier population on the local environment may be seen as occurring through a number of contextual or mediating factors representing political, human, and landscape ecology theory, as well as complexity theory. Political ecology is comprised of factors and forces that are imposed upon a system as context, and include socio-cultural norms (such as the drive to raise cattle and gender roles), the standard of living and technology, local and national government policies, commodity prices at the local to global scales, and the physical infrastructure. Human ecology argues that people are active agents on the landscape that shape and are shaped by it. Activities, such as the gathering of fuel wood or the out-migration of young adults because of excessive land fragmentation, are examples. Landscape ecology points to the interactions between landscape composition, spatial organization, and time as factors affecting human behavior (Walsh et al. 1999). Complex systems are systems that contain more possibilities than can be actualized. The goal of complexity theory is to understand how simple, fundamental processes can combine to produce complex holistic systems (Gell-Mann 1994).

## DATA AND METHODS

Using longitudinal household survey data collected in 1990 and 1999, a 2000 community survey, a multi-resolution remote sensing time-series, GIS coverages of resource potentials and endowments, and field verification, we analyze the determinants of change in LULC across a range of space-time scales. The central analytical methods include the use of multi-level models and spatial simulations achieved through cellular-automata (CA) techniques. The multi-level models are used to integrate household, community, and regional variables that impact household decision-making and hence the mapped LULC patterns at the farm level. Spatial simulations are developed through CA approaches that operate at the annual and decadal scales. Linear and non-linear responses or "critical landscapes" are studied to model the ecological responses to a range of spatial patterns of LULC, derived through hypothetical, modeled, and observed conditions. The multi-level models are assessed through statistical measures of model performance, whereas the derived CA patterns are compared to the actual patterns represented in the satellite time-series and assessed through image change-detections, change trajectories or pixel histories, summary correlations, and pattern metrics for comparisons of expected versus observed patterns.

### Multi-Level Models

Multi-level models are used to examine space-time variations in household decisions. An important issue is the estimation of zero-sum outcome variables (i.e., LULC) through the specification of effects at the individual/ household and community-levels, combined with the presence of hierarchical and spatial autocorrelation effects. To assess these effects, a generalization of the Laird and Ware (1982) Generalized Linear Model (also referred to as a Multi-level Model) is used, which can account for within-area heterogeneity (i.e., farms within *fincas* and *fincas* within communities), hierarchical effects (arises from small areas being grouped into larger areas), and spatial effects (i.e., between *fincas* and between communities).

*Cellular Automata*

A cell-based morphogenetic model (or CA) of LULC dynamics is used to generate spatial simulations of regional landscape dynamics, with emphasis on deforestation, agricultural extensification, secondary plant succession, and urbanization. A CA consists of a regular grid of cells, each of which can be in one of a finite number of $k$ possible states, updated synchronously according to a local interaction rule. The state of a cell is determined by the previous states of a surrounding neighborhood of cells (Wolfram 1984). The rule contained in each cell is specified in the form of a transition function or growth rule that addresses every possible neighborhood configuration of states. The ability of a system to grow and then alter its rate of growth and possibly reverse or "die" is a fundamental attribute. The system modeled by Clarke et al. (1997) and the modeling scheme proposed by Messina and Walsh (2001) attempt to follow biological patterns and processes.

## SOME OUTCOMES

CA models suggest that land conversion along roads, nearness to market centers and communities, household demographics and socio-economic conditions, resource endowments at local sites, and geographic access are the dominant factors affecting LULC dynamics. Pattern analysis suggests landscape cycles between one dominated by forest to one dominated by patches of agriculture and pasture. Further, the simulations suggest a more homogenous landscape with time, a scenario that fits the theoretical understanding of how in-migration of farmers into existing *fincas* alters the natural landscape. Regarding urban environments, the influence of Lago Agrio, the central market town in the region, accounts for LULC dynamics through diffusive processes as urban expansion occurs and nearby lands are enveloped into the urban landscape, creation of urban farm lots (*solares*) that are suggestive of incipient urbanization, and the creation of development corridors along primary roads that connect urban growth nodes and supply centers throughout the region.

System behaviors, defined through the multi-level models and simulated and visualized through the CA models, are interpreted within a policy-relevant context by comparing LULC scenarios to targeted land-management outcomes. Multiple LULC change scenarios are developed around defined policy goals. Model convergence and variable sensitivity are examined relative to the LULC patterns, model variables, and policy goals and expectations.

The basic intent of the research is to infuse an analytical perspective and a set of methods to integrate elements of population and the environment. The problem of deforestation of tropical rainforests is pervasive, which has implication in LULC dynamics, ecology, population migration, urbanization, and land fragmentation. Geographers are poised to make important contributions through their space-time frameworks and interdisciplinary perspectives.

## REFERENCES

Bilsborrow, R. 1987. Population Pressure and Agricultural Development in Developing Countries: A Conceptual Framework and Recent Evidence. *World Development* 15(2): 183–203.

Clarke, K., L. Gaydos, and S. Hoppen. 1997. A Self-Modifying Cellular Automaton Model of Historical Urbanization in the San Francisco Bay Area. *Environment and Planning B* 23: 247–261.

Fox, J., R. Rindfuss, S. Walsh, and V. Mishra, eds. 2002. *People and the Environment: Approaches for Linking Household and Community Surveys to Remote Sensing and GIS.* Boston: Kluwer Academic Publishers.

Gell-Mann, M. 1994. *The Quark and the Jaguar.* New York: Freeman.

Laird, N., and J. Ware. 1982. Random-Effects Models for Longitudinal Data. *Biometrics* 38: 963–974.

Messina, J. and S. Walsh. 2001. 2.5D Morphogenesis: Modeling Landuse and Landcover Dynamics in the Ecuadorian Amazon. *Plant Ecology* 156: 75–88.

Myers, N., R. Mittermeier, G. da Fonseca, and J. Kent. 2000. Biodiversity Hotspots for Conservation Priorities. *Nature* 403: 853–858.

Walsh, S., T. Evans, W. Welsh, B. Entwisle, and R. Rindfuss. 1999. Scale-Dependent Relationships Between Population and Environment in Northeastern Thailand. *Photogrammetric Engineering and Remote Sensing* 65: 97–105.

Walsh, S., J. Messina, K. Crews-Meyer, R. Bilsborrow, and W. Pan. 2002. Characterizing and Modeling Patterns of Deforestation and Agricultural Extensification in the Ecuadorian Amazon. In S. Walsh and K. Crews-Meyer, eds. *Linking People, Place, and Policy: A GIScience Approach*, 187–214. Boston: Kluwer Academic Publishers.

Wolfram, S. 1984. Cellular Automata as Models of Complexity. *Nature* 311: 419–424.

# 49

# FOREST DEGRADATION AND FRAGMENTATION WITHIN CELAQUE NATIONAL PARK, HONDURAS

JANE SOUTHWORTH
*University of Florida*

DARLA MUNROE
*University of North Carolina, Charlotte*

HARINI NAGENDRA and CATHERINE TUCKER
*Indiana University, Bloomington*

Finding effective means to halt or slow deforestation in natural areas is of great concern throughout the world. In the past few decades, the increase in the number of protected areas in Latin America has been particularly dramatic (Bates and Rudel 2000). Indeed, the tropical rain forests located within these protected areas have increased from less than 200,000 km$^2$ in 1965, to over 1,100,000 km$^2$ in 1990 (Harcourt and Sayer 1996). We chose one of these protected areas, Celaque National Park in Honduras, to conduct an empirical case study on the effectiveness of park establishment in halting deforestation. Established in 1987, Celaque National Park encompasses ~266 km$^2$. Much controversy exists regarding parks as management regimes (Bruner et al. 2001). Although the effects of protected-area establishment on local livelihoods is well documented, there are few case studies that examine whether parks experience different rates and patterns of land-cover change compared to non-protected areas. Geographers have the perfect suite of tools to undertake such studies.

The area where Celaque National Park was established faces a high demand for land; it typifies the conflict in many developing nations between land for development and conservation. The park contains the highest mountain peak in Honduras; but it has been little studied due largely to its inaccessibility. Prior research has explored land-use changes in this region, which helps inform our interpretations of land-cover change (Southworth et al. 2002; Munroe et al. 2002; Nagendra et al. 2002). The park was established relatively recently and so we are able to use satellite imagery of constant spatial, spectral and radiometric resolution (Landsat TM imagery) to examine whether landscape fragmentation and land-cover change differs within and outside park boundaries since the park's creation, and thus gain fundamental information to

*D. G. Janelle et al. (eds.), WorldMinds: Geographical Perspectives on 100 Problems, 305–310.*
© 2004 *Springer. Printed in the Netherlands.*

evaluate park effectiveness. Most protected areas are ecosystem remnants of a limited size. Few, if any, represent intact ecosystems; therefore it has become increasingly important to locate each protected area as a functional component of a larger landscape (Mladenoff et al. 1993). Remotely sensed data are particularly effective for such geographic studies and are the most frequently used technique for the mapping of tropical forest change. Data from several time points allow the creation of land-cover maps over greater spatial extents and more frequent time steps than is possible with expensive field studies (Nagendra 2001).

## MATERIALS AND METHODS

The digitized park boundary and core zone (> 1,800 m elevation) were obtained from the Friends of Celaque (FOC) research group (Figure 49.1a). The park boundary provides a protective area around the core except for a region on the west side, where a settlement exists. We created a GIS buffer region extending 5 km beyond the outer-park boundary to allow comparison of land-cover change within the park boundary and park core to changes in the surrounding landscape. Coverages of roads, towns, elevation, and slope were also created.

Landsat TM images were obtained for 1987 (park created), 1991, 1996, and 2000 in March, the end of the dry season when fallow agricultural lands can be easily distinguished from forests. The images were registered to each other using geometric rectification, with an RMSE (Root Mean Square Error) of less than 0.5. Following rectification, the images underwent calibration procedures (radiometric calibration and atmospheric correction) to correct for sensor drift and variations in the solar angle and atmospheric conditions.

Training sample data were used to determine the land-cover classes on the ground and then train a supervised classification of the satellite image, generating two cover classes to simplify the change analysis. Classes for agriculture, young fallows (approximately 1–3 years), cleared areas, bare soil, water, and urban areas were aggregated in this analysis to create a non-forest class. Forest was defined as having a canopy closure of 25 percent or greater, based on results of forest mensuration. Land-cover maps for each date were derived by independent supervised classification of the Landsat images, using a Gaussian maximum-likelihood classifier (Table 49.1).

To detect change post-park creation, following classification, change detection analysis was undertaken for the years 1991, 1996, and 2000. Change detection is a technique used to determine the spatial change between two or more time periods or for a particular land-cover. It offers an important tool for monitoring and managing natural resources (Macleod and Congalton 1998). Changes in land-cover (forest and nonforest) across the latter three dates result in eight possible classes (Table 49.2). The three separate images are

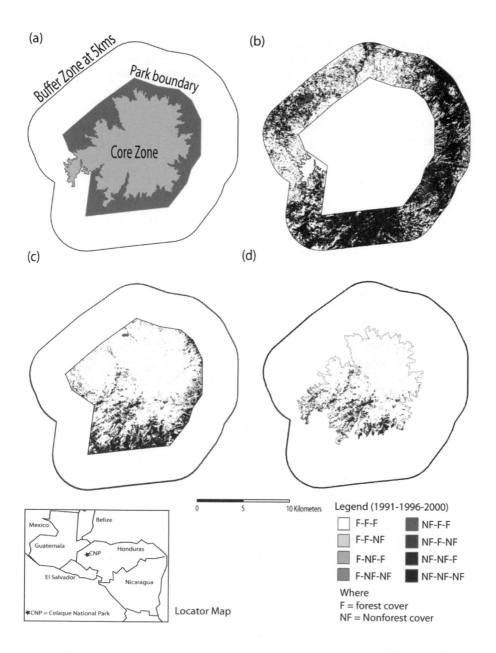

*Figure 49.1.*  Celaque National Park location and land-cover change images for the years 1991–1996–2000. (a) Park core, park boundary and surrounding buffer region used in the analysis. (b) Land-cover change within the buffer region. (c) Land-cover change within the park boundary. (d) Land-cover change within the park core zone.

*Table 49.1.* Extent of Forested and Nonforested Area (km$^2$) of Celaque National Park, for the Area Within the Core, the Boundary, and the Surrounding Landscape, Assessed Using Landsat TM Satellite Imagery from 1987, 1991, 1996, and 2000.

| Year | Class Name | Core Zone (km$^2$) | Park Boundary (km$^2$) | Surrounding Landscape (km$^2$) |
|------|-----------|------|------|------|
| 1987 | Forest | 144.75 | 217.07 | 198.95 |
|      | Nonforest | 20.79 | 46.68 | 197.79 |
| 1991 | Forest | 149.64 | 224.12 | 188.74 |
|      | Nonforest | 15.91 | 41.60 | 208.01 |
| 1996 | Forest | 144.55 | 218.44 | 199.42 |
|      | Nonforest | 20.99 | 47.31 | 197.33 |
| 2000 | Forest | 143.82 | 203.99 | 148.98 |
|      | Nonforest | 21.71 | 61.73 | 247.76 |

*Table 49.2.* Land-cover Change from 1991–1996–2000 for the Celaque National Park Core, Boundary, and Surrounding Buffer Region.

|  | f-f-f | f-f-nf | f-nf-f | f-nf-nf | nf-f-f | nf-f-nf | nf-nf-f | nf-nf-nf |
|--|-------|--------|--------|---------|--------|---------|---------|----------|
| *(a) in km$^2$* | | | | | | | | |
| Core | 133.60 | 5.98 | 5.97 | 4.08 | 2.40 | 2.59 | 1.95 | 8.97 |
| Park | 190.01 | 17.59 | 8.12 | 8.40 | 3.66 | 7.16 | 2.27 | 28.50 |
| Buffer | 111.11 | 41.94 | 14.28 | 21.37 | 15.34 | 30.99 | 8.22 | 153.50 |
| *(b) in %* | | | | | | | | |
| Core | 80.70 | 3.61 | 3.61 | 2.47 | 1.45 | 1.56 | 1.18 | 5.42 |
| Park | 71.51 | 6.62 | 3.06 | 3.16 | 1.38 | 2.70 | 0.85 | 10.73 |
| Buffer | 28.01 | 10.57 | 3.60 | 5.39 | 3.87 | 7.81 | 2.07 | 38.69 |

Where: f-f-f =Stable forest; f-f-nf = Recent forest clearing; f-nf-f = Old forest clearing with regrowth; f-nf-nf = Older, more permanent forest clearing; nf-f-f = Older, more permanent forest regrowth; nf-f-nf = Forest regrowth with new clearing; nf-nf-f = Recent forest regrowth; nf-nf-nf = Stable agriculture.

summed together to create one single image. In this "change image" each pixel (cell of information) now includes information on land-cover for all three dates (Table 49.2, Figures 49.1b–d).

## RESULTS

*Extent of Land-Cover Change*

Overall classification accuracies were above 85 percent, with kappa statistics greater than 0.75, for all dates. As depicted in Table 49.1, most of the area in the core, and to a lesser extent within the boundary, is forested for all dates. In contrast, nearly half of the area in the surrounding landscape is non-forested. In the core and boundary, the area under forest cover remained fairly static across all four dates of analysis. The core lost less than 1 km$^2$ of forest cover between 1987 and 2000, while the boundary lost about 13 km$^2$. In contrast, the forested area in the surrounding landscape remained fairly stable between 1987 and 1996, but then decreased sharply between 1996 and 2000.

*Land-Cover-Change Image*

The three-date land-cover change reveals the land-cover trajectories for every pixel on the landscape. While the absolute amount of forest or nonforest (Table 49.1) does remain constant within the park core, there is change occurring (Table 49.2). If the amount of reforestation and deforestation equal out however, the static land-cover change shown in Table 49.1 will make it appear as if no change has occurred. Through the use of change images we can see the trajectories of change much more clearly. Roughly 15 percent of the landscape within the core area changed land-cover types during the study period (Table 49.2b).

While the park core appears to be maintaining forest cover quite well (Table 49.1, Figure 49.1d), there is infringement into this region, which could indicate future problems. Analysis of the park boundary (20% change across the study period) and buffer region (30% change) finds that the changes within the core area are less than those within the park boundary (Table 49.1, Figure 49.1c), and the park experiences less change overall than regions outside the park (Table 49.1, Figure 49.1b). Therefore, we conclude that while the park is not entirely preventing land-cover changes within its boundaries, it appears to limit them.

DISCUSSION

Our results from the remote sensing and GIS analyses indicate there is increasing pressure on the park region, with growing deforestation in the 5-km landscape surrounding the park. Nevertheless, the area within the park boundary, particularly inside the core, appears to be maintaining forest cover. However, agricultural extensification and increasing conversion to mountain-grown coffee are placing pressure on the park, as also indicated by the change in spatial pattern within the boundary (Figure 49.1b).

It must be noted that analysis of remotely sensed data requires fieldwork to interpret human activities and incentives that relate to land-cover change. In short, our data are robust in showing that the park has been comparatively stable in forest cover as compared to the surrounding landscape. It does not indicate why the park has been effective; the steep slopes and complex topography and lack of roads within the core area of the park may constitute more important factors than park boundaries in preserving forest cover. To ensure park survival, there is a need to involve local inhabitants with conservation efforts, rather than simply excluding them from the park. Currently, the boundaries of the park are not well known by the population that lives in and around the park, and even the park rangers are not well informed about the protected areas of Honduras, or the flora and fauna within this region (Fonseca et al. 1999). While local non-government organizations are working with local populations, their efforts need to be supplemented by Honduran government agencies. The study of relationships across such boundaries is an integral component of geography, a discipline in which the human environment and the physical environment are studied interactively. This type of research, the methods used, and the tools it incorporates provide an example of geography's strengths, at the center of such interdisciplinary research.

## ACKNOWLEDGEMENTS

This research was supported by the National Science Foundation (NSF) (SBR-9521918) as part of the ongoing research at the Center for the Study of Institutions, Population, and Environmental Change (CIPEC) at Indiana University.

## REFERENCES

Bates, D., and T. K. Rudel. 2000. The Political Ecology of Conserving Tropical Rain Forests: A Cross-National Analysis. *Society and Natural Resources* 12: 619–634.

Bruner, A. G., Gullison, R. E. Rice, R. E. Gustavo and A. B. da Fonseca. 2001. Effectiveness of Parks in Protecting Tropical Biodiversity. *Science* 291: 125–128.

Fonseca, J. P., M. L. Moreno, and G. S. Padgett. 1999. *Estructura florìstica, uso de recursos y educación ambiental del parque nacional Montaña de Celaque.* Tesis Licenciatura en Biologia, Universidad Nacional Autónoma de Honduras, May 1999, p. 104.

Harcourt, C. S. and J. Sayer. 1996. *The Conservation Atlas of Tropical Forests: The Americas.* New York: Simon and Schuster.

Macleod, R. D. and R. G. Congalton. 1998. A Quantitative Comparison of Change-Detection Algorithms for Monitoring Eelgrass From Remotely Sensed Data. *Photogrammetric Engineering and Remote Sensing* 64: 207–216.

Mladenoff, D. J., M. A. White, and J. Pastor. 1993. Comparing Spatial Patterns in Unaltered Old-Growth and Disturbed Forest Landscapes. *Ecological Applications* 3: 294–306.

Munroe, D. K., J. Southworth, and C. M. Tucker, 2002. The Dynamics of Land-Cover Change in Western Honduras: Exploring Spatial and Temporal Complexity. *Agricultural Economics* 27(3).

Nagendra, H., J. Southworth, and C. M. Tucker. Accessibility as a Determinant of Landscape Transformation in Western Honduras: Linking Pattern and Process. Forthcoming in *Landscape Ecology.*

Nagendra, H. 2001. Using Remote Sensing to Assess Biodiversity. *International Journal of Remote Sensing* 22: 2377–2400.

Southworth, J., H. Nagendra, and C. M. Tucker. 2002. Fragmentation of a Landscape: Incorporating Landscape Metrics into Satellite Analyses of Land-cover Change. *Landscape Research* 27: 253–269.

# 50

# HUMAN INTERACTIONS WITH ECOSYSTEM PROCESSES: CAUSES OF ASPEN DECLINE IN THE INTERMOUNTAIN WEST

AMY E. HESSL

*West Virginia University*

Aspen is the most widely distributed tree species in North America and is the second most widely distributed tree species in the world. In the inter-mountain West of North America, aspen covers only a small percentage of the landscape. Despite its relative rarity, aspen is critical to the health of western landscapes. Aspen groves harbor a rich diversity of understory plants, butterflies, and birds not found in neighboring coniferous forest. Aspen also provide forage and cover for native and domestic ungulates and give us the spectacular fall vistas typical of the mountain west.

Since the 1930s, foresters and ecologists have noticed a "decline" in aspen across the west. Declining aspen stands are dominated by old-age classes, have many dead and dying stems, contain little understory regeneration, and may have coniferous trees or sagebrush invading the understory. Many hypotheses seek to explain the decline of aspen, including an overabundance of elk, fire suppression, and climatic variability. Though the total disappearance of aspen in the west seems unlikely given its wide distribution, the documented decline is alarming (National Research Council 2002). Given the value of aspen in western forests and the dramatic changes that have occurred in the west as a result of European-American settlement, we must ask: have human activities contributed to the apparent decline of aspen?

Geographers are uniquely qualified to answer questions of this nature. We have training in society and the environment, we like to attack complex inter-disciplinary problems with many confounding variables, and we approach environmental problems with an appreciation of the importance of space and time. In the case of aspen decline, human activities, such as fire suppression, grazing, hunting and predator control, have all had direct and indirect impacts on aspen forest ecosystems. Aspen are clonal organisms that produce new stems vigorously following fire, suggesting that fire suppression may have reduced opportunities for aspen to regenerate. Many grazing herbivores, both domestic and native (especially elk), eat young aspen sprouts, suggesting that management of grazing and hunting could play a role in the current

*D. G. Janelle et al. (eds.), WorldMinds: Geographical Perspectives on 100 Problems*, 311–316.
© 2004 *Springer. Printed in the Netherlands.*

"old" age structure of aspen stands. Finally, top-down trophic interactions suggest that the absence of major predators, such as wolves, may have created unusually large populations of elk that could prevent the regeneration of young-age classes of aspen. These factors, in combination with natural processes, such as climate variation and species interactions, make understanding aspen decline extremely complex.

By focusing on the spatial and temporal patterns of human activities, natural variability, and aspen forest structure, it is possible to disentangle many of the interacting factors that make aspen decline difficult to understand. Historically, fire and elk population dynamics operated over large spatial and long temporal scales, generating far reaching, pervasive impacts on regional ecosystem patterns, such as aspen forest dynamics (Figure 50.1). For example, while elk natality and mortality fluctuate on an annual basis, elk migrations change over decadal time scales. Similarly, an individual fire exists for a few days to a few months, but fire regimes (both frequency and extent) change at decadal to centennial or millennial time scales with changing climate (Millspaugh and Whitlock 1995). Aspen forest dynamics, affected by both fire regimes and elk populations, operate on annual to millennial timescales.

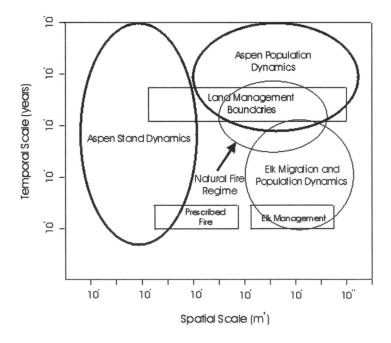

*Figure 50.1.* Spatial and temporal scales at which human institutions (boxes) and ecological processes (ellipses) operate with respect to aspen forest dynamics. Reprinted with permission from A. E. Hessl, 2002. Figure 3 in "Aspen, Elk, and Fire: Direct and Indirect Effects of Human Institutions on Ecosystem Processes." *BioScience* 52: 1011–1022. Copyright, American Institute of Biological Sciences.

Stem regeneration occurs annually, stand structure changes over decades, and recruitment of new aspen clones may take thousands of years. In contrast, 20th century human institutions that influence these ecosystem processes, such as prescribed burning and regulated hunting, have operated on short temporal and small spatial scales defined by agency policies and land ownership. For example, elk are managed by multiple agencies dealing with land areas ranging from the size of the National Elk Refuge, Wyoming (100 km$^2$) to Yellowstone National Park (8,900 km$^2$). The number of elk fed or hunted may change on an annual basis and may vary with land ownership. Similarly, prescribed fires may be planned on annual to decadal time scales, but to date many prescribed fires are less than 10,000 ha. These mismatches between human institutions and ecological processes manifest themselves as ineffective management policies and ecological surprises (Folke et al. 1998). The purpose of this paper is to describe the ways in which a geographical approach, focusing on the temporal and spatial scale of aspen dynamics, contributes to our understanding of a complex ecological problem: aspen decline.

## METHODS

I used tree-ring methods to date the history of aspen stem regeneration in the southern portion of the Greater Yellowstone Ecosystem (Hessl and Graumlich 2002), where aspen decline has been observed since the 1950s. Using age structures of aspen stands (including > 700 trees), I compared major episodes of recruitment at decadal time scales (1820–2000) with data on fire history, drought (Cook et al. 1999; Pisaric and Graumlich 2002), and elk population history. I also used published and unpublished sources that document the history of human institutions and their impacts on ecological processes.

## RESULTS

Aspen stem regeneration has occurred consistently for all plots but at low frequencies since 1830, with three peaks of regeneration: 1860–1885, 1915–1940, and 1955–1990 (Hessl and Graumlich 2002) (Figure 50.2). However, the history of stem recruitment varies spatially across different elk management areas. Plots with high elk use show a pulse of establishment during the 1870s and 1880s that is probably the result of fires during that period. Plots adjacent to elk feed grounds show a small pulse of regeneration during the 1920s and 1930s that is not clearly expressed in other areas. This event may be the result of elk feeding that began in 1912, reducing the effect of elk on aspen by concentrating herds near feed grounds. A recent period of recruitment in the 1990s is present in all levels of elk use, though this period

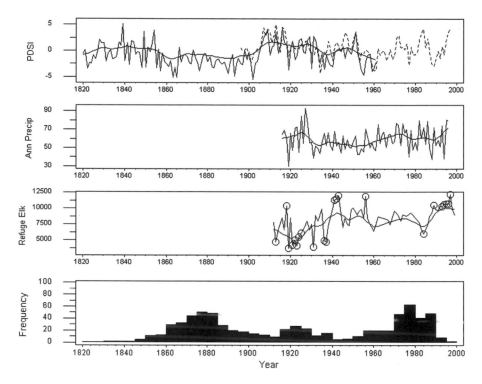

*Figure 50.2.* Raw data and loess-smoothed curves. (a) Douglas-fir tree-ring reconstruction of PDSI from Wind River, Wyoming, 1820–1997 (solid line) and observed PDSI (dashed line); (b) annual precipitation, 1915–1997; and (c) estimated elk population for the National Elk Refuge, 1912–1997. Estimates of elk population highs and lows are noted with black circles (population estimate > 1 standard deviation above or below the series mean); (d) Combined age structure of aspen stem regeneration frequency per five-year period (black bars) 1830–1997.

is most accentuated in the plots with low elk use, where prescribed burns have generated a pulse of regeneration that has not been browsed heavily by elk.

Aspen regeneration also varies over time and may be associated with changing numbers of elk and fire events. Periods of high aspen stem regeneration coincided with low to moderate elk populations, and aspen regenerated only sporadically when elk populations were high. Aspen regeneration was at its highest sustained level in the late 1800s when elk populations were low due to unregulated hunting. Aspen regeneration reached its lowest observed levels across all stands between 1940 and 1955, when elk population estimates were at a peak, equaled only by elk estimates in the 1990s. Long-term fire history indicates that fire frequency was higher during the period between 1840 and 1879 than during the 20th century when no fires greater than 16 hectares are known to have occurred in the study area. Forty percent of the mature aspen stems and 21 of the 29 dated cohorts of aspen regenerated during this period.

Interestingly, recent prescribed fires set to generate a flush of aspen recruitment had no effect on sprout density because of heavy elk browsing on post-fire stems. Drought appears to have little or no relationship to the temporal pattern of aspen stem regeneration.

## DISCUSSION AND CONCLUSIONS

The two major factors affecting aspen-stand age structures, elk-population dynamics and fire history, are associated either directly or indirectly with changes in human land use during the 19th and 20th century. Unregulated market hunting of elk in the mid- to late-19th century led to a reduction in elk populations and may have altered browsing patterns. Fire exclusion in the 20th century created an almost complete absence of fire and may have reduced opportunities for aspen regeneration. Elk feeding during the 20th century may have kept elk populations high and affected the distribution and behavior of elk in the study area. As a result, management efforts to regenerate aspen using prescribed fire have failed due to high elk numbers.

The results of this work demonstrate the importance of comparative temporal and spatial analysis for understanding both ecological relationships and human-environment relations (Hessl 2002). Human activities, such as the elimination of predators, fire suppression, ungulate feeding, and hunting have reduced variability in ecosystem processes (i.e., fire history and elk browsing patterns), have indirectly altered aspen regeneration patterns, and have potentially uncoupled complex relationships between predators, elk, fire, and aspen. These influences have created a landscape with a complex history of human impact that can only be understood through a geographical approach. Identifying the connections between historical land uses and their associated ecological consequences represents the value of a geographical approach to solving complex ecological problems. By integrating studies of human land use with ecology, geographical studies explore the many factors that conditioned ecosystem structure, function, and composition in the past and suggest which factors may be important in the future.

## REFERENCES

Cook, E. R., D. M. Meko, D. W. Stahle, and M. K. Cleaveland. 1999. Drought Reconstructions for the Continental United States. *Journal of Climate* 12: 1145–1162. (Data archived at the World Data Center-A for Paleoclimatology, Boulder, Colorado.)

Folke, C., L. Pritchard, F. Berkes, J. Colding, and U. Svedin. 1998. The Problem of Fit Between Ecosystems and Institutions. *IHDP Working Paper No. 2*. International Human Dimensions Programme on Global Environmental Change (IHDP), Bonn, Germany.

Hessl, A. E. 2002. Aspen, Elk, and Fire: Direct and Indirect Effects of Human Institutions on Ecosystem Processes. *BioScience* 52: 1011–1022.

Hessl, A. E., and L. J. Graumlich. 2002. Aspen: Ecological Processes and Management Eras in Northwestern Wyoming, 1807–1998. *Journal of Biogeography* 29: 889–902.

Millspaugh, S. H., and C. W. Whitlock. 1995. A 750-Year Fire History Based on Lake Sediment Records in Central Yellowstone National Park, U.S.A. *The Holocene* 5: 283–292.

National Research Council. 2002. *Ecological dynamics on Yellowstone's Northern range. Committee on Ungulate Management in Yellowstone National Park*, Board on Environmental Studies and Toxicology, Division of Earth and Life Sciences.

Pisaric, M. and L. Graumlich. 2002. The Frequency and Intensity of Droughts in the Greater Yellowstone Ecosystem During the Past 827 Years Reconstructed from Douglas-Fir Tree-Ring Chronologies. Poster 19th Annual Paclim Workshop, Pacific Grove, California.

# 51

# THE COLORADO RIVER DELTA OF MEXICO: ENDANGERED SPECIES REFUGE

JOHN ALL

*University of Arizona*

Ecological resources are often a source of competition and conflict between bordering countries. Ecosystem boundaries do not follow legal borders, and it is often difficult for stakeholders to view them holistically and to develop policies that are mutually beneficial with minimal negative impact on the natural resource. National competition to fully exploit shared resources often results in ecosystem integrity being compromised. Neighboring countries share equally in the burden of damaged ecosystems, even if the exploitation and benefits have not been equivalent. This practice may have long-term negative consequences for the natural environment and its ability to support economic activities. Legal instruments are currently inadequate to deal directly with these issues and piecemeal negotiation results. Holistic geographic analysis offers insight and potential solutions for these multidisciplinary problems.

This case study of international water use and shared freshwater resources focuses on the Colorado River and examines how upstream diversions have led to the degradation of the lower Colorado River Delta and upper Gulf of California in Mexico. Geographic analysis is assisting in the search for ways to remediate this area. Use of Colorado River water in agriculture and cities has greatly diminished freshwater flow into the Colorado River Delta since the completion of Glen Canyon Dam in 1968. Simultaneously, socio-economic changes related the use of natural resources have occurred in Mexico. These changes have negatively impacted the Delta and Upper Gulf, and alternative ways to support the ecosystem are being sought (Environmental Defense Fund 2000). Time will tell what long-term ramifications these events may have on the ecosystem, its inhabitants, and the people who depend directly or indirectly on the Colorado River for sustenance.

## ISSUES

The conservation group American Rivers declared the Mexican section of the Colorado River one of the ten most endangered stream reaches in North America in 1998 due to proposals in Arizona and Nevada to create programs

*D. G. Janelle et al. (eds.), WorldMinds: Geographical Perspectives on 100 Problems*, 317–320.
© 2004 *Springer. Printed in the Netherlands.*

that would impound excess flows during ENSO (El Nino/Southern Oscillation) floods (American Rivers 1998). The Secretary of the Interior is authorized to approve state storage off-site of Colorado River flows during flood years. Nevada and Arizona want to use these regulations to recharge groundwater aquifers, while California needs this water to continue its current usage as Upper Basin States begin to increase their demands for water. These actions will curtail large-scale flooding along the Colorado River basin caused by climate variability. This will mean that Mexico will receive little or no excess above the 1.5 million acre-feet required under a 1944 treaty with the United States dividing the flow of the Colorado River (Pontius 1997). The loss of the excess floodwaters would effectively eliminate any future freshwater flows to the Colorado's Delta in Mexico and the Gulf of California (cf. Snape et al. 1999). Any consideration of the Colorado River Delta in these maneuverings has been curtailed, partially due to the lack of adequate information. Geography offers the holistic research vision needed in this region.

The Colorado River Delta and Upper Gulf of California are home to a number of U.S. endangered and threatened species, including the vaquita (*Phocoena sinus*), the most endangered aquatic mammal in the world (Bureau of Reclamation 1996). The Delta ecosystem slowly died during the decades after Colorado River water diversions to the Imperial Valley in the United States began around the beginning of the 20th century. From an original area of 1.2 million acres, the Delta shrank to under 5,000 acres – this mainly watered from agricultural run-off. Beginning with the ENSO-driven floods in 1978 and 1983, freshwater from the Colorado began to flow back into the Delta. The ecosystem responded strongly as new cohorts of trees and other vegetation were established (Glenn et al. 1999). Geographers are analyzing the re-invigorated endangered-species habitat using remote sensing in conjunction with ground efforts.

The creation of a Biosphere Reserve around the Delta and northern end of the Gulf by Mexico in 1993 is evidence of a commitment by Mexico toward biodiversity and of her growing desire to protect the resource (See McGuire and Valdez-Gardea 1997 for a full discussion). But local management efforts are hampered by a lack of freshwater flows from the Colorado, due to management decisions made up-river by both governments. Protection of rare species could create potential for eco-tourism in some areas, but this is dwarfed under the relentless demand by agricultural and municipal groups to manage the river to serve their interests. There have been proposals for year-round flows to protect wildlife and fulfill Endangered Species Act (ESA) requirements (Wilson 1994). Including Mexico in the planning process might enhance cooperative efforts toward conservation of habitat in the Delta and Gulf. International cooperation might improve the potential for developing eco-tourism and for protecting of wildlife in the region.

## SOLUTIONS

The "Law of the River," a collection of federal and state statutes, Supreme Court cases, contracts, and treaties with Mexico developed over the last 100 years, governs the use of the Colorado River. The "Law of the River" and its focus on maximizing the "beneficial use" of the River's waters for human usage places little or no value on ecosystem conservation or enhancement. The ESA offers the best legal option for increasing movement of freshwater resources into the Delta and Gulf because it will force a re-ordering of beneficial use (Postel et al. 1998). Geographers are showing the benefits of freshwater flow for the ecosystem and endangered-species habitat in an effort to support ESA litigation. The ESA can be enforced in Mexico if the Mexican government does not interfere with flood flows or any other water releases intended to benefit endangered-species habitat. In addition the potential value of eco-tourism to replace extractive industry in the local economy is being examined using a variety of geographic field methods.

Local shrimp fishermen and some researchers have claimed that the resumption of Colorado water flow during the 1983 flood restored shrimp fisheries for a period. Geographers are attempting to validate the impacts of freshwater flow on the upper Gulf to establish the economic importance of Colorado River flows for this ecosystem. However, analysis shows that very little water actually reaches the Gulf and so other factors, such as overfishing, may have contributed to past declines in shrimp populations during non-flood years. Thus, better enforcement of fishing regulations could enhance the ecosystem of the upper Gulf.

Ducks Unlimited, an environmental organization, has proposed to re-build a natural earthen dam in the Delta that had been destroyed by the 1983 flood (Payne et al. 1992). This dam slowed the flow of water from the upper Delta into the lower Delta and Gulf and allowed the Rio Hardy wetlands to expand greatly. Its destruction by the immense power of the 1983 floods decreased the Rio Hardy wetlands, area by two-thirds. The Ducks Unlimited proposal was rejected by Mexico because it was concluded that such channel manipulations were not in keeping with the spirit of a Biosphere Reserve but efforts are underway to prove that the benefits outweigh the costs of such action. This proposal offers an easy inexpensive way to preserve endangered-species habitat and is likely the course of future management.

## SUMMARY

Intense political machinations have resulted as cities and farms have grown and demanded increasing amounts of freshwater. This discussion delineates the impacts of climate variability on a region highly vulnerable to slight changes in river-flow volumes and shows how geographers are researching the issues to provide insight and solutions. Since the Colorado is already heavily engineered and can be managed to capture the benefits and to mitigate the damages of climatic fluctuations, an understanding of ecological impacts caused by the current channel hydrology can lead to ideas and proposals for improving the management and use of the Colorado River. Enlightened river management benefits both the ecosystem and the inhabitants of the Delta and the upper Gulf of California. Results from such research serves an international audience, addressing key issues relating to water resource management in borderland regions.

## REFERENCES

American Rivers. 1998. See http://www.amrivers.org/20-6.html
Bureau of Reclamation. 1996. *Biological Assessment, Lower Colorado River.* (Prepared for the multi-species conservation planning efforts.)
Environmental Defense Fund. 2000. *A Delta Once More: Restoring Riparian and Wetland Habitat in the Colorado River Delta.* See http://www.edf.org/pubs/Reports/Delta/pagetwo.html.
Glenn, E. P., J. Garcia, R. Tanner, C. Congdon, and D. Luecke. 1999. Status of Wetlands Supported by Agricultural Drainage Water in the Colorado River Delta, Mexico. *Horticulture Science* 34(1): 39–45.
McGuire, T. R. and G. C. Valdez-Gardea. 1997. Endangered Species and Precarious Lives in the Upper Gulf of California. *Culture & Agriculture* 19(3): 101–107.
Payne, J. M., F. A. Reid, and E. C. Gonzalez. 1992. *Feasibility Study for the Possible Enhancement of the Colorado Delta Wetlands Baja California Norte, Mexico.* Report to Ducks Unlimited Inc. and Ducks Unlimited of Mexico, Sacramento, CA.
Pontius, D. 1997. *Colorado River Basin Study Final Report.* Report to the Western Water Policy Review Advisory Commission. National Technical Information Service, Springfield, VA.
Postel, S. L., J. I. Morrison, and P. H. Gleick. 1998. Allocating Fresh Water to Aquatic Ecosystems: The Case of the Colorado River Delta. *Water International* 23: 119–125.
Snape, B., K. Gillon, D. Hogan. 1999. *Notice of Intent to Sue for Violations of the Endangered Species Act Relating to Lower Colorado River Activities.* 14 December.
Wilson, F. S. 1994. A Fish out of Water: A Proposal for International Instream Flow Rights in the Lower Colorado River. *Colorado Journal of International Environmental Law & Policy* 5: 249–272.

# 52

# GEOGRAPHICAL PERSPECTIVES FOR TACKLING PROBLEMS IN WETLANDS

CHRISTOPHER F. MEINDL
*Georgia College and State University*

Wetland environments have long been shrouded in mystery. Historically, uninformed people viewed wetlands as slimy, stinky, insect-infested places whose full potential could be realized only if they were converted to some other use. Indeed, the United States government became so exasperated by wetlands during the mid-19th century that it gave more than 60 million acres of "swamp and overflowed lands" to several states hoping they could better promote wetland drainage and development. Because wetlands were so poorly understood, however, debate soon erupted over what actually constituted a wetland. Swamps, bogs, marshes, wetlands – call them what you will – these environments have eluded understanding for centuries and only since the second half of the 20th century have we begun to unlock the secrets of wetland patterns and processes (Vileisis 1997).

Only after we began to understand the many physical processes and ecological functions of wetlands did we begin to appreciate the many values wetlands provide people (Mitsch and Gosselink 2000). For example, many wetlands are capable of providing flood control and water purification, some provide timber and other economic benefits to local communities, and many coastal wetlands provide habitat for commercially valuable species of fish and shellfish. Because wetlands occur in all 50 states, and we now know that they frequently provide many important values to people, wetlands have assumed a position near the top of the environmental agenda in the United States. Today, government agencies and environmental groups around the nation are engaged in wetland education, preservation, and restoration.

This essay examines the following issues: (1) the development and refining of wetland inventories in the United States, (2) the discovery of selected physical processes and ecological functions in wetlands that effect people, and (3) the role of geographic information technology in tackling problems associated with wetlands. Resolving problems associated with wetland environments involves work at a variety of scales. Examination of broad patterns of change in wetlands over time is an example of macro-scale work that allows policy makers and private agencies to adjust wetland preservation and restoration efforts as needed. At the same time, micro-scale or detailed investigations

*D. G. Janelle et al. (eds.), WorldMinds: Geographical Perspectives on 100 Problems, 321–324.*
© 2004 *Springer. Printed in the Netherlands.*

in small areas are necessary to uncover specific ecological processes in wetlands. The results of these studies often provide decision makers with information they can use to develop strategies to combat specific problems. One of the best ways to illustrate the usefulness of work in wetlands at different scales is to examine the development and refinement of wetland inventories.

## WETLAND INVENTORIES AND GIS

Early 20th century inventories of wetlands in the United States were crude efforts that generated only the broadest of generalizations. Even modern wetland inventories of the United States, such as those produced by the National Wetlands Inventory (in St. Petersburg, Florida) are not thorough and precise accountings of absolutely all wetlands in the nation. This macroscale effort relies upon statistical extrapolation from geographically diverse samples of thoroughly investigated landscapes. It is useful for examining national trends in wetland changes over time (Dahl 1990). Meanwhile, workers at the National Wetlands Research Center (in Lafayette, Louisiana) use sophisticated geographic information technology, such as remotely sensed data (from air photos and satellite imagery) and computer based geographic information systems (GIS), to examine a variety of wetland-related problems, such as monitoring changes in the amount of wetlands in southern Louisiana (see Barras et al. 1994). A GIS is a computer software package that allows people to store, retrieve, and analyze large quantities of geographic data from a wide variety of sources. These intermediate or meso-scale studies have limitations, but they are useful in providing information on where wetland change is most significant. Furthermore, meso-scale studies provide field personnel with important information on where to look for wetlands so that they can be delineated precisely and examined more carefully to determine the physical processes at work. The bottom line is that micro-scale work is typically very expensive; therefore, focusing scarce resources on specific problem areas is far more cost effective than engaging in intensive (and expensive) study of all wetlands at the same time (Larry Handley, personal communication).

## PHYSICAL PROCESSES AND ECOLOGICAL FUNCTIONS

Reed (1995) has conducted a series of micro-scale investigations of sediment dynamics in coastal marshes, particularly along the southern Louisiana coast, where wetland losses have reached alarming proportions in recent decades. Specifically, she has outlined complexities associated with predicting the impact of sea-level rise on the marshes of the Mississippi deltaic plain and has called for further research (at micro- and meso-scales) into the interactions between vegetation, soil, and hydrologic processes. Reed and her colleagues have also engaged in a study intended to determine if the region's hydrology can be managed to promote increased sedimentation and consequent vertical accretion in the region's marshes to combat the influence of sea-level rise (Reed et al. 1997). They concluded that managed marshes in the region were receiving insufficient inorganic sediment deposition to keep pace with current rates of sea-level rise. Water managers have since built structures intended to divert sediment-laden water from the Mississippi River to some of the marshes at risk of being drowned by sea-level rise.

Tobin (1986) has contributed to our understanding of physical processes in wetlands by reviewing claims that wetlands help reduce flooding. Although this often-repeated generalization is attractive, Tobin points out that wetland hydrology is far too complicated for land managers to simply assume the mere presence of wetlands will always reduce flooding. Wetlands are a diverse set of environments: some wetlands frequently help reduce flooding but others do not. Furthermore, even individual wetlands may produce different hydrologic responses to precipitation events, depending on the time of year, the antecedent moisture conditions in the wetland, and the duration and intensity of precipitation. Reed (1993) adds that to adequately model wetland hydrology, we still need site-specific data to develop models capable of predicting the consequences of human alterations to hydrologic systems. In other words, both Tobin and Reed emphasize the need for micro-scale studies of wetland hydrology to determine wetland function and value at specific locations.

Meanwhile, Phillips (1989) argues that wetlands provide temporary sediment storage in fluvial (river) systems, a role that becomes more pronounced in progressively larger watersheds. He contends that land management programs should consider wetlands as components of larger fluvial systems, rather than as discrete and independent units within those systems – because fluvial processes at smaller scales are linked to similar processes at much larger scales. In other words, Phillips exposes the danger of remaining narrowly focused on small portions of fluvial wetlands (particularly those "up river") because, in reality, collections of fluvial wetlands produce cumulative effects in downstream portions of larger river systems.

As suggested earlier, wetlands have long suffered from an "image problem." Until recently, the only value attributed to many wetlands was based upon their potential to be converted to some other use – and this mindset encouraged drainage and development. Raphael and Jaworski (1979) helped improve our image of wetlands by estimating the dollar values of fish, wildlife, and recreational activities along Michigan's coastal wetlands. Their meso-scale research led people all over the country to consider the "hidden" values in many wetlands, such as the economic value of hunting, fishing, and non-consumptive recreation in wetlands.

We have already seen the value of GIS in efforts to create wetland inventories and monitor changes in those inventories over time. Geographic information systems also play a crucial role in ongoing efforts to restore portions of Florida's Kissimmee River-Everglades watershed to some semblance of its pre-drainage condition. A GIS allows land managers to store and retrieve huge amounts of spatially referenced data to monitor myriad ecological variables that change over time, and to simulate the effects of alternative courses of action (Lyon and McCarthy 1995). This is important because, as Gunderson (2001) points out, the Everglades is such a complex system that managers of the restoration effort need to be able to alter their strategies as ecologists and other scientists continue to learn how the system operates and reacts to restoration activities.

## CONCLUSION

Geographic perspectives are crucial in the investigation and resolution of problems associated with wetland environments. Geographic information technologies provide large quantities of data – and the ability to store, retrieve, and analyze that data. Macro-level studies help us discover broad patterns of change in wetlands across time and space. Micro-scale approaches help us solve problems in specific wetland environments. A geographic perspective suggests we need to look at the big picture – to examine all of the puzzle pieces in an effort to explain changes in wetlands, and to explain the bewildering diversity of wetland environments – recognizing that different wetlands provide people different suites of benefits. Indeed, it is this geographic perspective – to recognize simultaneously the trees AND the forest – that puts us in a position to tackle complex and multifaceted problems in wetlands.

## REFERENCES

Barras, J. A., P. E. Bourgeois, and L. R. Handley. 1994. *Land Loss in Coastal Louisiana, 1956–1990*. Lafayette, LA: National Biological Survey, National Wetlands Research Center Open-File Report, 94-01.

Dahl, T. E. 1990. *Wetlands Losses in the United States, 1780s to 1980s*. Washington, D.C.: U.S. Department of the Interior, Fish and Wildlife Service.

Gunderson, L. H. 2001. Managing Surprising Ecosystems in Southern Florida. *Ecological Economics* 37: 371–378.

Lyon, J. G. and J. McCarthy, eds. 1995. *Wetland and Environmental Applications of GIS*. Boca Raton, FL: Lewis Publishers.

Mitsch, W. J. and J. G. Gosselink. 2000. *Wetlands* (3rd edition). New York: John Wiley.

Phillips, J. D. 1989. Fluvial Sediment Storage in Wetlands. *Water Resources Bulletin* 25: 867–873.

Raphael, C. N. and E. Jaworski. 1979. Economic Value of Fish, Wildlife, and Recreation in Michigan's Coastal Wetlands. *Coastal Zone Management Journal* 5: 181–193.

Reed, D. J. 1993. Hydrology of Temperate Wetlands. *Progress in Physical Geography* 17: 20–31.

Reed, D. J. 1995. The Response of Coastal Marshes to Sea Level Rise: Survival or Submergence? *Earth Surface Processes and Landforms* 20: 39–48.

Reed, D. J., N. DeLuca, and A. L. Foote. 1997. Effect of Hydrologic Management on Marsh Surface Sediment Deposition in Coastal Louisiana. *Estuaries* 20: 301–311.

Tobin, G. A. 1986. Modification of Peak Flood Discharges by Wetland Environments: A Review. *Geographical Perspectives* Spring (57): 6–18.

Vileisis, A. 1997. *Discovering the Unknown Landscape*. Washington, D.C.: Island Press.

# 53

# MARINE GEOGRAPHY IN SUPPORT OF "REEFS AT RISK"

DAWN J. WRIGHT

*Oregon State University*

In the landmark publication, *Reefs at Risk: A Map Based Indicator of Threats to the World's Coral Reefs* (Bryant et al. 1998) of the World Resources Institute report the following sobering facts regarding the state of the world's coral reefs:
- Fifty-eight percent of the world's reefs (including 80 percent in Southeast Asia) are potentially threatened by human activity – ranging from coastal development and destructive fishing practices to overexploitation of resources, marine pollution, and runoff from inland deforestation and farming.
- Nearly half a billion people – 8 percent of the total global population – live within 100 kilometers of a coral reef.
- At least 40 countries lack any marine protected areas for conserving their coral reef systems.
- Scientists still do not know the actual condition of the vast majority of the world's reefs. In the Pacific, for example, 90 percent of them have never been mapped or assessed.

Clearly one of the most pressing environmental problems of the age revolves around the protection of these regions, often referred to as the "rainforests of the ocean." Reefs provide us with seafood, new medicines for treating everything from colds to cancer, recreation, protection of our shorelines from the impact of waves and storms, and billions of dollars in revenues each year to local communities and national economies. Geographers are now tackling the challenges of coral reef conservation and management mainly by providing the critical tools and methodologies for their mapping and assessment (Figure 53.1). Two of these technologies include "multibeam surveying" to derive the bathymetry (i.e., underwater topography) of coral reef environments, and geographic information systems (GIS, computerized mapping, and database systems) that can be used to map and integrate bathymetry with many other kinds of scientific and resource management data for purposes such as benthic habitat characterization, and managerial decision making (e.g., which reefs need to be designated as marine protected areas?). This essay briefly reviews results of shallow-water multibeam bathymetric surveys and GIS work conducted recently in support of the Fagatele (Fohng-ah-téh-leh) Bay National Marine Sanctuary in American Samoa.

*D. G. Janelle et al. (eds.), WorldMinds: Geographical Perspectives on 100 Problems*, 325–330.

*Figure 53.1.* Marine geographers solve critical environmental problems at sea as well as on land. Scientists prepare to survey with underwater videography the recovering corals within Fagatele Bay National Marine sanctuary, as part of a National Geographic Society Sustainable Seas expedition mission in March 2001. Photograph by the author.

There are currently thirteen sites in the U.S. National Marine Sanctuary System that protect over 18,000 square miles of American coastal waters. Fagatele Bay is the smallest, remotest, and least explored of these, the only true tropical coral reef in the sanctuary system. It is located at the southwest corner of the island of Tutuila in the United States' territory of American Samoa, 2,276 miles south of Hawaii. The bay is an ancient flooded volcano, with a thriving coral and calcareous algal reef community that is rapidly recovering from an infestation of crown-of-thorns starfish that devastated the corals in the late 1970s, and from the effects of two hurricanes in the early 1990s (Birkeland et al. 1987). Although much of the coral cover has been destroyed, fish populations still thrive, particularly surgeonfish, damselfish, and angelfish (Craig 1998). In addition, the steep slopes surrounding the bay contain some of the rarest paleo-tropical rainforests in the United States (http://www.fbnms.nos.noaa.gov). One of the greatest threats currently facing Fagatele Bay, as well as much of Samoa's coastal waters, is the rapid depletion of fish stocks by the illegal use of gill netting, spearfishing, poison, and dynamite (Sauafea 2001). In addition, the sanctuary staff is concerned about the potential for algal blooms with subsequent incidents of hypoxia (extremely low dissolved oxygen in the water) due to unchecked sewage outflow "upstream" from the bay. Before initial multibeam surveys of marine geographers in April-May 2001 (Wright et al. 2002), it was largely unexplored below depths of ~30 m, with no comprehensive documentation of the plants, animals, or bathymetry. The sanctuary's primary mission of protecting the coral reef terrace and broader marine ecosystem cannot be accomplished without adequate knowledge of the deeper environment.

## SCIENTIFIC AND MANAGEMENT OBJECTIVES

Prior to the April–May 2001 mission, two self-contained underwater breathing apparatus (SCUBA) surveys reached depths of ~43 m, but both were only brief, localized "snapshots": an algal reconnaissance in 1996 (N. Daschbach 2001: personal communication), and a rapid-assessment survey for fish and coral in 1998 (Green et al. 1999). Therefore, two primary surveying objectives during the 2001 mission were to obtain: (1) complete bathymetric coverage of the ocean floor; and (2) digital video and still photography of the biological habitats and physical features below 30 m via rebreather technology. In contrast to SCUBA, where the entire breath of a diver is expelled into the surrounding water when s/he exhales (open circuit), a rebreather apparatus is able to "reuse" the oxygen left unused in each exhaled breath (closed or semi-closed circuit), resulting in greatly extended dive times that are relatively quiet (little or no bubbles produced) and with much smaller tanks. Specific research questions that guided surveying included:

- What are the physical characteristics of the sanctuary (average depth, morphology and extent of reef structures, lava flows, and sand accumulation, etc.)? And further, what organisms and habitats currently reside within the sanctuary?
- In 1996, visiting divers to the sanctuary reported unusually large fleshy algal blooms at ~43 m depth, suggesting a nutrient source in the bay that should be identified and monitored, particularly if chronically harmful. Is it likely that the nutrient source is human-induced (e.g., sewage outfall carried to the bay along prevailing westerly currents or underground seepage into the bay from a landward watershed), and where are the most appropriate sites for long-term monitoring of water quality and ocean currents?
- What are the broader implications for coral reef conservation and management (Gubbay 1995; Allison et al. 1998)? For example, which sites should be of special biologic significance (such as a no-take zone or an area exhibiting a high degree of biological diversity)?

## CURRENT SOLUTIONS VIA SEAFLOOR MAPPING AND GIS

Depth soundings were gathered by the Kongsberg-Simrad EM-3000 multibeam acoustic mapping system, contracted from the University of South Florida's College of Marine Studies, and operated from a boat owned by the Department of Marine & Wildlife Resources of the American Samoa Government. The Kongsberg-Simrad EM-3000 system fans out up to 121 acoustic beams that quickly descend to the seafloor and return, capturing depths in the range of 3–150 m at survey boat speeds of 3–12 knots.

Maps created with data from this multibeam system are of excellent quality in provided detail about geological and manmade features (i.e., shipwrecks). Figure 53.2 gives an initial picture of the physical structure of the reef platforms in Fagatele Bay, which also helped to guide the location of a deep-diving photography mission in the bay on 16 May 2001, immediately after initial bathymetric surveys had been completed. Even though the rebreather diving mission was cut short by poor weather, several new species of fish were discovered and photographed in the deepest, previously unsurveyed and unexplored portions of the bay (see also http://www2.bishopmuseum.org/PBS/samoatz01). One cannot protect what one does not know about, so the discovery of these species within the sanctuary was important indeed.

Current research includes rigorous bathymetric terrain analysis and seafloor habitat classification of the multibeam bathymetry using GIS techniques (Wright et al. 2003) to provide among the first series of deeper-water (> 30 m) benthic habitat characterization maps of coral reefs in the region. These maps are greatly needed for ongoing studies in the bay, the National Park of American Samoa, and additional areas under the purview of the American Samoa Coastal Management Program. The studies include selecting sites for habitat class designation and protection (e.g., no-take marine protected areas, a major American Samoa initiative), developing of sanctuary program monitoring protocols, and developing a general understanding of species composition and abundance. These studies will also form the deep-water, multibeam bathymetric component of habitat assessment of shallow regions (~1–30 m), already begun by the National Oceanic and Atmospheric Administration's Biogeography Program (http://biogeo.nos.noaa.gov/), in support of the efforts of the aforementioned agencies. The assessment is based on 4-m multispectral (red, blue, green, and near infrared) imagery from Ikonos, launched in 1999 as the world's first commercial high-resolution satellite.

The Fagatele Bay National Marine Sanctuary GIS data archive at http://dusk.geo.orst.edu/djl/samoa/ provides current maps, images, digital photographs, and a wealth of GIS data, including the 2001 shallow-water multibeam bathymetric surveys, new bathymetry from a deep-water survey that circumnavigated the entire island of Tutuila in 2002, and a compilation of

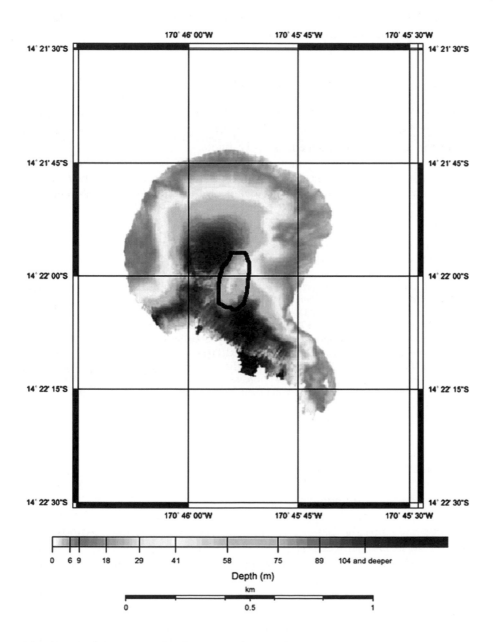

*Figure 53.2.* Gray-shaded bathymetric map of American Samoa's national marine sanctuary, created from a 1-m Kongsberg Simrad EM-3000 bathymetric grid. Black outline shows the region visited by a rebreather diving mission in the sanctuary, immediately following bathymetric surveying. Map projection is Mercator.

terrestrial GIS data layers obtained from the American Samoa Government, the National Park of American Samoa, and several other sources.

## REFERENCES

Allison, G. W., J. Lubchenco, and M. H. Carr. 1998. Marine Reserves are Necessary but not Sufficient for Marine Conservation. *Ecological Applications* 8(1): S79–S92.

Birkeland, C. E., R. H. Randall, R. C. Wass, B. Smith, and S. Wilkins. 1987. *Biological Assessment of the Fagatele Bay National Marine Sanctuary.* NOAA Technical Memorandum. Washington, D.C.: NOAA.

Bryant, D., L. Burke, J. McManus, and M. Spalding. 1998. *Reefs at Risk: A Map-Based Indicator of Potential Threats to the World's Coral Reefs.* Washington, D.C.: World Resources Institute.

Craig, P. 1998. Temporal Spawning Patterns for Several Surgeonfishes and Grasses in American Samoa. *Pacific Science* 52: 35–39.

Green, A. L., C. E. Birkeland, and R. H. Randall. 1999. Twenty Years of Disturbance and Change in Fagatele Bay National Marine Sanctuary, American Samoa. *Pacific Science* 53(4): 376–400.

Gubbay, S. ed. 1995. *Marine Protected Areas: Principles and Techniques for Management.* London: Chapman and Hall.

Sauafea, F. S. 2001. Community-based Fisheries Management in American Samoa. *Proceedings of the Fifth Regional Symposium, PACON 2001.* Burlingame, California: Pacific Congress on Marine Science and Technology.

Wright, D. J., B. T. Donahue, and D. F. Naar. 2002. Seafloor Mapping and GIS Coordination at America's Remotest National Marine Sanctuary (American Samoa). In D. J. Wright, ed. *Undersea with GIS,* 33–63. Redlands, CA: ESRI Press.

Wright, D. J., E. Lundblad, E. Larkin, and J. Miller. 2003. *Benthic Habitat Characterization for Coastal Management and Marine Protected Areas: American Samoa.* NOAA Coastal Services Center Special Projects for the Pacific Islands, manuscript in preparation.

# 54

# THE OCULINA BANKS EXPERIMENTAL RESEARCH RESERVE: A HABITAT ASSESSMENT USING MULTI-MEDIA AND INTERNET GIS

JOANNE N. HALLS
*University of North Carolina, Wilmington*

The Oculina Banks Experimental Research Reserve extends from Fort Pierce to Cape Canaveral off the eastern coast of Florida. This area contains a species of branching coral, *Oculina varicosa*, or the Ivory Tree Coral, which grows on high-relief pinnacles and ridges, 10–30 m in height. The reefs span over 90 nautical miles at depths ranging from 70–100 m and represent the only colony of deep water *O. varicosa* on earth (Reed 1980; Reed 2002). The colonies provide valuable habitat for a variety of fish species, including the economically important snapper/grouper fishery. However, as fishing pressures have increased, so have the impacts to the habitat resulting in a decline of the fisheries over the last twenty years (Koenig et al. 2000).

To monitor the habitats, the South Atlantic Fishery Management Council designated a 315 km$^2$ area (79°56′ W to 80°03′ W and 27°53′ N to 27°30′ N) as a Habitat Area of Particular Concern (HAPC) in 1984 (Figure 54.1). Ten years later, this area was designated as the Experimental Oculina Research Reserve (EORR), when all commercial fishing was band for a period of ten years. During this time, the reserve is closed to all trawling, anchoring, and bottom fishing. Expansion of the EORR continued in 1998, with legislation that now protects 1,029 km$^2$ of benthic habitat (Reed 2002). The extension of the reserve came as a result of the amendments to the Magnuson-Stevenson Fisheries Management Act in 1996, which states that Essential Fish Habitats must be identified in impacted areas to enhance overall conservation.

In 2001, development of a multi-media GIS began for visualizing the habitats of the EORR, analyzing historical fisheries data, and disseminating these data via cd-rom and the Internet. The EORR was given an initial ten-year period during which a no-take policy is in effect. This policy will be reviewed in 2004. To make an informed decision on the fate of the reserve, officials from the South Atlantic Fisheries Management Council are using the GIS to analyze data. The GIS enables researchers and managers to visualize the submersed habitats, incorporate data that otherwise would be inaccessible, investigate new hypotheses about the changing conditions of the reefs, and propose management alternatives.

*D. G. Janelle et al. (eds.), WorldMinds: Geographical Perspectives on 100 Problems*, 331–336.
© 2004 *Springer. Printed in the Netherlands.*

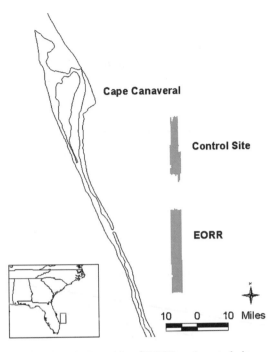

*Figure 54.1.* Location of EORR and control site.

## INCORPORATING DISPARATE DATA SOURCES

Until the development of a GIS for the Oculina Banks, EORR researchers and stakeholders had no method for analyzing the changing landscape of the reefs. With the advent of GPS technology, and the improvements made to this technology in recent years, it is now possible to accurately map submersed habitats both remotely and *in situ*.

In 1995, sidescan sonar imagery was collected for the EORR as well as for a control area located north of the EORR that is open to fishing and considered unprotected. Collection of sidescan sonar imagery is rather difficult in this area due to the fast currents of the Gulf Stream. Image processing software was used to post-process the imagery. Geographic distortion was further corrected using remotely operated vehicle (ROV) video footage and sediment samples from the bottom. Mosaicked images of the EORR and the control area were produced at a spatial resolution of 2 m by 2 m (Scanlon et al. 1999).

In September 2001, as part of the National Oceanographic and Atmospheric Administration's (NOAA) Ocean Exploration – Sustainable Seas Expedition, the habitat conditions were investigated throughout the EORR and the control site. During the seven-day mission aboard the research vessel Seward Johnson

II, researchers from Florida State University, Harbor Branch Oceanographic Institute, National Undersea Research Center, and University of North Carolina at Wilmington employed submersibles to map and quantify the fisheries and habitats. An ROV was used for investigating the bottom habitat, and the manned submersible "Clelia" was deployed to carry observers to critical sites within the EORR and within the control areas for analysis of the habitat. Using constant communications between the submersible and the ship, GPS locations were obtained and used to georeference the field data. High definition video cameras logged fish aggregations, habitat conditions, and apparent anthropogenic impacts. Tapes from a total of fourteen dives have been analyzed and a database has been created to document the dives with video clips, images, and descriptions of each dive.

To investigate the condition of the Oculina Banks study area, additional historical data were incorporated into the GIS. These data included fish surveys from the 1980s and 1990s and bathymetry soundings collected at various times by NOAA. The earlier fish surveys were georeferenced using the Loran A system, which is rather crude by today's mapping standards, however these data are still useful for providing a guide to the past conditions of the area. The bathymetry soundings (points) are useful for generating topographic surfaces using various spatial interpolation algorithms. The result is a grid-based map of the surface that can be overlaid with the other data layers.

The final GIS contains habitat maps based on the sidescan sonar images, multiple dates of bathymetry data surveyed by NOAA, high-resolution multi-beam backscatter topographic data obtained in October 2002, multiple dates of fish-count surveys, and digital video conducted at various locations within the study area.

## HABITAT ANALYSIS

Several methods were used to characterize the condition of the reef habitat. First, the habitat types classified from the sidescan sonar imagery were compared with the digital video surveyed at various locations within the EORR and the unprotected control area. To help with this analysis, a shaded-relief image was created using the bathymetric soundings and the habitat data. The georeferenced video clips were superimposed on the topographic and habitat data to document differences seen in the video from the earlier sidescan sonar imagery (Figure 54.2).

Second, fisheries data (species type and abundance) gathered in the 2001 survey were compared with earlier fish surveys (1980 and 1995). While specific methodologies differ over time, the surveys performed in 1995 showed disturbing shifts in the ecology of the reef, as smaller fish were more prevalent and overall diversity declined (Gilmore and Jones 1992; Koenig et al. 2000). The 2001 mission also tabulated fish abundances by species for specific

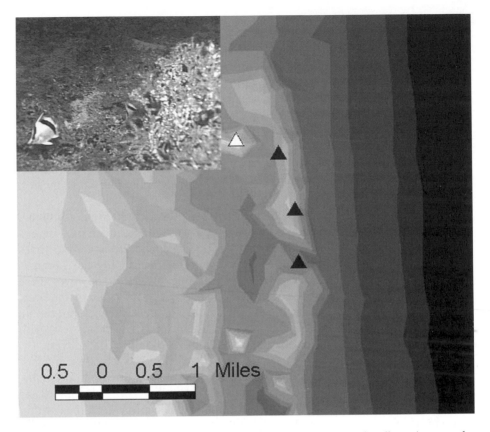

*Figure 54.2.* Example digital elevation model with locations of submarine dives. An example picture is given for the dive location highlighted.

geographic locations within the reserve. The resulting database allows users to query individual species by dive location. Generally, these surveys show a continued shift in the community towards non-fishery species. Further investigations are planned to gauge the spawning aggregations that develop during the spring and that are essential to maintaining populations of gag and scamp within the reef.

## MULTI-MEDIA GIS

With the computing advances made in recent years, it is now possible to create complex databases that contain a variety of data formats. GIS is well suited to embrace this computing technology and enables users to interact with a wide variety of data. The Oculina GIS project has used this technology to develop a database structure that incorporates traditional spatial (raster and vector) data and attribute information as well as multi-media digital data, such as videography. The GIS database structure enables users to select locations, see how many fish have been observed there, the types of species, the type of habitat, and video clips of the actual sites.

Outreach is a fundamental component to a successful GIS project. An Internet GIS website (www.uncw.edu/oculina) was built where users can access the data and perform querying and spatial analysis functions that are also available in the more traditional GIS application delivered via CD-Rom. ArcView™ was chosen as the GIS software because of its functionality and popularity. Within the GIS, hotlinks were created for linking the sampling sites to photographs, video, and Adobe PDF files containing detailed descriptions of the habitats, species, and fish abundances. For more robust analysis, a relational database was constructed to contain the species and abundance data.

## CONCLUSIONS

Through an analysis of historical information and more recently collected data, results indicate that the closure of the Oculina EORR has not restored the area to pre-trawling conditions. However, it does appear that some areas within the EORR are experiencing an increase in the numbers of fish and that this is a positive impact from resource protection. Conversely, in some areas there are recent trawl scars on the sea floor that, although the area is closed to fishing, is evidence that this activity is continuing.

The development of the GIS has enabled researchers, stakeholders, and resource managers to visualize, analyze, and explore the submersed landscape. Additionally, the GIS now houses a basemap of the area and corresponding georeferenced data layers that can be updated in the future. Unlike conventional change-detection techniques in terrestrial environments, where aerial photography and satellite imagery are primary data sources, the multi-media sources obtained for this project were used to assess changes in the habitats quantitatively and qualitatively.

## REFERENCES

Gilmore, R. G., and R. S. Jones. 1992. Color Variation and Associated Behavior in the Epipheline Groupers, *Mycteroperca microlepis* (Goode and Bean) and *M. phenax* (Jordan and Swain). *Bulletin of Marine Science* 51: 83–103.

Koenig, C. K., F. C. Coleman, C. B. Grimes, G. R. Fitzhugh, K. M. Scanlon, C. T. Gledhill, and M. Grace. 2000. Protection of Fish Spawning Habitat for the Conservation of Warm-Temperate Reef-Fish Fisheries of Shelf-Edge Reefs of Florida. *Bulletin of Marine Science* 66(3): 593–616.

Reed, J. 1980. Distribution and Structure of Deep-Water *Oculina varicosa* Coral Reefs Off of Central Eastern Florida. *Bulletin of Marine Science* 30(3): 667–677.

Reed, J. 2002. Deep-Water *Oculina* Coral Reefs of Florida: Biology, Impacts, and Management. *Hydrobiologia*, in press.

Scanlon, K. M., P. R. Briere, and C. C. Koenig. 1999. Oculina Bank: Sidescan Sonar and Sediment Data From a Deep-Water Coral Reef Habitat Off East-Central Florida. *USGS Open File Report 99-10*.

Stanbury, K. B. and R. M. Starr. 1999. Applications of Geographic Information Systems (GIS) to Habitat Assessment and Marine Resource Management. *Oceanologica Acta* 22(6): 699–703.

Waltenburger, B. and M. Pickett. 1999. Integration of NOA Aerial Monitoring and GIS Programs for Research, Education, and Management Within the Channel Islands Marine Sanctuary. *Proceedings of the 5th California Islands Symposium*. Santa Barbara, CA, 29 March–1 April 1999: 469–471.

Wright, D. J., B. T. Donahue, and D. F. Naar. 2001. Seafloor Mapping and GIS Coordination at America's Remotest Marine Sanctuary (American Somoa). In D. J. Wright, ed. *Undersea with GIS*. Redlands, CA: ESRI Press.

PART IV

# BALANCING ENVIRONMENT WITH ECONOMY

Balance is often difficult to achieve in a world that privileges accumulated individual wealth and the corporate quest for profit. Yet, a sustainable resource base requires measured moderation in the access to and use of water, animal, mineral, and plant resources. The often-incompatible desires of human agents and the vulnerability of the environments they seek to dominate necessitate perspectives from geographical science to help balance economic well-being against exploitation and long-term survival against short-term greed.

*Kathy Hansen*

# 55

# GEOGRAPHERS AND SUSTAINABILITY:
# A MISSING CONNECTION?

JAMES C. EFLIN
*Ball State University*

As geography matured during the 20th century, it followed a course that might appear, to outsiders at least, as a series of distributary channels – rather than coming together in a coherent and unified flow. Schisms among geographers and competing identities for their discipline risk impairing geography's effectiveness to provide tangible solutions to problems plaguing Planet Earth (Hanson 1999). Although this parallels changes in some other disciplines, it also carries with it the potential for unraveling the sinews that give strength to its disciplinary project.

## CHARTING A USEFUL COURSE

To be effective and be taken seriously, geographers must re-awaken the minds of others – politicians, scientists in other disciplines, and society generally – by charting useful courses. Current debates involving global climate change, population pressures on natural resources, world trade, sustainable development, desertification, biodiversity, and wetlands losses are signals that geographers should take seriously. Can geography help interpret their meanings? Geographers are adept at understanding spatial patterns and processes within economies, the roots of socio-spatial inequities, and complex and interacting biophysical systems. But, an enduring integration of these interlinking processes – sustainability of economic, social, and ecological processes – is, arguably, the most significant problem facing planet Earth for humans who wish to be here indefinitely. How may this triad be maintained, integrated, and kept in balance for an indefinite future? What connection might geography make in sustainability?

## SUSTAINABILITY FOR WHOM?

Sustainability for humans in concert with the earth is at question, here. Sustainability is a hot-button topic elsewhere, yet voices within geography have

*D. G. Janelle et al. (eds.), WorldMinds: Geographical Perspectives on 100 Problems*, 339–343.
© 2004 *Springer. Printed in the Netherlands.*

been strangely silent on it. For those who paid attention to his speech in New York (1 March 2001), United Nations Secretary-General Kofi Annan appealed to members of the Association of American Geographers to "take advantage of [the] close affinity" between professional geographers and the United Nations:

> I would like to suggest to you today that we take advantage of this close affinity, and work together to tackle some of the greatest challenges facing the human community: climate change, the perilous state of the global environment, and the long-term goal of achieving truly sustainable development (Annan 2001: 10).

Annan was referring to an evolution of thinking among world leaders in policy, academic, business, and other communities to build on efforts forged at the United Nations Conference on Environment and Development (or Earth Summit) in Rio de Janeiro in 1992. The Earth Summit continued a thrust that emerged when the World Commission on Environment and Development released *Our Common Future* in 1987, with its oft-repeated definition of sustainable development as ". . . development that meets the needs of the present without compromising the ability of future generations to meet their own needs" (WCED 1987: 43). From that base, world leaders at the Earth Summit crafted the Rio Declaration and Agenda 21, designed as a road map to move nations toward sustainable development in the current century. Its many chapters gave evidence of one thing: the sustainability of the human prospect cannot ignore the grave disparities among people of the world nor the delicate interlinkages between material development and environmental quality.

Why were generational equity, economic development, and environmental quality at issue? And why haven't geographers been listening?

## PIECING IT ALL TOGETHER

If we look at the evolution of geography during the 20th century, we see that geographers *were* listening – just not seeing the pieces uniting to form the complete puzzle. Environmental degradation, linked to the development of modern economies and their industrial growth complexes, as well as earlier transformations of the earth by humans, was long recognized by geographers (Thomas et al. 1956). Similarly, geographers sought to become "relevant" to society by interpreting unequal distribution of the access to material goods and services since the early 1970s (Harvey 1973). More recently, social equity implications of natural resource bases and the human dimensions of global change became key themes within geography (Turner 1990.)

It may help geographers to picture the integration of three key elements of sustainability using what is commonly viewed by the metaphor of a "three-legged stool." Sustainability requires each of its legs – economy, society, and environment – to be firmly connected and *integrated* with the others in

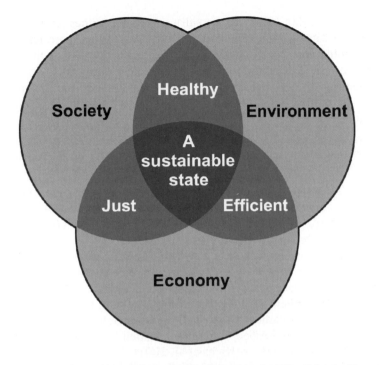

*Figure 55.1.* Overlapping and integrative realms leading to sustainability. Printed with permission of New Jersey Future (www.njfuture.org). Accessed on 13 March 2003.

order that together the whole might stand. In this way, overlaps in which geographers fit themselves *between* the three realms can be clearly seen (Figure 55.1).

Much has changed in the focus of other disciplines and in the focus of progressive societies since the Earth Summit. One example was the development of The Natural Step framework and its "systems conditions" for sustainability, based on scientific principles of thermodynamics:

> Metaphorically, the current situation for people on the Earth can be viewed as a funnel with diminishing room to maneuver. . . . In the sustainable society, nature is not subject to systematically increasing (1) concentrations of substances extracted from the Earth's crust, (2) concentrations of substances produced by society, (3) degradation by physical means and, in that society (4) human needs are met worldwide (The Natural Step 2001/2002).

Another example is the increasing use of indicators of sustainability to help understand the "wellbeing of nations" or the interconnection between people and ecosystems. An example with special appeal to geographers is the concept of the "ecological footprint," an indicator that expresses ". . . the land and water area that is required to support indefinitely the material standard of living of a given human population, using prevailing technology"

(Chambers et al. 2000: 177). Conceptually appealing, further refinement should make this indicator a robust measure for empirical geographic research with practical applications for policy-makers.

Today, geographers continue to search for the identity of their discipline. In commenting on our investment of "large amounts of intellectual energy in search of its identity," Turner (2002: 63) takes us forward toward a new reconciliation:

> If geography is to gain a full seat at the academy's head table and retain the breadth of its traditions (practice) it must seek to reunite its two main identities in a way that is congruent with the prevailing logic by which the academy partitions knowledge.

Here, Turner speaks of the prospect for a union between disparate sides of geography, a merger that ". . . would enable the retention of geography's breadth and bridging qualities and avoid the transaction costs of creating new fields of study" (Turner 2002: 64).

Our attention has recently begun to grow toward that integration (Kates 2002). Notably, the American Association for the Advancement of Science (AAAS) used "Science in a Connected World" as the theme for its 2002 meeting after the evident disintegration of trust within the global community following 9/11. This came on the heels of an article titled "Sustainability Science" that AAAS published in *Science*, written by a collaboration of geographers and non-geographers (Kates et al. 2001). "Sustainability science," we are told by Kates, et al., is a focus "on the dynamic interactions between nature and society." Why does that sound so familiar? Perhaps it is because sustainability science was advanced as a holistic approach to address ". . . meeting human needs while preserving the life support systems of the earth" (p. 1), a new synthesis and integration, the very heart of what geography was and can be once more.

Kates (2002) and Turner (2002) point geographers toward a new connection that has been missing within geography for some time, a connection to be made because geography gives its practitioners the insights and skills to synthesize and integrate for the good of meeting human needs and maintaining their environmental resource base. But, we do not have to await the emergence of another, separate "science" of sustainability; we must regain our soul, our focus.

## REGAINING OUR PLACE?

In September 2002, the United Nations convened the World Summit on Sustainable Development in Johannesburg to review the status of ten years of progress toward sustainability. Looking back on the decade, Kofi Annan saw hope in the results, concluding the conference with these words: "This Summit makes sustainable development a reality. . . . This Summit will put us on a path that reduces poverty while protecting the environment, a path that works for all peoples, rich and poor, today and tomorrow" (Annan 2002). Whether Annan was right that sustainable development is a reality, it is clear that nations of the world view the path ahead as being one of importance. The time has come for geographers to recognize this course and to come together to lend the strength of their discipline while preserving geography's own place – and importance – at the tables of academe and governance.

## REFERENCES

Annan, K. 2001. Address to the Association of American Geographers, 1 March 2001 in New York City, Association of American Geographers. *AAG Newsletter* 36(4): 10–12.

Annan, K. 2002. Quoted in *Johannesburg Summit*, 2002: United Nations Website for the World Summit on Sustainable Development; see http://www.johannesburgsummit.org/html/whats_new/feature_story39.htm. Accessed 4 October 2002.

Chambers, N., C. Simmons, and M. Wackernagel. 2000. *Sharing Nature's Interest: Ecological Footprints as an Indicator of Sustainability*. London: Earthscan.

Hanson, S. 1999. Isms and Schisms: Healing the Rift Between Nature-Society and Space-Society Traditions in Human Geography. *Annals of the Association of American Geographers* 89(1): 133–143.

Harvey, D. 1973. *Social Justice and the City*. Baltimore: The Johns Hopkins University Press.

Kates, R. W., W. C. Clark, R. Corell, J. M. Hall, C. C. Jaeger, I. Lowe, J. J. McCarthy, H. J. Schellnhuber, B. Bolin, N. M. Dickson, S. Faucheux, G. C. Gallopin, A. Gruebler, B. Huntley, J. Jäger, N. S. Jodha, R. E. Kasperson, A. Mabogunhe, P. Matson, H. Mooney, B. Moore III, T. O'Riordan, and U. Svedin, 2001. Sustainability Science. *Science* 292: 641–642. Also published as Kennedy School of Government Faculty Research Working Paper RWP00-018, December 2000; see http://sust.harvard.edu.

Kates, R. W. 2002. Humboldt's Dream, Beyond Disciplines, and Sustainability Science: Contested Identities in a Restructuring Academy. *Annals of the Association of American Geographers* 92(1): 9–81.

The Natural Step, 2001/2002. *Intro to Sustainability: What is The Natural Step Framework? and The Natural Step Four System Conditions*. See http://www.naturalstep.org/framework/framework_overview.html. Accessed 9 March 2003.

Thomas, W. L., C. O. Sauer, M. Bates, and L. Mumford, eds. 1956. *Man's Role in Changing the Face of the Earth*. Chicago: University of Chicago Press.

Turner, B. L. II, ed. 1990. *The Earth as Transformed by Human Action: Global and Regional Changes in the Biosphere Over the Past 300 Years*. Cambridge: Cambridge University Press.

Turner, B. L. II. 2002. Contested Identities: Human-Environment Geography and Disciplinary Implications in a Restructuring Academy. *Annals of the Association of American Geographers* 92(1): 52–74.

WCED World Commission on Environment and Development. 1987. *Our Common Future*. Oxford/New York: Oxford University Press.

# 56

# THE GREAT PLAINS AND THE BUFFALO COMMONS

DEBORAH E. POPPER

*City University of New York/College of Staten Island*

FRANK J. POPPER

*Rutgers University*

The Great Plains region, North America's grassland and breadbasket, is a vast, beautiful, charismatic place with a volatile settlement history. Since the end of the Civil War, the Plains has suffered three large cycles of population, economic, and environmental boom and especially bust. Seventeen years ago the authors (Popper and Popper 1987) proposed a solution for the region's difficulties, a vision of the Plains' land-use future that we called the Buffalo Commons. Our idea took hold in the Plains, has lasted, and promises to affect the region's development.

The Great Plains occupies almost a sixth of the lower 48 United States' land area. By conventional definition, its east edge is the 98th meridian, where Oklahoma City and San Antonio lie (Webb 1981 [1931]: 3–9). Its west edge is the Rockies. The region consists of parts of ten U.S. states, from North Dakota and Montana in the north to Texas and New Mexico in the south, plus parts of three Canadian provinces, Alberta, Manitoba, and Saskatchewan.

The Plains are the burnt right flank of the Western desert. The region has high winds and few trees, and the climate is semiarid, with average rainfall of 12 to 20 inches annually. The vegetation is mainly midgrass in the east, shortgrass in the west. The topography rolls in the north, flattens in the south. The prime crops are cattle, corn, wheat, sheep and, in the south, cotton. The prime minerals are coal, oil, and natural gas. The Plains once were nearly entirely deep-rural and small-town, but these areas have long been losing population to cities in the Plains and elsewhere. Still, the largest city in the northern half of the U.S. Plains is Billings, Montana, with a population of only 90,000 in 2000. The region's total United States population is under 6 million people, about that of a mid-size state such as Indiana.

The Plains are the land of the Big Sky and the Dust Bowl, one-room schoolhouses and settler homesteads, straight-line Interstates and custom combines, and *Little House on the Prairie* and *Dances with Wolves*. The oceans-of-grass vistas of the Plains induce the somber-serene awe that enthralled early literary observers like Washington Irving and George Catlin and modern-day

*D. G. Janelle et al. (eds.), WorldMinds: Geographical Perspectives on 100 Problems, 345–349.*
© 2004 *Springer. Printed in the Netherlands.*

ones like Sharon Butala and Wallace Stegner. All of them knew the difficulties the settlers would find there.

## THE PROBLEM

Large-scale Euroamerican habitation, which began soon after the Civil War, shows a basic pattern: federally subsidized settlement and cultivation produce a boom, which then leads to overgrazing and overplowing, which then leads to a bust that features heavy depopulation, especially in the region's most rural sections. Plains settlement has repeatedly displayed what University of North Dakota historian Elwyn Robinson called the "Too-Much Mistake" – too many people, farms, ranches, towns, railroads, and roads for the land to take (Robinson 1966:VII). Nature and the economy inevitably rebelled.

The three great cycles of short-boom/long-bust have now played out. The first originated with the 1862 Homestead Act, which gave settlers 160 acres of free federal land if they could live on it for five years. By 1890, a series of hard winters and cattle die-ups forced a bust, with failed homesteading, widespread starvation, and large convoys of fully loaded wagon trains headed east, out of the Plains. The second cycle began in the early 1900s, with additional homesteading laws that allowed claimants up to 640 acres of free federal land. Plains locusts and a national rural depression hit in the early 1920s, followed by the 1929 Great Depression and the ecological catastrophe of the 1930s Dust Bowl, when huge clouds of Plains soil blew away across large parts of the region. By the late 1930s, the federal government prohibited new homesteading, and John Steinbeck's *Grapes of Wrath* Okies were driving, hitchhiking, or rail-hopping west to California.

As a result of the cycles, many deep-rural Plains towns and counties had their largest populations early in the last century and have been steadily emptying since. Young people are particularly prone to leave, and so the local population ages. Maps of the Plains show many towns that have been essentially unoccupied for decades – Keota, Colorado, say, or Hoover, South Dakota. Kansas alone has hundreds of ghost towns. Nebraska has 5,000 to 10,000 deserted farmhouses.

Since the late 1980s, the present authors have argued that a third great boom/bust cycle was well under way and likely to go farther than the other two (Popper and Popper 1987). It began with the New Deal advent of large modern federal subsidies from the Agriculture Department. It peaked in the 1970s, when the Department encouraged fencepost-to-fencepost cultivation, and the energy crisis made Plains coal, oil, and natural gas particularly profitable. By the middle 1980s the environmental, economic, and demographic toll was clear, and the bust phase set in. Soil erosion approached Dust Bowl rates, although it was less visible than in the 1930s, because it was mainly water – rather than wind-borne. Despite wider use of water-conservation methods than in earlier periods, the Ogallala Aquifer, the source of agricultural and urban groundwater for the southern two-thirds of the Plains, was dropping fast, especially under its largest consumer, Texas (Opie 2000 [1993]).

Much of the region's farm, ranch, energy, and mining economies suffered near-depression. Americans ate less beef, substituting lower-cholesterol and lower-fat chicken and fish, which the Plains did not produce. The Interior Department's Bureau of Reclamation stopped building big dam-and-irrigation projects, such as the 1940s and 1950s Missouri River ones that underwrote large chunks of Plains economic development. As in the other two cycles, failing farmers and ranchers mostly sold their holdings to more successful ones, but now depopulation reached the point where entire counties lost their churches, doctors, hospitals, and banks.

The middle-1980s third-cycle pressures have persisted, and others emerged. The large population losses, especially of young people, continued in the 1990 and 2000 Censuses (Popper and Popper 2002: 21). All Plains states and provinces suffered severe drought, and some saw at least ten years of it over the last fifteen. Friendlier American-Russian relations and bigger federal budget deficits downsized many military bases and eliminated missile silos in the Dakotas, Montana, and Wyoming that gave localities much-needed subsidies. Deregulation of the airline, trucking, railroad, bus, and telephone industries further isolated many remote Plains communities, as did cutbacks in medical, educational, and other public services. The United States and Canada, increasingly urban and suburban, became ambivalent about big federal farm subsidies that rose as farm populations fell and farm overproduction swelled. Beginning in the late 1990s, white Plains Nebraska cattle counties supplanted Indian Plains South Dakota ones as the poorest in the nation (*Rural Action* 2000: 1). Global warming seemed likely to put new stresses on Plains' agriculture.

## BUFFALO COMMONS

In the late 1980s, the present authors offered the Buffalo Commons as an alternative to the boom-and-bust cycles (Popper and Popper 1987, 2002). The Buffalo Commons suggests ecologically and economically restorative possibilities for large stretches of the Plains. We foresaw a Plains with new land uses that fell somewhere between traditional agriculture and pure wilderness. Environmental protection and ecotourism would supplement existing agricultural and resource-extraction methods. Buffalo and other native animals and grasses would in some places replace cattle, a nonnative species. The shift from corn-fed cattle to grass-fed buffalo would diminish the overall environmental pressures on Plains' agricultural land.

The Buffalo Commons proposal created an ongoing national debate about the future of the Plains (Matthews 2002 [1992]; Wheeler 1998; and Williams

2001). Since the late 1980s, a robust range of Buffalo Commons initiatives has emerged on private, public, Native American, and nonprofit holdings. The total number of buffalo on private and public lands in the United States and Canada approaches 400,000 – a remarkable figure for a large species that nearly went extinct less than a century ago. A thriving buffalo industry appeared, touting the animal's high-protein, low-fat, low-cholesterol, no-preservative, no-biocide features. The industry had three main sectors. Ranchers, such as the members of the North-Dakota-based North American Bison Cooperative, were usually fleeing the boom-and-bust cycles of running cattle. Indian tribes and ranchers, exemplified by the 51 tribes, primarily from the Plains, that formed the South Dakota-based InterTribal Bison Cooperative, often sought spiritual, psychological, self-determination, and cultural values from buffalo, as much as financial ones. Ted Turner, the Atlanta billionaire, bought nearly two million acres of Plains cattle ranches, converted them to buffalo ones, primarily for their environmental values, and now owns a tenth of U.S. buffalo. Buffalo prices rose at first and then, in the last few years, dropped sharply because the industry expanded too fast. But the industry, tiny half a generation ago, today is not.

The Buffalo Commons idea flourishes elsewhere. Beginning in the early 1990s, the Nature Conservancy, the leading U.S. land-preservation organization, made big Plains purchases and restored buffalo and other native animals and plants on them. Starting in 2000, the Sierra Club began talking of doing likewise. Environmental groups promote the Buffalo Commons – for instance, the Ft. Worth- and Denver-based Great Plains Restoration Council, whose main goal is to create it. Government agencies have taken Buffalo Commons steps. Saskatchewan created Grasslands National Park, eventually to encompass 350 buffalo-filled square miles, but already open to visitors. Several state and provincial governments now offer ranchers loans and technical assistance to run buffalo, and private banks lend for buffalo as well. In 1996, North Dakota's agriculture commissioner said the state would one day have more buffalo than cattle. By some estimates, buffalo are already its second-leading agricultural product. In 2001 and 2002, the state hosted two federally sponsored national conferences on Plains rural depopulation, and the Buffalo Commons figured prominently. In May 2001, the state's new governor, several of whose predecessors had attacked the concept, told the *New York Times*, "There is a lot of that Buffalo Commons idea that's probably true. . . . It's never going to look like it did before, when all the farms and ranches were healthy" (Egan 2001: 20). In August 2001, South Dakota's Rosebud tribal government became the first public body formally to endorse the Buffalo Commons. More such public-sector actions are predictable in coming years. The issue is no longer whether the Buffalo Commons will happen, but how.

# REFERENCES

Egan, T. 2001. As Others Abandon Plains, Indians and Bison Come Back. *New York Times* 27 May: 1, 20.

Matthews, A. 2002 [1992]. *Where the Buffalo Roam: Restoring America's Great Plains.* Chicago: University of Chicago Press.

Opie, J. 2000 [1993]. *Ogallala: Water for a Dry Land.* Lincoln: University of Nebraska Press.

Popper, D. E. and F. J. Popper. 1987. The Great Plains: From Dust to Dust. *Planning* 53(12): 12–18.

Popper, D. E. and F. J. Popper. 2002. Small Can Be Beautiful: Coming to Terms with Decline. *Planning* 68(7): 20–23.

Robinson, E. B. 1966. *History of North Dakota.* Lincoln: University of Nebraska Press.

Rural Action. 2000. *New Data on Rural Poverty.* September, 1. Walthill, Nebraska: Center for Rural Affairs.

Webb, W. P. 1981 [1931]. *The Great Plains.* Lincoln: University of Nebraska Press.

Williams, F. 2001. Plains Sense: Frank and Deborah Popper's 'Buffalo Commons' is Creeping Toward Reality. *High Country News*: 15 January. See www.hcn.org.

Wheeler, R. S. 1998. *The Buffalo Commons.* New York: Forge.

# 57

# REDUCING ENERGY SHORTAGES AND POLLUTION IN CHINA: THE WORLD BANK'S COAL TRANSPORT STUDIES

MICHAEL KUBY
*Arizona State University*

ZHIJUN XIE
*Henwood Energy Services, Inc.*

China, like many developing countries, faces a difficult tradeoff between economic development and environmental protection. Being the world's largest producer and consumer of coal places a massive burden on China's transportation and environmental systems. In 1989, energy shortages and runaway pollution prompted the World Bank and the Chinese State Planning Commission (SPC) to jointly undertake the China Coal Transport Study (CTS). At the time, China relied on coal for 73 percent of commercial energy, while coal accounted for 42 percent of rail freight. Rail transport, coal, and electricity were all suffering from debilitating shortages, which threatened China's 10 percent annual economic growth. Meanwhile, $SO_2$ and particulate levels were among the highest in the world, and $CO_2$ emissions reached 11 percent of the world total – second after the United States.

The goal of the Coal Transport Study was to "develop and analyze investment strategies and associated policies for China's coal and electricity delivery system, while also taking into account the air pollution effects of the different strategies" (World Bank 1995: xi). Initially, the focus was on the transportation supply side – railways, ports, ships, barges, and slurry pipelines. China's biggest and best coal deposits are in the north central and northwest regions, but its economy is concentrated on the coast, thus creating a huge transport demand. China's coal transport demand in 1990 was 533 billion ton-kilometers (tkm). But as the international team of geographers, economists, engineers, and planners studied the problem, it became obvious that China could not simply build its way out of this predicament – and even if it could, the environmental impacts would be disastrous. Instead of building more infrastructure to move more tkm, China could invest in reducing the tkm that need to be moved by reducing the tons and/or the kilometers (Kuby et al. 1993). Investment strategies for reducing transport distance included mining higher-

*D. G. Janelle et al. (eds.), WorldMinds: Geographical Perspectives on 100 Problems*, 351–356.
© 2004 *Springer. Printed in the Netherlands.*

cost coastal coal deposits and building minemouth power plants and electricity transmission lines, i.e., "coal by wire." Strategies for reducing tons included washing coal to remove waste and substituting hydro or nuclear power. Later on, another important substitute – energy conservation investment – was added to the model as a third category of tonnage-reducing strategies (Xie and Kuby 1997).

## METHODOLOGY

The methodology adopted for this project – mixed-integer, multi-objective programming – is widely used for optimizing facility location and transportation networks. This branch of geography overlaps with engineering, business, and economics. Known alternatively as operations research or management science, it is an interdisciplinary field with real-world applicability. Its geographical roots date back to Alfred Weber's (1909) *Theory of the Location of Industries*. The techniques used have evolved from Weber's simple isocost curves and Varignon's pulleys-and-weights to mathematical solutions, simulated annealing, and integration with GIS in spatial decision-support systems. Geographers have published a number of seminal papers over the past three decades in journals such as *Geographical Analysis* (ReVelle and Swain 1970; Church and ReVelle 1976; O'Kelly 1986). Models have been formulated and solved for a wide range of applications: locating fire stations, schools, health clinics, and banks; locating undesirable facilities like waste dumps; moving passengers, goods, and waste over a network; locating distribution facilities like ports, warehouses, or airline hubs; and targeting links to expand (or contract) a transportation network. Additionally, optimization is increasingly being used to promote sustainability in applications such as nature reserve design, hazardous materials transportation, and in our case, cost-pollution tradeoffs.

The best investment strategy was determined using a model to minimize the costs of satisfying projected coal and electricity demands over three Five-Year Plans (1990–2005). The CTS-C model (with Conservation) represented the energy production, conversion, transportation, consumption, and environmental impact system as a chain in which the output of one sector is the input to the next (Figure 57.1). The underlying physical network consisted of more than 50 coal supply and demand nodes, 300 railway arcs, 30 ports, 100 long-distance transmission lines, 166 hydropower sites, plus several slurry pipelines, inland waterways, and nuclear sites. The CTS-C model contained 12,000 constraints and 31,000 variables. Key inputs included existing infrastructure; candidate new projects; cost data; technical coefficients such as coal types and conversion efficiencies; and policy assumptions, like demand forecasts and budget constraints. The model's main decisions were how much new capacity to build, where, and when; how much to produce or transport with the built capacity; and how much demand remains unsatisfied. Other

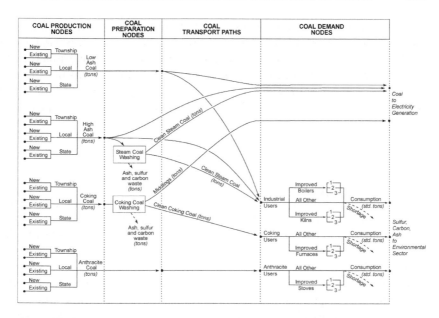

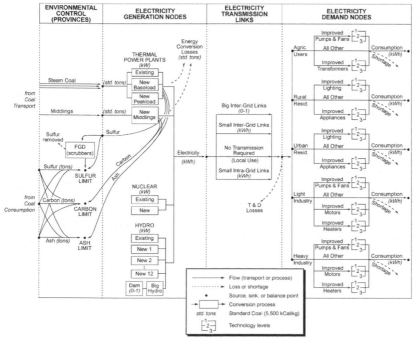

*Figure 57.1.* The supply, delivery, and utilization system for coal (top) and electricity (bottom) in the CTS-C model. Reprinted from Xie Zhijun and Michael Kuby, 1997. "Supply-side – demand-side optimization and cost-environment tradeoffs for China's coal and electricity delivery system," *Energy Policy* 25: 313–326. Reprinted with kind permission from Elsevier Science.

outputs related to costs, bottlenecks, and pollution, while a GIS module mapped coal and electricity flows, bottlenecks, and shortages. Constraints in the model, which ensure a realistic solution, imposed capacity limits, investment budgets, energy conversion balances, and restrictions on the ash, sulfur, and carbon content of the delivered coal. These environmental limits could be loosened or tightened to explore the tradeoffs between minimizing costs and minimizing pollution.


## RESULTS


It is sometimes said by or of developing countries that they cannot afford the luxury of protecting their environment because of more pressing basic needs. Our tradeoff analysis tried to quantify just how much different levels of environmental protection would actually cost the Chinese economy. In Figure 57.2, look at the tradeoff curve for the CTS Base Case, which included coal washing, hydro, and transmission investments, but not energy conservation. Solving the model with no limitation on pollution generated Point A, which represents the least-cost supply-side solution for China. One of the CTS's most important findings is shown by Point B, which demonstrates that a 10 percent cut in ash and sulfur pollution can be achieved for only a 3 percent

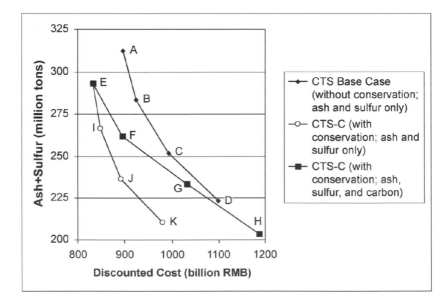

*Figure 57.2.* Non-inferior tradeoff curves between cost and environmental goals for three sets of assumptions. Reprinted from Xie Zhijun and Michael Kuby, 1997. "Supply-side – demand-side optimization and cost-environment tradeoffs for China's coal and electricity delivery system," *Energy Policy* 25: 313–326. Reprinted with kind permission from Elsevier Science.

cost increase. The cheapest strategy for doing so involves substituting hydro for thermal power, building some sulfur scrubbers, and nearly doubling the washing of steam coal. Coal washing, a mechanical procedure for cleaning waste from coal before it is burned, is a win-win technology that can reduce pollution *and* save money on transportation. Using a combination of crushers, jigs, and flotation tanks, washing can reduce ash content from as much as 40 percent to as little as 10 percent, cut sulfur by half, and reduce weight by up to a quarter. The least-cost solution (Point A) included 200 million tons of steam coal washing fully paid for by the transportation cost savings, increasing to 350 million tons for Point B. Points C and D on the CTS-Base-Case curve show that the second 10-percent reduction in pollution drives up national costs by 7 percent, while the third 10-percent cut adds 10 percent to the cost. In keeping with economic theory, each successive unit of pollution reduction costs more than the last.

Typically, multi-objective tradeoff curves are viewed as "frontiers." The points on the curve represent "noninferior" solutions such that there exist no solutions that perform better on both objectives. Only the decision maker, not the analyst, can say whether a low-cost, high-pollution solution on the frontier outranks a high-cost, low-pollution alternative. But that is true only for the specific investment options provided to the model. Our further study with conservation showed that when demand-side options are added to the model, it is possible to reduce both cost and pollution through energy conservation (Points E, I, J, K). Compared with the supply side base case in Point A, Point F satisfies the same final energy demands with 16 percent less ash and sulfur and *275 million tons less coal*. The greatest gains come from efficient lighting, boilers, stoves, pumps, and fans.

Finally, the middle tradeoff curve demonstrates the economic impact of cutting carbon as well as ash and sulfur. Costs rise significantly with each 10 percent cut from Point E to Points F, G, and H. For carbon reduction, coal washing can no longer provide low-cost environmental benefits: carbon cannot be removed without losing energy. More expensive conservation and hydropower investments must be substituted, driving up costs.

## IMPACTS

Gui Shiyong, Vice Chairman of the State Planning Commission, wrote "the CTS results and conclusions are the first of their kind used in China, and are also very realistic, which have been taken seriously by related departments" (Kuby et al. 1995: 68). Many of the strategies recommended by the study were implemented. For example, in 1990 only 18 percent of Chinese coal was washed. We recommended in 1994 that over 40 percent of coal should be washed based on transport cost savings alone, and more if China were willing to pay for environmental benefits. By 2001, China was washing 36 percent of its coal, and the government announced a target of 50 percent for 2005. More importantly, national coal use had actually declined in 1998 and 1999.

*Editor's Note*

The CTS project described in the chapter was the recipient of an Applied Geography Citation Award of the Association of American Geographers, and was one of six finalists for the 1994 Franz Edelman award from the Institute for Operations Research and Management Science.

## REFERENCES

Church, R. L. and C. S. ReVelle. 1976. Theoretical and Computational Links Between the P-Median Location Set-Covering and Maximal Covering Location Problems. *Geographical Analysis* 8: 406–415.

Kuby, M., S. Qingqi, T. Watanatada, S. Xufei, C. Wei, X. Zhijun, Z. Dadi, Z. Chuntai, Y. Xiaodong, L. Fatang, P. Cook, T. Friesz, S. Neuman, R. Jiang, W. Xuesheng, and G. Shenhuai. 1995. Planning China's Coal and Electricity Delivery System. *Interfaces* 25: 41–68 (Special Issue: Franz Edelman Award Papers).

Kuby, M., S. Neuman, Z. Chuntai, P. Cook, Z. Dadi, T. Friesz, S. Qingqi, G. Shenhuai, T. Watanatada, C. Wei, S. Xufei, and X. Zhijun. 1993. A Strategic Investment Planning Model for China's Coal and Electricity Delivery System. *Energy – The International Journal* 18: 1–24.

O'Kelly, M. E. 1986. Activity Levels at Hub Facilities in Interacting Networks. *Geographical Analysis* 18: 343–356.

ReVelle, C. S. and R. Swain. 1970. Central Facilities Location. *Geographical Analysis* 2: 30–42.

Weber, A. 1909. *Ueber den Standort der Industrien* (Theory of the Location of Industries). Erster Teil. Reine Theorie der Standorte. Mit einem mathematischen Anhang von G. PICK (in German). J. C. B. Mohr, Tübingen, Germany: Verlag.

World Bank. 1995. *Investment Strategies for China's Coal and Electricity Delivery System.* Report No. 12687-CHA (Gray Cover). The World Bank, China and Mongolia Department, Transport Operations Division, 8 March.

Xie, Z. and M. Kuby. 1997 Supply-Side – Demand-Side Optimization and Cost-Environment Tradeoffs for China's Coal and Electricity Delivery System. *Energy Policy* 25: 313–326.

# 58

# UNINTENDED CONSEQUENCES: THE WAR ON DRUGS AND LAND USE AND COVER CHANGE IN THE ECUADORIAN AMAZON

JOSEPH P. MESSINA

*Michigan State University*

It is now well-recognized that, at the local, regional, and global scales, land-use changes are significantly altering land cover, perhaps at an accelerating pace. This transformation of the Earth's surface, particularly through deforestation, is, in turn, linked to a variety of scientific and policy issues affecting climate change and environmental sustainability. The world's scientific community is increasingly recognizing what, in retrospect, should have been obvious: that human behavior, as exacerbated by government policy, is a critical driver of global change.

The Ecuadorian Amazon (Figure 58.1) remains one of the few critical habitats left in the world. It is believed to be a Pleistocene refugia, or a place that has existed essentially unchanged for hundreds of thousands of years. This part of the Amazon lies in the classic humid tropics and is a true non-seasonal rainforest environment. It possesses several major centers of endemism (Whitmore and Prance 1987). Despite being one of Myers' (1988) global "hot spots" of biodiversity, the region is threatened by agricultural settlement. Settlers in the region are generally poor farmers, who originally lived on 50 hectare plots called *fincas*, clearing primary forest to grow subsistence crops (e.g., yucca) as well as commercial crops (e.g., coffee), and to create small pastures for cattle. High fertility and mortality rates characterize this rural population of in-migrant settlers. Among the Amazon-basin countries, Ecuadorian forests are most rapidly disappearing (FAO 1995). Oil exploration, road construction, and in-migration threaten globally significant conservation areas and national parks.

This paper introduces the use of geo-spatial data combined with expert knowledge and a dynamic spatial simulation model to describe, explain, and explore land use and cover change (LUCC) in the Ecuadorian Amazon, specifically along the border between Ecuador and Colombia, as both cause and consequence of the theorized direct and indirect impacts of Plan Colombia (the U.S.-based program to eradicate drug production in Colombia).

*D. G. Janelle et al. (eds.), WorldMinds: Geographical Perspectives on 100 Problems, 357–362.*
© 2004 *Springer. Printed in the Netherlands.*

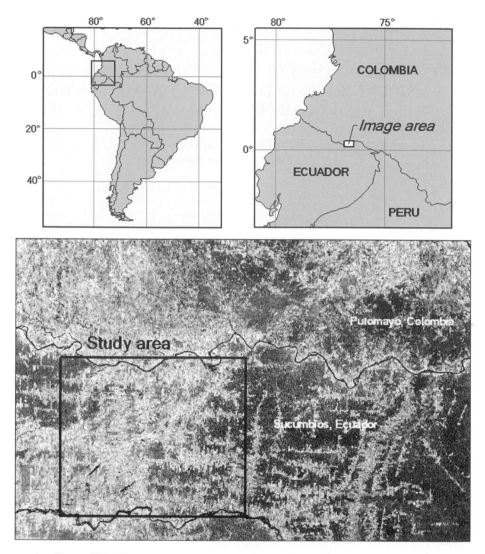

*Figure 58.1.* The study site in the northwestern Amazon Basin. Source: author.

## PLAN COLOMBIA

Plan Colombia was created during the Clinton administration to reduce the impact of drugs in the United States. (Crandall 2002). The specific goals of the plan are as follows:

> . . . one is a push into Southern Colombia; second is support for narcotics interdiction efforts; third is support for the Colombian National Police; fourth is support for developmental and particularly alternative development programs and approaches; fifth is support for justice

and other social sector reform. A sixth element, which does not directly affect Colombia itself, is our support for issues outside of Colombia, forward-operating locations, as well as several other countries that are affected by what happens in Colombia (U.S. Dept. of State, Office of the Spokesman, 12 March 2001).

As an example of the currency of this situation and its unintended consequences for Ecuador, excerpted below is a wire report from Reuters (6 Sept. 2001):

> QUITO, Ecuador: Colombia's defense minister traveled to Ecuador . . . to talk with President Gustavo Noboa about boosting border security as fears grow of a regional spillover from Colombia's war on rebels and drug traffickers. . . . Brief clashes on the border between Ecuadorian security forces and Colombian Marxist rebels raised worries Colombia's intensifying battles with guerrillas, paramilitaries and drug dealers will spill into neighbors like Ecuador. A U.S.-backed anti-drug offensive in Colombia . . . has been criticized for adding tension along the borders, with reports of rebels and refugees crossing into Ecuador for safety.

Plan Colombia has influenced the northern Oriente of Ecuador to the point where it is no longer safe to farm or conduct research in some of these areas. The author's fieldwork along the border in 1999 and 2000 revealed that large groups of farmers have been displaced, crops have been damaged, and access to markets severely limited, and the drug trade infrastructure has moved into the northern edges bordering the Putomayo River. The question remains, however, whether it is possible to map and model the landuse changes that have been seen and are expected to appear in the region over time. The goal of the research is to build predictive models that incorporate the social and the physical drivers of landscape change to characterize and simulate future landscape alterations due at least in part to the Plan Colombia.

## COMPLEXITY THEORY AND GEOGRAPHY

The goal of the science of complexity is to understand how simple, fundamental processes can combine to produce complex holistic systems (Gell-Mann 1995). This approach is based on the contention that some systems become complex, and for which computer simulation modeling is appropriate to capture the dynamics that reductionist-based, process-driven science cannot appropriately explain. Recognition of complex dynamics challenges some of the conventional geographic notions of stability and change (Manson 2000).

Cellular Automata (CA) models were conceived originally by Ulam and Von Neumann in the 1940s to provide a formal framework for investigating the behavior of complex, extended systems (von Neumann 1958). CA models are dynamic, discrete space-and-time systems and, as such, fall neatly within the purview of the discipline of geography. Geographers have long sought to describe and discover the relationships among the various elements that make up the population-environment matrix, and geographers have readily adopted

CA to give a conceptual cognizance with traditional geographic theory (Couclelis 1997).

A cellular automaton (CA) system is a raster data model. Each cell can be in one of a finite number of possible states, updated in discrete time steps according to a neighborhood function. Transition probabilities for the typical CA model depend on the state of a cell, the state of its surrounding cells, the physical characteristics of the cell (e.g., terrain, soil quality, vegetation, hydrology, and demographic characteristics), and the weights associated with the neighborhood context of the cell (e.g., proximity to the border). These weights and neighborhood conditions are determined from empirical analyses of LUCC based on social survey data, a comprehensive geographic information system (GIS) database of natural resources, and the linkages between villages, land parcels, and other landscape features.

The CA model developed for this research works by: (1) simulating the present by extrapolating from the past using LUCC base-maps derived from satellite imagery; (2) validating the simulations through field observations; (3) allowing the model to predict the future; and (4) comparing model outputs to traditional statistical tests (Messina and Walsh 2001). The Plan Colombia scenario was run in comparison with a control image, specifically, an image created by allowing the model to run into the future as if Plan Colombia did not exist. The basic assumptions as placed within the model include: (1) Plan Colombia had no effect on Ecuador earlier than 1999; (2) starting in 1999, increased urbanization due to more remote farmers being forced off their land, due to threats by drug lords, and to the current process of demographics in Latin America, where urbanization occurs along with population growth; (3) an increase in the effect of transportation on urban processes, due to greater constraints upon agricultural production and increased smuggling activities; (4) an even split in agriculture and secondary forest in the typical flux class due, to less controls in either functional direction; and (5) a reduction in relief influence, as drug production is likely to occur in any available area.

The results of the model run show dramatic increases in the amount of urban and a significant decrease in the amount of dense forest (Table 58.1). One of the interesting simulation results centers on the increase in urban associated with the enhanced transportation effect. The change in transition probabilities had little affect on the total amount of land in agriculture or

*Table 58.1.* Simulation Results (in Hectares).

|  | Control | Plan Colombia | Difference |
| --- | --- | --- | --- |
| Primary Forest | 21,762 | 18,875 | −2,887 |
| Secondary Forest | 73,452 | 72,022 | −1,430 |
| Agriculture | 45,049 | 45,622 | +573 |
| Urban | 3,154 | 6,910 | +3,756 |

secondary forest as compared to the reference image. With the 50/50 split though, patchiness should increase. Given these results, it is reasonable to summarize the impacts of Plan Colombia on the region as causing significant urban expansion and increasing deforestation. The first outcome was expected, but the second was not.

## CONCLUDING COMMENTS

The true measure of spatial complexity as applied in a CA context is one not yet fully realized in the literature. The tools exist to build complex (in the complicated not complexity sense) models to predict LUCC in a variety of environments and time periods. By integrating new computational methods, data sources, and theory, the underpinnings of LUCC may be explored and future scenarios predicted. One of the great challenges to the wider geospatial use of cellular systems specifically and complex dynamics generally is that very few mathematical and computational tools exist to deal with non-linearity. Currently, there are no workable alternatives that allow geographic models to function in a meaningful spatio-temporal sense. The advantage of the computer-driven dynamic spatial simulation is that it is a good middle ground between theory and experiment. It offers the rigor of mathematics without the generality, and the selection and repeatability of good experiments without the enforced connection to reality.

Spatial simulation models provide dynamic realizations of events occurring in time and space and allow policy development to proceed on an informed basis. Complex systems theory is an inherently spatial theory and, as such, is well utilized by geographers with disciplinary experience, relating theory to practice in a naturally spatial context. Geo-spatial tools, while computationally significant, should also be considered in terms that address general applicability to real-world problems. The drug war impacts many people directly and millions of people indirectly. Complex systems, computer modeling, and geography combine to place a very real face on the unintended consequences of this spatial issue and, over time, may influence the course of human events.

## REFERENCES

Couclelis, H. 1997. From Cellular Automata to Urban Models: New Principles for Model Development and Implementation. *Environment and Planning B* 24: 163–174.

Crandall, R. 2002. *Driven by Drugs: U.S. Policy towards Colombia*. Boulder: Lynne Reinner Publishers.

FAO. 1995. *State of the World's Forest*. Rome: Food and Agricultural Organization.

Gell-Mann, M. 1995. *The Quark and the Jaguar: Adventures in the Simple and the Complex*. New York: W. H. Freeman.

Manson, S. 2000. Simplifying Complexity: A Review of Complexity Theory. *Geoforum* 32: 405–414.

Messina, J. and S. Walsh. 2001. 2.5D Morphogenesis: Modeling Landuse and Landcover Dynamics in the Ecuadorian Amazon. *Plant Ecology* 156: 75–88.

Myers, N. 1988. Threatened Biotas: Hotspots in Tropical Forests. *The Environmentalist* 8: 1–20.

U.S. Department of State. 2001. Office of the Spokesman, 12 March.

von Neumann, J. 1958. *The Computer and the Brain.* New Haven, Yale University Press.

Whitmore, T. and G. Prance, eds. 1987. *Biogeography and Quaternary History in Tropical America.* Oxford Monographs on Biogeography No. 3. Oxford, Clarendon Press.

# 59

# A GEOGRAPHICAL PERSPECTIVE ON COCA/COCAINE IMPACTS IN SOUTH AMERICA

KENNETH R. YOUNG

*University of Texas*

Coca is the common name of a shrub, *Erythroxylum coca*, domesticated millennia ago in the Andes Mountains of South America. Its leaves are picked and transformed into various products, one of which is cocaine. The result is to spatially link the locations of coca cultivation in South America to the places in the world where demand for cocaine exists. There are only a handful of plants that directly affect both the welfare of millions of people and the world economy, including food species such as rice, wheat, and maize. The two globally significant plants that are transformed into important illicit drugs are poppy and coca. In this essay, I present a geographical perspective on the negative impacts associated with the latter.

## THE COCA/COCAINE COMMODITY CHAIN

The coca plant grows about one to two meters tall, and has elliptical leaves and thin branches. There are wild populations growing in humid tropical forests in the foothills of the eastern Andes from Colombia to Bolivia. Domesticated coca is similar in appearance, but has been brought into cultivation over a broader region, including parts of lowland Amazonia and up into the Andes (Plowman 1984). The principal source countries are Colombia, Peru, and Bolivia.

The traditional use of coca is for leaf chewing. Mastication of leaves with lime liberates a variety of alkaloids that counteract feelings of hunger and tiredness. There is no addictive effect at those dosage levels. Instead, coca is a useful product for people working many hours in their agricultural fields or walking long distances across the mountains. The coca leaf, itself, is symbolic of ancient and deeply held spiritual beliefs concerning the Andes (Allen 1988). Even some secular groups in Bolivia and Peru use the coca leaf to symbolize nationalism and Andean values.

However, only a few percent of harvested coca leaves are destined for this benign use. Most leaves are collected and sold to intermediaries who transform the raw leaves into cocaine paste by mixing with sulfuric acid and

*D. G. Janelle et al. (eds.), WorldMinds: Geographical Perspectives on 100 Problems, 363–367.*
© 2004 *Springer. Printed in the Netherlands.*

other chemicals (Morales 1989). Cocaine paste is bulky and impure, but is currently a favored drug of the underclass of cities in the Andes and Amazon because is relatively cheap, can be smoked, and is highly addictive. The remainder of the paste is refined in laboratories into cocaine. This final step typically takes place far from the areas of cultivation of coca. From there most cocaine is sent to the United States and Europe, where its value becomes much greater than what was paid for the leaves. However, even the small amount received by the coca farmer is hundreds of times more than would be earned from maize, sugarcane, or coffee.

## ENVIRONMENTAL AND SOCIAL IMPACTS

Probably several million hectares of tropical forests have been cut by people growing coca. This would rank coca/cocaine as one of the world's major causes of tropical deforestation. The foothill forests being cut are some of the world's most biodiverse (Young 1996). There are plant and animal species present both from the higher-elevation cloud forests and also from the Amazonian lowlands. There are also numerous birds, frogs, spiders, snails, ferns, and orchids found nowhere else.

The response to this threat has been inadequate on the part of the international environmental community. The nature reserves in these regions of South America are only protected by virtue of their inaccessibility and those located near deforestation fronts soon lose much of their biodiversity value. In addition, deforestation in areas such as the Putumayo in Colombia, the Huallaga and Ene River valleys in Peru, and the Chapare in Bolivia has fragmented the once near-continuous forest corridor along the base of the tropical Andes.

One of the ironies of this deforestation is that the expected spatial patterns are reversed. Most colonists in the humid tropics cut forest in areas near to roads or rivers, and choose places with the best soils and gentle slopes. When the goal is to plant coca, this must be done in isolated places, far from military and civil control. The more intense the drug enforcement, the more likely it is that remote, steep hill slopes will be deforested. Because cocaine demand is inelastic, eradication of coca in one area simply shifts cultivation to other places. When the enforcement involves aerial spraying of herbicides or military tactics, it becomes controversial on environmental and humanitarian grounds.

The deforested sites where coca is grown include some of the most dangerous places on Earth. The money brought in to pay for the leaves, paste, and cocaine also buys weapons and produces corruption. The national government of Colombia has lost effective control of these and nearby areas. This was the case also for much of Peru and parts of Bolivia from 1985 to 1995 (e.g., Poole and Rénique 1992). Deforestation frontiers attract new, often young, colonists who are more likely to assume the inherent risks, more likely to cut forests at or beyond the end of the road system, and least likely to worry about long-term sustainability.

This puts long-term settlers, and in particular the original inhabitants of these lands, the indigenous peoples, directly in the path of deforestation, violence, land speculation, inflation, and dollarization. It is interesting to note that some places, where indigenous people, such as the Aguaruna or Machiguenga, are isolated from roads and the market economy, and where they have traditional rights to large tracts of land, have proven resistant to the entrance of coca/cocaine (Young in press).

## SUBSIDIES FOR DEFORESTATION

The cocaine-consuming countries have reacted to coca in a harsh, although misdirected manner. The United States and many European countries condition their foreign assistance and diplomatic ties on perceptions of Latin American countries' willingness to combat in the "drug war." For the transshipment countries, this implies making money laundering laws and enforcement efforts. For the source countries, this has required a series of both enforcement and alternative development activities. Cynically, the visible corruption under the regimes of Noriega in Panama and Fujimori in Peru was tolerated for many years for other geopolitical reasons.

The humid tropical foothills where coca is cultivated share similar biophysical environments. They also often have a colonial or even pre-colonial history of coca cultivation. But the most important spatial connection to coca expansion is the presence of roads, which are used to bring in the processing chemicals and to take out the products of deforestation (timber, crops, coca paste). Thus, an expansion of coca cultivation accompanies roads and colonization.

Without exception, the current coca-growing areas are past tropical forest colonization projects. This began in the 1960s and continued into the 1990s despite a near universal failure of these projects. Currently the United States is funding the construction of roads through tropical forests in Peru under the guise of "road improvements," which escape mandated environmental oversights. These roads go to places where coca is cultivated. Thus, the very source areas of illicit drugs are rewarded with better infrastructure and lower transportation costs. There have been cases in both Peru and Bolivia where farmers have repeatedly received payments for supposedly abandoning their coca crops.

The alternative development programs typically consist of subsidies to produce other crops, such as citric crops, coffee, or cacao (OTA 1993). None of these have been successful. More to the point, these subsidies favor the same land-use practices that have resulted in so much forest loss.

## NEW POLICIES FOR COCA/COCAINE

By being sensitive to the spatial relationships and to the underlying environmental and socioeconomic processes, geographers can help reorient policy for illicit drugs, such as cocaine, and its precursor, coca. The places at risk can be identified, mapped, and studied in terms of biodiversity loss, environmental justice, and land-use change.

There are two land-tenure systems in the western Amazon that have protected tropical forests and indigenous lands despite coca/cocaine: nature reserves and the lands that indigenous people control. Granted this protection is often passive, soon pushed aside when the deforestation front approaches. But active conservation practices might slow or even halt that advance, given interest, funding, and the removal of the perverse subsidies (Myers and Kent 2001) that favor deforestation.

The resistance to coca/cocaine shown by indigenous people is probably best facilitated by legal and social assistance. Land titles and territorial demarcation need to be made available (Stonich 2001). Exploration for petroleum and natural gas continues throughout the region; when poorly planned or implemented, this can result in informal (and illegal) colonization. In addition, there are determined efforts throughout tropical South America to cut the last stands of mahogany. These efforts always pull along and expand crude road systems and new colonization.

Asháninka warriors in central Peru showed the most dramatic resistance to outside influences from 1986 to 1993 (e.g., Brown and Fernández 1991). They fought with bows and arrows against determined groups of Shining Path guerillas armed with automatic weapons and financed with cocaine money. Hopefully, less vivid but more long-term solutions can be found to the environmental degradation and social disruption caused by coca/cocaine. Geographers could and should be a part of the search for explanation and sustainable alternatives.

## REFERENCES

Allen, C. J. 1988. *The Hold Life Has: Coca and Cultural Identity in an Andean Community.* Washington, D.C.: Smithsonian Press.
Brown, M. F. and E. Fernández. 1991. *War of Shadows: The Struggle for Utopia in the Peruvian Amazon.* Berkeley: University of California Press.
Morales, E. 1989. *Cocaine: White Gold Rush in Peru.* Tucson: University of Arizona Press.
Myers, N. and J. Kent. 2001. *Perverse Subsidies: How Tax Dollars Can Undercut the Environment and Economy.* Washington, D.C.: Island Press.
OTA. 1993. *Alternative Coca Reduction Strategies in the Andean Region. Office of Technology Assessment.* Washington, D.C.: Congress of the United States.
Plowman, T. 1984. The Ethnobotany of Coca (*Erythroxylum* spp., Erythroxylaceae). *Advances in Economic Botany* 1: 62–111.
Poole, D. and G. Rénique. 1992. *Peru: Time of Fear.* London: Latin America Bureau.

Stonich, S. C. ed. 2001. *Endangered Peoples of Latin America: Struggles to Survive and Thrive.* Wesport, CT: Greenwood Press.

Young, K. R. 1996. Threats to Biological Diversity Caused by Coca/Cocaine Deforestation in Peru. *Environmental Conservation* 23: 7–15.

Young, K. R. In press. Environmental and Social Consequences of Coca/Cocaine in Peru: Policy Alternatives and a Research Agenda. In M. K. Steinberg, J. J. Hobbs, and K. Mathewson, eds. *Dangerous Harvest: Psychoactive Plants and the Transformation of Indigenous Landscapes.* Oxford: Oxford University Press.

60

# GLOBAL ENVIRONMENTS AND RURAL COMMUNITIES: ENHANCING COMMUNITY CONSERVATION INITIATIVES IN EAST AFRICA

JEFFREY O. DURRANT and MARK W. JACKSON
*Brigham Young University*

When Hans Meyer stood on the summit of East Africa's legendary Mount Kilimanjaro in 1889 he most likely didn't consider the possibility that his Chagga guide, Yohani Kinyala Lauwo, would still be alive to participate in the one-hundredth anniversary of this initial ascent. Yet amazingly Lauwo attended the celebration seven years before his death in 1996 at age 125.

During the century between Lauwo's initial ascent and his death, Kilimanjaro, like many other global environments, has changed dramatically. Concern over these changes has been the driving force in the worldwide establishment of a dominant preservation strategy based on establishing formal protected areas, such as Kilimanjaro National Park. While few would argue against the value of preserving these spectacular and ecologically important areas, there has been concern and even criticism of the "fences-and-fines" approach typical of many protective area designations. Criticism has centered on both the effectiveness and appropriateness of a strategy that often focuses on ecological preservation, without regard to the socio-cultural elements of nearby human populations (Hughes 1996; Pimbert and Pretty 1997). The result of this failure to integrate the local human realm into conservation planning and management, critics argue, has typically led to counter-productive and adversarial relations with rural residents instead of enhanced conservation (Neumann 1998).

On Mount Kilimanjaro, thousands of people from across the globe come hoping to catch a glimpse of the roof of Africa. Lauwo's Chagga people are now more numerous than ever, and those who fail in the increasingly difficult task of obtaining fertile land on Kilimanjaro's lower slopes must either follow Lauwo's legacy and join a new generation of porters and guides assisting climbers or leave the mountain in search of opportunities elsewhere. There is also concern about environmental change on the mountain in the form of a diminishing forest, reduced stream flow, and the gradual disappearance of the summit glaciers encountered by Meyer and Lauwo.

Global environmental change during the past century is certainly not limited

369

*D. G. Janelle et al. (eds.), WorldMinds: Geographical Perspectives on 100 Problems, 396–374.*
© 2004 *Springer. Printed in the Netherlands.*

to Kilimanjaro. It extends to the rainforests of Borneo and the Amazon, South Pacific coral reefs, Alaskan tundra, and other places. Yellowstone became the world's first National Park 17 years before Lauwo's ascent of Kilimanjaro, initiating a trend that has gained substantial momentum during the past several decades. In 1997, the World Conservation Union listed over 50,000 National Parks, Wilderness Areas, Conservation Areas, and other protective designations worldwide (Green and Paine 1997). During the past five years the trend has continued, with numerous protected areas designated worldwide, including a new national park in central Borneo, a dozen conservation areas and National Monuments in the Western Untied States, and additional designations in Tanzania, where 12 national parks and other relatively restrictive conservation areas already cover nearly 15 percent of the country.

Criticism of the dominant protected-area approach has led to the creation of strategies that attempt to integrate the social and ecological processes. These programs, known by names such as "conservation within development," "community-based conservation," and "integrated conservation and development programs," all attempt to connect or tie local community "development" and interests with ecological conservation (Little 1994; Alpert 1996). At Kilimanjaro National Park (KINAPA) and other Tanzania Parks these efforts have been formalized in the creation of a "Community Conservation Service" that has the mandate of working with local communities surrounding Park boundaries, including revenue sharing (TANAPA 2000).

As with previous conservation efforts, community-based approaches have not escaped criticism. Apart from the difficulty of combining ecological processes and the oft-contentious concept of "development," these programs have recently had their purported success seriously questioned. Specifically, critics point out that their conclusions are formed with little or no basis on sound ecological data (Newmark and Hough 2000).

The common flaws of failing to effectively take into account biophysical and social-cultural processes and working from sparse or even absent ecological data are serious threats to the long-term success of preserving the earth's ecological heritage. While geography has not been completely absent from conservation efforts, an increased effort to use geographical approaches offers interesting possibilities for working with rural populations to save important global environments. In particular, geography's long tradition of engaging both human and environmental realms and geographical data collection and analysis methods provide both theory and methods that can bring biophysical and socio-cultural processes together based on more rigorous data collection.

On Mount Kilimanjaro, we are using a geographical approach to develop better understanding of connections between the social and ecological realms in and around the protected area boundaries and to improve the management capabilities of the Park, particularly in relation to the efforts of an under-resourced Community Conservation Service (CCS) department.

Kilimanjaro National Park and Forest Reserve encompass the upper slopes of the mountain, and there are 80 villages adjacent to the Park boundaries that are the focus of CCS efforts. The villages are rural agrarian-based communities that have historically accessed the upper forest for livestock forage, firewood collection, and other uses. While several paved roads run from the base of the mountain to a few villages, these roads are primarily associated with the more popular tourist trekking routes, and most villages are relatively isolated and difficult to access.

The Kilimanjaro CCS began activities in 1994 and received a permanent CCS warden in June 1995. As of 2002, the CCS staff stood at two – the warden and an assistant. In 1999, KINAPA developed a "Strategic Action Plan" for the CCS. Among other findings listed was the obvious weakness of inadequate staff, equipment, and financial resources, all critical shortcomings. The key issues facing the CCS are outlined in the report as: wildfires; cattle grazing; inadequate communication facilities; lack of current information; poaching; poor farming practices; environmental education program development; and transboundary issues (KINAPA 1999).

Like many protected areas, KINAPA is facing a formidable set of issues. To help the CCS address these issues, we have begun the development of a

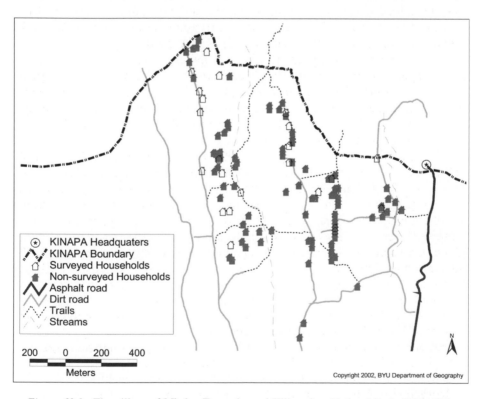

*Figure 60.1.* The village of Mbahe, Tanzania, and Kilimanjaro National Park (KINAPA).

geographic information system (GIS) for use by KINAPA, particularly the CCS. The backbone of this initiative is an extensive effort to field-map the surrounding villages, including roads and paths into the protected area boundaries using global positioning system (GPS) technology. The mapping will help KINAPA better understand the villages they are working with, while also laying the foundation for ecological data collection that can be linked to social data on the communities. Figure 60.1 illustrates initial field mapping of Mbahe, a village near KINAPA headquarters. It is estimated that it will take 3–5 years to map all 80 villages and routes across the Park to this level of detail.

Each household in upper Mbahe village was geolocated. A random sample of households was then selected and located (Figure 60.1) for participation in an extended survey and interview. Surveyed households are connected to data gathered from the survey on size of plot, household size, livelihood sources, and land-use. Extended local-resident comments and observations,

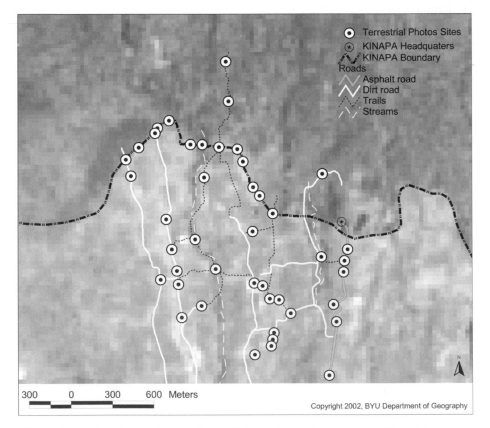

*Figure 60.2.* Ground control points for rectifying and classifying Landsat ETM+ data on Mt. Kilimanjaro, Tanzania.

such as opinions on the purpose of KINAPA and past experiences with the CCS outreach program, are also tied to these points.

In addition to providing a detailed and relatively accurate map of these communities, the mapping efforts allow visualization and integration with environmental data, such as land-cover change, as detected by satellite imagery. A subset of Landsat 7 data, acquired in 2000, is overlayed with elements of Mbahe field mapping data (Figure 60.2). When imagery, cartographic, and survey data are combined digitally in this manner, it becomes possible to monitor land use/cover inside and adjacent to the Park effectively by providing not only change information but also clues as to why it has changed. In addition to household survey data, georeferenced terrestrial photographs (Figure 60.2) are also connected to social and ecological field-note observations in a larger database. Though the Landsat 7 imagery provides an initial view of vegetation change along the Park boundary region, the data coarseness (30 m pixels) is only marginally useful in land use/cover change on a scale as fine as the villages along the border. For socioeconomic data analysis, a resolution of 0.25–5 m is necessary (Jensen 2000), and this is increasingly available through data sets from higher-resolution satellite sensors.

While the prohibitive cost of some aspects of newer GPS technology and satellite imagery would appear to make this approach impractical on a widespread basis, these costs are diminishing, and it will undoubtedly not take another Lauwo lifetime before they become practical. Combining the advances in technology used by geographers with time-tested field observations and interviews offers exciting possibilities for advancing knowledge in many realms of geographic inquiry, particularly in community conservation efforts. These techniques provide more accurate and integrated data and analysis methods that address the common failings of efforts to preserve the earth's environment. A century after a young Lauwo scaled Kilimanjaro and ushered in a new era on the mountain, geography's insight and approach are improving efforts to conserve the mountain's environment for future generations.

## REFERENCES

Alpert, P. 1996. Integrated Conservation and Development Projects. *Bioscience* 46: 845–855.

Green, M. and J. Paine. 1997. *State of the World's Protected Areas at the End of the Twentieth Century.* Paper presented at the IUCN World Commission on Protected Areas on "Protected Areas in the 21st Century: From Islands to Networks" Albany, Australia: 24–29 November.

Hughes, D. 1996. When Parks Encroach Upon People: Expanding National Parks in the Rusitu Valley, Zimbabwe. *Cultural Survival Quarterly* Spring: 39–40.

Jensen, J. 2000. *Remote Sensing of the Environment: An Earth Resource Perspective.* Upper Saddle River, NJ: Prentice Hall.

KINAPA (Kilimanjaro National Park). 1999. *Community Conservation Service Strategic Action Plan,* February.

Little, P. 1994. The link Between Local Participation and Improved Conservation: A Review of Issues and Experiences. In D. Western and P. Wright, eds. *Natural Connection: Perspectives in Community-Based Conservation*, 347–372. Washington D.C.: Island Press.

Neumann, R. 1998. *Imposing Wilderness: Struggles over Livelihood and Nature Preservation in Africa.* Berkeley, CA: University of California Press.

Newmark, W. and J. Hough. 2000. Conserving Wildlife in Africa: Integrated Conservation and Development Projects and Beyond. *Bioscience* 50: 585–592.

Pimbert, M. and J. Pretty. 1997. Parks, People and Professionals: Putting 'Participation' into Protected Area Management in Social Change and Conservation. In K. Ghimire and M. Pimbert, eds. *Environmental Politics and Impacts of National Parks and Protected Areas*, 297–332. London: Earthscan Publications Limited.

TANAPA (Tanzania National Park Authority). 2000. *Strategic Action Plan for Community Conservation Services* (Draft Report).

# 61

# WATER MANAGEMENT IN THE WEST: CONTROLLING THE IMPACT OF CUMULATIVE POND DIVERSIONS

TERESA L. BULMAN
*Portland State University*

Small ponds are generally unregulated under western water law, meaning owners can divert and store water without a permit. In areas of intensifying agriculture, the number of newly created small ponds has led to cumulative water diversions in excess of flow capacity. At the same time, analysis of the location of ponds, diversion points, and of timing of diversions has created a new understanding of the impact of pond diversions. The result has been increased regulation of pond diversions to meet water needs while maintaining instream flows for wildlife and downstream users.

## THE CONTEXT OF SMALL PONDS IN WESTERN WATER MANAGEMENT

Small agricultural ponds have played an important role in the development of western agriculture, especially in water-depleted areas such as California and Arizona. Such storage diversions were initially hailed as a solution to the massive direct diversions, with instantaneous water pumping by hundreds of users that frost events and mini-droughts prompted. Storage diversions to small reservoirs allow for slower rates of water withdrawal and storage in anticipation of extreme weather events. The small reservoirs usually are not subject to regulation because they fall under the regulatory standard that exempts ponds of less than approximately 10 acre-feet. As a result, little attention has been paid to the cumulative impact of small storage. However, a geographic analysis completed in the early 1990s identified the growing impact of small reservoir diversions in Napa County (Bulman 1994). Since that time, regulations in several western states have begun to address the problem.

*D. G. Janelle et al. (eds.), WorldMinds: Geographical Perspectives on 100 Problems*, 375–379.

## THE NAPA CASE STUDY

With growth in urban areas and agriculture during the latter half of this century, water has been in increasingly short supply in Napa County, California, where the dominant economic activity is wine making. Natural precipitation is usually sufficient for grape growing and, as a result, dry-farming of vineyards is the predominant practice. However, a serious threat to grapevines is early-spring frost and the probability of damage from frost was one of the factors initially limiting the areas suitable for grape growing (Winkler et al. 1974).

In their efforts to defeat the threat of frost damage, grape growers have used several active frost protection systems brought about by advances in technology, including orchard heaters, vineyard fans, and sprinkler systems, with sprinklers being the most effective in deep or prolonged frost events. The benefit of sprinkler protection systems was vividly demonstrated in 1971 when grapevine leafing and budding came late after a severe frost. Fruit on frost-protected vines matured normally, while those on unprotected vines did not. Today, wind machines, heaters, sprinkler systems, or a combination of these protects most vineyards. Subsequently, as the demand for grape production increased, and the price of grapes increased, large investments in irrigation and frost-protection systems began, including construction of private reservoirs to provide a reliable water supply.

By 1990, over 600 small, private reservoirs diverted large amounts of water to storage for vineyard uses (Figures 61.1 and 61.2), and served over 17,000 hectares of Napa County land, an increase of 555 percent over hectares served in 1972. The diversion of over 30,000 cubic decameters (dam$^3$) of water to private reservoirs has a discernible impact on the seasonality of water supply, on annual streamflow, and on ground-water storage. These impacts, in turn, affect public water supply and the urban and economic growth of the county (Bulman 1994).

The expanding use of water for frost-protection purposes, beginning in the 1960s, gave the California State Water Resources Control Board (the "Board") special concern because it feared that high instantaneous demand for water during a frost event in Napa Valley might drain the Napa River dry. Hence, construction of reservoirs was initially encouraged so water could be stored prior to frost events. Reservoirs under the 10 acre-feet exemption did not require permits, however, so the total number and volume of reservoirs could not be monitored. The growing number of reservoirs, combined with the instantaneous water diversion during severe frost events, prompted the Board to address the problem. Ironically, their attempt to control diversions from the river further promoted the growth of reservoir construction.

On 13 March 1974, the Board initiated an action against "Boots" Forni, one of the original users of sprinkler frost protection in Napa Valley, to enjoin Forni and others from drawing water from the Napa River to their vineyards for frost protection – *People ex rel. State Water Resources Control Board v.*

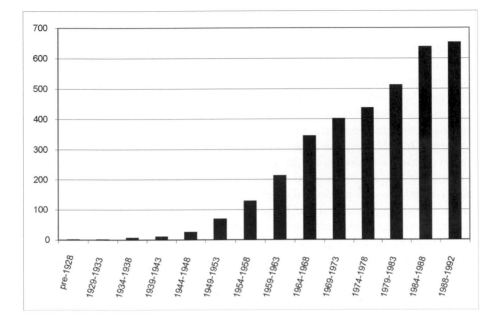

*Figure 61.1.* Cumulative number of reservoirs in Napa County. Reprinted with permission of the American Water Resources Association from T. L. Bulman, 1994. "When the River Runs Dry: Management Responses to Water Diversions." In *Effects of Human-Induced Changes on Hydrologic Systems, Proceedings of the American Water Resources Association*, 707–716. Jackson Hole, WY: American Water Resources Association.

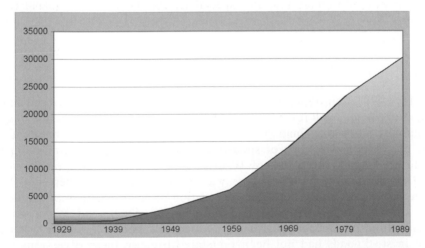

*Figure 61.2.* Total volume of reservoirs in Napa County (cubic decameters). Reprinted with permission of the American Water Resources Association from T. L. Bulman, 1994. "When the River Runs Dry: Management Responses to Water Diversions." In *Effects of Human-Induced Changes on Hydrologic Systems, Proceedings of the American Water Resources Association*, 707–716. Jackson Hole, WY: American Water Resources Association.

*Forni*, 54 Cal.App. 3d 743, 126 Cal. Rptr. 851 (1976). The *Forni* case was settled in 1976, requiring the defendants to comply with the orders of a Board-appointed watermaster. The watermaster may allot Napa River water and control the amount, rate, and times of pumping by the defendants, and require installation of meters. The watermaster may determine each grower's allotment and, when the available water supply is less than ample to satisfy all growers' allotments, the watermaster will allot available water correlatively in proportion to the total riparian acreage under sprinkler frost-protection systems. In addition, the watermaster can order the construction of small reservoirs to store diversions in anticipation of frosts. The Board wanted to ensure that the general provisions of the judgment would apply to all Napa *River* riparians, and proceeded to effect changes in California water regulations aimed specifically at the Napa River situation. The general provisions of the *Forni*-stipulated judgment are in effect codified in Section 735 *of* Title 23, "Napa River, Special," of the California Code of Regulations and apply to all diversions of water from the Napa River stream system between 15 March and 15 May.

The *Forni* case has been cited as an example of a line of cases in water law, which put limits on the waste or misuse of water (Dunning 1987). Reservoir construction has in fact solved the water-scarcity problem for some Napa County grape growers, particularly riparian owners who previously had no legal right to store water for long periods of time. However, as vineyard and urban expansion continue in Napa Valley, the failure to consider the cumulative impact of private reservoir construction on water supply in Napa County has led to more lawsuits, special legislation, and water-management crises.

## RECENT SMALL POND REGULATION

By the 1990s, California, like other western states, recognized that the cumulative impact of small reservoirs needed to be evaluated. However, the biggest stumbling block to evaluation and regulation was a lack of information about the existence and volume of such reservoirs. Because reservoirs under the statutory exemption volume of 10 to 15 acre-feet (depending on the state) had never required permits, there was no record of such reservoirs. Now states have begun to require the recordation of small water impoundments. By the late 1970s, Arizona had adopted the Stockpond Water Rights Registration Act, recognizing that ". . . many diversions of surface water for use in stockponds had not been registered because these diversions were thought to be too small to be of concern. With the growing recognition of the competition for surface water, the . . . Act was enacted to grant statutory recognition of previously unrecorded stockponds" (Arizona Department of Water Resources 2002). The Act also gives water managers vital information about volume and location of small pond diversions. California adopted regulations (Water Code 1226.1) governing impoundments under 10 acre-feet, which give valid water rights to only those who register their impoundments with the Board. The regulations give the Board data on the impoundments, such as volume and surface area. Even Oregon, a state whose western region is well endowed with water, has had to address ponds. In 1995, the Oregon Legislature enacted regulations providing exempt status to ponds less than 9.2 acre-feet if the owners applied for exemption, thereby registering the pond. Within four years, Oregon had issued exemptions for more than 6,500 ponds (State of Oregon 1999).

But registering ponds may not be enough to regulate these impoundments and reduce their cumulative impact. While registration provides state water managers baseline data on small reservoirs, it may not give enough regulatory control. As recent litigation in California has shown, water managers may need to regulate numerous aspects of pond diversions. That case, State Water Resources Control Board Water Right Decision 1636 (1997), held that the Board could not only regulate the quantity and time of diversion of water to a 9 acre-feet pond, but could also require stream bypass minimums, erosion control, and pesticide management. Such increasing regulatory scope is likely to be the trend in the Western United States as the demand for scare water resources increases.

## REFERENCES

Arizona Department of Water Resources. 2003. *Summary of Statewide Surface Water Management: The Stockpond Water Rights Act of 1977.* Available online at: http://www.water.az.gov/adwr/Content/Publications/files/SurfMgt.pdf.

Bulman, T. L. 1994. When the River Runs Dry: Management Responses to Water Diversions. In *Effects of Human-Induced Changes on Hydrologic Systems, Proceedings of the American Water Resources Association*, 707–716. Jackson Hole, WY: American Water Resources Association.

Dunning, H. C. 1987. State Equitable Apportionment of Western Water Resources. *Nebraska Law Journal* 66: 76–119.

Hadley, R. F. 1980. The Role of Small Reservoirs in Hydrologic and Geomorphic Studies. In H. G. Stefan, ed. *Proceedings of the symposium on surface Water Impoundments, June 2–5, 1980, Minneapolis, Minnesota.* New York: American Society of Civil Engineers.

State of California Water Resources Control Board. 1976. *People ex rel. State Water Resources Control Board v. Forni*, 54 Cal.App. 3d 743, 126 Cal. Rptr. 851.

State of California Water Resources Control Board. 1997. Water Right Decision 1636.

State of Oregon. 1999. Strategic Plan 1999–2001. See http://www.wrd.state.or.us/publications/stratplan99/supply2.html.

Winkler, A. J., J. A. Cook, and L. A. Lider. 1974. *General Viticulture.* Berkeley: University of California Press.

# 62

# THE CONSERVATION RESERVE PROGRAM:
# A SOLUTION TO THE PROBLEM OF AGRICULTURAL
# OVERPRODUCTION?

PHILIP J. GERSMEHL and DWIGHT A. BROWN
*University of Minnesota*

American agriculture has long had a problem of overcapacity – the country is capable of producing far more food than it can sell at the prices its citizens and foreign customers are willing to pay. The reasons for this overcapacity are multiple and complex, including exceptionally favorable environment (in some areas), impressive technology (in some areas), families who undervalue their labor to gain a rural lifestyle (in some areas), government subsidies that encourage production (in some areas), and the ability to externalize major costs of nutrient management and sediment control (in some areas).

Geographers have examined many facets of agriculture, including patterns of food production, at a variety of scales from global to local (Hart 1991; Furuseth 1997), the effects of agricultural innovations such as hybrid corn or no-till farming (Hudson 1994), the movement of water, sediment, nutrients, and pesticides to and from fields and feedlots (Kromm and White 1992; Trimble 1999), the effects of different political economies, ownership patterns, and management systems (Archer and Lonsdale 1997), and the integration of environmental and agricultural information into a regional framework for planning purposes (Omernik 1987). And some have translated these findings into educational materials (Gersmehl et al. 1998). Many of these studies were part of the milieu of scientific opinion that led to the passage of the 1985 Farm Bill, one of the landmark pieces of land-use legislation in the history of the United States.

## SCALE PROBLEMS IN THE ANALYSIS OF THE 1985 FARM BILL

The direct effects of the 1985 Farm Bill included the paid idling of about 15 million hectares of cropland and the imposition of federally approved land-use plans on more than 80 million hectares of land still used for crops. Together, these two programs had a direct impact on nearly 70 percent of all cropland in the United States. Scholars from many disciplines have tried to assess the

*D. G. Janelle et al. (eds.), WorldMinds: Geographical Perspectives on 100 Problems*, 381–386.
© 2004 *Springer. Printed in the Netherlands.*

effects of this legislation. Unfortunately, the vast majority of those studies commit an elementary geographic fallacy: they scale the results of a field test up to a national estimate without checking whether national totals derived from other sources also support the conclusions derived from that scaling.

Here is a simple example of that fallacy. At a local scale, a wildlife biologist studies a 40-hectare field that was "removed" from crop production under the Conservation Reserve Program. She observes that the field now supports 10–12 pheasant coveys under its new cover of perennial grasses and forbs, whereas it had at most one or two when it was used for annual crops. Meanwhile, at a state scale, an analyst for the Farm Service Agency reports a total of 400,000 hectares of land enrolled in the CRP in that state. Hunters also notice an increase in pheasant harvest. A third analyst then takes those three bits of information – the field study, a state total, and some anecdotal observations – and draws the plausible conclusion that the CRP is "responsible" for an increase of 90,000 pheasant coveys in the state (an average of nine "new" coveys per 40 hectares, multiplied by 400,000 hectares). The logic sounds plausible, but it may also be wildly erroneous. It is worth noting that the 1985 Farm Bill was passed at the beginning of a decade that had six of the mildest winters in the century. When a "normal" winter returned, the number of pheasants dropped sharply. That one year suggests that the original causal analysis did not include all of the variables that might have been causally related and, as a result, the CRP was given credit that it did not deserve. Unfortunately, it is difficult to find solid evidence either way, because the apparent plausibility of the first conclusion led to a reduction of efforts to find a link between pheasant density and other environmental variables.

Other studies noted a dramatic reduction in dust storms that occurred in the years following the passage of the 1985 Farm Bill. They attributed this improvement in air quality to the enrollment of land in the CRP. One analysis by geographers, however, noted a significant change in tillage practices during the same years, and the analysis revealed that the effect of the tillage change was greater than the effect of the CRP (Ervin and Lee 1994). This cautionary stance was the exception, however; scale-up fallacies plagued a large proportion of the studies that tried to find links between CRP enrollment and other environmental conditions, such as water quality, lake eutrophication, pesticide concentrations, and predator efficiency.

Amid this flurry of field-scale studies, few people bothered to consult the Census Bureau, which gathers an independent form of data that can shed light on land-use trends. The first release of national totals from the Census of Agriculture gave no cause for concern; they showed a decline of about 14 million hectares of cropland between 1982 and 1992, which corresponded with the total of slightly less than 15 million hectares enrolled in the CRP in that time interval (Osborn et al. 1990; Kellogg et al. 1994). Subsequent state-by-state analysis, however, revealed huge and disturbing discrepancies.

In seven states of the Great Plains, the region with the most CRP participation, landowners enrolled 7.0 million hectares of land in the CRP between 1985 and 1991. The Census, however, reported that the total amount of harvested cropland in those states declined by only 1.1 million hectares between the 1982 and 1992 census years. After adjusting for mismatched reporting years and for differences in paid diversion programs and designated natural disaster areas, we are left with the clear indication that someone must have plowed between four and five million hectares of "new" cropland in the region. The identity of that "someone" is not well known, because that particular aspect of the issue has received little attention. The lack of rigorous study is perhaps understandable because the decision to embark on, or even to fund, a study of cropland "creation" is predicated upon acknowledging that such creation is occurring, and that is something the proponents of the CRP are reluctant to do.

This point emerges even more dramatically when we compare the 1992 and 1997 Censuses of Agriculture. The top 15 states had 7.6 million hectares of land enrolled in the CRP in 1992. Many of the ten-year contracts from the first CRP were scheduled to expire shortly after 1995. Overwhelmingly, the owners of CRP land wanted to "re-up," to sign another agreement to keep their land out of crop production in exchange for another ten years of payments; in most cases, CRP payments had been at least 50 percent higher than cash rent for similar land. Meanwhile, many other landowners also wanted to get into the reserve program, and by 1997 the CRP acreage in these 15 states had increased by more than 25 percent, to 9.5 million hectares (U.S. Department of Agriculture 1997).

If one employs a "pie-slice" view, in which different kinds of land use are depicted as slices on a pie graph, and removal of land from one slice results in a decrease in its size and an increase in at least one of the other slices, one would expect that the amount of harvested cropland would have declined by about 2 million hectares during this period. The Census, however, revealed an *increase* of nearly 4 million hectares of cropland in those states between 1992 and 1997. The state with the largest increase in CRP acres, Texas, also had the largest increase in harvested cropland. The next four states put 0.6 million additional hectares in the CRP, but the amount of harvested cropland in those states increased by nearly 1.6 million acres. Indeed, the positive correlation between the change in CRP acreage and the change in harvested cropland is high enough to justify an utterly counter-intuitive statement: "one fairly good predictor of the *increase* in harvested cropland between 1992 and 1997 is the amount of additional land that the government paid landowners to *remove* from production in that state during roughly the same time period" (Figure 62.1); for a rigorous study of this "slippage" process, see Leathers and Harrington 2000).

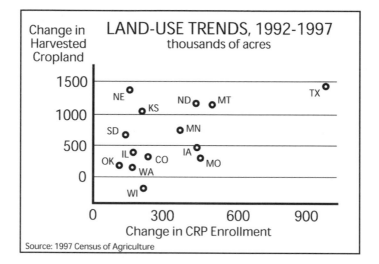

*Figure 62.1.* Land-use trends, 1992–1997.

## CONCLUSION

The Conservation Reserve Program on the Great Plains is a well-regarded failure. It is well regarded because most analysts who have studied its effects at a field scale tend to draw favorable conclusions about the effects of retiring erosion-prone land from crop production (Kantrud 1993; Young and Osborn 1990). It is a failure because these favorable effects are offset by the widespread creation of "new" cropland during the same time frame. In the absence of data to the contrary, one must assume that the new cropland would be at least as erosion-prone and wildlife-unfriendly as the cropland that was being retired. A few geographers have studied the effects of the CRP at an intermediate scale, such as a geomorphic region or multi-county watershed, and not surprisingly they report no significant effect.

The tragedy is that the flurry of geographically suspect field-scale analyses has overwhelmed other opinions and, as a result, the federal response to continuing food surpluses has been to expand the scope of the Conservation Reserve Program, to pay farmers to "idle" even more cropland that can be converted to wetlands, buffer strips along streams, grassland preserves, and so forth. Unfortunately, as the amount of paid diversion increases, the incentives to create additional cropland also seem to increase. As farmers we interviewed in at least 14 states told us, "I may have missed it this time around, but I'll be ready for the next buyout program."

Solving this knotty problem will require a multi-scale approach. Analysts will have to use state-scale analyses to assess the cumulative impacts of field-scale studies. Policy-makers will have to recognize that a change in farm-scale incentives is essential to the solution of national-scale problems. And, at some point, the United States must recognize that its county-scale administration of flawed land-use policy has an impact on its international-scale ability to promote a reduction in trade barriers (other countries rightfully point to the CRP as a form of subsidy for farmers and therefore a presumptive violation of free-trade agreements). It requires no great flash of insight to predict that this problem of farm surpluses and the impacts of policies to deal with them will continue to attract the attention of scholars from many disciplines. One would hope that these scholars would continue to insist on a multi-scale frame of reference, even as they conduct empirical research at a field scale.

## ACKNOWLEDGEMENTS

The authors acknowledge support of the National Science Foundation for early stages of the CRP research, and extend their thanks to Fraser Hart, Glenn Johnson, Bryan Baker, Tim and Sheryl Beach, Nick Dunning, Bill Casey, J. Fonkert, Phil Heywood, Dianne Vosick, Barb Weismann, Althea Willette, and the hundreds of farmers, District Conservationists, NRCS technicians, and other county officials who shared their time and expertise so willingly during our 404-county field study.

## REFERENCES

Archer, J. and R. Lonsdale.1997. Geographical Aspects of US Farmland Values and Changes During the 1978–1992 Period. *Journal of Rural Studies* 13: 399–413.

Davie, D. and C. Lant. 1994. The Effect of CRP Enrollment on Sediment Loads in Two Southern Illinois Streams. *Journal of Soil and Water Conservation* 49: 407–412.

Ervin, R. and J. Lee. 1994. Impact of Conservation Practices on Airborne Dust in the Southern High Plains of Texas. *Journal of Soil and Water Conservation* 49: 430–437.

Furuseth, O. 1997. Restructuring of Hog Farming in North Carolina: Explosion and Implosion. *Professional Geographer* 49: 391–403.

Gersmehl, P., A. Hoehn, L. Kigin, and C. Ruprecht, eds. 1998. *Food for Thought: A Geography of Minnesota Agriculture.* Minneapolis: Minnesota Alliance for Geographic Education and the Minnesota Department of Agriculture.

Hart, J. 1991. *The Land that Feeds Us.* New York: W.W. Norton.

Hudson, J. 1994. *Making the Corn Belt.* Bloomington: Indiana University Press.

Kantrud, H. 1993. Duck Nest Success on Conservation Reserve Program Land in the Prairie Pothole Region. *Journal of Soil and Water Conservation* 48: 238–242.

Kellogg, R., G. TeSelle, and J. Goebel. 1994. Highlights from the 1992 National Resources Inventory. *Journal of Soil and Water Conservation* 49: 521–527.

Kromm, D. and S. White. 1992. *Groundwater Exploitation in the High Plains.* Lawrence: University Press of Kansas.

Leathers, N. and L. Harrington. 2000. Effectiveness of Conservation Reserve Programs and Land "Slippage" in Southwestern Kansas. *The Professional Geographer* 52: 83–93.

Omernik, J. 1987. Ecoregions of the Conterminous United States. *Annals of the Association of American Geographers* 77: 118–125.

Osborn, C., F. Llacuna, and M. Linsenbigler. 1990. The Conservation Reserve Program: Enrollment Statistics for Signup Periods. *U.S. Dept. of Agriculture Statistical Bulletin* 811: 1–9.

Trimble, S. 1999. Decreased Rates of Alluvial Sediment Storage in the Coon Creek Basin of Wisconsin, 1975–1993. *Science* 285: 1244–1246.

U.S. Department of Agriculture National Agricultural Statistics Service. 1997. *Census of Agriculture Volume 2 Subject Series Part 2 Ranking of States and Counties.* Table 5: Harvested Cropland AC97-S-2. Washington, D.C.

Young, C. and C. Osborn. 1990. Costs and Benefits of the Conservation Reserve Program. *Journal of Soil and Water Conservation* 45: 370–373.

# 63

# IT'S THE OVERGRAZING, STUPID!
# DESTRUCTION OF THE GLOBAL RANGELANDS

PETER VINCENT
*Lancaster University*

The interaction between global environments and the ever-growing demands of the world's burgeoning population is of major interest to geographers. No other discipline has such a broad, integrating perspective when dealing with such issues and herein lies the importance and vitality of much modern geography. In this essay I shall attempt to illustrate this richness of perspective by reference to one particular example; namely, the possible socio-economic causes for the deterioration of the global rangelands.

Rangelands occur throughout the developed and undeveloped world and, because of their physical limitations, such as steep topography, low and erratic precipitation, poor drainage, and low or high temperatures, are unsuited for cultivation (Stoddart et al. 1975). This open, unfenced, terrain is dominated mostly by grasses and shrubs that support grazing and browsing agricultural systems. It is estimated that about half of the world's usable surface is rangeland that supports about 360 million cattle, half of which are in the humid savannas, and over 600 million sheep and goats, mostly in the arid areas.

Animal stocking rates on these pastures often far exceed rangeland carrying capacities, leading to vegetation impoverishment, soil erosion, and eventually to barren wasteland. The problem of rangeland degeneration is genuinely global and critically important when viewed in the context of a growing demand for meat by an ever-increasing world population. The scale of degeneration is truly staggering. For example, Dregne et al. (1991) estimate that some 73 percent of the world's 4.5 billion hectares of rangeland is moderately or severely degraded, leading to a loss of biological productivity and eventually to barren wasteland and desert.

The situation in China illustrates the point vividly. According to official sources, throughout China about 100,000 square kilometers of rangeland has become desert over the last 50 years and since the mid-1980s, the desert area has been expanding by about 2,500 km$^2$ per year, particularly in eastern Inner Mongolia, where the once-green pastures now resemble a scene from the American Dust Bowl of the 1930s. The impacts of this range destruction often go well beyond the overgrazed rangelands, and dust storms downwind of the deflation areas are now a major hazard in cities, such as Beijing. Indeed,

*D. G. Janelle et al. (eds.), WorldMinds: Geographical Perspectives on 100 Problems*, 387–390.
© 2004 *Springer. Printed in the Netherlands.*

dust from Inner Mongolia has now been detected in Hawaii, several thousand kilometers away.


## SELFISH NEIGHBORS?

In addition to pressures on rangeland resources brought about by rapid pop-ulation growth and the possible role of global climatic change, the decline of global rangeland productivity can be partly understood in terms of the social rules and institutions of the pastoralists themselves and the way these may or may not control access to, and use of, range resources. These institutions, also known as property regimes, are often classified into one of four basic types: state, private, common, and open-access. It is important to make clear the distinction between the last two categories in this list. In an open-access range, authority rests with no one and pastoralists are free to use the range as they think fit without the mediating forces of social interaction. In a common property regime, the resource is used by an identifiable community who can exclude others and regulate use (Berkes et al. 1989; Ostrom 1999).

In his now classic paper, "The Tragedy of the Commons," Garrett Hardin (1968) argued that the eventual fate of all resources held 'in common' is over-exploitation because access is unrestricted and there is no incentive among individuals toward resource protection since the gain to each individual user from overusing the commons will always outweigh the individual losses they have to bear due to its resulting degradation. It is further argued that individuals act selfishly, that there is no communication among resource users, and that no social norms mediate their actions. As Hardin put it:

> Therein is the tragedy. Each man is locked into a system that compels him to increase his herd without limit – in a world that is limited. Ruin is the destination toward which all men rush, each pursuing his own best interest in a society that believes in the freedom of the commons (Hardin 1968: 1244).

Hardin later acknowledged that such "free riding" and resource over-exploitation actually refers to open-access rangeland rather than common property rangeland and a failure to understand the difference by governments has, in several cases, resulted in even more range deterioration.


## ARABIAN HIMA

A case in point is the more or less complete collapse of the traditional hima system controlling semi-arid rangelands in the Arabian Peninsula. In the case of Saudi Arabia, a misunderstanding of a Royal Decree in 1953 brought about the demise of the hima system. A hima (pl. ahmyah or ahmia) is a protected grazing reserve belonging to a tribal group or village and is one of the oldest common property rangeland management systems in the world. The system is actually pre-Islamic in origin but was adopted by the Muslim community because the benefits of this approach to range management do not contradict any fundamental Islamic principles.

The hima system, or variants on it, was once widely practiced in the Arabian Peninsula as a whole (Grainger and Llewellyn 1994). In Oman, Wilkinson (1978) describes the practice of communal range control by villages of eastern oases and Thesiger (1959) remarks on the areas called hawtah, where hunting, cutting, or grazing were proscribed. Similar regulations governing rangeland have been recorded in Syria (locally called mahmia or ma'a), from the Kurdish areas of Iraq and Turkey, where they are referred to as koze (Draz 1978), and from Tunisia, where they a referred to as ghidal or zenakah (Hobbs 1985). Hima can also be found in northern Nigeria, Algeria, and Syria. With the almost complete demise of the several thousand himas, much of the rangeland in western Saudi Arabia has reverted from a common property to an open-access rangeland, and overstocking and overgrazing are now rife. For example, Shaltout et al. (1996) indicate that about 60 percent of the rangelands of Saudi Arabia are seriously degraded due to overgrazing. They estimate that the expected optimal carrying capacity is 1.04 million Livestock Grazing Units (LSUs) compared with an actual number of about 6.7 million. One LSU requires about 14 kg of forage (dry biomass weight) in a Saudi

*Figure 63.1.* Overgrazed hill slopes in the Middle Atlas Mountains, Morocco. Source: author.

Arabian-type environment and is equal to one camel or 9–10 sheep or goats. Furthermore, for many pastoral societies, herd size is also a sign of social status, regardless of the quality of the animals themselves. As a consequence, overstocking of rangeland is an endemic phenomenon in many parts of Africa and the Middle East.

As some researchers have noted, the Tragedy of the Commons (Hardin 1968) is often better described as the Tragedy of the Enclosure. Where the range has been enclosed, it has been observed that the enclosers rarely have the same relationship with their environment. In many parts of the Middle East, and particularly in Saudi Arabia, nomadic pastoralists (Johnson 1969) are no longer able to follow the ephemeral rains to fresh grazing as so much of the range-land has been fenced in for vast center-pivot irrigation schemes using precious fossil groundwater. This is not a case of "making the desert green" but rather "making the desert surface irreversibly saline" since the ground water is high in dissolved salts as a result of its long contact time with the bedrock.

The global rangelands that we observe today have developed over many thousands of years and have been subject to several climatic oscillations. Left to themselves, they are resilient ecosystems adapted to environmental stresses. Left to a burgeoning, meat eating, global population, the prospect for the global rangelands looks bleak indeed. It is only by appreciating all the various factors at play that we can hope to save this precious resource.

## REFERENCES

Berkes, F., D. Feeny, B. J. McCay, and J. M. Acheson. 1989. The Benefits of the Commons. *Nature* 340: 91–93.

Draz, O. 1969. The Hema System of Range Reserves in the Arabian Peninsula. Its Possibilities in Range Improvement and Conservation Projects in the Middle East. FAO/PL: PFC/13.11, Rome: FAO.

Dregne, H. E., M. Kassas, and B. Rozanov. 1991. A New Assessment of the World Status of Desertification. *Desertification Control Bulletin* 20: 6–18.

Johnson, D. L. 1969. *The Nature of Nomadism.* University of Chicago, Department of Geography Research Paper No. 118.

Granger, J. and O. Llewellyn. 1994. Sustainable Use: Lessons From a Cultural Tradition in Saudi Arabia. *Parks* 4(3): 8–16.

Hardin, G. 1968. The Tragedy of the Commons. *Science* 162: 1243–1248.

Ostrom, E. 1999. *Governing the Commons. The Evolution of Institutions for Collective Action.* Cambridge, U.K.: Cambridge University Press.

Shaltout, K. H., E. F. El-Hahwany, and H. F. El-Kady. 1996. Consequences of Protection From Grazing on Diversity and Abundance of the Coastal Lowland Vegetation in Eastern Saudi Arabia. *Biodiversity & Conservation* 5: 27–36.

Stoddart, L. A., A. D. Smith, and T. W. Box. 1975. *Range Management.* New York: McGraw-Hill.

Thesiger, W. 1959. *Arabian Sands.* London: Penguin.

Wilkinson, J. 1978. Islamic Water Law With Special Reference to Oasis Settlement. *Journal of Arid Environments* 1: 87–96.

# 64

# FORESTS AND MANAGEMENT: A CASE STUDY IN NEPAL USING REMOTE SENSING AND GIS

HARINI NAGENDRA
*Indiana University, Bloomington*

CHARLES M. SCHWEIK
*University of Massachusetts, Amherst*

The past decades have witnessed increasing awareness of forest degradation, with a concurrent rising interest in alternative methods of forest management. Among developing countries, Nepal has proved an enthusiastic leader in experimenting with participatory systems of forest governance (Agrawal et al. 1999). A careful look at the measures that Nepal has put into place, and analysis of their successes and limitations, will help enhance our understanding and ultimately facilitate implementation of effective forest policy.

Starting in the early 1970s, efforts at maintaining forest cover and biodiversity in Nepal have given rise to a variety of programs, including establishment of a network of protected areas under the control of the national government. Community forestry is another program that began in 1993, in an attempt to strengthen local capabilities in forest management, and to encourage sustainable use of forests (Agrawal et al. 1999). However, most community forests are located in the mountains. While community forestry is generally declared successful in improving the conditions of people and forests in the Nepal middle hills (Chakraborty 2001), reservations have been expressed about its outcome in the southern lowland Terai regions (Brown 1998). While the middle hills of Nepal have supported local populations for centuries, there has been extensive migration into the Terai from the hills after malarial eradication in the 1960s. As a result, there has been far greater recent deforestation in the Terai, compared to the middle hills (Schweik et al. 1997).

Concerns about forest degradation in the Terai have led to legislative initiatives by the Nepal government, which propose to bring Terai community forestry under partial control of the National government's Forest Department. These initiatives have, understandably, created high levels of concern in the user communities. Consequently, there is a strong need for broad-scale geographical studies that evaluate the impact of community forestry on forests in the Nepal Terai. Remote sensing is a potentially useful tool for this purpose,

*D. G. Janelle et al. (eds.), WorldMinds: Geographical Perspectives on 100 Problems, 391–396.*
© 2004 *Springer. Printed in the Netherlands.*

providing a spatial synoptic view of changes in forest condition and cover over time (Nagendra 2001). This research analyzes the changes in the forests around the Royal Chitwan National Park, before and after implementation of the buffer-zone community forestry development programs. This Park is the oldest protected area in Nepal, and a site of much publicized initiatives towards buffer-zone community forestry (Nepal and Weber 1994). A combination of information from Landsat satellite images and forest plots is used to evaluate the success of these efforts within a spatially and temporally explicit framework.

## STUDY AREA

The Terai District of Chitwan in southern Nepal is the location for our study. In the northern part of the district lie the foothills of the Himalayas, in the south, relatively flat terrain. As late as the 1950s, this region was covered with tropical moist deciduous forests, dominated by *Shorea robusta* (*Sal*), and infested with malaria-carrying mosquitoes, keeping the population low. By 1970, 91 percent of the previously affected region was declared malaria-free. This led to a rapid settlement of the region and to large increases in population. A sizable portion of the available flat terrain in the region is now cleared for agriculture or other development (e.g., settlements), with the notable exception of the parkland in the south.

The Royal Chitwan National Park, the oldest national park in Nepal, was established in 1973. The Royal Nepal Army now strictly protects this area, formerly managed as a hunting reserve for the royal family. Local people are allowed to enter the park only for 10 days every year, to collect building material (Stræde and Helles 2000). However, a population of nearly 300,000 surrounds the park, and park authorities have expressed concern about the effect of this illegal extraction on forest conditions. Some efforts at conflict resolution have been made in recent years.

## METHODS

We use U.S. Landsat Thematic Mapper™ images in this study. Landsat images cover a broad spatial extent (approximately 185 km × 185 km), and there are seven sensors that enable us to distinguish broad types of land cover, such as forest, agriculture, soil, and water. The spatial resolution is relatively fine, with a pixel size of 28.5 m, allowing us to detect changes in forest cover at a fairly detailed scale. We selected two nearly cloud-free Landsat™ images: 24 January 1989 (Figure 64.1) and 27 March 2000 (Figure 64.2) as this allows us to evaluate efforts of community forestry in the Terai, which have been strongest in the last decade. Images were collected during the pre-monsoon season, to ensure comparability across years.

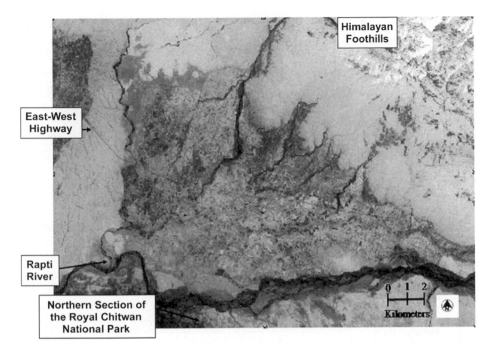

*Figure 64.1.* Landsat TM NDVI image of the East Chitwan, Nepal, 24 January 1989. Bright areas indicate vegetation.

The 1989 image was georeferenced using 1:25,000 scale topographic maps, and the 2000 image was georeferenced to the 1989 image with a root mean square error of less than 0.5 pixels. This enables us to overlay information from different images within a geographical information system (GIS) to evaluate forest change. Radiometric calibration, atmospheric correction, and radiometric rectification procedures were used to ensure image comparability. Without such calibration, change detection analysis may evaluate differences at the sensor level rather than changes at the Earth's surface.

The forest-plot data used for this study come from part of a larger set of studies in Nepal, conducted using common research protocols developed by the International Forestry Resources and Institutions (IFRI) research program, and coordinated by Indiana University (Ostrom 1998). Three locations in the Chitwan district were selected for analysis, with a total of seven forest patches under a variety of management regimes. The forests are dominated by relatively similar *Shorea robusta*, tropical moist deciduous hardwood trees. A total of 69 plots were laid in the community forests, 102 plots in national forests, and 45 plots in the national park, using a systematic random sampling to cover the entire elevation gradient within the forest. Nested circular plots were laid, with the outermost plot being 10 m in radius. Within this, the species and the diameter at breast height (dbh) were recorded for all trees with dbh

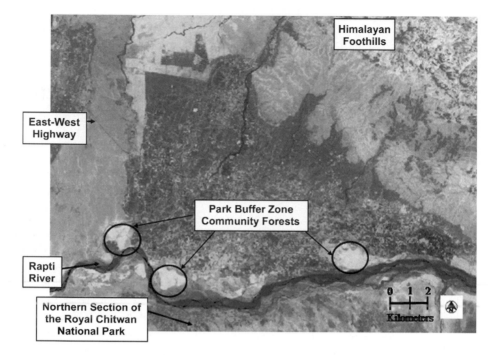

*Figure 64.2.* Landsat TM NDVI image of the East Chitwan, Nepal, 27 March 2000. Bright areas indicate vegetation. Note the increase in vegetation in the buffer-zone areas, between the 1989 image (Figure 64.1) and the 2000 image.

greater than 10 cm. A nested sub-plot of 3 m radius was used to record species, dbh, and height for all saplings with dbh less than 10 cm but greater than 1 cm. Data from all three research sites were pooled to examine whether there were significant differences in forest quality (tree and sapling density, biomass, species richness, and diversity) between forest plots located in the three different management regimes.

## RESULTS AND DISCUSSION

As debate about the effectiveness of community forest management becomes increasingly vocal, the need for geographers who can contribute to empirical evaluations is increasing. Although there is much current controversy on the effectiveness of community forestry in the Nepal Terai, there are few empirical investigations of this issue (Nagendra, in press). A Mann-Whitney pair-wise analysis of differences between management regimes, using quantitative data on forest condition, indicates that the forests within the Royal Chitwan National Park are in noticeably better condition compared to the community forests and national forests located outside the park (Nagendra, in press). This is no surprise, given the resources and manpower devoted to guarding the park. However, protection is provided through strict monitoring of the borders by armed guards and is not participatory in the least, although some efforts at conflict resolution have been made in recent years. Thus, while these efforts may be successful at maintaining biodiversity in limited protected park areas, they are clearly not sustainable efforts that can be broadened to cover the entire Terai region.

The factors impacting forest cover change need to be examined at multiple scales for an integrated understanding (Nagendra and Gadgil 1999). It is essential to broaden the spatial extent of this study to arrive at a complete understanding of the effect of tenure on forest condition, not just within a single forest patch, but also for the entire region. Our research, using two Landsat™ satellite images for 1989 and 2000, indicated that several forest patches just adjoining the Park, that fall within the purview of the buffer-zone community forestry program, show a substantial increase in forest cover over this decade (Schweik et al. in press). This can be seen from the Normalized Difference Vegetation Index (NDVI) images of 1989 (Figure 64.1) and 2000 (Figure 64.2). Thus, the buffer-zone community forestry programs appear to be successful at regenerating forest cover in several patches located within our study area.

Within our region of study, we find that at broad-scale, efforts to involve local communities in somewhat more participatory efforts at forest management have increased forest cover. Nevertheless, while on paper there is much enthusiasm for encouraging participatory management in Nepal, in practice, the level of devolvement is yet to match the rhetoric (Chakraborty 2001). There is a real need for geographers to provide empirical evaluations of such complex policy issues. Complemented by a fine-scale plot-based analysis of changes in forest condition over time, this research will allow comprehensive evaluation of the effect of Nepal's changing forest policies on forest sustainability. A two-pronged approach of this kind, using a combination of field data and remote sensing techniques at multiple scales, is useful for geographers wishing to address multi-dimensional and complex problems of this nature.

## ACKNOWLEDGEMENTS

This research was supported by the National Science Foundation (NSF) (SBR-9521918) as part of the ongoing research at the Center for the Study of Institutions, Population, and Environmental Change (CIPEC) at Indiana University. We thank Deb Sinha, Mukunda Karmacharya, and Birendra Karna for invaluable assistance with field investigations.

## REFERENCES

Agrawal, A., C. Britt, and K. Kanel. 1999. *Decentralization in Nepal: A Comparative Analysis.* Oakland, CA: ICS Press.

Brown, K. 1998. The Political Ecology of Biodiversity, Conservation and Development in Nepal's Terai: Confused Meanings, Means and Ends. *Ecological Economics* 24: 73–87.

Chakraborty, R. N. 2001. Stability and Outcomes of Common Property Institutions in Forestry: Evidence from the Terai Region of Nepal. *Ecological Economics* 34: 1–353.

Nagendra, H. and M. Gadgil 1999. Biodiversity Assessment at Multiple Scales: Linking Remotely Sensed Data with Field Information. *Proceedings of the National Academy of Sciences USA* 96: 9154–9158.

Nagendra, H. 2001. Using Remote Sensing to Assess Biodiversity. *International Journal of Remote Sensing* 22: 2377–2400.

Nagendra, H. 2003. Tenure and Forest Conditions: Community Forestry in the Nepal Terai. *Environmental Conservation*, in press.

Ostrom, E. 1998. The International Forestry Resources and Institutions Program: A Methodology for Relating Human Incentives and Actions on Forest Cover and Biodiversity. In F. Dallmeier and J. A. Comisker, eds. *Forest Biodiversity in North, Central and South America, and the Caribbean: Research and Monitoring, Man and the Biosphere Series* (Vol. 1: 1–28). Paris: UNESCO; New York: Parthenon.

Schweik, C. M., K. Adhikari, and K. N. Pandit. 1997. Land-Cover Change and Forest Institutions: A Comparison of Two Sub-Basins in the Southern Siwalik Hills of Nepal. *Mountain Research and Development* 17: 99–116.

Schweik, C., H. Nagendra, and D. R. Sinha. 2003. Using Satellites to Search for Forest Management Innovations in Nepal. *Ambio*, in press.

Stræde, S. and F. Helles. 2000. Park-People Conflict in Royal Chitwan National Park, Nepal: Buying Time at High Cost? *Environmental Conservation* 27: 368–381.

PART VII

# UNMASKING DANGEROUS ENVIRONMENTS

The world can be a dangerous place. But, through understanding and careful planning, humankind can cope and adapt in ways to reduce the toll on life and resources. Environmental contamination and waste disposal highlight the risks that human agents pose on themselves and their neighbors, while weather, water, and geological events reflect risks associated with human activities that clash with and sometimes reinforce the forces of nature. Unmasking these complexities is increasingly essential in a world of significant population growth that has extended human occupancy beyond the margins of the most easily occupied regions.

*Kathy Hansen*

# 65

# EXCURSIONS INTO THE TOXIC PAST

## CRAIG E. COLTEN
*Louisiana State University*

The rediscovery of more than 20,000 tons of toxic wastes buried in a Niagara Falls suburban school yard in the late 1970s captured public attention and directed the media spotlight on hazardous wastes, an issue industry experts and government officials had been grappling with for decades. Congress had already passed legislation in 1976 (Resource Conservation and Recovery Act) that sought to regulate hazardous waste management, but the act neglected to deal with long-abandoned waste dumps such as Love Canal (New York). Consequently in 1980, Congress passed the Superfund legislation (The Comprehensive Environmental Response, Liability, and Compensation Act), which employed the principle: the polluter pays. Inherent in this approach was a retrospective perspective.

## A QUESTION FOR HISTORICAL GEOGRAPHY

Identifying and assigning liability to former operators or owners of abandoned dumps demanded historical investigation. As the United States Environmental Protection Agency (USEPA) attempted to wrest payment from those it identified as the responsible parties, litigation spiraled out to ensnarl many other participants in the nation's hazardous waste legacy. One key question that arose out of that litigation was "what did the waste managers know about the potential environmental impact of their actions?" This is a question historical geographers have helped answer.

There were three basic phases of liability litigation spawned by Superfund. The first, was the PRP (potentially responsible party) phase. In this phase, the USEPA identified the most obvious and deep-pocket PRPs and sued them for the clean-up costs. Those corporations targeted by the federal government then conducted further research to identify more PRPs. Historians often assisted by conducting land-use/ownership and corporate histories to expand the net of financial responsibility (Bookspan 1991). A second phase turned the litigation around and involved suing the federal government as a responsible party. In numerous Superfund cases the federal government had owned or played a role in site operations, and litigants sought to tap the U.S. Treasury's

*D. G. Janelle et al. (eds.), WorldMinds: Geographical Perspectives on 100 Problems,* 399–403.
© 2004 *Springer. Printed in the Netherlands.*

resources. Similar to PRP litigation, the historical research involved documenting federal ownership of former military facilities or contractual oversight of private companies supplying goods to the government. In both of these overlapping phases, the initial PRP group often successfully reduced its financial liabilities by enlarging the pool of responsible parties through historical documentation.

The third phase, which this essay focuses on, involved suits by responsible parties against their insurance companies. Manufacturing firms have sought coverage for damages caused by hazardous wastes that they had released, sometimes in the distant past, under provisions of general liability policies. Historical geography came to play a role in this litigation in two key ways. First, by establishing the historical state-of-knowledge about the potential for groundwater contamination (a form of what John K. Wright (1948) termed geosophy), historical geographers countered claims by engineers and physical scientists that at the time of disposal technical knowledge was inadequate to anticipate damages from hazardous waste dumps (Colten 1991, 1998a, 1998c). Second, by reviewing state laws and groundwater pollution litigation, historical geographers were able to provide a portrait of past public policies and their enforcement, which demonstrated concern with the issue from coast to coast and at a time when industry was releasing wastes that are now embroiled in Superfund-spawned litigation (Colten 1998b).

## DOCUMENTING PAST GEOGRAPHICAL KNOWLEDGE

In order to draw on past general liability policies and spread the financial liabilities, companies facing clean-up costs needed to convince the courts that the damages from past waste management practices were unanticipated. Key terms in former policies were "expected" or "intended." From the manufacturers' standpoint, this involved demonstrating that waste managers would not have had sufficient comprehension of groundwater processes to anticipate the movement of wastes from pits or ponds or the waste's long-term persistence in the environment. With most of the practitioners who had overseen waste management planning in the 1940s and 1950s deceased, historical documents became the linchpin for constructing arguments one way or the other. The role geographers could play was to investigate the state-of-knowledge about groundwater contamination at the time of waste disposal. Unlike engineers or physical scientists, who all too often used current expertise as the baseline for historical actions and supported their positions with limited historical research (Mutch and Eckenfelder 1993; Hart 1995), historical geographers, uninfluenced by current standards in waste management that could lead them to impose an unrealistic standard on past events, approached the question dispassionately. They could attack the research problem with a basic comprehension of physical geographic processes and, thereby, understand the technical literature. In addition, using fundamental historical research skills, they could trace the evolution of understanding forward in time and, also, reconstruct the literature available to practitioners in the past along with the sources actually used by authorities in the field. By blending their basic training in physical geography with their research skills, historical geographers offered a unique and unbiased perspective. The ultimate contribution was to document the state-of-knowledge, rather than attempting to argue the absence of knowledge.

For example, general engineering texts and handbooks indicated the appearance of discussions about groundwater processes by 1900. As part of general practice, civil engineers had developed expertise in managing groundwater to protect mines, railroads, highways, and dams. Textbooks showed that an engineer's education encompassed basic principles of groundwater movement and that the texts' authors drew on the work of geologists and hydrologists – indicating dissemination from basic researchers to practitioners (Colten 1991, 1998a). Building construction also demanded manipulation of groundwater to avoid leaks or foundation failure. Groundwater knowledge was fundamental to an engineer's training and rudimentary to one's practice by the 1950s (Colten 1998a). Most sanitary engineers were trained as civil engineers and, thus, the practitioners engaged in managing wastes received exposure to concepts of groundwater hydrology.

In terms of groundwater contamination, waste managers and public authorities had access to and acknowledged basic groundwater-movement principles by the 1950s, as well. Early twentieth-century examples of groundwater contamination lawsuits demonstrated the presumption that contaminants could flow from a waste source to down-gradient wells (Colten 1991, 1998c). Federal investigators had employed a network of wells down gradient from waste pits in the 1920s and 1930s to document contaminant movement in soils (Colten 1998c). When industrial wastes contaminated groundwater in Long Island and Colorado, geologists sampled wells to map contaminant plumes in the 1940s and early 1950s (Colten 1998c). When investigating the extent of contamination issues, waste managers in the 1950s turned to previously published technical literature about groundwater contamination, demonstrating the linkages to prior developments and the perpetuation of knowledge developed in the past (Colten 1998c). Also, the consistent use of down-gradient monitoring wells in groundwater-pollution incidents illustrated the general belief that wastes typically entered groundwater and that contaminants moved as part of the groundwater flow. Although some consultants have argued in litigation that there was insufficient comprehension of groundwater processes in the 1940s and 1950s (Mutch and Eckenfelder 1993), the extensive body of literature indicates otherwise.

Frequently, records of these insurance cases that never reach trial are protected from public view, but public records that surfaced in investigative efforts reveal many specific examples of historical corporate knowledge. In the early 1940s, New Jersey officials convinced manufacturers in Parlin to stop dumping acidic wastes on the ground after this practice caused groundwater contamination (Barksdale 1943). Groundwater pollution from an aluminum maker's disposal pit prompted a cooperative monitoring program, starting in 1946 (Price 1967). Chemical maker DuPont borrowed 1930s oil-field technology to inject saline wastes into a geologic formation already containing brines to prevent water pollution (de Ropp 1951). In the early 1950s, Thomas Powers, a waste management official with Dow Chemical Company, served on a national panel that reviewed the extent of the problem and concluded that groundwater contamination was to be "expected" in areas where industry operated in groundwater recharge areas (AWWA 1953). These examples illustrate considerable sophistication on the part of industry personnel dealing with actual groundwater problems and their access to scientific knowledge about the issue.

Beyond the community of waste managers and hydrologists, there is ample evidence that society in general accepted the notion that wastes could foul down-gradient water sources. Basic sanitation texts had warned against placing wells down gradient from outhouses. This idea of contaminant movement is also amply supported by public policy and legal actions throughout the twentieth century. In 1905, the U.S. Geological Survey reported that 31 states specifically referred to wells, springs, or groundwater in their water-pollution laws. Individuals filed suits under those laws to protect private and public groundwater supplies. By 1956, cases in 18 state appellate courts had ruled on groundwater-pollution cases – and the fact that the legal research tool included a specific category for this type of case, underscores its commonality. State laws tended to expand their reach from surface to groundwater as the century progressed. By the late 1950s, 37 states either protected "all" waters or specified groundwater in their water-pollution laws. A survey of public health officials indicated that at least 27 states had regulations concerning groundwater contamination. By reviewing state policies, legal remedies, and enforcement actions, geographers were able to document an extensive apparatus for dealing with groundwater issues by the 1950s (Colten 1998b).

## BENEFITS OF THE HISTORICAL GEOGRAPHIC PERSPECTIVE

In addition to historical geographers, engineers, hydrologists, geologists, and historians have all entered the arena of insurance litigation, where millions of dollars are at stake. Each brings a particular perspective and expertise. Their fundamental task has been to offer a well-documented perspective on past understanding of physical processes and technical knowledge among waste-management practitioners. While physical scientists bring an abundance of sophisticated knowledge of recent developments in groundwater hydrology to the table, they appear to be trapped within that very expertise. When they find historical actors lacking in the same level of knowledge that current practitioners possess, these modern experts tend to dismiss them as completely unprepared to comprehend groundwater processes. Yet, by seeking the knowledge available in the past, rather than trying to demonstrate that historical actors were unsophisticated by today's standards, historical geographers have been able to reconstruct the state-of-knowledge, which shows that historical waste managers would have anticipated wastes moving with groundwater and the potential for down-gradient contamination. This approach has helped clarify issues before jurists (Chemical Leaman Tank Lines v. Atena et al. 1995).

## REFERENCES

American Water Works Association. 1953. Findings and Recommendations on Underground Waste Disposal. *Journal American Water Works Association* 45(12): 1295–1297.

Barksdale, H. C. 1943. *The Ground-water Supplies of Middlesex County New Jersey.* Trenton, NJ: New Jersey State Water Policy Commission.

Bookspan, S. 1991. Potentially Responsible Party Searches: Finding the Cause of Urban Grime. *Public Historian* 13: 25–34.

Chemical Leaman Tank Lines, Inc. v. Aetna Casualty and Surety Company, et al. 1995. Decision by Circuit Judge McKee. U.S. Court of Appeals for the Third Circuit, No. 93-5777 and 93-5794, p. 35.

Colten, C. E. 1991. A Historical Perspective on Industrial Wastes and Groundwater Contamination. *Geographical Review* 81: 215–228.

Colten, C. E. 1998a. Industrial Topography, Groundwater, and the Contours of Environmental Knowledge. *Geographical Review* 88: 199–218.

Colten, C. E. 1998b. Groundwater and the Law: Records v. Recollections. *Public Historian* 20: 25–44.

Colten, C. E. 1998c. Groundwater Contamination: Reconstructing Historical Knowledge for the Courts. *Applied Geography* 18: 259–273.

de Ropp, H. W. 1951. Chemical Plant Disposal at Victoria, Texas, Plant of Du Pont Company. *Sewage and Industrial Wastes* 23(2): 194–197.

Hart, F. C. 1995. Superfund Reauthorization: It's Not Time to Revise History. *Mealey's Litigation Reports: Insurance* 9(29): 17–28.

Manufacturing Chemists' Association. 1959. *Water Pollution Abatement Manual: Compendium of Water Pollution Laws.* Washington, D.C.: Manufacturing Chemists' Association.

Mutch, R. and W. W. Eckenfelder. 1993. Out of the Dusty Archives: The History of Waste Management becomes a Critical Issue in Insurance Litigation. *Hazmat World* 6 (October): 59–68.

Wright, J. K. 1947. Terrae Incognitae: The Place of the Imagination in Geography. *Annals of the Association of American Geographers* 37: 1–15.

# 66

# NON-POINT SOURCES: HISTORICAL
# SEDIMENTATION AND 20TH CENTURY GEOGRAPHY

L. ALLAN JAMES
*University of South Carolina*

Geographers have traditionally studied spatial relationships, human-land interactions, and the synthesis of this information on a regional basis. How better to integrate these foci than to study regional impacts of soil erosion and sedimentation following the introduction of European and African agricultural technology to North America? Physical geography has contributed greatly by elucidating concepts of fluvial sediment transport, non-point source (NPS) pollution, and channel-morphological changes that effect flooding. These intellectual traditions arose as federal government priorities were changing from soil conservation, to resource conservation and to modern concerns over public health and environmental preservation and restoration. Since passage of the Clean Water Act (CWA) Amendments of 1987, study of NPS pollution has grown in relevancy. This chapter puts early geographic research into a historical context and shows its importance to modern NPS pollution control and to a great body of on-going research by a new generation of geographers.

In spite of the early concern of physical geography for fluvial systems, few early 20th century North American geographers studied contemporary or historical soil erosion and sedimentation. As measured by a sampling of the research by past presidents of the Association of American Geographers (AAG), early geographers were interested in erosion primarily from a perspective of landform evolution (Table 66.1). By the 1920s, however, Marbut and Bennett (1943) began to note anthropogenic soil erosion and massive downstream effects. In the second half of the 19th century, physical geography focused keenly on these impacts as a natural extension of traditional concerns. These studies were ahead of their time in respect to the approach taken by most other environmental sciences.

Outside of geography, water-quality studies in North America initially focused on the reach (local) scale of stream channels and on point sources of pollution with little regard for sediment or integrated basin processes. Recently, focus has progressed from a non-spatial perspective to practices that increasingly rely on geographic notions of space and scale as applied to watersheds (USEPA 1995). In response to a growing awareness of the importance of diffuse sediment to the production and storage of nutrients,

*D. G. Janelle et al. (eds.), WorldMinds: Geographical Perspectives on 100 Problems, 405–410.*

*Table 66.1.* Research of Selected Early AAG Presidents.

| Name | Year of Term | Focus of Writing |
|------|-------------|------------------|
| W. M. Davis | 1905 & 1909 | Landform Evolution. |
| G. K. Gilbert | 1908 | Landform Evolution; Theories of Sediment Production and Transportation. |
| R. S. Tarr | 1911 | Landform Evolution and Glacial Erosion. |
| R. D. Salisbury | 1912 | Landform Evolution; Textbook on Physiography Estimated Denudation Rates From Sediment Concentrations. |
| N. M Fenneman | 1918 | Landform Evolution; Textbook on Physiography Mentioned Historical Soil Erosion and Guiles in Southern Piedmont. |
| C. F. Marbut | 1924 | Father of Modern Soil Classification. |
| D. W. Johnson | 1928 | Landform Evolution. |
| Lawrence Martin | 1930 | Landform Evolution. |
| Isaiah Bowman | 1931 | Landform Evolution. |
| Francois Matthes | 1933 | Landform Evolution. |
| Wallace Atwood | 1934 | Landform Evolution. |
| Carl O. Sauer | 1941 | Advocating Study of Soil Erosion |
| Hugh J. Bennett | 1943 | Soil Erosion and Conservation (Son of H.H. Bennet) |

metals, pesticides, and toxins, the 1987 CWA Amendments emphasized NPS sediment and associated pollutants. Consequently, the methods and concerns of other environmental sciences are coming into harmony with past geographic research that has gained substantially in stature and relevance.

## ACCELERATED EROSION AND CONSERVATION MOVEMENTS

Rates of erosion and sedimentation prior to anthropogenic disturbances, often referred to as the *geologic norm*, are far less than rates after human disruptions, referred to as *accelerated erosion* (Strahler 1956). Accelerated erosion followed the advent of agriculture and, therefore, has been substantial in most of North America. In the western world, it began with Neolithic agriculture in Mesopotamia, expanded through the Mediterranean region to woodland clearance in northern Europe, and followed the introduction of European agricultural practices to North America. Soil erosion was such a threat to land viability in the Colonies, that it was noted as a menace by Thomas Jefferson, and by the end of the 19th century, vast tracts of land were permanently damaged.

In the 1920s and 1930s a *soil conservation* movement initiated by Hugh Bennett focused farm policies on soil erosion. From the 1940s through 1970s, this concern was expanded to *land conservation*, emphasizing reduction of on-site damages and losses to natural resources. Not until the environmental movement gained momentum in the 1970s, however, did off-site damages of erosion begin to receive attention. The Section 319 NPS Management Program of the 1987 CWA Amendments called for studies of NPS pollution that represent the first substantial federal mandate to study, manage, or mitigate off-site impacts of land use. Fortunately, geographers had not waited for government sanction to engage in research on the downstream effects of accelerated erosion. Arising from long-held traditions in human-land inter-actions and fluvial geomorphology, geographers were trained and motivated to examine these questions. Their predisposition to historical landscapes and spatial analysis anticipated by two decades the recent move towards a watershed-scale approach by the broader scientific community.

## GEOGRAPHIC CONCERNS WITH SEDIMENTATION

At the turn of the 20th century, geomorphology in North America was exam-ining long-term landform-evolution processes (Davis 1900). Erosion and sedimentation were emphasized as processes explaining the formation of drainage systems, hill-slope profiles, and regional topography in general. While landform evolution dominated geomorphology for the first half of the 20th century, the stage was being set for studies of erosion and aggradation on historical time scales. G. K. Gilbert (1914, 1917) was an early contributor to the study of accelerated aggradation. He wrote two classic monographs on the transport of hydraulic mining sediment. Data from his flume experiments are still used in hydraulic calculations of sediment transport and spatial-temporal concepts of long-distance sediment transport from the other study are still prevalent in fluvial geomorphology.

Neglect of downstream impacts of historical erosion by geographers during the first half of the 20th century was in accord with the national policy of ignoring off-site impacts of soil erosion. It also reflects a shift in North American geography away from physical geography at the time. Sauer (1941: 2) referred to the period beginning in the 1920s as "the Great Retreat" when geography separated from geology departments and moved away from physical and historical geography. He lamented that, in seeking an area of study independent of geology, "American geography gradually ceased to be a part of Earth Science." Ironically, few geologists in North America were inter-ested in human impacts on the environment until recent decades, so geographers' apparent perception of redundancy with geology was distorted. Sauer lauded the early work of George Marsh and advocated the study of humans as an agent of physical geography. Although he stopped short of following erosion impacts downstream, he signaled the need to study soil erosion as an agent of change:

The incidence of soil erosion may be a major force in historical geography. Did soil losses
sap the Mediterranean civilizations? Were the Virginians great colonizers because they were
notable soil wasters? Geographical field work should embrace thorough search for full, original
soil profiles and note the characteristic diminution or truncation of soil profiles in fields
and pastures. (Sauer 1941: 18).

This call to arms heralded a prolific new era of research soon to be initiated
by geographers.

In the 1950s, physical geography began to experience a metamorphosis
and resurgence propelled by the quantitative revolution. The mathematical
concepts of Robert Horton's hydrology were enhanced to incorporate concepts
of erosion and equilibrium from geomorphology (e.g., Strahler 1956; Figure
66.1). The emphasis during this period was on process geomorphology,
however, and there remained a dearth of historical geomorphology studies
concerned with human impacts on hill-slope or river sedimentation. This was
soon to change as geographers began to raise questions about human impacts
on the Earth. For example, Karl Butzer, recognizing the importance of human
disturbance to Quaternary stratigraphic sequences in the Mediterranean region,
criticized Vita-Finzi's prominent explanation that was based on a simple model
of two climatic intervals and little human impact (Butzer 1969).

In the 1970s and 1980s a new movement in fluvial geomorphology took
root that merged traditions in historical geography and geomorphology to
examine downstream consequences of accelerated erosion. Spatial and temporal

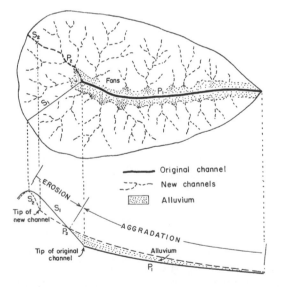

*Figure 66.1.* Interactions between gully erosion extending a drainage network and aggradation
downstream in larger channels. Reprinted with permission from the University of Chicago
Press from A. N. Strahler, 1956. "The Nature of Induced Erosion and Aggradation." In W. L.
Thomas, Jr., C. O. Sauer, M. Bates, and L. Mumford, eds. *Man's Role in Changing the Face
of the Earth*, Vol. 1, 448, 621–638. Chicago: University of Chicago Press.

patterns of anthropogenic hydrologic changes, soil erosion, sediment delivery, and channel morphologic change began to emerge. In the upper Midwest, Knox (1972, 1977, 1987) presented detailed evidence from field and documentary data to show the impacts of European agricultural and mining technology on floodplain stratigraphy. Knox's biogeomorphological response model (Figure 66.2) has been widely influential and reproduced in the literature. In the southern Piedmont, Trimble (1974) examined the history of soil erosion and stream sedimentation. He identified several spatial patterns, including the progression of plowing up hill slopes and the timing of stream aggradation related to stream order. In the Southwest, Graf examined arroyo cutting in response to boom-and-bust mining activities (Graf 1979) and documented spatial patterns of radionuclides in fluvial systems related to stream power (Graf 1990).

Currently, a large new generation of fluvial geomorphologists is combining physically based process geomorphology, stratigraphy, and historical research to the study of human-induced aggradation (cf., Abrahams and Marston 1993). The convocation of fluvial geomorphologists at the annual AAG meetings has flourished for more than 25 years and continues to examine questions of fluvial sedimentation. These on-going studies are contributing to a much-needed understanding of watershed dynamics and potential sources of non-point source pollution.

## CONCLUSION

Geographic traditions spawned important research on downstream effects of sediment production that are now relevant to major policy initiatives mandated

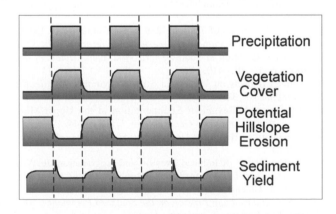

*Figure 66.2.* Knox's biogeomorphic response model shows non-linear complexities in sediment yields resulting from a simple step-function change in precipitation regime. These concepts can explain channel morphological responses to Holocene climate change or NPS pollution responses to environmental change. (Adapted from Knox 1972.)

by the federal government. Knowledge of the extensive historical deposits left by episodic erosion is essential to watershed management aimed at protecting water quality and water-resources infrastructure. These deposits may be unstable and can introduce toxic, hazardous, or carcinogenic materials, such as metals, radio nuclides, and agricultural chemicals adsorbed on sediment grains. They are also essential to understanding pre-historic channel conditions for reconstructing reference channels in aquatic restoration projects. The demand for this expertise is growing, as is interest by non-geographers in this research and in the perspective of physical geographers. As the skills of physical geography are solicited by outside interests, however, it is important to remember and cherish the deep geographic traditions that brought us to this juncture.

## REFERENCES

Abrahams, A. D. and R. A. Marston. 1993. Drainage Basin Sediment Budgets: An Introduction. *Physical Geography* 14(3): 221–224.

Bennett, H. H. 1943. Adjustment of Agriculture to its Environment. Presidential address. *Annals of the Association of American Geographers* 33: 163–198.

Butzer, K. W. 1969. Changes in the Land (review of C. Vita-Finzi: The Mediterranean Valleys: Geological Changes in Historic Times). *Science* 165: 52–53.

Davis, W. M. 1900. The Physical Geography of the Lands. *Popular Science Monthly* 57: 157–170; reprinted in D. W. Johnson, ed. *Geographical Essays by William Morris Davis; 1909 & 1954.* Dover Pubs.

Gilbert, G. K. 1914. Transport of Debris by Running Water. *U.S. Geological Survey Professional Paper* 86.

Gilbert, G. K. 1917. Hydraulic-Mining Debris in the Sierra Nevada. *U.S. Geological Survey Professional Paper* 105.

Graf, W. L. 1979. Mining and Channel Response. *Annals of the Association of American Geographers* 69: 262–275.

Graf, W. L. 1990. Fluvial Dynamics of Thorium-230 in the Church Rock Event, Puerco River, New Mexico. *Annals of the Association of American. Geographers* 80: 327–342.

Knox, J. C. 1972. Valley Alluviation in Southwestern Wisconsin. *Annals of the Association of American Geographers* 62: 401–410.

Knox, J. C. 1977. Human Impacts on Wisconsin Stream Channels. *Annals of the Association of American Geographer* 67: 323–342.

Knox, J. C. 1987. Historical Valley Floor Sedimentation in the Upper Mississippi Valley. *Annals of the Association of American Geographers* 77: 224–244.

Sauer, C. O. 1941. Foreword to Historical Geography. *Annals of the Association of American Geographers* 31 :1–24.

Strahler, A. N. 1956. The Nature of Induced Erosion and Aggradation. In W. L. Thomas, Jr., C. O. Sauer, M. Bates, and L. Mumford, eds. Man's Role in Changing the Face of the Earth; 621–638, Vol. 1; Chicago: University of Chicago Press. 448.

Trimble, S. W. 1974. *Man-Induced Soil Erosion on the Southern Piedmont, 1700–1970.* Ankeny, Iowa: Soil Conservation Society of America.

U.S. Environmental Protection Agency, 1995. *Watershed Protection: A Statewide Approach.* Office of Water. EPA 841-R-95-004. Washington, D.C.: U.S. Government Printing Office.

# 67

# SPATIAL ANALYSIS OF HAZARDOUS FUELS AND ECOLOGICAL DECADENCE

DENISE TOLNESS
*Bureau of Land Management*

On 8 August 2000, President Clinton asked Interior Secretary Bruce Babbitt and Agriculture Secretary Dan Glickman to prepare a report that would recommend how best to respond to the severe fires of 2000, specifically how to reduce the impacts of these wildland fires on rural communities. The President also asked for short-term actions that federal agencies, in cooperation with states, local communities, and tribes, could take to reduce immediate hazards to communities in the wildland-urban interface and use to ensure land managers and firefighters are prepared for the extreme fire conditions expected in the future. A key point was the investment in projects to reduce fire risk (hazardous-fuels-reduction programs). Addressing fire risk and wildland fire issues will require significant investments in time, manpower, and money. It is through multiple-treatment types, including chemical applications, mechanical thinning, seeding efforts, and prescribed fire that this is to be accomplished. Since 1994, the Forest Service and the Bureau of Land Management (BLM) have increased the number of acres treated to reduce fuel buildup, from fewer than 500,000 acres in 1994 to more than 2.4 million acres in 2002 (Executive Summary 2000).

The BLM Upper Snake River District (USRD) manages approximately 5.4 million acres of public land in southern Idaho, of which the Southern Idaho Interagency Fire Center (SIIFC) is responsible for fire suppression and fuels reduction for approximately 3.3 million of those acres in the Shoshone and Burley Field Offices Figure 67.1). The region offers a wide diversity of landscapes, from the Basin and Range topography and the grotesque, eroded granite formations of the City of Rocks in the south, the sagebrush-grasslands of the Snake River Plains, to the volcanic Craters of the Moon and Great Rift area, and the forested mountains of Sun Valley. Such a large region, composed of a diversity of landscapes, can pose unique problems for an effective hazardous-fuels-reduction program. The SIIFC Fuels Program has been aggressively working on a hazardous-fuels-reduction project since 1999 in an effort to assess fuel loading, vegetation characteristics, and fire hazard, and to assist with proper land-management needs.

411

*D. G. Janelle et al. (eds.), WorldMinds: Geographical Perspectives on 100 Problems*, 411–417.
© 2004 *Springer. Printed in the Netherlands.*

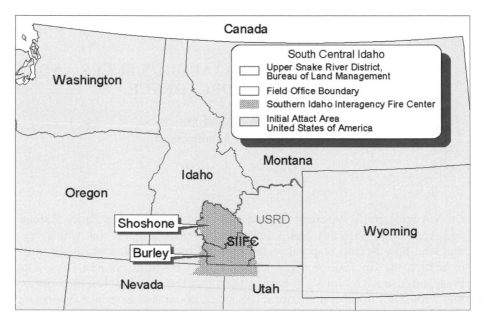

*Figure 67.1.* Southern Idaho Interagency Fire Center area of concern.

SCOPE OF PROJECT

Primarily, the goal of the fuels-reduction project is to target the reduction of hazardous-fire conditions within the BLM-administered lands. A secondary effect of the project is to achieve ecological balance of the land through the reintroduction of fire on the landscape. Some additional achievements include Quaking Aspen (*Populus tremuloides*) regeneration, Cheatgrass (*Bromus tectorum*) and noxious-weed eradication, a return to native perennial vegetation, and the "ground-truthing," or verification of multispectral imagery that is used to create a District vegetation-classification scheme. Geographical information systems (GIS) technology is used to gather, correct, and project future sites for fuels-reduction projects throughout the USRD.

METHODS

*Digital spatial data were gathered.* To accomplish the GIS side of the project, standards, processes, guides, data dictionaries, and datasets were acquired, created, or converted before anything else could be done digitally or spatially. Coverages from other agencies were acquired and converted. Where data was missing, resource specialists' field maps were digitized, scanned, and attributed. Over a year was put into the initial effort to develop GIS layers that are consistent on an interagency, interstate, and statewide basis. This provided the baseline information for evaluation and for the fire/fuels planning process.

*Field data.* Fuel specialists designate large polygon areas for the collection of field data. Field crews then travel to random survey plots within the polygon project perimeter and conduct a field-based vegetation survey. The survey plot is a circle with a radius of 1 chain (66 feet). The number of plots surveyed varies depending on the number and areas with different vegetation types within designated polygons. The vegetation types are determined using maps from a modified version of the Utah Gap database (Edwards and Homer 1996). Two transects are run off the diameter of the survey plot. Transects run 50 feet up slope and 50 feet down slope. Along these transects, species identification and canopy-coverage type and amount are collected at 10-foot intervals and logged into the survey forms. On average, approximately 500 surveys can be taken across the SIIFC within one field season.

*Post processing the field data.* The collected data are downloaded, differentially corrected, and exported to a GIS format. The GIS coverage for the points is hotlinked to the images and placed in a usable format for the fuels specialists and resource managers.

*Modeling.* Specific data fields were assigned ranks to allow for mathematical computations. Attribute values in the range of 0 to 200 were assigned to each data field to create a series of consistent ordinal scales (Zar 1999) that can be used to perform additional functions (Figure 67.2 and Figure 67.3).

## WHY?

In a natural fire regime, fires will generally burn quickly through underbrush or grass and shrubs. The fuel load will remain low because periodic fires clean house regularly and prevent the fuel build up. Therefore, fires do not burn with great heat or intensity, established plants are not adversely affected, and those species dependant on fire retain their natural cycles. However, due to the continued fire-suppression efforts of the last 50 years, ecosystems formerly adapted to frequent fire occurrence are now ecosystems with high hazardous fuel density (Varga 2000). It is the communities adjacent to such hazardous fuels that are at the greatest risk to wildfires. Most of the communities in Idaho fall within the Federal Register's definition of a Community at Risk or Wildland-Urban interface area, leaving land managers with the problem of where to start. Modeling the hazardous-fuels rankings for known areas of concern allows a risk factor to be assigned to the adjacent communities. So even though most of Idaho is at risk of wildfire, areas with the highest fuel buildups can be targeted for immediate action.

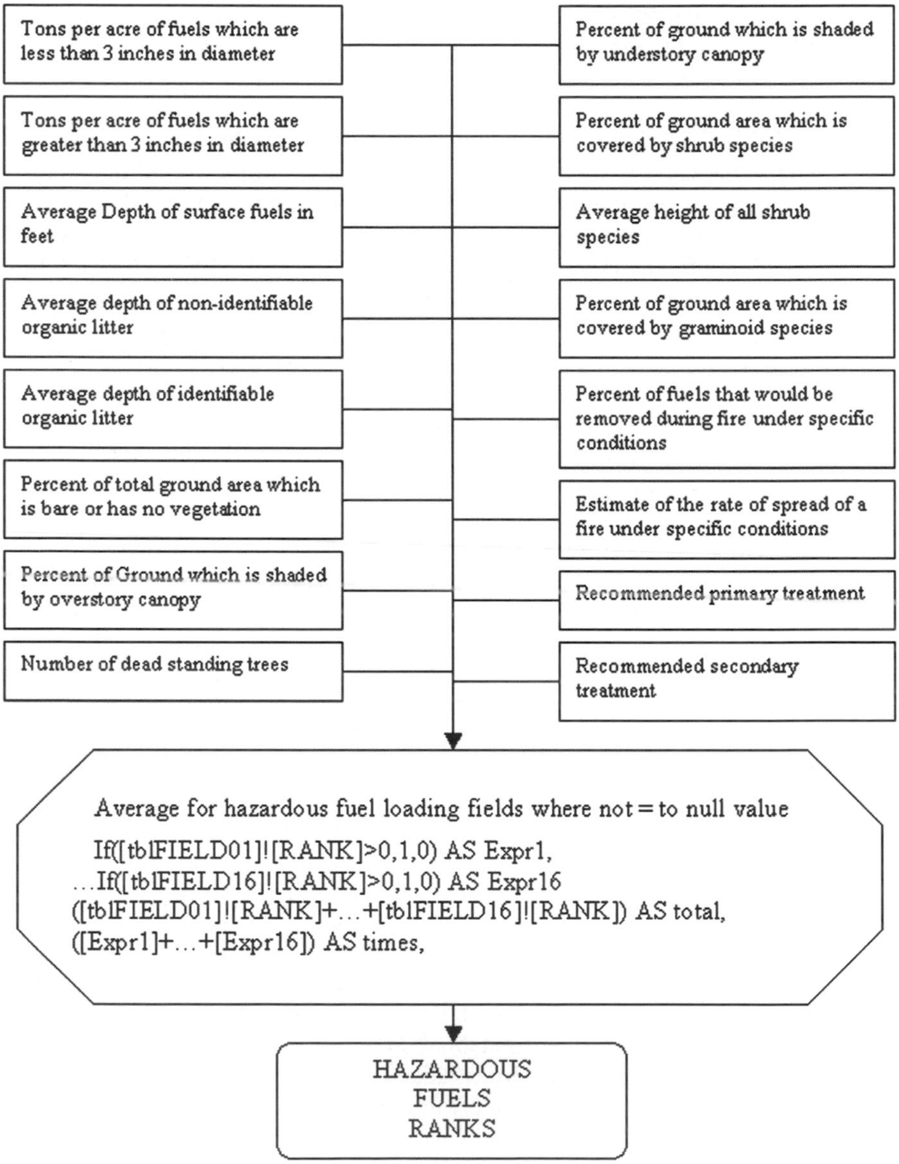

| Tons per acre of fuels which are less than 3 inches in diameter | Percent of ground which is shaded by understory canopy |
| Tons per acre of fuels which are greater than 3 inches in diameter | Percent of ground area which is covered by shrub species |
| Average Depth of surface fuels in feet | Average height of all shrub species |
| Average depth of non-identifiable organic litter | Percent of ground area which is covered by graminoid species |
| Average depth of identifiable organic litter | Percent of fuels that would be removed during fire under specific conditions |
| Percent of total ground area which is bare or has no vegetation | Estimate of the rate of spread of a fire under specific conditions |
| Percent of Ground which is shaded by overstory canopy | Recommended primary treatment |
| Number of dead standing trees | Recommended secondary treatment |

Average for hazardous fuel loading fields where not = to null value

If([tblFIELD01]![RANK]>0,1,0) AS Expr1,
...If([tblFIELD16]![RANK]>0,1,0) AS Expr16
([tblFIELD01]![RANK]+...+[tblFIELD16]![RANK]) AS total,
([Expr1]+...+[Expr16]) AS times,

HAZARDOUS
FUELS
RANKS

*Figure 67.2.* Hazardous fuels model.

Idaho has been in drought-like conditions for over three years now. Cheatgrass invasion is a big problem in many of the western states, including Idaho. Cheatgrass allows for the fire to move quickly and with higher heat intensity than native grasses. Cheatgrass is an invasive grass that is more

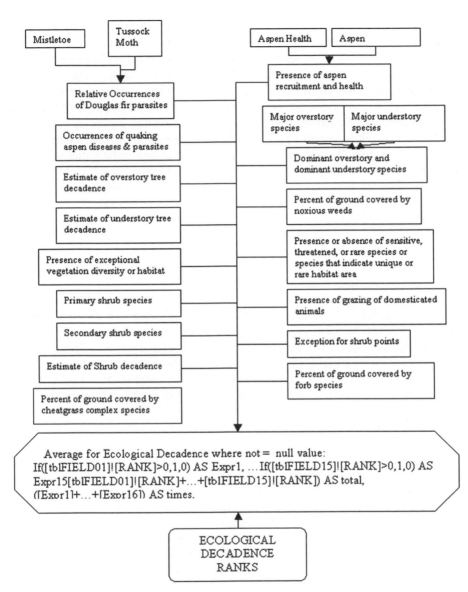

*Figure 67.3.* Ecological decadence model.

aggressive than most native-grass species and often takes over in areas after wildfires occur. Thus, when cheatgrass moves into an area and a wildfire occurs, the result is a hot, fast, and large fire.

Unlike wildfires, prescription fires can be done at times when weather conditions are good, and maximum control of the fire is the result. Through vegetation surveys and monitoring, sound management decisions can be made as to the type of treatments to use. Often, the result is multiple treatments over the same piece of ground, as illustrated in the following example. In the summer, a project area is surveyed and determined to be in a degraded state of mostly cheatgrass. The following spring a prescription fire is done to remove the invasive species from the project area. In the fall a chemical application is applied to slow the cheatgrass reinvasion. That winter, a seeding is applied to the same area to reintroduce the native grasses and shrubs. Other treatment methods may be used where fire is deemed inappropriate, due to safety, animal-habitat endangerment, etc. A targeted area surrounding a community is designated for a prescription burn. When the time comes for the planned burn, it may be deemed unfeasible due to weather constraints. As time goes by, it may be determined that other methods must be taken. A firebreak may be bulldozed to slow or stop future wildfire spread. Green strips of native grasses and shrubs are also planted to slow cheatgrass reinvasion. Finally continual monitoring is used to assess the effectiveness of results and the need for future treatments.

Some plant species are dependent on fire for reproductive success. For example, Quaking Aspen needs the intense heat of fires for the seeds to be released. Identification of areas of aspen that are ecologically decadent result in planned treatment projects. Through the use of mechanical thinning and prescription fires, these areas are regenerated. Areas that are degraded can be fenced off, allowing for a time of rest. However, over such a large expanse of land, it is impossible to visit all the areas of aspen. Therefore through the modeling techniques discussed above, the conditions of areas not visited can be inferred based on similar surrounding areas.

## RESULTS

The fuels specialists and resource managers use the gathered information to assess fuel loading, vegetation characteristics, and fire hazard. These surveys help to identify areas of fire exclusion and changed fire regimes, to plan priorities, and to develop a baseline inventory. With a good and current baseline inventory, managers can compare post-treatment monitoring data to determine community change after natural disturbance, changes over the years, pre-treatment and post-treatment achievements, and failures. Software used to maintain the massive fuels data dictionary and tabular data allows for specialized queries, weighting assignments, and modeling. It also allows ease of use for the non-GIS user when spatial information is not necessary. The final result is a series of management tools that are easy to use, updateable, and modifiable, and capable of demonstrating the dynamics of an ever-changing landscape.

## PROJECT CONTRIBUTORS

Glen Burkhardt, Fuels Specialist, South Central Idaho, Burley Field Office, Bureau of Land Management; Brandon Brown, Fire Use Specialist, South Central Idaho, Shoshone Field Office, Bureau of Land Management; Denise Tolness, Geographic Information Systems Specialist, Burley Field Office, Bureau of Land Management; Patrick Kennedy, GIS/GPS Support, Private Contractor; and Steve Popovich, Quality Control/Training, Private Contractor.

## REFERENCES

Edwards, Jr., T. and C. G. Homer. 1996. Idaho and Western Wyoming USDA Forest Service Inventory and Assessment/Gap Vegetation Analysis Mapping Project – *Idaho/Western Wyoming LandcoverClassification.* Remote Sensing/GIS Laboratories, Department of Geography and Earth Resources, Utah State University.

Executive Summary, Report to the President, 8 September 2000. Retrieved 25 January 2002 from http://www.fs.fed.us/fire/nfp/president.html.

Cohen, J. 1998. *Humans and Fire.* Retrieved 25 January 2002 from http://www.micro.utexas.edu/courses/mcmurry/spring98/10/jerry.html.

Federal Register. 2001. Volume 66, Number 3, Notices 751-777, 4 January 2001. Retrieved 4 April 2001 from http://www.nifc.gov/fireoplan/fedreg.htm.

Ferry, G. W. 2002. Bureau of Land Management, Division of Fire and Aviation Policy and Management. *Altered Fire Regimes Within Fire Adapted Ecosystems.* Retrieved 18 December 2002 from http://biology.usgs.gov/s+t/noframe/m1197.htm.

Fighting Fire With Fire: Sustainable Ecosystems and Fire Management in Florida. Retrieved 25 January 2002 from http://www.islandpress.com/ecosystems/forest/florifire.html.

Varga, E., 2000. Coos SWCD EQIP Outreach Coordinator, Coos Soil & Water Conservation District. *The Role of Fire in Ecosystems.* Retrieved 18 December 2002 from http://coosswcd.oacd.org/fire.htm.

Zar, J. H. 1999. *Biostatistical Analysis,* Fourth Ed. Englewood Cliffs, NJ: Prentice-Hall.

## 68

# CLOSED MUNICIPAL LANDFILLS IN TEXAS: USING GIS TO STUDY THEIR HEALTH, SAFETY, AND ENVIRONMENTAL RISKS AND IMPACTS

ROBERT D. LARSEN, RONALD J. STEPHENSON, and
JAMES VAUGHAN
*Texas State University – San Marcos*

Closed municipal solid-waste landfill-disposal sites represent a potential threat to health, safety, and welfare, and to the environment (U.S. Congress 1991b). Significant problems associated with old landfills include: accumulations of hazardous wastes; generation of potentially explosive methane gas that threatens enclosed structures; leachate resulting from the percolation of water through waste that can contaminate groundwater for several miles; and adverse effects of differential settling to overlying land development (U.S. Congress 1976).

Most of the closed and abandoned landfill sites that operated in the United States and Texas prior to the sweeping federal and state regulatory changes did not have liners, impermeable covers, or monitoring wells to detect and protect against potential groundwater contamination and off-site gas migrations (U.S. Congress 1991b). Many landfills were poorly sited, with communities using old sand and gravel borrow-pits, ravines, and floodplains for their waste disposal sites. Hundreds of landfill sites were selected because of their location in these "marginal-value" lands. A major problem with these sites was that leachate and landfill gases could rapidly dissipate through porous soils into adjacent "sensitive" surroundings, such as streams, water intakes, schools, apartment complexes, hospitals, residences, etc.

Concerns over documented problems caused by methane gas, leachate production, differential subsidence, and property abandonment served as catalysts for the passage of Texas state laws, requiring an official inventory of closed municipal solid-waste landfills. New rules specified that closed landfill information be property-deed recorded with current land use and ownership identified (TNRCC 1995).

*D. G. Janelle et al. (eds.), WorldMinds: Geographical Perspectives on 100 Problems*, 419–424.
© 2004 *Springer. Printed in the Netherlands.*

## DEVELOPING A THREE-PHASED INVENTORY

A three-phased Closed Landfill Inventory (CLI) has been ongoing under interagency agreements between the Texas Natural Resource Conservation Commission or TNRCC (name was changed to Texas Commission on Environmental Quality or TCEQ on 1 September 2002), regional Councils of Governments (COGs), and the Department of Geography at Southwest Texas State University (SWT; now Texas State University) since August 1995 (TNRCC 1978). In the initial phase of the inventory, a then state-of-the-art Geographic Information System (GIS) technology was used to locate, map electronically, and compile attribute files on nearly 4,200 closed municipal solid-waste landfills in Texas (TNRCC 1995), shown in Figure 68.1.

The two major categories of the nearly 4,200 identified sites are the PERMAPP (from PERMit APPlication) facilities and the UNUM (from Unauthorized NUMbered) sites. PERMAPP sites consist of officially recognized landfill facilities that were: permitted; had applied for a permit but one was never granted by the state; were officially recognized by the state as illegal sites; or had been grandfathered into the records when the permitting process began in the late 1970s. Dumps and landfills that were never intended to be permitted, their owners instead choosing to operate them as unauthorized/ unpermitted sites, became classified as UNUMs.

The processes developed to conduct the three phases of the inventory are summarized below. Through these procedures, landfills were identified and located, and data were verified. Several additional relational databases were developed regarding natural and cultural features that might be adversely impacted by their close proximity to a closed municipal landfill. Based on prior experiences, SWT researchers and the State agreed that for data-management and ease of interpretation, we should digitally record all data in Environmental Systems Research Institute Inc.'s (ESRI) ArcView GIS. Also, a methodology was developed whereby a GIS-based CLI might be used to undertake assessments of closed landfill risks associated with leachate, methane, and other closed landfill-induced problems.

## INITIAL INVENTORY PROCESS

Phase I of the CLI involved building a Geographic Information System using an interactive, computerized "point and click" map and attribute file system of all known closed or abandoned municipal solid-waste landfill facilities in Texas. This inventory was completed using mostly secondary sources of data and information. The basic steps in Phase I of the CLI process included: (1) data acquisition; (2) geocoding; (3) database development; (4) data verification; (5) mapping; and (6) building the GIS.

In the initial inventory, 1,415 closed landfill facilities were located, docu-

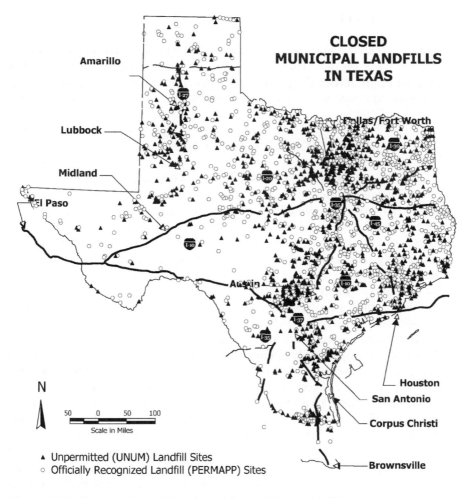

*Figure 68.1.* Concentrations of closed municipal solid waste landfills can be found in close proximity to urban areas and along major state and interstate highways throughout the State of Texas.

mented, and placed in the PERMAPP dataset of the GIS. The dataset of UNUM sites was generated mainly from an exhaustive, 18-month search of city and county site inspection reports, complaint letters, notices of violations, Attorney General Referral Letters, and other correspondence on file at the central records of the TNRCC in Austin, Texas. We located and digitally mapped 2,772 closed landfill sites in the UNUM dataset.

The inherent problems in the Phase I GIS were the limitations at that time of having only the data and GIS technology to support point-data identification rather than shape files of the boundaries for the landfills. Also, it became evident that there was a need for stronger, more encompassing legislation to govern closed landfills (Texas 1993). The landfill sites, represented in the GIS only as points, could not be deed recorded. Fortunately, as Phase-I inventory efforts were completed, a new Texas state law was passed that required additional work on the statewide inventory of closed landfills (Texas 1999). New state requirements, improved data collection methods, and advances in GIS technology lead to Phase II of the inventory.

## BOUNDARY COVERAGES AND SITE ATTRIBUTES

By TNRCC's adopting of this SWT Phase I GIS database as its "official" inventory, it became the statewide basis for all continuing inventory work under Phase II. Any changes, deletions, or additions to the Phase-I inventory had to be justified by written documentation developed in Phase II. The key new elements of Phase II, instituted under the amended legislation, included not only locating, but also mapping the exact or approximate boundaries of the closed landfills, statewide (Texas 1999). Six of the 24 COGs contracted with SWT to complete their Phase-II Inventories.

It then became possible to locate and map the legal boundaries of many of the known PERMAPP and UNUM landfill sites previously represented only as points. Perhaps even more significant, in many cases it becomes possible to locate the legal point of beginning (POB) of the traverse of the site (a corner of the site) in a metes-and-bounds description given as a latitude/longitude point. This can give a very high degree of accuracy as to the site's specific location and where to begin the legal description of its boundary. In addition, the present owners and current uses of the land containing the closed landfills can often be identified, and recorded along with other attribute information.

The inventory of all closed landfills in each COG planning region must be included in their Amended Regional Solid Waste Management Plan (Texas 1993,1999; TNRCC 2000). Closed or abandoned municipal landfills must also be recorded officially in the property-deed documents by the county clerk of the county where the closed landfills are located. Also, the chief planning officer of each city or county where the closed landfill sites are situated must be notified and provided an official copy of the inventory for their jurisdiction (Texas 1999).

Based on the innovative structure of the CLI, and the unique capabilities of the GIS that was used to process those data, the CLI can be viewed at varied scales, from street-corner locations, to census tracts, to legislative districts, etc. This process is easily used and interpreted, creating an extremely powerful research and decision-making tool for various federal, state, and local regulatory agencies, and private research organizations.

## SOME REMAINING QUESTIONS

What was of profound research interest in conducting the CLI was whether closed landfills could pose problems to public health, safety, and the state's environment. As the CLI neared Phase-II completion, any geographic relationships that might exist between closed landfills and "sensitive" features that could be negatively impacted by their proximity to those closed landfills became foci for further research efforts. We are in the process of developing this research as Phase III of the project. Many questions need answers. For example: are any of these old disposal sites generating leachate that has moved off-site, threatening possible contamination of ground and/or surface waters; because of their explosive characteristics, is the migration of landfill gases threatening public health and safety? Concerns are magnified because once-remote lands that were used for landfills have often been subsequently encroached upon and covered by urban sprawl.

To enable the GIS-based CLI to give researchers more specific geographic and scientific information for determining which landfill sites pose the greatest existing or potential problems, it was necessary early in Phase III to begin developing additional relational databases for the location of aquifers, public water-supply intakes, floodplains, schools, hospitals, population concentrations, etc. It is then possible to locate specific "sensitive" land uses and environments, and to electronically overlay them upon landfill data in the inventory. In this way, schools, public water supplies, hospitals, etc. that are over, under, or too near to closed landfills can be identified as relatively high-risk problem areas. The TCEQ now recommends that a closed landfill risk assessment be conducted for each planning region in each COG's Amended Plan (TNRCC 2000).

With many more questions than answers, there is a critical need for additional geographic research to address the global issues of landfilling municipal solid wastes.

## REFERENCES

Texas. 1993. *Use of Land Over Municipal Solid Waste Landfills.* Seventy-Third Texas State Legislative Session, H.B. No. 2537. General and Special Laws of the State of Texas. Amendment of Section 1, Chapter 361, Health and Safety Code, Subchapter R. Austin.

Texas. 1999. *Contents of Regional or Local Solid Waste Management Plan.* Seventy-Sixth Texas State Legislative Session, S.B. 1447. General and Special Laws of the State of Texas. Amendment of Section 1, Chapter 361, Health and Safety Code, Subchapter T. Austin.

Texas Natural Resource Conservation Commission (TNRCC). 1978. *Regional and Local Solid Waste Management Planning and Financial Assistance General Provisions.* Texas Administrative Code, Title 30. Environmental Quality Chapter 330, Subchapter O. Austin, Texas. See http: //www.tnrcc.state.tx.us/RuleS/tac/30/I/330/O/index.html.

Texas Natural Resource Conservation Commission (TNRCC). 1995. *Use of Land Over Closed Municipal Solid Waste Landfills.* Texas Administrative Code, Title 30. Environmental Quality

Chapter 330. Subchapter T. May. Austin, Texas. See. http: //www.tnrcc.state.tx.us/RuleS/tac/30/I/330/Y/330.953.html.

Texas Natural Resource Conservation Commission (TNRCC) *Strategic Assessment Division.
2000. Solid Waste Management in Texas, Strategic Plan 2001–2005*. SFR-042/01. Report to the Seventy-Seventh Legislature, Appendix 3. December. Austin, Texas.

U.S. Congress. 1976. *Resource Conservation and Recovery Act of 1976* (RCRA) (Solid Waste Disposal Act). U.S. Code 42 § 6901 et seq., October. Washington, D.C.

U.S. Congress. 1991a. *Resource Conservation and Recovery Act of 1976* (RCRA). Subtitle D, 40 CFR § 257 (Subtitle D Amendments). October. Washington, D.C.

U.S. Congress. 1991b. *Resource Conservation and Recovery Act of 1976* (RCRA). Subtitle D, 40 CFR § 258.40 (a) (2) (Subtitle D Amendments). October. Washington, D.C.

# 69

# UNDERSTANDING PESTICIDE MISUSE IN DEVELOPING COUNTRIES

## LAWRENCE S. GROSSMAN
*Virginia Tech*

Pesticide misuse is a widespread problem in developing countries (Weir and Schapiro 1981; Wright 1990; Murray 1994; Thrupp 1995; Arbona 1998). It can lead to severe health problems, contaminate the environment, facilitate resistance to pesticides, and create financial burdens for farmers.

Analyzing the causes of pesticide misuse is essential for understanding the difficulties in eradicating the problem. Researchers have focused on two types of causes. Some highlight the influence of political-economic constraints on farmers, including poverty, powerlessness, unequal control over land, biased government policies that favor large-scale growers and export agriculture, and the marketing policies of unscrupulous transnational corporations selling pesticides (e.g., Weir and Schapiro 1981; Wright 1990; Thrupp 1995). Others blame behavioral patterns at the local level, portraying farmers as being "careless" and "indiscriminate" in their use of pesticides (e.g., Black et al. 1987; Guan-Soon and Seng-Hock 1987).

The solutions implied in these contrasting analyses differ. The former approach indicates the need to alter political-economic constraints. The latter implies the need to eradicate "ignorance" and educate farmers to make them more careful in their application of pesticides. Changing the political-economic context of agriculture and educating farmers can alleviate agrochemical abuse to some extent. But another dimension makes pesticide misuse more difficult to solve than either perspective indicates (Grossman 1998).

A weakness in both approaches is that they portray pesticide use at the local level as being relatively uniform. Only the dimension of misuse is highlighted (e.g., Arbona 1998). In geographic analyses of environmental problems, we need to discard assumptions of homogeneity at the local level (Bassett and Zueli 2000). In the case of pesticides, we need to explore the extent to which patterns of agrochemical use are highly variable within communities.

To understand such variability, including the coexistence of both cautious and careless use of pesticides, it is helpful to consider Johnson's (1972) concepts of individuality and experimentation. "Individuality" refers to the tendency for villagers to make agricultural decisions based on their own unique circumstances, experiences, needs, and perceptions, which vary from farmer

425

*D. G. Janelle et al. (eds.), WorldMinds: Geographical Perspectives on 100 Problems*, 425–429.
© 2004 *Springer. Printed in the Netherlands.*

to farmer. "Experimentation" refers to the tendency for farmers to experiment with new techniques and crops in an attempt to improve agricultural output. Johnson employed these concepts to challenge the widespread belief that traditional agricultural practices were uniform. He asserted that individuality and experimentation generated considerable diversity in farming practices among members of the same community. Although Johnson was concerned with patterns in traditional agriculture, I assert that his concepts are also relevant to understanding variability in contemporary pesticide use in developing countries; these influences contribute not only to problems of misuse but also help account for the coexistence of varying degrees of caution and carelessness in pesticide use.

Individuality and experimentation are adaptations to the "environmental rootedness" of agriculture (Grossman 1998). Agriculture is anchored in the environment, and the environmental conditions of production – including rainfall, humidity, winds, temperatures, pests, soils, aspect, and slopes – are inherently variable in space and time. Similarly, household land and labor resources. and the crop combinations planted often change from year to year. In such a context of diversity, a simplified, uniform set of production strategies and practices among members of the same community would clearly be inappropriate for all farming conditions, and consequently farmers traditionally devise particular solutions to their own unique circumstances – patterns reflected in individuality and experimentation.

To illustrate these issues, I explore the case of pesticide use and misuse on the small island of St. Vincent in the Eastern Caribbean. I conducted fieldwork there from August 1988 to August 1989 and returned for several shorter visits between 1990 and 1995.

## PEASANT AGRICULTURE AND PESTICIDE USE ON ST. VINCENT

St. Vincent is the largest island in St. Vincent and the Grenadines. Measuring 17.7 km wide and 29.0 km long, this rugged, volcanic island has 101,000 inhabitants. Agriculture is the mainstay of the Vincentian economy, and bananas have been the island's major export cash crop since the mid-1950s, though such exports have declined substantially since the 1990s due to changing import regulations and intensifying market pressures in the European Union. The large majority of growers have been peasants, who cultivate bananas in small, scattered plots on rugged terrain. In addition to bananas, peasants grow a variety of local food crops.

A key feature of peasant agriculture on St. Vincent is increasing dependence on synthetic pesticides, a trend paralleling that in much of the rest of the Third World. Banana growers obtain "official" information about pesticides from the St. Vincent Banana Growers' Association (SVBGA), the statutory corporation that regulates the island's industry. Farmers also obtain advice about pesticides for local food crops from government agricultural extension officers.

Pesticide use is extensive. In one Vincentian village in which I conducted research, 97 percent of the agricultural households used at least one synthetic pesticide, with the average being four pesticides (Grossman 1998: 197). Peasants apply insecticides, herbicides, and nematicides in both local food production and banana export agriculture, with the latter receiving the majority of agrochemicals.

Pesticide misuse certainly occurs. Examples include failing to use protective clothing, applying agrochemicals with bare hands (Figure 69.1), determining the amount of water and pesticide to mix together based on a solution's color or odor instead of precise measurement, and formulating "pesticide cocktails." The latter are created when farmers mix more than one pesticide together to spray on their fields, and their use can be extremely hazardous to human health and the environment.

But it is misleading to focus only on the dimension of misuse. In fact, many villagers are more cautious in their use of agrochemicals than would be expected from reading the literature on pesticide use in developing countries. The spatial organization of crops is one example. When cultivating food crops separately from bananas in steeply sloping fields, peasants plant food crops on the top portion of the hills and the bananas below. They fear that if bananas were planted above the food crops on the hillsides, runoff during rainfall would carry pesticides originally applied in banana fields to the food

*Figure 69.1.* Farmer mixing pesticide and seedlings with bare hands. Source: author's fieldwork on St. Vincent.

crops downhill and thus contaminate them. Frequency of applications is another. Many prefer to wait for longer intervals between reapplications of pesticides than those recommended by SVBGA and government agricultural extension officers. Such farmers believe that longer intervals reduce the likelihood of pesticides contaminating their food crops. Caution is also revealed in the gender distribution of pesticide users. Because villagers believe that men are more resistant than women to the dangers from pesticides, women are much more hesitant to apply them, with some women, especially those who are pregnant, refusing to use them at all.

Thus, both caution and carelessness coexist. Other examples of variability in agrochemical use include inter-personal differences in the timing of pesticide applications, strength of dosages, and choices of which agrochemicals to apply to particular crops at each stage of cultivation.

## INDIVIDUALITY AND EXPERIMENTATION

Individuality and experimentation generate much of the variability in pesticide use. Individuality helps account for both caution and carelessness. For example, some farmers refuse to use pesticides at all, whereas a few others apply them with their bare hands. Farmers also experiment with pesticides in an attempt to solve problems and improve yields. They, thus, sometimes use pesticides in a manner unintended by the officials importing them. One example is the creation of "pesticide cocktails," which growers devise on their own initiative – a pattern found elsewhere in the Caribbean and Third World (Murray 1994). Another involves the case of the insecticide Primicid, which the SVBGA imports to sell to banana growers to control the banana borer. The SVBGA instructs that Primicid should be applied only on bananas, but villagers experiment by also applying it on food crops. But they do not use the same strength solution of Primicid on food crops as they do on bananas, believing that the strong mixtures typically applied on bananas would contaminate their food crops and harm crop growth. Instead, they experiment with weaker solutions, which they formulate in varying ways – by looking at the color of the resulting liquid mixture, by measurement, or by smell.

Certainly, some variations in agrochemical use among households, such as amounts applied, reflect differences in incomes and land holdings. But the majority of variation is independent of such economic conditions.

## CONCLUSION

Individuality and experimentation, as sources of innovation and change, were adaptive in traditional agriculture but create problems in the context of contemporary pesticide use. They combine to produce considerable variations in pesticide-use patterns, some of which are safe for human health and the environment while others have harmful consequences. Homogenizing away such variability in our analyses masks the complexity of the dynamics that are involved.

This analysis indicates that we cannot solve the problem of agrochemical misuse by only changing the political-economic context of agriculture or by focusing on educating farmers – though such efforts are beneficial. Individuality and experimentation are additional forces that facilitate pesticide misuse. They make the problem more difficult to eradicate because they are longstanding, fundamental characteristics of farming communities in developing countries.

## REFERENCES

Arbona, S. I. 1998. Commercial Agriculture and Agrochemicals in Almolonga, Guatemala. *Geographical Review* 88: 47–63.

Bassett, T. and K. B. Zueli. 2000. Environmental Discourses and the Ivorian Savanna. *Annals of the Association of American Geographers* 90: 67–95.

Black, R., N. Jonglaekha, and V. Thanormthin. 1987. Problems Concerning Pesticide Use in Highland Agriculture, Northern Thailand. In J. Tait and B. Napompeth, eds. *Management of Pests and Pesticides*, 28–37. Boulder, CO: Westview.

Grossman, L. 1998. *The Political Ecology of Bananas: Contract Farming, Peasants, and Agrarian Change in the Eastern Caribbean.* Chapel Hill: University of North Carolina Press.

Guan-Soon, L. and O. Seng-Hock. 1987. Environmental Problems of Pesticide Usage in Malaysian Rice Fields – Perceptions and Future Considerations. In J. Tait and B. Napompeth, eds. *Management of Pests and Pesticides*, 10–21. Boulder, CO: Westview.

Johnson, A. 1972. Individuality and Experimentation in Traditional Agriculture. *Human Ecology* 1: 149–159.

Murray, D. L. 1994. *Cultivating Crisis: The Human Cost of Pesticides in Latin America.* Austin: University of Texas Press.

Thrupp, L. 1995. *Bittersweet Harvests for the Global Supermarket.* Washington, D.C.: World Resources Institute.

Weir, D. and M. Schapiro. 1981. *Circle of Poison: Pesticides and People in a Hungry World.* San Francisco: Institute for Food and Development Policy.

Wright, A. 1990. *The Death of Ramon Gonzalez: The Modern Agricultural Dilemma.* Austin: University of Texas Press.

# 70

# THE BIOGEOGRAPHIC RESPONSE TO ACID RAIN

SARAH FINKELSTEIN
*University of Toronto*

Although acid rain is now a well-documented environmental issue, it was not always recognized as a problem. The work of biogeographers, analyzing the changing distributions of aquatic organisms through space and time, played a vital role in demonstrating that the acidification of lakes was occurring at a rapid rate and could be linked to human emissions of sulfur and nitrogen. This research compelled governments and individuals to reduce emissions leading to acid deposition.

Acid rain is caused by the release of sulfur dioxide ($SO_2$) and nitrogen oxides ($NO_x$) into the atmosphere. Major sources are the burning of coal and oil and the smelting of sulfur-bearing metal ores. These pollutants react in the atmosphere to form sulfuric and nitric acids, which are then deposited through all forms of precipitation. One of the earliest warnings about the phenomenon of acid rain was a report presented by Swedish scientists in 1972 at the United Nations Conference on the Human Environment, detailing the demise of aquatic life in dozens of Swedish lakes. They attributed this to air pollution from industrial areas in Britain and Northern Europe. Despite this early warning, very little action was taken anywhere in the world.

Awareness of the consequences of acid rain grew in the early 1980s. In addition to the appearance of clear and fishless "dead lakes", other impacts included detrimental effects on terrestrial vegetation and deterioration of buildings and monuments. Eastern North America, Scandinavia, and Northern Europe emerged as highly sensitive areas.

Although there was growing concern about this issue, there remained uncertainty about its causes and what could be done to stop it. It had long been recognized that some lakes acidify naturally and so, some researchers suggested that lake acidification was not ultimately controlled by human activity. Mechanisms for natural acidification include the development of peat-forming wetlands around a lake or the gradual leaching of base cations from catchment soils over time. Krug and Frink (1983) suggested that the benefits of reducing sulfur emissions would be small since ongoing soil-mediated processes of acidification would carry on, especially in temperate, humid regions on resistant, base-poor rocks.

Biogeographic research was essential in demonstrating that although some

*D. G. Janelle et al. (eds.), WorldMinds: Geographical Perspectives on 100 Problems*, 431–434.
© 2004 *Springer. Printed in the Netherlands.*

lakes do acidify naturally, the increase in acid deposition linked to human activity was causing unprecedented changes in aquatic ecosystems. This research also suggested that lakes could recover if emissions of $SO_2$ and $NO_x$ were reduced. Studies on long-term changes in populations of aquatic organisms provided data on background rates of lake water pH change and showed that the current process was unparalleled.

To gauge the severity of recent acidification in lakes, modern pH values need to be compared to historical or prehistoric values. Historical data on environmental change are often lacking in detail and accuracy and generally do not extend back further than the past century in North America. The use of paleoecological methodology has enabled biogeographers to look at changes in species distribution and abundance on time scales of tens of thousands of years and in some cases, even longer. This methodology involves recovering sediment cores in which records of ecological change are recorded in the form of micro- and macrofossils and chemical and physical signatures. These sedimentary records are typically found in lake basins or in wetlands where waterlogging of the sediments prevents their decomposition or dispersion. Sediments may be dated using the decay of Carbon-14 or other radioactive isotopes, or, by the presence of marker horizons.

One of the most powerful indicators used by biogeographers interested in paleolimnology is diatom analysis. Diatoms are a class of single-celled algae that secrete siliceous cell walls (Figure 70.1). The elaborate ornamentation on the surface of these frustules permits the identification of fossilized diatoms to the species or variety level under a microscope. Because diatoms are found in all aquatic environments, preserve well in sedimentary records, and are typically associated with a particular set of environmental conditions, they are excellent paleoecological indicators.

Geographer Richard Battarbee from University College London was a leader in early paleolimnological research using diatoms to trace long-term trends

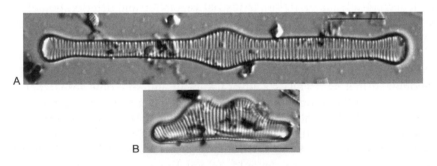

*Figure 70.1.* (a) *Tabellaria quadriseptata Knudson.* This species is typically found in acidic bogs and has been widely used as an indicator of recent acidification. (b) *Eunotia diodon Ehrenberg.* This species, like many in the genus Eunotia, indicates acidic conditions (see Patrick and Reimer 1966). Scale bars represent 10 μm. Photographs by Sarah Finkelstein.

in lake acidity. Battarbee (1984) summarized the work on lake acidification that had been completed up to that date. He presented findings from over 30 studies in Scandinavia, the United Kingdom, the United States, and Canada, with many showing the onset of significant and rapid acidification beginning anywhere from 1850 to 1970. The range of pH declines was from 0 to 1.7 units in that time span. To determine whether recent lake acidification was a natural or anthropogenic phenomenon, he contributed to the establishment of several large research projects to coordinate and standardize data collection and analysis. These projects had two components. The first was to create calibration sets to relate a particular assemblage of diatoms to a narrow pH range. This was done by studying a large number of lakes, determining the pH of each one, and inventorying the diatom species present, along with the abundance of each taxon. This gives an optimum pH for each taxon. In the second component, sediment cores were collected, dated usually by the Lead-210 method, and analyzed for fossil diatoms. The fossil diatom assemblages were used by way of the calibration sets to infer past pH levels.

Very large datasets emerged from these projects. The results confirmed those of the earlier studies. Widespread lake acidification was occurring in Eastern North America, Scandinavia, and Britain (reviewed in Battarbee et al. 1999). Typical changes in the diatom flora associated with increased acidity include a decrease or complete elimination of euplanktonic species and an increase of acid-tolerant taxa, such as *Tabellaria quadriseptata* Knudson, *Eunotia exigua* (Brébisson ex Kützing) Rabenhorst or *Fragilaria acidobiontica* Charles.

To compare recent declines in pH with natural rates of change, Whitehead et al. (1986) used diatom analysis to gain a better understanding of acidification in lakes prior to human perturbation. This study considered natural acidification in three Adirondack mountain lakes. These lakes were situated in the High Peaks region and, due to resistant anorthosite bedrock, were sensitive to acidification. They found that for about 2,000 years following deglaciation, lake pH was circumneutral. It then began to drop, reaching stable levels of about 5.5 by 3,000 years ago in Heart Lake. Pollen analysis and geochemical testing revealed that acidification here was caused by the postglacial spread of conifers into the region, the associated increase in availability of organic acids, the development of histosols, and increased cation exchange capacity of soils. These factors all resulted in the gradual depletion of the base cations in the watershed needed to maintain a circumneutral pH in lake basins. One of the most important findings of this study was that the maximum rate of acidification in the Holocene was 0.2–0.4 pH units in 100 years. A paleo-biogeographic approach was able to put the magnitude of recent changes (e.g., decline of 1.2 pH units in about 50 years at Loch Grannoch, SW Scotland; Battarbee 1984) into the context of natural variation. This served to emphasize the unprecedented speed of recent acidification.

The work of these biogeographers helped galvanize governments and individuals to action on acid rain. Most European nations adopted the Helsinki

Protocol on the Reduction of Sulphur Emissions in 1985; the U.S. government adopted the Clean Air Act in 1990; in March 1991, Canada and the United States adopted an air-quality agreement to lower $SO_2$ and $NO_x$ emissions that has since been expanded to include a post-2000 strategy. Since emission reductions have begun to take place, there is some indication that water quality is being restored in heavily damaged lakes, and, in some cases, biological recovery is already occurring. Gunn and Keller (1990) document recovery in Whitepine Lake, near Sudbury, Ontario, in the nickel-smelting region that was one of the most severely damaged in the world. The authors discuss increases in acid-sensitive species such as lake trout and crayfish. The reversibility of the acidification process once $SO_2$ emissions decline is further evidence that it was anthropogenically driven. In other cases, however, recovery is not yet occurring. Ek and Korsman (2001) showed that despite 50 percent reductions in $SO_2$ emissions in Sweden, diatom-inferred lake water pH has not yet recovered in seven study lakes, probably due to local biogeochemical processes.

The use of diatom analysis in sediment cores helped to show that recent lake acidification is taking place at an unprecedented rate. Acid rain remains a problem, but now, it is better understood and the steps required to solve it are better known. The results from paleo-biogeographic studies are now being similarly used to demonstrate that the ecological changes associated with anthropogenically driven climatic change are unprecedented.

## REFERENCES

Battarbee, R. W. 1984. Diatom Analysis and the Acidification of Lakes. *Philosophical Transactions of the Royal Society of London Series B* 305: 451–477.

Battarbee, R. W., D. F. Charles, S. S. Dixit, and I. Renberg. 1999. Diatoms as Indicators of Surface Water Acidity. In E. F. Stoermer and J. P. Smol, *The Diatoms: Applications for the Environmental and Earth Sciences*, 85–127. Cambridge: Cambridge University Press.

Ek, A. S. and T. Korsman. 2001. A Paleolimnological Assessment of the Effects of Post-1970 Reductions of Sulfur Deposition in Sweden. *Canadian Journal of Fisheries and Aquatic Sciences* 58: 1692–1700.

Gunn, J. M. and W. Keller. 1990. Biological Recovery of an Acid Lake After Reductions in Industrial Emissions of Sulfur. *Nature* 345: 431–433.

Krug, E. C. and C. R. Frink. 1983. Acid Rain and Acid Soil – A New Perspective. *Science* 221: 520–525.

Patrick, R. and C. W. Reimer. 1966. *The Diatoms of the United States Exclusive of Alaska and Hawaii 1*. Philadelphia: The Academy of Natural Sciences of Philadelphia, Monograph No. 13.

Whitehead, D. R., D. F. Charles, S. T. Jackson, S. E. Reed, and M. C. Sheehan. 1986. Late Glacial and Holocene Acidity Changes in Adirondack (N.Y.) Lakes. In J. P. Smol, R. B. Davis, and J. Meriläinen, eds. *Diatoms and Lake Acidity*, 251–274. Dordrecht: Junk.

# 71

# THINKING OUTSIDE THE CIRCLE:
# USING GEOGRAPHICAL KNOWLEDGE TO FOCUS
# ENVIRONMENTAL RISK ASSESSMENT
# INVESTIGATIONS

JAYAJIT CHAKRABORTY
*University of South Florida*

MARC P. ARMSTRONG
*The University of Iowa*

The perception of inequity in the distribution of technological hazards in the United States has generated headlines, stimulated scientific investigations, and fostered public policy initiatives during the past decade. The empirical evidence for such inequities comes from numerous case studies that have attempted to establish associations between hazards and the characteristics of at-risk populations. These studies have employed a variety of methods, and have stirred epistemological debates that are often rooted in the approach used to geographically define exposure to a polluting facility.

## THE PROBLEM

Most early studies to identify areas and populations exposed to environmental risk used administrative zones (e.g., zip-codes, census tracts) (e.g., United Church of Christ 1987; Anderton et al. 1994). The logic of this approach is that people are at risk if they reside in the same enumeration zone as a hazardous facility. While this "spatial coincidence" approach (Sheppard et al. 1999) facilitates statistical comparisons, it rests on the unsupportable assumption that pollution effects are equal everywhere within the administrative unit containing a hazardous facility. The approach also ignores the possibility that the facility could be located near the boundary of its "host" unit, thereby exposing the population in neighboring areas to its effects.

Several studies have attempted to address these limitations by creating GIS-based circular buffers around polluting facilities to determine at-risk populations (e.g., Glickman 1994). While a circular buffer approach may be viewed in the abstract as an improvement, several limitations remain. Circle

*D. G. Janelle et al. (eds.), WorldMinds: Geographical Perspectives on 100 Problems, 435–442.*
© 2004 *Springer. Printed in the Netherlands.*

radii are typically selected arbitrarily (e.g., as 1,000 yards, or one mile) and do not reflect the quantity, volatility, and toxicity of substances stored at each facility. The approach also assumes that exposure remains invariant with the direction and speed of prevailing winds; a person residing one mile upwind from a polluting facility faces the same exposure as someone living one mile downwind. Although physical processes, in general, do not operate in a perfectly symmetrical manner, meteorological conditions are rarely used in geographic definitions of exposure to environmental hazards.

## THE SOLUTION: GEOGRAPHIC PLUME ANALYSIS

One solution to these problems is provided by geographic plume analysis (GPA), a methodology that integrates physical dispersion modeling with GIS and demographic information to estimate areas and populations exposed to airborne releases of toxic substances (Chakraborty and Armstrong 1997). Dispersion models combine data on the quantity and physical properties of a released chemical with site-specific information and atmospheric conditions to estimate the shape and size of the area likely to be affected by a spreading plume. Our application of GPA uses ALOHA (Areal Locations of Hazardous Atmospheres), a popular dispersion model designed to support emergency responses to hazardous chemical accidents. ALOHA is particularly useful for estimating plume extent and concentration for short-duration chemical releases (NOAA and U.S. EPA 1996).

ALOHA output includes a top view (footprint) of the plume calculated by the model (Figure 71.1). The shaded area inside a footprint is predicted to have ground-level concentrations higher than a threshold value, usually the Immediately Dangerous to Life and Health (IDLH) concentration. A chemical's IDLH is the maximum airborne concentration to which a healthy person could be exposed for a short duration without suffering permanent health effects (NIOSH 1997). The dotted lines bounding the footprint in Figure 71.1 indicate a 95 percent confidence interval based on shifts in wind direction.

An ALOHA footprint can be imported by GIS software and superimposed on other map layers (e.g., street centerlines, census blocks) to examine affected areas and their demographic characteristics. It is important to consider, however, that a single footprint, as shown in Figure 71.1, is based on meteorological conditions at the time of a chemical release. A change in any input parameter (e.g., wind direction) will alter the size and shape of the at-risk area. The reliability of exposure estimates can be improved with a composite plume model that considers local climatic variability in a study area (Chakraborty and Armstrong 1997). A composite plume is created by using long-term average weather conditions to compute multiple-dispersion footprints for a set of different time-periods (e.g., months). Unlike circular buffer analyses, GPA can be used to model anisotropic distributions of exposure to airborne hazards. The

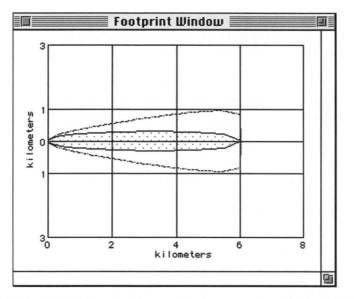

*Figure 71.1.* The ALOHA footprint. Source: http://www.nwn.noaa.gov/sites/hazmat/cameo/alotech/inout.html.

integration of meteorological and facility-specific chemical information, when linked with dispersion models and GIS technology, significantly improves the ability of researchers to identify populations that are potentially exposed to environmental hazards.

## AN ILLUSTRATION OF GEOGRAPHIC PLUME ANALYSIS

A single facility, the Winn-Dixie Tampa Warehouse in Tampa, Florida, is used to demonstrate and compare the circular buffer and GPA methodologies for risk assessment. This facility is used to store various perishable and non-perishable items for distribution to local Winn-Dixie supermarkets. According to the Risk Management Plan submitted to the Environmental Protection Agency (EPA) in 1999, the warehouse contains two separate ammonia refrigeration systems that are capable of storing 14,763 lbs. of anhydrous ammonia (http://d1.rtk.net/rmp/fac.php). Ammonia is the only chemical at this facility that appears on EPA's list of extremely hazardous substances.

Three methods are used in this demonstration to delineate the area at risk around the facility:

*Circular buffer with arbitrarily chosen radius:* Previous research on environmental risk assessment suggests that radii of uniform circular buffers around hazardous facilities typically range from one-half to two miles. In our illustration, the radial buffer methodology was implemented using a one-mile radius around the warehouse (Figure 71.2).

*Circular buffer with radius based on worst-case assumption:* In this extension of the circular-buffer approach, the radius of the circle reflects the area potentially exposed to a worst-case release at the facility. This methodology is consistent with the latest EPA guidelines (U.S. EPA 1998) for modeling worst-case exposure to toxic substances (Chakraborty 2001). Following EPA guidelines, the total quantity of anhydrous ammonia was assumed to be released in 10 minutes, at a rate of 1,476.3 lbs. per minute. This information was used in combination with EPA-recommended worst-case atmospheric assumptions (U.S. EPA 1998) in ALOHA to estimate the plume endpoint (impact distance) for a worst-case ammonia release. This distance (1.8 miles) was then used as a radius to construct a circular buffer around the facility (Figure 71.2).

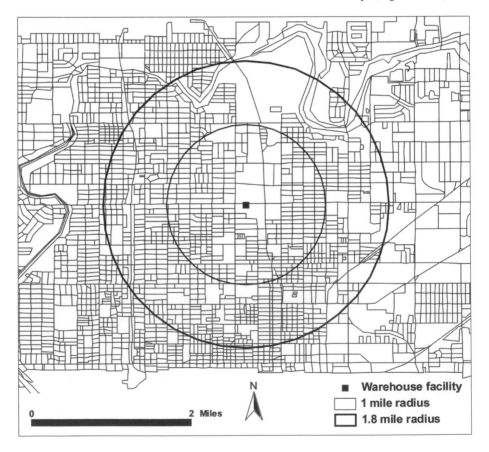

*Figure 71.2.* GIS-based circular buffers around Winn-Dixie Tampa warehouse facility.

*Composite plume footprint based on local weather and wind patterns:* A composite plume footprint was centered on the warehouse through the computation of 12 (monthly) footprints. Long-term average weather data for each month (Table 71.1) are used by ALOHA to estimate 12 footprints for a worst-case ammonia release (14,763 lbs. in 10 minutes) at the facility. The analytical capabilities of GIS are used to consolidate these footprints into a single composite footprint (Figure 71.3). In addition to the quantity and toxicity of chemicals stored at the facility, the size and shape of this composite footprint reflects seasonal variations in local weather conditions. Given the predominant wind directions in Tampa, neighborhoods directly north or south of the warehouse facility are unlikely to be exposed to a potential ammonia release.

Demographic characteristics of the at-risk population were computed on the basis of information from the 2000 U.S. Census. Population data from census blocks within the circular- and plume-based buffer zones were used to estimate the number of potentially exposed residents. Since block boundaries do not coincide exactly with the buffer zones, a method of areal interpolation (Goodchild and Lam 1980) was used to develop these estimates. Our method consists of summing the populations of all census blocks, weighted by the fraction of the area of the block that falls inside the buffer (Chakraborty and Armstrong 1997).

The results (Table 71.2) indicate important differences in the total population at risk and its racial composition, when prevailing wind patterns are incorporated in the analysis. Neighborhoods south of the facility contain a high proportion of African-American and low-income residents. Since the composite footprint does not extend to the south, the proportion of the African-American

*Table 71.1.* Atmospheric Data: Long-Term Averages for Tampa, Florida.

| Month | Average Temperature (degrees F) | Relative Humidity (percent) | Average Wind Speed (miles/hour) | Prevailing Wind Direction |
|---|---|---|---|---|
| January | 60.4 | 75 | 8.6 | ENE |
| February | 62.3 | 73 | 8.9 | ENE |
| March | 66.6 | 72 | 9.5 | E |
| April | 71.6 | 69 | 9.2 | E |
| May | 77.6 | 70 | 8.6 | W |
| June | 81.3 | 74 | 7.9 | W |
| July | 82.3 | 77 | 7.2 | W |
| August | 82.4 | 78 | 7.0 | E |
| September | 81.0 | 78 | 7.6 | E |
| October | 75.0 | 74 | 8.3 | ENE |
| November | 67.6 | 75 | 8.3 | ENE |
| December | 62.2 | 75 | 8.2 | ENE |

Source: National Climatic Data Center 2001.

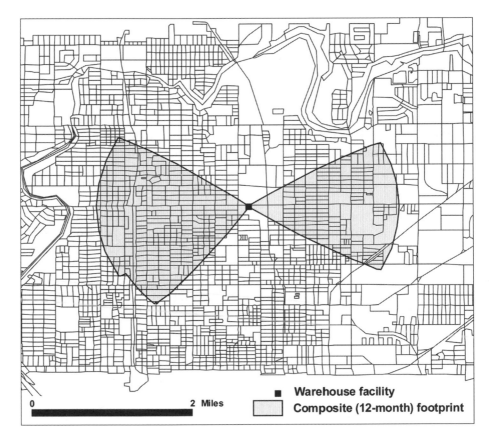

*Figure 71.3.* ALOHA-based composite (12-Month) footprint at facility.

population inside this zone is 54 percent, compared to 61 percent within the one-mile buffer and 66 percent within the worst-case buffer. An analysis of environmental justice based on the isotropic circular approaches, therefore, could lead to incorrect inferences regarding racial inequity in the distribution of potential exposure.

Though the GPA results show clear differences when compared to circular buffers, additional enhancements can be made. A more complete set of climatic indicators (> 12 monthly observations) would support the development of a probabilistic assessment procedure. Additional improvements would focus on the areas immediately surrounding each site, which serve as foci for conic sections that extend in different directions. Under calm-wind conditions, however, plumes assume more circular forms, and the population in all directions is affected over shorter distances.

*Table 71.2.* Percentage of Population at Risk from Winn-Dixie Warehouse Facility, 2000.

|  | Circular Buffer (Arbitrary: 1-mi. radius) | Circular Buffer (Worst-case: 1.8-mi. radius) | Composite ALOHA (12-month) Footprint |
|---|---|---|---|
| Total Population | 12,176 | 36,359 | 15,389 |
| Race and Ethnicity: |  |  |  |
| White | 27.6 | 32.1 | 39.6 |
| Black | 66.4 | 61.2 | 53.6 |
| Asian | 0.6 | 0.5 | 0.7 |
| American Indian/Eskimo | 0.4 | 0.4 | 0.5 |
| Other Race | 3.1 | 3.3 | 3.3 |
| Multi-racial | 1.9 | 2.5 | 2.4 |
| Hispanic origin | 12.1 | 12.8 | 14.5 |
| Age: |  |  |  |
| Under 5 | 6.2 | 6.9 | 7.9 |
| Above 65 | 15.3 | 13.4 | 12.2 |

## CONCLUSION

Previous research on environmental justice and risk assessment has adopted assumptions that were based on convenience and inappropriate GIS-based tools. Though the spatial-coincidence and circular-buffer approaches are easy to implement, they unfortunately fit the old saw: "when the only tool you have is a hammer, everything looks like a nail." GPA provides an alternative assessment methodology that attempts to incorporate knowledge about dynamic spatial processes into the calculation of affected areas and populations. Though additional improvements remain to be implemented, the use of geographic knowledge, when combined with appropriate geographic information technologies, can lead to a better understanding of complex, dynamic urban problems.

## REFERENCES

Anderton, D. L., A. B. Anderson, J. B. Oakes, and R. M. Fraser. 1994. Environmental Equity: The Demographics of Dumping. *Demography* 31(2): 229–248.

Chakraborty, J. 2001. Acute Exposure to Extremely Hazardous Substances: An Analysis of Environmental Equity. *Risk Analysis* 21 (5): 883–894.

Chakraborty, J. and M. P. Armstrong. 1997. Exploring the Use of Buffer Analysis for the Identification of Impacted Areas in Environmental Equity Assessment. *Cartography and Geographic Information Systems* 24(3): 145–157.

Glickman, T. S., 1994. Measuring Environmental Equity With Geographic Information Systems. *Renewable Resources Journal* 12(3): 17–21.

Goodchild, M. F. and N. S. N. Lam. 1980. Areal Interpolation: A Variant of the Traditional Spatial Problem. *Geoprocessing* 1: 297–312.

National Climatic Data Center. 2001. *Local Climatological Data: Annual Summary With Comparative Data: Tampa, Florida.* Washington D.C.: National Oceanic and Atmospheric Administration.

National Institute of Occupational Safety and Health. 1997. *NIOSH Pocket Guide to Chemical Hazards.* NIOSH Publication No. 97-140-640. Springfield: U.S. Department of Health and Human Services.

National Oceanic and Atmospheric Administration and the U.S. Environmental Protection Agency. 1996. *ALOHA User's Manual.* Washington D.C.: National Safety Council.

Sheppard, E., H. Leitner, R. B. McMaster, and H. Tan. 1999. GIS-Based Measures of Environmental Equity: Exploring Their Sensitivity and Significance. *Journal of Exposure Analysis and Environmental Epidemiology* 9(1): 18–28.

United Church of Christ. 1987. *Toxic Wastes and Race in the United States: A National Report on the Racial and Socio-Economic Characteristics of Communities With Hazardous Waste Sites.* New York: United Church of Christ Commission for Racial Justice.

U.S. Environmental Protection Agency. 1998. Risk Management Program Offsite Consequence Analysis Guidance. *Risk Management Programs 40 CFR Part 68.* Washington D.C.: Office of Emergency and Remedial Response.

# 72

# PALEOTEMPESTOLOGY: GEOGRAPHIC SOLUTIONS TO HURRICANE HAZARD ASSESSMENT AND RISK PREDICTION

KAM-BIU LIU

*Louisiana State University*

Hurricanes are a meteorological phenomenon that can severely impact both the natural and socio-economic systems of the world (Elsner and Kara 1999). In terms of impacts on the natural environmental system, hurricanes and their associated storm surges and overwash processes can cause significant changes in coastal landforms and processes. Hurricanes can also cause perturbations in the hydrological and geomorphic systems, resulting in excessive rainfall, flooding, landslides, and storm deposition in lakes and waterways. As an ecological agent, hurricanes are an important disturbance mechanism that has long-term effects on the patterns of vegetation composition, biocomplexity, and successional pathways. In terms of societal impacts, hurricanes rank at the top of all natural disasters in terms of the number of deaths and the economic losses they caused in the United States. The Galveston (Texas) Hurricane of 1900, for example, resulted in more than 8,000 deaths – the deadliest natural disaster in U.S. history. The costliest hurricane in U.S. history – Hurricane Andrew of 1992 – devastated Miami and then southern Louisiana, resulting in economic losses totaling more than $30 billion, about $16 billion of which are paid out by insurance companies. As a natural hazard, hurricanes account for 62 percent of all catastrophic insurance losses, hence a major concern for the property catastrophic insurance industry (Elsner and Kara 1999). As population and property along the hurricane-prone U.S. coastal areas were growing rapidly during the last three decades, there has been increasing societal concerns over a number of hurricane-related issues, such as emergency preparedness, evacuation strategies, disaster mitigation, building codes, urban zoning and physical planning, and post-disaster health hazard assessments.

## HURRICANES AS A MULTI-FACETED RESEARCH PROBLEM

Hurricanes, then, represent a multi-faceted research problem that integrates both the biophysical and social dimensions of science. Multi-faceted problems

*D. G. Janelle et al. (eds.), WorldMinds: Geographical Perspectives on 100 Problems*, 443–448.
© 2004 *Springer. Printed in the Netherlands.*

require a multi-disciplinary or multi-dimensional approach to tackle them effectively. This problem is well suited for study by geographers who, as individuals or as collaborative research teams, can offer an integrative approach to the understanding of not only the environmental impacts of hurricanes but also the interactions between hurricanes and society. Indeed, one of the areas in hurricane research where geographers have already played a major role in developing is paleotempestology – a new field of science that studies past hurricane activities by means of both geological and historical-archival techniques (Liu in press).

## PALEOTEMPESTOLOGY AND ITS APPLICATIONS IN HURRICANE RISK ASSESSMENT

The vulnerability of the U.S. coastal areas to catastrophic hurricane strikes has increased significantly during the last few decades due to population migration, income redistribution, growth in coastal property and wealth, and tourism development. Therefore, the assessment of hurricane risk can have significant implications for the insurance structure, real estate values, tourism development, and regional planning for coastal locations. In the light of this, a question of major scientific and practical significance is: How likely is any given coastal location (such as New Orleans or Miami) to be directly hit by a catastrophic hurricane of Saffir-Simpson category 4 or 5 intensity? Obviously, this question is important to a wide range of stakeholders from the climate scientists to the risk managers working for the reinsurance industry.

This question cannot be easily answered by examining the historical record of hurricane activities in the United States. The period of instrumental observations is very short – essentially the past 150 years (since 1850). Within the 20th century, the U.S. coasts were hit directly by only three hurricanes of category-5 intensity – the Labor Day Hurricane of 1935, Hurricane Camille of 1969, and Hurricane Andrew of 1992 – and 15 hurricanes of category-4 intensity (Elsner and Kara 1999). Considering the typical radius of maximum wind and the storm surge height around the eye of the hurricane ((50 km), most U.S. coastal locations have not been directly hit by a category-4 or -5 hurricane during the past 150 years. Therefore, a longer empirical dataset is needed for estimating the return period or landfall probability of these rare but most destructive hurricanes.

Paleotempestology can provide such a long-term perspective that would encompass a wider range of variability in the hurricane climate regime. Two main sources of data are available in paleotempestology – historical documents and geological proxy records. In certain coastal regions of the United States historic, information about hurricanes dating back to the 17th century can be extracted from old newspapers, travelers' logbooks, and plantation diaries. In other regions of the world, such as China, historical documentary records

in the form of county gazettes and official histories have permitted the reconstruction of a 1,000-year time-series of typhoon landfalls that reveal trends and
cycles occurring at multi-decadal timescales (Liu et al. 2001). On the other
hand, proxy evidence of past hurricane strikes can potentially extend the
empirical record back to the past several thousand years. Several geological
or biological proxies have been attempted that offer potential in paleotempestology, but so far the most useful proxy has been overwash sand layers
found in the sediments of coastal lakes and marshes (Liu, in press; Liu and
Fearn 1993, 2000a, b; Donnelly et al. 2001). These sand layers were deposited
when waves and currents overtopped a sandy barrier (beach, dunes, beach
ridges) during storm surges generated by landfalling hurricanes, causing successive overwash fans to form behind the sandy barrier (Liu and Fearn 2000a,
b). In addition, microfossils from pollen, diatoms, phytoliths, and foraminifera
can be used as supplementary evidence to confirm the occurrence of marine
incursion into otherwise freshwater or brackish coastal lakes and marshes. Once
the past hurricane landfall events have been identified from the sedimento-
logical evidence, their ages can be estimated by means of radiocarbon ($^{14}$C)
dating or other dating techniques, such as Lead-210 ($^{210}$Pb) and Cesium-137
($^{137}$Cs) dating. For each location, a dated record of past hurricane strikes can
then be translated into an estimate of hurricane return periods and landfall
probabilities (Figure 72.1).

During the past decade, several sites from the Gulf of Mexico coast have
yielded proxy records of catastrophic hurricane landfalls spanning the last 5,000
years (Liu 1999; Liu and Fearn 2000b). Each of these sites was directly
struck by catastrophic hurricanes about 9–12 times during the last 3,800
calendar years, thus implying a return period of approximately 350 years, or
a landfall probability of about 0.3 percent per year. Another striking discovery is that the landfall probabilities have varied on the millennial timescale.
At Western Lake, northwestern Florida, for example, the recent millennium
seems to be a relatively quiescent period, but it was preceded by a hyperac-
tive period about 1,000–3,800 years ago during which the Florida Panhandle
was hit frequently by catastrophic hurricanes, resulting in a landfall probability
of approximately 0.5 percent per year (Liu and Fearn 2000a). The millennial-scale variability observed along the Gulf coast has been attributed to
long-term shifts in the mean position of the Bermuda High and to long-term
changes in the North Atlantic Oscillation (NAO) (Liu and Fearn 2000a; Elsner
et al. 2000).

The discovery of an approximately 350-year return period for catastrophic
hurricanes has significant implications for risk assessment along the Gulf coast.
Previously, no empirical evidence existed to show whether the decidedly rare
landfalling catastrophic hurricanes like Camille or Andrew were 100-, 500-,
or 1,000-year events. With the newly discovered empirical estimate of 0.3
percent per year, climate scientists and risk managers can calibrate their
computer models to come up with better risk assessment for coastal commu-

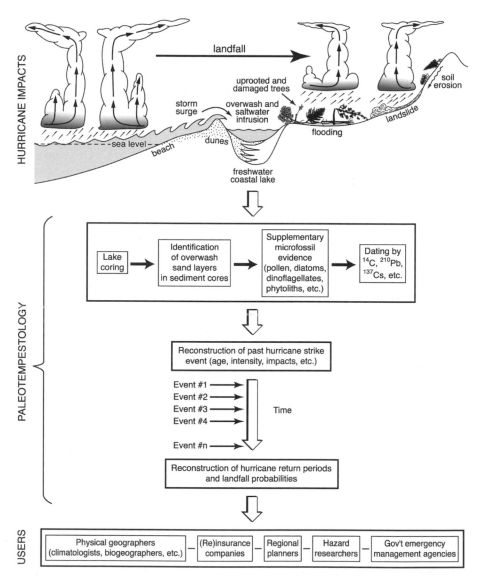

*Figure 72.1.* Schematic diagram showing major hurricane impacts and the principles, methods, and applications of paleotempestology.

nities. This information will be very useful for the insurance/reinsurance industry, real estate developers, emergency management agencies, and regional planners in their decision-making processes. In fact, the reinsurance industry has already incorporated the results of paleotempestology into their risk prediction models (Murnane et al. 2000).

## CONCLUSIONS

Paleotempestology is a multi-disciplinary solution to a multi-faceted problem in hurricane hazard assessment and risk prediction. Paleotempestology is an integrative science that requires knowledge in climatology, coastal geomorphology, limnology, sedimentology, biogeography, and paleoecology. The results of paleotempestology have significant socio-economic implications and applications, thus offering ample opportunities for paleotempestology researchers to collaborate with social scientists, regional planners, risk managers, business leaders, and policy makers in both government and private industry. By looking at the past, paleotempestology provides us with a long-term perspective that is vital for understanding the present and predicting the future. In addition, paleotempestology has a spatial dimension that can be effectively studied by the use of GIS and spatial analytical techniques, because multiple cores from within a lake and multiple lakes from coastal areas are necessary for deciphering the depositional pattern of each past overwash event and for reconstructing the spatial variability of hurricane landfall probabilities between regions. Because of their broad, multi-disciplinary training that typically integrates the principles and methods of both the physical and social sciences, geographers are well positioned to tackle the multi-faceted research problems posed by hurricanes by means of the emerging science of paleotempestology.

## ACKNOWLEDGEMENTS

This research has been supported by grants from the National Science Foundation (SES-8922033, SES-9122058, ATM-9905329; BCS-0213884) and the Risk Prediction Initiative of the Bermuda Biological Station for Research (RPI-96-015, RPI-96-048, RPI-99-1-002, RPI-00-1-002).

## REFERENCES

Donnelly, J. P., S. S. Bryant, J. Butler, J. Dowling, L. Fan, N. Hausmann, P. Newby, B. Shuman, J. Stern, K. Westover, and T. Webb III. 2001. A 700 yr Sedimentary Record of Intense Hurricane Landfalls in Southern New England. *Geological Society of America Bulletin* 113: 714–727.

Elsner, J. B. and A. B. Kara. 1999. *Hurricanes of the North Atlantic: Climate and Society.* New York: Oxford University Press.

Elsner, J. B., K-b. Liu, and B. Kocher. 2000. Spatial Variations in Major U.S. Hurricane Activity: Statistics and a Physical Mechanism. *Journal of Climate* 13: 2293–2305.

Liu, K-b. 1999. *Millennial-Scale Variability in Catastrophic Hurricane Landfalls Along the Gulf of Mexico Coast.* Preprint Volume of the 23rd Conference on Hurricanes and Tropical Meteorology, January 10–15, 1999, Dallas, Texas, 374–377. Boston: American Meteorological Society.

Liu, K-b. (in press). Paleotempestology: Principles, Methods, and Examples From Gulf Coast Lake-Sediments. In R. J. Murnane and K-b. Liu, eds. *Hurricanes and Typhoons: Past, Present, and Future.* New York: Columbia University Press.

Liu, K-b. and M. L. Fearn. 1993. Lake-Sediment Record of Late Holocene Hurricane Activities From Coastal Alabama. *Geology* 21: 793–796.

Liu, K-b. and M. L. Fearn. 2000a. Reconstruction of Prehistoric Landfall Frequencies of Catastrophic Hurricanes in Northwestern Florida From Lake Sediment Records. *Quaternary Research* 54: 238–245.

Liu, K-b. and M. L. Fearn. 2000b. Holocene History of Catastrophic Hurricane Landfalls Along The Gulf of Mexico Coast Reconstructed from Coastal Lake and Marsh Sediments. In Z. H. Ning and K. Abdollahi, eds. *Current Stresses and Potential Vulnerabilities: Implications of Global Change for the Gulf Coast Region of the United States,* 38-47. Baton Rouge: Franklin Press.

Liu, K-b., C. Shen, and K. S. Louie. 2001. A 1000-year History of Typhoon Landfalls in Guangdong, Southern China, Reconstructed from Chinese Historical Documentary Records. *Annals of the Association of American Geographers* 91: 453–464.

Murnane, R. J., C. Barton, E. Collins, J. Donnelly, J. B. Elsner, K. Emanuel, I. Ginis, S. Howard, C. Landsea, K-b. Liu, D. Malmquist, M. McKay, A. Michaels, N. Nelson, J. O'Brien, D. Scott, and T. Webb III. 2000. Model Estimates Hurricane Wind Speed Probabilities. *EOS, Transactions of the American Geophysical Union* 81: 433–438.

# 73

# SOCIETAL IMPACTS OF BLIZZARDS IN THE CONTERMINOUS UNITED STATES, 1959–2000

ROBERT M. SCHWARTZ
*Ball State University*

Many individuals think of tornadoes and hurricanes when considering weather-related storms. However, severe winter storms are significant natural hazards as they cause disruptions to transportation, damage to buildings, closure of schools and businesses, loss of electricity, and hazards to human health (Rooney 1967; Helburn 1982). Geographers have a strong tradition of examining various natural hazards. Mileti (1999: 302) updated the hazards assessment in the United States pioneered by White and Haas (1975), and stated that "a blizzard map" was needed for the United States. Additionally, research regarding climate change identified a need for a reference baseline for climate model validation to identify trends in weather-related hazards frequencies and severities (Mileti 1999: 304). The wide and severe economic and societal impacts associated with blizzards demonstrate the value of studying the societal impacts by examining the spatial and temporal patterns of blizzards along with the societal impacts. This included the population in affected counties, fatalities, injuries, economic damage, and the number of federally declared disaster areas.

The most extreme form of winter storm is the blizzard, combining strong winds and falling or blowing snow to cause low visibility, deep snowdrifts, and extreme wind chill. According to the National Weather Service (NWS 1999b), a blizzard is defined as having falling or blowing snow with winds in excess of 35 mi h$^{-1}$ (15 m s$^{-1}$) and visibility of less than 1/4 mi (0.4 km) for a minimum of 3 hours.

## METHODS

Since there was no database of blizzards, it was necessary to establish a climatology of blizzards. This was accomplished for state and county levels of analysis from 1959 to 2000 using the publication *Storm Data (STORM DATA)*. A map and corresponding attributes were produced for each blizzard through GIS, allowing patterns to be examined on monthly, annual, decadal, and the overall 41-year period (Schwartz 2001; Schwartz and Schmidlin 2002).

*D. G. Janelle et al. (eds.), WorldMinds: Geographical Perspectives on 100 Problems*, 449–454.
© 2004 *Springer. Printed in the Netherlands.*

The population in counties affected by each blizzard during 1959–2000 was extracted from United States Census Bureau data and correlated with the database.

The social and economic impacts were obtained from *STORM DATA*, which lists fatalities, injuries, property damage, and crop damage for all storms. Fatalities and injuries were recorded for each blizzard and summarized in SPSS statistical software.

Property and crop damage was adjusted to the mean of the damage category, as *STORM DATA* originally reported in one of nine classes. Damage after 1994 was given in dollars rather than a category. Property and crop damage was listed in both unadjusted dollars and 2001 standardized dollars adjusted by the Consumer Price Index from the Department of Labor.

Declared disaster areas were obtained from the Public Papers of the Presidents from 1959 through 1991. After 1992, the Federal Emergency Management Agency (FEMA) listed all disaster and emergency declarations.

RESULTS

There were 438 blizzards identified for the 41 winters from 1959–1960 to 1999–2000. The annual number of blizzards (by winter) averages 10.7 and ranged from 1 blizzard in 1980–1981 to 27 blizzards in 1996–1997. The frequencies of blizzards by county over the study period were developed (Figure 73.1). Forty of the 48 states reported a blizzard during the period. The "Blizzard Zone" consisted of North Dakota, South Dakota, and 34 counties in western Minnesota with an average of one or more blizzards per winter (Schwartz 2001; Schwartz and Schmidlin 2002).

The mean population was 2,462,949 people affected per blizzard. Additionally, the average total population affected per winter during the study period was 26,311,505. The population in affected counties each winter season was regressed against the year to determine if there was linearity. A correlation of 0.078 was not significant ($p = 0.313$) and the slope of 168,646 people per year was not significant ($p = 0.627$). Population in the continental United States, for 1960 through 1990, rose by 68,728,426 persons. In spite of the increased population, the population affected by blizzards did not increase. Although the number of blizzards increased from 69 in the 1960s to 147 in the 1990s, the affected population did not increase. The likely reason for no linear time trend is that the population decreased by 30,737 in the core blizzard region.

There were 679 reported fatalities directly associated with blizzards during the 41-year period, which is a mean of 1.55 fatalities per blizzard. Fatalities ranged from zero in several cases to 73 during the Midwest Blizzard of January 1978.

During the study, 2,011 injuries were reported as direct results of bliz-

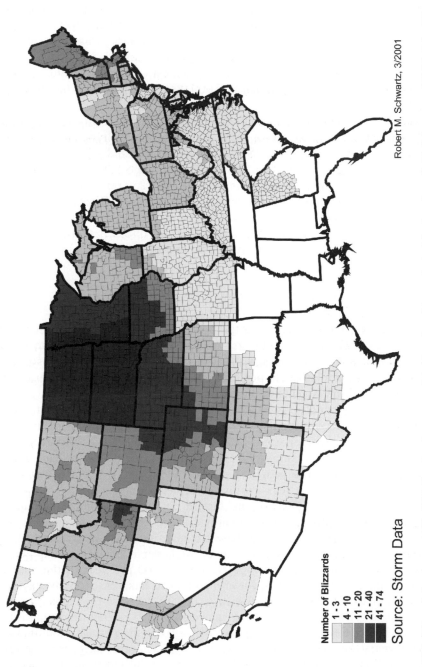

*Figure 73.1.* Blizzards by county, 1959–2000. Reprinted, courtesy of the American Meteorological Society, from R. M. Schwartz and T. W. Schmidlin. 2002. "Climatology of Blizzards in the Conterminous United States, 1959–2000." *Journal of Climate* 15: 1765–1772.

zards. The mean was 4.59 per event and the minimum number was zero, while the maximum was 426, with the Superstorm of March 1993.

Total property damage adjusted to 2001 dollars was reported at $22,600,000,000 or a mean of $51,593,998 per blizzard. No damage was reported for several blizzards. The greatest property damage was $4,550,000,000 (unadjusted) or $12,300,000,000 (adjusted) from the Midwest Blizzard of January 1978.

Even though many agricultural areas are affected by blizzards, crop damage was often not reported by *Storm Data*. In unadjusted dollars, $439,542,000 in crop damage was reported ($1,080,000,000 adjusted). The mean crop damage per blizzard was $1,003,520 unadjusted and $2,466,324 in adjusted dollars. Maximum reported crop damage was $227,250,000 (unadjusted) and $613,575,000 (adjusted) from the January 1978 Midwest Blizzard.

There were 25 Federal Disaster declarations associated with blizzards in the study period. The 1960s had no declared blizzard disasters while the 1970s had eight declarations, the 1980s had three, and the 1990s had 14.

## DISCUSSION AND CONCLUSIONS

Although most blizzards occur in the northern plains, this is not the most populated area. The blizzard core region has lost population over the study period. Blizzards affect an average of 26,311,505 people per winter season and the major population impacts per blizzard occur in the populated areas of the Midwest and the Northeast United States.

There are vulnerability issues and blizzards are a significant hazard in the United States. It is interesting to note that fatalities are decreasing while injuries are increasing. Most winter storm fatalities occur with vehicles and exposure. Weather forecasts are becoming more accurate and, with longer lead times, warnings can lead to more people staying off the roads. It is not known why injuries are increasing but it is worthy of further investigation.

Blizzards are not the most serious hazard but still illustrate vulnerability. Comparing impacts from the threatening hazards of floods, tornadoes, and hurricanes, blizzards rank behind these hazards. Yet, with an average of 11 blizzards per year, these impacts can be high per event, especially when compared to tornadoes that average around a 1,000 per year in the United States. Mileti (1999) stated that for climatological hazards, property damage ranged from $4.8 billion to $196 billion in 1992 dollars for floods, tornadoes, tropical cyclones, and snow and ice events for 1975–1994. In comparison, blizzards had $16.6 billion in property damage for the period. Examination of these annual damages from these weather-related hazards range from around a half a billion dollars from blizzards to almost $6 billion from flooding.

Property and crop damage has significant costs from natural hazards that are paid by individuals, insurance, or government subsidies. An average per winter of $551 million in property damage and $26 million in crop damage in 2001 dollars ($51.5 million in property damage and $2.4 million in crop damage per blizzard) was caused by blizzards according to *STORM DATA*.

The number of declared disasters or emergencies due to blizzards has been increasing, especially in the 1990s. This may be due to the upward trend in the number of blizzards during 1959–2000. Other questions, such as the political motivation in the declarations, are not addressed in this study.

While most blizzards are in the lesser-populated areas of the country, the largest numbers of affected population live in the states surrounding the Great Lakes in the Midwest and along the Eastern coast of the United States. The core blizzard area has a smaller economy and a less complex transportation infrastructure than heavily populated areas. Also, frequent blizzard exposure may lead to higher preparedness, which leads to less vulnerability in these core blizzard regions. Perhaps the most vulnerable locations of the United States for blizzards are around the Midwestern Great Lakes and the Northeast regions. This is due to the larger population, complex economies, larger ground transportation network, and airport hubs. These areas may not have as much blizzard experience, which could lead to greater vulnerability.

Besides the social impacts studied, an additional major impact is one of disruption. Transportation can have severe interruptions from blizzards, as highways and airports may be closed, potentially affecting people and businesses on a global scale. This could happen when international airports are closed, as from the March 1993 Superstorm that closed the entire Eastern Seaboard and took days for recovery. Disruption is very difficult to measure but is a key variable from a blizzard, and additional research is warranted.

It may help to reduce vulnerability from blizzards by increasing awareness of the potential impacts from these severe storms. When the NWS issues a watch or warning, the general population should follow basic preparation measures, such as staying indoors and off the roads. Additionally, emergency management and other agencies could initiate procedures to restore infrastructure and clear roads as soon as it is practical. Another mitigative measure would be planning for these events by having personnel and procedures ready for implementation. This could offset the lack of blizzard experience for the rare events.

## REFERENCES

Helburn, N. 1982. Severe Winter Storms as Natural Hazards. *Great Plains-Rocky Mountain Geographic Journal* 10: 86–95.

Mileti, D. S. 1999. *Disasters by Design, A Reassessment of Natural Hazards in the United States.* Washington, D.C.: John Henry Press.

*Storm Data*. 1959–2000. Ashville, NC: National Climatic Data Center.

National Weather Service (NWS). 1999. *National Weather Service Says: Know Your Winter Weather Terms.* Washington, D.C.: National Weather Service. http://www.nws.noaa.gov/om/winter/wntrtrms.htm (last accessed 13 September 2002).

Rooney, J. F. 1967. The Urban Snow Hazard in the United States: An Appraisal of Disruption. *Geographical Review* 57: 538–559.

Schwartz, R. M. 2001. *Geography of Blizzards in the Conterminous United States, 1959–2000.* Ph.D. dissertation, Kent State University.

Schwartz, R. M. and T. W. Schmidlin. 2002. Climatology of Blizzards in the Conterminous United States, 1959–2000. *Journal of Climate* 15: 1765–1772.

White, G. F. and J. E. Haas. 1975. *Assessment of Research on Natural Hazards.* Cambridge, MA: The MIT Press.

# 74

# COASTAL HAZARDS AND BARRIER BEACH DEVELOPMENT

STEPHEN P. LEATHERMAN and KEQI ZHANG
*Florida International University*

Barrier beaches along the U.S. East and Gulf Coasts are some of the most popular recreational areas and most expensive real estate in the United States. For those people who take vacations, more than 70 percent will visit a beach (www.aaa.com). According to insurance industry reports, there is more than $3.1 trillion at risk to coastal storms (Institute for Business and Home Safety).

The primary issues in coastal geomorphology, all having profound implications for society, are as follows:

- Coastal Erosion – 80 to 90 percent of U.S. sandy beaches are presently eroding (Leatherman 1988; Heinz Center 2000). This demonstrates the scale of the problem, and the nearly ubiquitous nature of beach erosion points to rising sea level as the underlying driver.
- Coastal Storms – property damage from winter nor'easters and especially from hurricanes has been rapidly increasingly during the last 100 years along the barrier beach coasts. Surprisingly, there has been no increase in storminess (Zhang et al. 2000), although there are significant interdecadal changes in storm frequency. There has been rapid urbanization of coastal areas in response to the coastward migration of the population (Figure 74.1).
- Sea Level Rise – there is a worldwide (eustatic) rise of sea level that has averaged 1.8 mm/yr for the past century, based on tide-gauge measurements; land subsidence along the U.S. East and Gulf Coasts has resulted in more than double this amount of relative sea-level rise (Douglas et al. 2001). Rising sea level has been linked directly to coastal erosion (Leatherman et al. 2000).

Coastal physical geographers have made significant progress in quantifying coastal hazards with direct implications for beachfront development. There has literally been a revolution in our knowledge base and in the evolution of geomorphic concepts, which have been greatly aided by technological advancements in imaging and mapping; notably geographic information systems (GIS), global positioning system (GPS), and light detection and ranging (LIDAR).

D. G. Janelle et al. (eds.), WorldMinds: Geographical Perspectives on 100 Problems, 455–460.
© 2004 Springer. Printed in the Netherlands.

*Figure 74.1.* Fire Island is a barrier island along the south shore of Long Island, New York. This house, built behind protective dunes in the late 1890s, was finally toppled by the December 1992 Nor'easter.

## COASTAL POLICY

Coastal-erosion hazards are a major scientific and public policy-issue because of their scope; almost 90 percent of U.S. sandy beaches are presently eroding at various rates on a geographical basis (Heinz Center 2000). At the same time, there is a coastward migration of people, and this burgeoning population means more coastal infrastructure and building, especially for the most desirable real estate at the water's edge. We are by definition on a collision course at the shoreline – the critical interface between land and water (Figure 74.1).

Chains of barrier islands provide the buffer between high-energy oceanic waves and the more placid waters of bays and lagoons that front the mainland shores of the U.S. East and Gulf Coasts. Barrier islands are composed almost entirely of loose, easily eroded sand. People want to live here because they are some of America's best beaches, but there is little understanding of barrier-island dynamics, in general, and beach erosion rates, in specific. In fact, coastal realtors do not like to discuss the issue of beach-erosion – it is bad for business. Therefore, there is a total disconnect between building beachfront houses and even mansions along a shoreline with little to no information on erosion rates, storm-induced impacts, and long-term sea-level trends.

## COASTAL EROSION MAPPING AND ANALYSIS

Leatherman (1983) developed the first computer-based historical shoreline-change mapping technique (termed Metric Mapping) that allowed for the generation of highly accurate beach-erosion rates, which is paramount for developing major databases for scientific research and coastal management purposes. The Metric Mapping technique uses and rectifies shoreline data from NOS "T" sheets that range to the mid-1800s, aerial photography (1940s to present), field-based kinematic GPS surveys (1990 to present), and LIDAR (1999 to present). These data are incorporated into a GIS/computerized system for manipulation and automated plotting of historical shoreline-change maps and histograms of erosion rates for various time periods.

The Metric Mapping technique allowed for the compilation of over 150 years of shoreline position data along the U.S. East Coast, yielding small error bars for the derivation of the long-term trend for scientific analysis (Crowell et al. 1993). Shoreline-change data and erosion-rate analyses are the most important data sets in coastal-zone management for establishment of building setback lines and potentially for insurance rates (E-zones) by FEMA (Federal Emergency Management Agency).

## COASTAL STORMS

The amount of storm damage along the U.S. East Coast has been increasing dramatically in the last 100 years, leading some researchers to speculate that increased storminess is a major cause. Damage estimates are highly biased by the amount of shorefront development, which boomed in the past half century. Zhang et al. (2000) compiled a storm surge data set using hourly tide-gauge measurements from stations along the U.S. East Coast. Surprisingly, there has been no change in storminess during the 20th century in spite of global warming. There are interdecadal variations in storm frequency, meaning that certain time periods are characterized by more and/or by more powerful storms, but there is no trend.

There is a functional relationship between coastal damage and the storm tide for nor'easters along the U.S. East Coast (Figure 74.2). This is a remarkable finding that provides a cornerstone for estimating the potential damage of certain sized nor'easters and for rating such events. The storm-erosion potential index is based on quantitative data – hourly water levels from tide-gauge data (Zhang 2001). All other indices of storm damage are qualitative in nature and dependent upon hindcasts. The other important parameter is the duration of the event during which a certain size storm tide occurs. This explains why the Ash Wednesday Storm of March 1962 caused such devastation (e.g., the greatest measured beach and dune erosion as well as unprecedented wave and surge damage to buildings along the U.S. East Coast

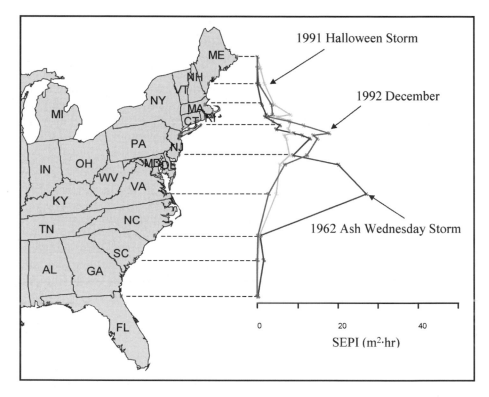

*Figure 74.2.* Spatial variation of the storm erosion potential index (SEPI) for the 1962 Ash Wednesday, 1991 Halloween (e.g., the "Perfect Storm"), and December 1992 storms (from Zhang et al. 2001). The 1962 Nor'easter caused the most erosion and damage along the U.S. mid-Atlantic Coast.

barriers) – it lasted through five high tides and occurred during a perigean spring tide.

## SEA LEVEL RISE

Sea level is known to control the position of the shoreline over geological time frames, and the Holocene rise in sea level has caused barrier islands to migrate landward many kilometers. It has long been postulated that rising sea level would cause coastal erosion at decadal time scales, but the proof of this causal relationship has been elusive. Many coastal scientists and engineers (e.g., Leatherman 1991) have tried to find this relationship without success. Although most coastal geomorphologists and even the informed public and journalist (e.g., Cory Dean, science editor of the *New York Times*) tacitly assume that rising sea level is the underlying driver of contemporary beach erosion, there has been no consensus. Establishing such a relationship is of extraordinary importance from a public-policy viewpoint because global warming will almost certainly cause an accelerated rate of sea-level rise, which will exacerbate the existing erosional problems along the U.S. coasts and indeed worldwide.

The U.S. East Coast provides a unique place to investigate the relationship between sea-level rise and shore position. Sea-level rise was shown to drive coastal erosion such that horizontal beach retreat is about 150 times the rate of vertical rise (Leatherman et al. 2000) – this is the "Holy Grail" of coastal geomorphology. There is no energy in sea-level rise itself, but changes in water level put the beach profile out of equilibrium, which is related to sediment size and wave energy. Coastal storms accomplish the geologic work of moving sediment and re-establishing equilibrium beach profiles in response to rising water levels, which translates to shoreline recession.

## CONCLUSIONS

It is an exciting time to be a coastal geomorphologist. The scientific discoveries and breakthroughs have been made possible by analyzing quantitatively the geographical variations in coastal processes along the U.S. East Coast and by taking advantage of the tremendous advancements in technology, especially GIS, GPS, and LIDAR (e.g., airborne laser mapping).

Geographers are playing the key role in addressing the coastal erosion problem by making quantum advancements in the understanding of coastal geomorphology and by applying this science to the real-world problems of establishing building setbacks and E (erosion) zones for insurance purposes. Our work is not yet complete, as FEMA and the U.S. Congress are still wrestling with the implementations of coastal erosion mapping and management for the geographically diverse U.S. coasts.

## ACKNOWLEDGEMENT

Funding from The Andrew W. Mellon Foundation is gratefully acknowledged.

## REFERENCES

Crowell, M., S. P. Leatherman, and M. K. Buckley. 1993. Shoreline Change Rate Analysis: Long Term Versus Short Term Data. *Shore and Beach* 61: 13–20.

Dean, C. 1999. *Against the Tide: The Battle for America's Beaches.* Columbia University Press.

Douglas, B. C., M. Kearney, and S. P. Leatherman. 2001. *Sea Level Rise: History and Consequences.* New York: Academic Press.

Heinz Center. 2000. *Evaluation of Erosion Hazards.* Washington, D.C.: The H. John Heinz III Center for Science, Economics, and the Environment.

Leatherman, S. P. 1983. Shoreline Mapping: A Comparison of Techniques. *Shore and Beach* 51: 28–33.

Leatherman, S. P. 1988. *Impact of Climate-Induced Sea-Level Rise on Coastal Areas.* Washington, D.C.: U.S. Senate Publication.

Leatherman, S. P. 1991. Modeling Shore Response to Sea-Level Rise on Sedimentary Coasts. *Progress in Physical Geography* 14: 447–464.

Leatherman, S. P., K. Zhang, and B. C. Douglas. 2000. Sea Level Rise Shown to Drive Coastal Erosion. *EOS* 81: 55–59.

Zhang, K., B. C. Douglas, and S. P. Leatherman. 2000. Twentieth Century Storm Activity along the U.S. East Coast. *Journal of Climate* 13: 1748–1761.

Zhang, K., B. C. Douglas, and S. P. Leatherman. 2001. Beach Erosion Potential for Severe Nor'Easters. *Journal of Coastal Research* 17: 309–321.

# 75

# BRIDGING HAZARDS GEOGRAPHY AND POLITICAL GEOGRAPHY: A BORDERLAND VULNERABILITY FRAMEWORK

LYDIA L. BEAN
*Southern Illinois University*

FRED M. SHELLEY
*Texas State University – San Marcos*

Historically, political boundaries have been analyzed as static barriers to flows of people, goods, capital, and ideas. However, geographers are now re-examining borders within a new, dynamic context (Bean 2002). In this chapter, we present a conceptual framework in which boundaries are categorized on the basis of symmetry/asymmetry and levels of interaction. We then illustrate the utility of this conceptual framework in understanding the impacts of boundaries on cultures, economies, and environments in border regions.

## SYMMETRY AND ASYMMETRY

Boundaries can be seen as geographical expressions of the territorial limits of sovereignty. Land and people on either side of a boundary are recognized as under the jurisdiction of different states. Geographers have long examined the physical forms of borders, but much of this analysis has been based on examining their spatial forms. For example, geographers have distinguished boundaries that follow physical features from those that are geometric lines or those that separate cultures, religious communities, or ethnic groups. However, this conceptualization fails to take into account the dynamics of interaction across the border (Bean 2002).

The dynamics of boundaries and cross-boundary interactions depend on differences between the two states and the effects of the border on interaction between them. Many boundaries separate states that have very similar economies, cultures, and standards of living. Other boundaries separate states characterized by considerable economic differences. For example, the standard of living in the United States is comparable to Canada but much higher than that in Mexico.

*D. G. Janelle et al. (eds.), WorldMinds: Geographical Perspectives on 100 Problems, 461–466.*
© 2004 *Springer. Printed in the Netherlands.*

Boundaries separating states with similar standards of living have been termed symmetric boundaries, while those separating states with very different conditions have been termed asymmetric boundaries (Bean 2002). Thus the Canada/United States boundary is relatively symmetric, while the United States/Mexico border is relatively asymmetric (Figure 75.1). This concept of symmetry and asymmetry has both economic and socio-cultural overtones. For example, the boundaries between India and Pakistan and between Québec and the northeastern United States are associated with economic symmetry, but not cultural symmetry. In both cases, economic conditions and standards of living are similar, but cultures are very different. In contrast, the border between Texas and Mexico is characterized by cultural symmetry (Arreola 1995) but less economic symmetry.


## INTERACTION

A second dimension along which borders can be compared is that of levels of interaction. Spatial interaction theory is based on the premise that levels of interaction are affected by the degree to which interaction opportunities exist. For example, the gravity model predicts that levels of interaction between two places are affected by the size of the two places and the distance between them. In this context, borders have often been conceptualized as barriers or impediments to spatial interaction. The degree to which a boundary impedes spatial interaction can be determined by comparing actual cross-border interaction relative to the amount expected based on the gravity model or other spatial interaction models. A completely closed border is one in which there is no interaction whatsoever. On the other hand, a border that has no effect on the level of interaction might be conceptualized as a completely open border. Thus, the level of interaction across the border can range from 0 percent across a completely closed border, for example between North Korea and South Korea, to approaching 100 percent across an open border, such as between the United States and Canada or between neighboring states within the European Union.

Cohen (1990) pointed out that some boundary communities are located in such a way as to magnify interaction beyond what would be expected. Such places, including Estonia, Slovenia, and Singapore, are found along boundaries whose locations encourage interaction between major trading blocs or large culture regions and are termed gateways. Interaction across gateways is more than 100 percent of what might be expected based on the gravity model. It should be pointed out that interaction levels over 100 percent generally would occur only under conditions of asymmetry, because of the principle of comparative advantage, which would be absent under symmetry (Bean 2002).

Interaction levels can also be distinguished on two dimensions: quantity and quality. The quantity dimension measures volumes of trade flows, border cross-

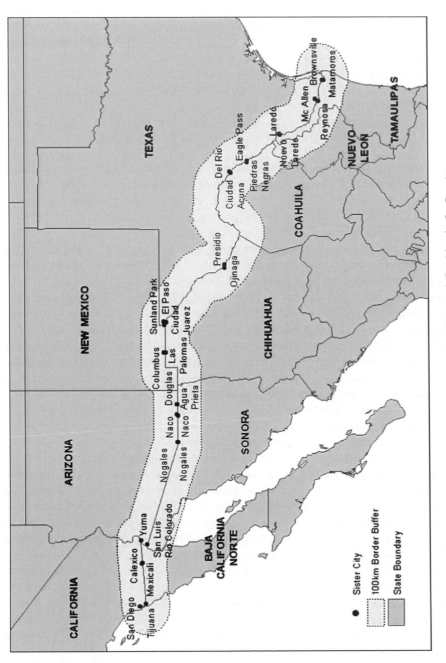

*Figure 75.1.* United States-Mexico border region, as defined by the La Paz Agreement.

ings, and other numbers of interactions between countries. However, inter-action can be of high or low quality. Exchanges of people, goods, and ideas that benefit both parties and contribute to enhancing the welfare of both societies might be considered high-quality interaction. On the other hand, smuggling, piracy, drug trafficking, illegal immigration, and similar activi-ties constitute low-quality interaction that is generally detrimental to persons on one or both sides of the boundary (Bean 2002).

## TOWARD A CONCEPTUAL FRAMEWORK

Together, the dimensions of symmetry/asymmetry and interaction levels allow the development of a conceptual framework, within which individual bound-aries can be located. Considering that levels of asymmetry and interaction are relative, suppose that for any boundary, levels of symmetry and interac-tion can be categorized as high, medium, or low. This yields 27 possible combinations of symmetry and interaction levels (Figure 75.2). For example, the boundary between the United States and Mexico would be characterized by high interaction and high asymmetry. The United States-Canada and France-Germany border, are characterized by high interaction and low asymmetry. A closed border such as that between North and South Korea, yields low interaction and high asymmetry. On the other hand, the India-Pakistan border is characterized by low – interaction and low asymmetry. The conceptual framework can be fine-tuned in more complex terms by bringing in addi-tional dimensions, for example by distinguishing low and high quality interaction or economic and cultural asymmetry as described above.

This conceptual framework fills a gap in the literature identified by Scott et al. (1997), who argued that studying border regions at a microscale, inde-pendent of global economic and cultural contexts, gives a reductionist view of border dynamics. It provides a holistic, integrated method by which researchers can evaluate the effects of boundaries on local and regional environments, economies, and populations. Its effectiveness can be illustrated by using it to examine how boundaries affect vulnerability to natural and technological hazards (Bean 2002). Vulnerability refers to the potential effects of disasters, hazards, or other unanticipated shocks on resident populations. Blaikie et al. (1994) cite gaps in knowledge concerning linkages and causes of vulnerability at all scales. The conceptual framework presented in this paper provides a basis for meaningful theoretical and empirical investigation of how vulnerability is affected by boundaries associated with differing levels of symmetry and interaction. It provides a basis for making predictions con-cerning the relationships between asymmetry, interaction, and vulnerability. For example, it is expected that enhanced vulnerability will be associated with increased asymmetry. However, with enhanced levels of quality cross-border interaction, the framework predicts that the degrees of asymmetry

ASYMMETRY

| INTERACTION | QUANTITY | QUALITY | HIGH (H) | MODERATE (M) | LOW (L) |
|---|---|---|---|---|---|
| | H | H | | | **POSSIBLE REALM OF THE FRANCE-GERMANY BORDER REGION** |
| | H | M | **POSSIBLE REALM OF THE U.S-MEXICO BORDER REGION** | | |
| | H | L | | | |
| | M | H | | | |
| | M | M | | | |
| | M | L | | | |
| | L | H | | | |
| | L | M | | | |
| | L | L | **POSSIBLE REALM OF THE NORTH KOREA-SOUTH KOREA BORDER REGION** | | **POSSIBLE REALM OF THE INDIA-PAKISTAN BORDER REGION** |

*Figure 75.2.* Borderland vulnerability framework matrix with border examples.

become lessened and that some border region populations enjoy the greater capacity to mitigate their individual and collective vulnerabilities.

This has been demonstrated, for example, in the border cities of Laredo, Texas and Nuevo Laredo, Mexico. In these historically impoverished and rapidly growing communities, enhanced cross-border interaction has fueled the development of an emerging middle class that is less vulnerable to natural and technical hazards; however, the poor and less educated on both sides of the boundary may be becoming increasingly marginalized by this growth (Bean 2002). As well, large numbers of skilled young adults are moving from rural interior Mexico to the border communities, leaving the young, elderly, poor, and unskilled in the rural areas.

The conceptual framework recognizes the spatial proximity of places separated by boundaries, yet it is also not dependent on the existence of these boundaries. In other words, it is equally applicable to examining interaction between places or societies that interact with one another but are not directly connected by boundaries, for example the United States and countries such as Cuba, Japan, or the European Union. It is applicable to examining a large variety of impacts of borders, not only the impacts of vulnerability but also the effects of short- and long-term political and economic changes on border regions. The ability to make predictions about how vulnerability is affected by interaction and asymmetry helps policy makers to properly direct resources toward mitigating potential vulnerability and planning for long-term development.

## REFERENCES

Arreola, D. D. 1995. Mexican Texas: A Distinctive Borderland. In J. F. Petersen and J. A. Tuason, eds. *Pathways: A Geographic Glimpse of Central Texas and the Borderlands: Images and Encounters*, 3–9. Indiana, PA: National Council for Geographic Education.

Bean, L. L. 2002. *Bridging Hazards Geography and Political Geography: A Borderland Vulnerability Framework With a Case Study along the United States-Mexico Border.* Unpublished Doctoral Dissertation. Department of Geography, Southwest Texas State University.

Blaikie, P., T. Cannon, I. Davis, and B. Wisner. 1994. *At Risk: Natural Hazards, People's Vulnerability and Disasters.* New York: Routledge.

Cohen, S. B. 1990. The World Geopolitical System in Retrospect and Prospect. *Journal of Geography* 89(1): 2–14.

Scott, J., A. Sweedler, P. Ganster, and W. Dieter-Eberwein. 1997. Dynamics of Transborder Interaction. In P. Ganster, A. Sweedler, J. Scott, and W. Dieter-Eberwein, eds. *Borders and Border Regions in Europe and North America*, 3–23. San Diego, CA: San Diego State University Press.

United States Environmental Protection Agency and *Secretaría de Medio Ambiente, Rescursos, Naturales, y Pesca*. 2002. U.S.-Mexico Border XXI interactive maps; accessed on 26 January 2002 from http://www.epa. gov/usmexico border/map/index.htm.

PART VIII

# ACTIVATING PHYSICAL GEOGRAPHY

Numerous complex forces shape the earth's landscape and its atmosphere. Geographers uncover the science of these landscape and atmospheric changes. They are also instrumental in putting this knowledge to use in planning for extreme events, restoring ecosystems, and evaluating protective strategies for living with natural changes. Increasingly, the monitoring and modeling of physical processes has included the role of humans as active agents of environmental transformation.

*Kathy Hansen*

# 76

# ECOLOGICAL RESPONSE TO GLOBAL CLIMATIC CHANGE

GEORGE P. MALANSON
*University of Iowa*

DAVID R. BUTLER
*Texas State University – San Marcos*

STEPHEN J. WALSH
*University of North Carolina, Chapel Hill*

Climate change and ecological change go hand in hand. Because we value our ecological environment, any change has the potential to be a problem. Geographers have been drawn to this challenge, and have been successful in addressing it, because the primary ecological response to climate changes in the past – the waxing and waning of the great ice sheets over the past 2 million years – was the changing geographic range of the biota. Plants and animals changed their location. Geographers have been deeply involved in documenting the changing biota of the past, and today we are called upon to help assess the possible responses to ongoing and future climatic change and, thus, their impacts. Assessing the potential responses is important for policy makers to judge the outcomes of action or inaction and also sets the stage for preparation for and mitigation of change.

## QUATERNARY STUDIES

Many studies have documented ecological responses to climatic change during the Quaternary. The basis for these studies is a record of fossil and quasi-fossil information consisting of pack rat middens, diatoms, animal remains, plant macrofossils, and, primarily, plant pollen. Observations of patterns of these data can be correlated with climate in the present and then used to reconstruct the past. For example, working with such data, Whitlock and Bartlein (1997) found good correspondence between vegetation change and a number of indicators of climate change in the northwestern United States at millennial time scales. Many such studies demonstrate the importance of location and scale to the geographical responses of ecosystems to climatic

*D. G. Janelle et al. (eds.), WorldMinds: Geographical Perspectives on 100 Problems, 469–473.*
© 2004 *Springer. Printed in the Netherlands.*

change (e.g., Lavoie and Filion 2001). Three regions, the tropics, temperate zone, and arctic treeline, encompass most of the work, but individual workers study regions spanning the globe. These studies are the foundation of our understanding of the nature of global environmental change.


## GREENHOUSE POTENTIAL

Interest in ecological response to climatic change has moved beyond the study of the past. Beginning in the 1980s, the potential of anthropogenic climate change in the near future was recognized as an impact on ecosystems. Projections of such impacts were initially based on simple geographic projections of the climates in which species now live onto maps of future climate scenarios. Shortcomings in these models were quickly recognized by geographers, however, who realized that the basis for such a projection relies on species locations being in equilibrium with present climate and that other factors, such as soil, either do not matter or will change in the same way (Malanson 1993). Newer models have been developed to incorporate more mechanistic and spatially explicit approaches. We now know more about barriers and time-lags and the significance of dynamics during a period of climatic change.

More detailed spatial studies have focused on ecotones, the transition zones between adjacent biomes, because it is hypothesized that the range limits of biomes will be most sensitive to climate change and, thus, will be indicators or early warning systems for more extensive change. Studies of recent change of arctic and alpine treeline overlap with studies extending back into the past (e.g., MacDonald et al. 1998). These studies indicate that ecological responses to climate change are somewhere between loosely and tightly linked: i.e., there are scale-dependent spatial and temporal lags leading to complex relations (Malanson 1999). The examination of ecotones as potential indicators has led to increased understanding of the spatial dimensions of ecological dynamics and, while raising caveats, the potential still exists in some cases.

Much current interest is on montane ecotones, but their role as indicators depends on their rates of response and the multiplicity of important niche dimensions (Kupfer and Cairns 1996). Here, spatial pattern may be a useful indicator to which geographic analyses are relevant (e.g., Mast and Veblen 1999). Treeline in specific locations may be determined by multiple interacting abiotic factors, such as wind, moisture, snow, substrate, biotic competition, and feedbacks (e.g., Cairns and Malanson 1998). While the response of treeline to climatic change may not be a good indicator of climatic warming alone (e.g., Butler et al. 1994), it may yet be sensitive to combinations of change, such as warming and increased precipitation. Variability in its sensitivity has multiple possible causes. Hypothetically, it may be sensitive because it may now not be in equilibrium with climatic change over the past two centuries, and it may be sensitive because feedbacks may induce nonlinear responses not taken into account in earlier critiques. Also, its response may be important because any advance of treeline would be at the expense of tundra, which may have no equivalent response due to boundaries with rock, ice, or simply a mountaintop.

## GEOGRAPHICAL APPROACHES

Geography brings particular strengths to the study of ecotones and climate change because geographers usually consider the interaction of spatial and temporal scales explicitly. Ecotones have characteristics wherein the scale affects interpretation; over large areas, simple coupling of climate and vegetation is relatively strong, while nonlinear feedbacks are at the scale of individual plants and patches. Moreover, geography has a tradition and associated strong analytics in three areas. Physical geography concentrates on biophysical interactions in the processes and patterns of the atmosphere, biosphere, hydrosphere, and lithosphere; the spatial tradition in geography links pattern to process using spatial statistics and simulations); and GIScience brings remote sensing and GIS to bear. These three aspects of geography coincide at treeline.

In addition to tree establishment and growth responses to temperature and moisture, trees respond to geomorphology and pedogenesis, such as the soil development on solifluction forms that is also linked to climate (needle-ice action, desiccation and aeolian deflation, fluvial activity, and biological agents have all been identified as potential causal agents necessary to disrupt the dense mat of tundra on solifluction risers for exposure of soil and subsequent successful germination of tree seeds). Treeline environments are also affected by feedbacks among these interactions (such as canopy structure and temperature via albedo) that have strong spatial pattern-process relations – distance decay being dominant. These observations are linked to advances in GIScience, which are often problem-dependent; e.g., advances in spatial resolution are important for capturing the interactions and feedbacks of solifluction forms and establishment (Walsh et al. 2003). Our study in Glacier National Park (GNP) illustrates this approach (Figure 76.1). We can identify particular results that are due to combinations of the three influences at different levels.

We have identified solifluction turf exfoliation as sites for tree establishment; this finding is most strongly based in the study of biophysical interactions from fieldwork, but has been reinforced by pattern detection using remote sensing and analyses using geostatistics. We have identified linear fingers of trees advancing into tundra as evidence of positive feedback; this relation between pattern and process is part of the spatial tradition of geography but has relied on fieldwork (e.g., dendrochronology), spatially explicit simulations, and mapping from multispectral aircraft imagery. On a broader scale we have

*Figure 76.1.* Trees are advancing into tundra in an area of relict solifluction, Lee Ridge, Glacier National Park, MT. While disrupted solifluction steps provide establishment sites, wind and feedback play a dominant role in creating linear features.

identified multi-scale pattern development of soil and vegetation overlaying a pattern of relict solifluction using primarily image interpretation but with ties to our understanding of the spatial patterns of biophysical processes revealed in simulations and which in turn leads to additional fieldwork.

We can identify areas that are most susceptible to a rapid advance of trees into tundra that depends on the exfoliation of relict solifluction for advance beyond pre-Little Ice Age extent. Although receding glaciers in GNP have caught the attention of the public, the park staff also sees the ecological response as a resource issue that needs attention. Resource and information managers in GNP present this information to the public to educate all about the dynamic nature of the environment. While the National Park Service will not mitigate the advance of trees into tundra, ecological responses may yet play a role in raising awareness of climatic change.

Geographers have not solved the problem of ecological response to climatic change, but they have made significant contributions to its definition, to our understanding of it, and, at this time, to identifying the locations where it may be of most consequence. It is likely that geographers will contribute in the future to efforts to mitigate the consequences.

## ACKNOWLEDGEMENTS

This work was supported by NSF grants SBR-9714347 and SBR-0001738 and by the USGS Global Change Program. This is a contribution from the Mountain GeoDynamics Research Group.

## REFERENCES

Butler, D. R., G. P. Malanson, and D.M. Cairns. 1994. Stability of Alpine Treeline in Northern Montana, USA. *Phytocoenologia* 22: 485–500.

Cairns, D. M. and G. P. Malanson. 1998. Environmental Variables Influencing Carbon Balance at the Alpine Treeline Ecotone: A Modeling Approach. *Journal of Vegetation Science* 9: 679–692.

Kupfer, J. A. and D. M. Cairns, 1996. The Suitability of Montane Ecotones as Indicators of Global Climatic Change. *Progress in Physical Geography* 20: 253–272.

Lavoie, M. and L. Filion. 2001. Holocene Vegetation Dynamics of Anticosti Island, Quebec, and Consequences of Remoteness on Ecological Succession. *Quaternary Research* 56: 112–127.

MacDonald, G. M., J. M. Szeicz, J. Claricoates, and K. A. Dale. 1998. Response of the Central Canadian Treeline to Recent Climatic Changes. *Annals of the Association of American Geographers* 88: 183–208.

Malanson, G. P. 1993. Comment on Modeling Ecological Response to Climatic Change. *Climatic Change* 23: 95–109.

Malanson, G. P. 1999. Considering Complexity. *Annals of the Association of American Geographers* 89: 746–753.

Mast, J. N. and T. T. Veblen. 1999. Tree Spatial Patterns and Stand Development Along the Pine-Grassland Ecotone in the Colorado Front Range. *Canadian Journal of Forest Research* 29: 575–584.

Walsh, S. J., D. R. Butler, G. P. Malanson, K. A. Crews-Meyer, J. P. Messina, and N. Xiao. 2003. Direct and Indirect Influences of Geomorphic Processes on the Alpine Treeline Ecotone, Glacier National Park, Montana, USA: Application of Geographic Information Science Technologies in Mapping, Modeling, and Visualization. *Geomorphology*, in press.

Whitlock, C. and P. J. Bartlein. 1997. Vegetation and Climate Change in Northwest America during the Past 125 kyr. *Nature* 388: 57–61.

# 77

# GLOBAL CHANGE AND ASSESSMENT OF THE ONSET OF SPRING

MARK D. SCHWARTZ
*University of Wisconsin, Milwaukee*

Atmosphere-biosphere interactions are a key component of the earth's physical systems. Living organisms, especially plants, are active agents in the production and transfer of important atmospheric gases, as well as regulators of the surface energy balance and solar-energy absorption. Understanding these interactions is crucial to accurate prediction and tracking of global climate-change impacts on the biosphere (Schwartz 1999), which may eventually include stress to world agricultural systems, and disruption to natural ecosystems.

Obviously, data gathering is a necessary step to better understanding any problem. Rapid and accurate daily weather forecasts, and ever-improving General Circulation Models (GCMs), are the results of over a century of coordinated climate station observations around the world. However, unlike the atmosphere, no global biosphere information network exists (Schwartz 1998). Satellite-derived vegetation information, with its frequent worldwide coverage can help, but satellite products have limitations (Schwartz et al. 2002). Collection of vegetation-condition and development-stage data through surface networks, coordinated with satellite-based observations, is likely the most promising route to effective biospheric monitoring, although functional ways to integrate all these measures are still being developed (Schwartz 1998). Such data are the realm of phenology, an integrative environmental science that concentrates on the observation and study of specific stages of plant and animal development (for example, first leaf, first bloom, insect hatches, and bird nesting) as related to changing environmental factors, especially temperature.

## ONSET OF SPRING

Although much work remains, considerable progress has been made in understanding some stronger atmosphere-biosphere interactions. Most of these efforts have concentrated on spring events. The relationship between weather events and plant development during spring is stronger than other seasons, making it sensible to study first. For example, multiple studies have shown that large

*D. G. Janelle et al. (eds.), WorldMinds: Geographical Perspectives on 100 Problems*, 475–480.
© 2004 *Springer. Printed in the Netherlands.*

parts of North America and Europe have experienced earlier arrivals of spring over the last 40 years (e.g., Menzel and Fabian 1999; Schwartz and Reiter 2000). Such information is invaluable in explaining the potential impacts of global climate change to the public. The implications of plants leafing and blooming a few days earlier is much easier to understand than the effects of a small change in mean temperature, as the connections among plant development, the growing season, and agriculture are generally understood. Yet monitoring of change for selected plant species is only the simplest of phenological applications to global change research.

This problem is worthy of additional study, as understanding the onset of spring at an ecosystem and global scale may facilitate better agricultural planning, and is a crucial part of efforts to understanding modifications to natural ecosystems. A few locations around the world have considerable phenology records, extending back 50–100 years or more, but widespread observation of the same event and species are rare (Chmielewski and Rötzer 2002; Menzel 2002). Development of global and regional phenology data-networks are slowly moving forward, and will be of great benefit to these research efforts in the future (Schwartz 1999).

A geographical research perspective has helped recent development of techniques designed to integrate vegetation data sets, gas/energy exchange data, and remote sensing products, to further interpretation of the first appearance of spring foliage, commonly called the "green wave." This term was apparently coined in response to sequential satellite views of vegetation development that "moved" like a wave across mid-latitudes (Rouse 1977). The green wave is crucial for accurate computation of net primary productivity, a sensitive measure of the impacts of global change, and intimately connected with springtime changes in ecosystem energy/carbon fluxes (Schwartz 1999; Schwartz and Crawford 2001).

## MODEL DEVELOPMENT

Initial research efforts concentrated on the development of simple models capable of simulating plant responses based on daily maximum-minimum temperature data. The first issue was spatial – how to build a model that was suitable for comparison to plant phenology over a large geographical area (Schwartz 1999). Conventional wisdom suggested that such models would not be effective because the environmental variation was too great. Fortunately, phenology data on cloned species of lilac and honeysuckle had been collected since the late 1950s and early 1960s across the United States, as part of regional agricultural experiment-station networks. Such data were ideal for large-area model development, and gathered for the cloned lilac *Syringa chinensis* "Red Rothomagensis" and the two honeysuckle varietals *Lonicera tatarica* "Arnold Red" and *L. korolkowii* "Zabeli."

The first continental-scale models concentrated on "first leaf," as this is the initial sign of plant growth. Models were based on a theory of plant development that linked growth to the number of times when large amounts of energy (essentially higher temperatures) occurred before the event, rather than the total accumulation of energy itself. Further, that there were two last surges of these high-energy events (capstones) that tended to occur about one week before and at the time of first leaf. These models proved superior to the conventional degree-day total-accumulation approach. In subsequent refinements, chilling requirements, and a model to simulate first bloom were added to what is now called the Spring Indices (SI) models (Schwartz and Reiter 2000). This "suite of measures" has been successfully tested against lilac data in the eastern and western United States, Germany, Estonia, and China. The robustness of the models in these tests has led to current work, which will apply the models to the largest daily maximum-minimum temperature base possible in mid-latitudes (both in space and in time) to provide a baseline of comparison for assessment of past, present, and future changes to the onset of spring (e.g., Figure 77.1).

## EXPERIMENTS WITH MODEL OUTPUT

With the operational SI models in hand, experiments began using these simulated phenological output data. Ultimately such "synthetic" data are not as rich as "real" multi-species data, but sufficient to make a start. The first studies examined changes in lower-atmosphere variables associated with the onset of spring. This approach turned the previous model-building research on its head. Now the question was not how climate affects plants, but how plants might affect climate. As a climatologist, I first assumed that plant impacts on the atmosphere must be tiny. Considerable examination of the topic now suggests that the effect is indeed small, but may be large enough to be important during spring. Early research results suggested that the impact of plant transpiration on the lower atmosphere is reflected in conventional variables like diurnal temperature range. Later direct examination of gas and energy exchange data have shown that they also change at the onset of spring greenness (Fitzjarrald et al. 2001; Schwartz and Crawford 2001). Complex issues, such as site-to-site comparability and inter-annual variation, show promise of being addressed in simple and straightforward ways with phenology. These potential benefits allow phenology to enhance the work of global-change scientists trying to understand the planet's carbon balance.

The SI models also provided a systematic way to evaluate the effectiveness of satellite-derived products used to measure the onset of the spring season. Cloud-cover interference problems severely restrict the temporal resolution of these products at one to two weeks. Also, some satellite onset-of-greenness or start-of-season (SOS) measures may reflect understory

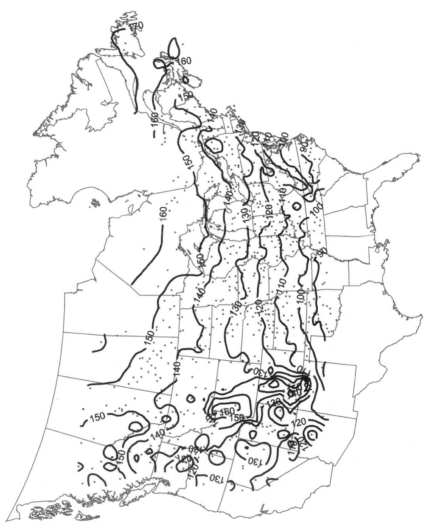

*Figure 77.1.* Average 1961–1990 Spring Indices (SI) first bloom date (January 1st = 1).

or non-biological changes (such as snow cover melt). Despite these limitations, SOS measures can still detect year-to-year variations at some specific sites, and often follow a reasonable correlation with surface phenology (Schwartz et al. 2002). Their main advantage is providing a data values for all portions of the earth's surface. This perspective is critical for whole earth/ecosystem studies, and confronts a variety of issues that are distinct from analyses based on site data. Newly launched/future sensors aboard satellites will improve, but probably never completely overcome the cloud-cover issue. So as mentioned before, a combined observation system using satellite and ground observations is the prudent course (Schwartz 1998, 1999).

## CONCLUSIONS

The work has a long way to go before completion, but a geographic perspective has helped to promote the framing and pointed the path toward resolution of this research question, because of geography's ability to address complex multi-scale and interdisciplinary problems. Neither the perspectives of the remote-sensing specialist nor the biologist alone are sufficient to address the full range of ideas needed to frame the issues. As with many such problems, geographers are well situated to act as facilitators for the diverse set of experts and perspectives necessary to make progress in planning a research strategy. Assessment of twentieth-century changes in the timing of the onset of mid-latitude spring (currently underway) will be a reference for future multidisciplinary research. As new global phenology data become available and analyses continue, better understanding of interconnections among the various atmospheric and biospheric changes associated with the green wave will emerge, and contribute to improvements in global-change science.

## REFERENCES

Chmielewski, F. M. and T. Rötzer. 2002. Annual and Spatial Variability of the Beginning of Growing Season in Europe in Relation to Air Temperature Changes. *Climate Research* 19: 257–264.

Fitzjarrald, D. R., O. C. Acevedo, and K. E. Moore. 2001. Climatic Consequences of Leaf Presence in the Eastern United States. *Journal of Climate* 14: 598–614.

Menzel, A. and P. Fabian. 1999. Growing Season Extended in Europe. *Nature* 397: 659.

Menzel, A. 2002. Phenology, its Importance to the Global Change Community. *Climate Change* 54: 379–385.

Rouse, J. W. Jr. 1977. *Applied Regional Monitoring of the Vernal Advancement and Retrogradation (Green Wave Effect) of Natural Vegetation in the Great Plains Corridor*, Final Report, Contract No. NAS5-20796. Greenbelt, Maryland: Goddard Space Flight Center.

Schwartz, M. D. 1998. Green-wave Phenology. *Nature* 394: 839–840.

Schwartz, M. D. 1999. Advancing to Full Bloom: Planning Phenological Research for the 21st Century. *International Journal of Biometeorology* 42: 113–118.

Schwartz, M. D. and T. M. Crawford. 2001. Detecting Energy Balance Modifications at the Onset of Spring. *Physical Geography* 21: 394–409.

Schwartz, M. D. and B. E. Reiter. 2000. Changes in North American Spring. *International Journal of Climatology* 20: 929–932.

Schwartz, M. D., B. C. Reed, and M. A. White. 2002. Assessing Satellite-derived Start-of-Season Measures in the Conterminous USA. *International Journal of Climatology* 22: 1793–1805.

# 78

# UNDERSTANDING URBAN CLIMATES

SUE GRIMMOND

*Indiana University, Bloomington*

Globally, cities are reaching an unprecedented size, in terms of their number, area, and population. By the end of the twentieth century, approximately half of the world's population, over three billion people, lived in urban areas. By 2025, the United Nations predicts that this number will double, and the proportion of the global population who are urban residents will rise to two-thirds (Uitto and Biswas 2000). Urbanization dramatically alters the land surface and converts pre-urban micro, local and regional climates into distinctively "urban climates." Probably the best-known and intensively studied urban-climatic feature is the "urban heat island," although considerable attention has also been directed to the effects of urbanization on precipitation, humidity, wind, and air quality (Lowry 1998). Urban climate effects have many implications; for example, they influence human comfort, heat related illnesses and mortality, energy consumption, as well as ecological and hydrological systems within and beyond city boundaries. Geographers, through measurements and modeling, have been at the forefront of assessing urban influences, especially the effects of surface materials and morphology, on local and regional climates.

## SURFACE-ATMOSPHERE EXCHANGES

Ultimately, urban climate effects are due to differences in the exchanges of heat, mass, and momentum between the city and its pre-existing landscape (Oke 1988). Thus, the understanding, prediction, and mitigation of urban climate effects are intricately tied to knowledge of surface – atmosphere exchanges in cities and their spatial variability. Geographers have conducted many studies of these exchanges within the framework of the surface energy balance, defined as:

$$Q^* + Q_F = Q_H + Q_E + \Delta Q_S + \Delta Q_A \qquad \text{[units: W m}^{-2}\text{]}$$

where $Q^*$ is the net all-wave radiation (the net balance of the incoming and outgoing radiative fluxes), $Q_F$ is the anthropogenic heat flux (the energy released by human activities), $Q_H$ is the turbulent sensible heat flux (the energy

*D. G. Janelle et al. (eds.), WorldMinds: Geographical Perspectives on 100 Problems,* 481–486.

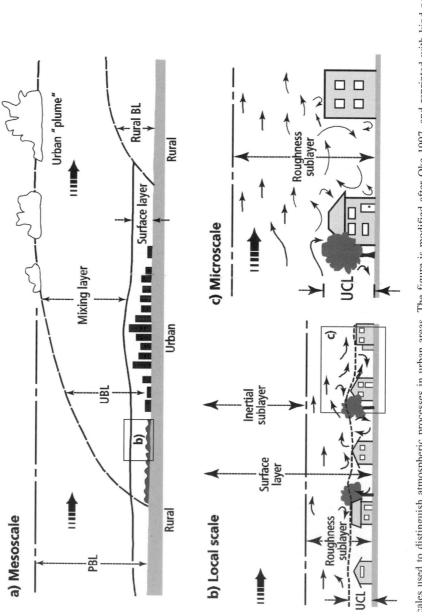

*Figure 78.1.* Scales used to distinguish atmospheric processes in urban areas. The figure is modified after Oke 1997, and reprinted with kind permission of Kluwer Academic Publishers from Piringer et al. 2002 (Figure 1, p. 3).

that heats the air), $Q_E$ is the latent heat flux (the energy taken up/released with the phase change of water, i.e., with evaporation and condensation), $\Delta Q_S$ is the net storage heat flux (the energy that heats and is stored in the urban fabric) , and $\Delta Q_A$ is the net horizontal heat advection (the lateral movement of energy into or out of an area).

## SCALE

Geographers have stressed the critical importance of scale, both spatial and temporal, in the understanding of urban climates. Based on Oke (1984), three spatial scales, micro-, local-, and meso- are commonly recognized. At the micro-scale, each individual element (buildings, trees, roads, etc.) creates its own microclimate. Because the city usually possesses repetitive structures, such as building lots and streets, these elements are combined into micro-scale units, such as street canyons, which in turn generate their own features (Figure 78.1). These are effects that exist beneath roof-level, and by analogy to plant canopies, are within the urban canopy layer (UCL). A larger neighborhood, comprising several similar street canyons plus intervening buildings, gardens etc., creates a local-scale climate that extends horizontally. The influence of each element extends above roof-level as a series of plumes and wakes in the roughness sub-layer (Figure 78.1). These eventually merge to form a more homogeneous surface layer, in which micrometeorological theory for extensive homogeneous surfaces apply (Figure 78.1). In turn, these are mixed together to form the urban boundary layer (UBL) of the entire city, a meso-scale phenomenon, within which the atmosphere shows a response to the integrated presence of the city. While understood well conceptually, those interested in urban form and atmospheric processes are still studying the fundamental dimensions associated with each of these scales, spatial and temporal.

## UNDERSTANDING URBAN CLIMATE

Urban areas represent one of the most challenging environments for meteo-rologists. The three-dimensional structure of cities, combined with the complex mix of surface types, with contrasting radiative, thermal, and moisture con-ditions make the application of standard measurements difficult. Geographers have focused much attention on how to collect representative meteorological observations in urban areas at different scales (e.g., Oke et al. 1989; Voogt and Oke 1997; Grimmond and Oke 1999; Spronken-Smith et al. 2000; Rotach 2001). Criteria have been developed for the siting of equipment on towers, both in terms of height above the roughness elements (buildings and trees) that make up the urban surface, and the areal extent of homogeneous land cover

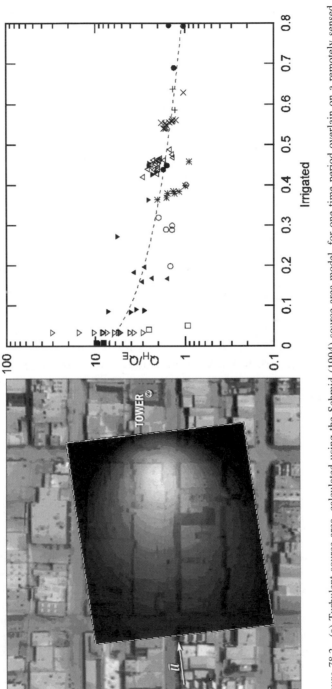

*Figure 78.2.* (a) Turbulent source area, calculated using the Schmid (1994) source area model, for one time period overlain on a remotely sensed thermal image for an industrial area of Vancouver, B.C. The brightest area has the greatest influence on the turbulent flux measurement (adapted from Voogt and Grimmond 2000, with courtesy from the American Meteorological Society). (b) Relation between the surface characteristics (fraction of area irrigated) and the partitioning of the turbulent heat fluxes during the middle of day for a number of different urban areas in North America (adapted from Grimmond and Oke 2002, with courtesy from the American Meteorological Society).

around the measurements site. Flux source-area models, for example FSAM of Schmid (1994), have been integrated to define the surface footprint of the measurements, based on the height of the observations, the roughness of the surface, and meteorological conditions, such as wind direction, speed, and atmospheric stability. The output of flux footprint models, when overlain on a georeferenced database (Grimmond and Souch 1994, Grimmond 1996) or a remotely sensed image (Voogt and Grimmond 2000) (Figure 78.2), allow relations between surface cover/properties and fluxes to be explored.

Emerging from this work is a greater understanding of surface energy balance fluxes in urban areas and their spatial variability. The turbulent sensible, latent, and storage heat fluxes all have been shown to represent important terms in the surface energy balance of most cities. Each of the heat fluxes varies both spatially and temporally. Under low-wind conditions, storage heat flux is most important at downtown and light industrial sites (at least 50 percent of daytime $Q^*$), and the sensible heat flux is most important at residential sites (40 to 60 percent of daytime $Q^*$) (Grimmond and Oke 2002). At residential sites, latent heat flux, if sustained by garden irrigation and/or frequent rainfall, is also significant (20 to 40 percent of daytime $Q^*$). Surface cover, notably the fraction of the surface vegetated and irrigated, exerts an important control on $Q_H$ and the relative heat partitioning between sensible $(Q_H)$ and latent $(Q_E)$ heat (Figure 78.2). At all sites there is distinct hysteresis in the diurnal course of the storage heat flux; much more of the net radiation is used to heat the urban fabric in the morning. In addition, the sensible heat flux remains positive after the net all-wave radiation turns negative at night. This has importance implications for the stability (mixing) of the urban atmosphere, with implications for urban air quality.

## FUTURE WORK

While advances have been made in the study of urban climates, the range of conditions studied, both in terms of meteorology (seasonal and synoptic conditions) and surface cover (building styles – sizes, shapes and arrangements; and vegetation cover) remain limited. Geographers will play a key role in the next generation of urban climate research, particularly as issues related to the integration of processes and effects at different scales are addressed.

## REFERENCES

Grimmond, C. S. B. 1996. Dynamically Determined Parameters for Urban Energy and Water Exchange Modelling. In M. F. Goodchild, L. T. Steyaert, B. O. Parks, C. Johnston, D. Maidment, D. Crane, and S. Glendinning, eds. *GIS and Environmental Modeling: Progress and Research Issues*, 305–309. GIS World Books.

Grimmond, C. S. B. and T. R. Oke. 1999. Aerodynamic Properties of Urban Areas Derived From Analysis of Surface Form. *Journal of Applied Meteorology* 38: 1262–1292.

Grimmond, C. S. B. and T. R. Oke. 2002. Turbulent Heat Fluxes in Urban Areas: Observations and Local-Scale Urban Meteorological Parameterization Scheme (LUMPS). *Journal of Applied Meteorology* 41: 792–810.

Grimmond, C. S. B. and C. Souch. 1994. Surface Description for Urban Climate Studies: A GIS Based Methodology. *Geocarto International* 9: 47–59.

Lowry, W. R. 1998. Urban Effects on Precipitation Amount. *Progress in Physical Geography* 22: 477–520.

Oke, T. R. 1984. Methods in Urban Climatology and in Applied Climatology. *Zürcher Geographische Schriften* 14: 19–29.

Oke, T. R. 1988. The Urban Energy Balance. *Progress in Physical Geography* 12: 471–508.

Oke, T. R. 1997. Urban Environments. In W. G. Bailey, T. R. Oke, and W. R. Rouse, eds. *The Surface Climates of Canada*, 303–327. Montreal: McGill-Queens University Press.

Oke, T. R., H. A. Cleugh, C. S. B. Grimmond, H. P. Schmid, and M. Roth. 1989. Evaluation of Spatially-Averaged Fluxes of Heat, Mass and Momentum in the Urban Boundary Layer. *Weather and Climate* 9: 14–21.

Piringer, M., C. S. B. Grimmond, S. M. Joffre, P. Mestayer, D. R. Middleton, M. W. Rotach, A. Baklanov, K. De Ridder, J. Ferreira, E. Guilloteau, A. Karppinen, A. Martilli, V. Masson, and M. Tombrou. 2002: Investigating the Surface Energy Balance in Urban Areas – Recent Advances and Future Needs. *Water, Air and Soil Pollution: Focus* 2(5–6): 1–16.

Rotach, M. W. 2001: Simulation of Urban-Scale Dispersion Using a Lagrangian Stochastic Dispersion Model. *Boundary-Layer Meteorology* 99: 379–410.

Schmid, H. P. 1994. Source Areas for Scalars and Scalar Fluxes. *Boundary Layer Meteorology* 67: 293–318.

Spronken-Smith, R. A., T. R. Oke, and W. P. Lowry. 2000. Advection and the Surface Energy Balance Across an Irrigated Urban Park. *International Journal of Climatology* 20: 1033–1047.

Uitto, J. I. and A. K. Biswas. 2000. *Water for Urban Areas: Challenges and Perspectives*, 245. New York: United Nations University Press.

Voogt, J. A. and T. R. Oke. 1997. Complete Urban Surface Temperatures. *Journal of Applied Meteorology* 36: 1117–1132.

Voogt, J. A. and C. S. B. Grimmond. 2000. Modeling Surface Sensible Heat Flux Using Surface Radiative Temperatures in a Simple Urban Area. *Journal of Applied Meteorology* 39: 1679–1699.

# 79

# DEVELOPING HEAT-WARNING SYSTEMS FOR CITIES WORLDWIDE

SCOTT SHERIDAN
*Kent State University*

Natural hazards and their impact upon humans have long been the focus of curiosity by researchers and the general public. Geologic hazards, such as earthquakes and volcanoes, and atmospheric phenomena, such as tornadoes and hurricanes, receive considerable attention, especially in light of the dramatic impact they may have upon the landscape.

## HEAT AS A HAZARD

In contrast, to tornados and hurricanes, the awareness of "heat" as an atmospheric hazard has been largely understated over time. This occurs despite the fact that heat is as deadly a phenomenon as all other atmospheric phenomena combined (National Weather Service 2002). For example, over the ten years from 1992 to 2001, on average 219 deaths per year were directly attributable to the heat, though with a lack of consensus on the definition of a heat-related death, the actual toll is far higher. While the Galveston hurricane of September 1900 is often cited as the deadliest atmospheric-phenomenon disaster in United States history, claiming 6,000 lives, the hot summer of 1980 is believed to be associated with 10,000 deaths (National Climatic Data Center 2002). Future scenarios suggest that, in a warmer world, heat vulnerability could increase significantly. Worldwide, Tol (2002) estimates that an additional 350,000 people could die from heat-related cardiovascular and respiratory problems per 1 °C increase in the global mean temperature. Yet it is arguable that only after the heat wave of 1995, during which the nation witnessed graphic images of several hundred Chicagoans perishing during a week of punishing heat and humidity, that the will and the resources were galvanized to attempt to mitigate the deadly hazard of excessive summertime heat.

*D. G. Janelle et al. (eds.), WorldMinds: Geographical Perspectives on 100 Problems*, 487–492.
© 2004 *Springer. Printed in the Netherlands.*

## COMBATING HEAT VULNERABILITY

The effort into combating the effects of excessive heat has been forged in two directions: better forecasting methodologies, and better mitigation strategies once oppressive weather has been forecast to occur (e.g., Kalkstein et al. 1996). It is the former in which I have participated, studying the relationship between heat and mortality. My work has been incorporated into improved forecasting systems for several worldwide locations, including Rome, Shanghai, and Toronto, as well as for several regions in the United States.

The appeal of the heat-health relationship to a geographer is that it is not static. Thresholds of heat vary from place to place, both on the large scale (e.g., Kalkstein and Davis 1989) and the small scale; they depend upon the time of year as well. Understanding heat vulnerability is more than just the ambient afternoon temperature or a "heat index": wind desiccates skin more rapidly, less cloud cover heats up buildings faster, and high overnight temperatures do not allow them to cool off. It is important to assess the heat-health relationship as precisely as possible. Announcing too few warnings would not protect the population, as many hazardous days would be ignored; calling too many days, on the other hand, would result in the population ignoring the warnings.

For all of these reasons, the National Weather Service's official criterion for an excessive heat warning – a heat index above 105°F on two consecutive days – is inadequate. It does not account for whether one is in Duluth or Miami, in May or August. A more holistic approach, used by many applied climatologists and geographers, is the "synoptic climatological" method. Synoptic climatology's main goal is to link the atmosphere and a surface "response," in this case, human health. It does so by categorizing the atmosphere holistically, viewing all components together, rather than independently. This method thus identifies the "air mass" or "weather type" over a particular location at a particular time.

## "SYNOPTIC" SYSTEM DEVELOPMENT

By identifying a weather type, one can account for all atmospheric conditions at once, and this is the "umbrella" of conditions to which we respond. The conditions associated with a weather type – temperature, humidity, and so on – vary by location, just as one would expect: a cold front advancing polar air southward never brings temperatures to Miami nearly as cold as those it brings to Minneapolis. The weather types that affect human health actually vary according to location as well. The system I have used in analyzing the heat-health relationship is the Spatial Synoptic Classification (SSC, Sheridan 2002a), which categorizes each day at a location into one of eight weather types. The two most commonly associated with heat-related health problems are, unsurprisingly, the two hottest: Dry Tropical (DT), hot and dry, with little cloud cover; and Moist Tropical Plus (MT+), an oppressively humid weather type with high overnight temperatures.

Up to 25 years of daily mortality data, standardized for population growth and seasonal migration, have been analyzed. As is expected, humans deal worst with conditions that are outside of their accustomed range. Across much of the middle latitudes, both weather types mentioned above occur infrequently, less than one day in ten. Mortality rates on these days generally rise 5 to 10 percent as a result. At locations further from the poles, such as Phoenix and New Orleans, often only the most extreme conditions evoke any response, and these are generally lower in magnitude (Table 79.1). These mean responses do not capture all the variability observed in the human response. That is, on some oppressive days there is a large increase in mortality; on others, none at all. What causes this variability? In many locations, particularly those farther poleward, seasonality is important. The same weather conditions evoke a stronger response earlier in the season than later in the season, after the population has acclimatized to the summer. The length of time that oppressive conditions have persisted is also crucial – previous forecasting methods never accounted for this obvious factor – the 5th day of a heat wave will be

Table 79.1. Mean Mortality Response to Different Weather Types by Location.

| Dry Tropical | | | Moist Tropical Plus | | |
|---|---|---|---|---|---|
| Frequency | Excess Mortality | Percentage Increase | Frequency | Excess Mortality | Percentage Increase |
| CINCINNATI, OHIO, USA (25 years) | | | | | |
| 1.9 | +4.4 | +15.6 | 6.5 | +1.8 | +6.4 |
| NEW ORLEANS, LOUISIANA, USA (25 years) | | | | | |
| | | | 2.5 | +1.9 | +6.2 |
| PHOENIX, ARIZONA, USA (25 years) | | | | | |
| 1.3* | +2.7 | +6.0 | | | |
| ROME, ITALY (11 years) | | | | | |
| 6. 8 | +6.2 | +12.1 | 3.9 | +5.0 | +9.8 |
| SHANGHAI, P.R. CHINA (10 years) | | | | | |
| | | | 11.0 | +42.4 | +11.3 |
| TORONTO, ONTARIO, CANADA (17 years) | | | | | |
| 3.4 | +2.4 | +7.7 | 3.9 | +2.2 | +7.1 |

Frequency is the percentage of days in the period studied that are classified as the weather type (from 15 May–30 September); excess mortality is mean total deaths per day greater than normal; and percentage increase represents this increase in mortality as a percentage of the above mean value. Blank indicates the air mass does not cause an increase in mortality.
* "Dry Tropical Plus," defined only for Phoenix, to separate it from the extremely common Dry Tropical weather type.

more unbearable than the first, especially inside, where interior temperatures will keep rising each day. Even within a weather type, certain characteristics are important, such as cloud cover and overnight temperatures, as mentioned above.

## TORONTO HEAT HEALTH ALERT SYSTEM
*Afternoon Forecast*
*Issued 8/7/2001 15:13:49*

**Forecast for 8/8 - 8/9/2001**

| DAY | 08/08 | | | | 08/09 | | | |
|---|---|---|---|---|---|---|---|---|
| HOUR | 05 | 11 | 17 | 23 | 05 | 11 | 17 | 23 |
| TEMPERATURE | 23 | 31 | 35 | 29 | 25 | 29 | 31 | 25 |
| DEW POINT | 22 | 22 | 23 | 23 | 22 | 23 | 23 | 22 |
| CLOUDINESS | | 4 | | | | 5 | | |
| AIR MASS | | MT+ | | | | MT+ | | |
| DAY IN ROW | | 3 | | | | 4 | | |

Forecast data provided by Meteorological Service of Canada - Ontario Region

## 8/8: HEAT EMERGENCY
Conditions oppressive - with a 97% chance of excess mortality

## 8/9: HEAT EMERGENCY
Conditions oppressive - with a 92% chance of excess mortality

## SYSTEM LEVELS

### HEAT EMERGENCY
The likelihood of weather-related excess mortality occurring exceeds 90 percent.

### HEAT ALERT
The likelihood of weather-related excess mortality occurring exceeds 65 percent.

### ROUTINE MONITORING
Conditions do not suggest excess mortality is likely.

*Figure 79.1.* The Toronto Heat Watch-Warning System web page.

All of these relationships between mortality, weather, and other parameters are ultimately quantified statistically, and equations are developed that can be used to relate forecast weather conditions to a likelihood of excess mortality occurring, based on past analogous conditions. These relationships appear on interactive websites (Figure 79.1) for use by the forecaster, as well as by health and other community officials. Forecasts are produced automatically twice a day based on computer-model output, prognosticating two days into the future. The forecasts can also be updated manually, as often as desired, should a forecast change. Using this output, local agencies then have the ultimate decision on whether to call attention to the oppressive conditions. Most agencies have more than one level of advisory. Toronto, for example, has a *heat emergency*, which represents a day whose weather conditions in the past are associated with a greater-than-90 percent chance of excess mortality, and a lower-level *heat alert*, where this likelihood exceeds 65 percent. Mitigation strategies vary according to location and type of advisory, but include media announcements, opening of cooling shelters, and additional emergency medical-services staffing, among other local community-action programs.

## FUTURE DIRECTIONS

While this work has focused on the metropolitan-area level thus far, much research remains. What makes a particular person vulnerable is not yet fully understood. It has often been surmised that the urban population is more vulnerable, due to the heat island and building type. Interestingly, initial analysis on behalf of the state of Ohio shows the percentage increase in mortality during oppressive heat is similar across rural, suburban, and urban areas (Sheridan 2002b). This suggests that more than the physical location, perhaps other socioeconomic factors are important, as Smoyer (1998) has suggested. As outdoor conditions can only serve as a proxy for the conditions we personally endure on any given hot day, be it indoors or outdoors, with or without a cooling system, vulnerability is an individual-level issue that demands further examination.

## REFERENCES

Kalkstein, L. S. and R. E. Davis. 1989. Weather and Human Mortality: An Evaluation of the Demographic and Interregional Response in the United States. *Annals of the Association of American Geographers* 79: 44–64.

Kalkstein, L. S., P. F. Jamason, J. S. Greene, J. Libby, and L. Robinson. 1996. The Philadelphia Hot Weather – Health Watch/Warning System: Development and Application, Summer 1995. *Bulletin of the American Meteorological Society* 77: 1519–1528.

National Climatic Data Center. 2002. Billion-dollar Weather Disasters, 1980–2002. See: http://www.ncdc.noaa.gov/ol/reports/billionz.html. Accessed 19 September 2002.

National Weather Service. 2002. *Natural Hazard Statistics*. See. http://www.nws.noaa.gov/
  om/hazstats.shtml. Accessed 22 September 2002.
Sheridan, S. C. 2002a. The Redevelopment of a Weather-Type Classification Scheme for North
  America. *International Journal of Climatology* 22: 51–68.
Sheridan, S. C. 2002b. Using a Synoptic Classification Scheme to Assess Rural-Urban Differences
  in Heat Vulnerability. *Proceedings, 13th Conference on Applied Climatology*, 323–325.
Smoyer, K. E. 1998: Putting Risk in its Place: Methodological Considerations for Investigating
  Extreme Event Health Risk. *Social Science and Medicine* 47: 1809–1824.
Tol, R. S. J. 2002: Estimates of the Damage Costs of Climate Change: Part I. Benchmark
  Estimates. *Environmental and Resource Economics* 21: 47–73.

# 80

# USING GEOMORPHOLOGY TO ASSESS AND ENHANCE BEACH HABITAT FOR HORSESHOE CRABS

NANCY L. JACKSON
*New Jersey Institute of Technology*

DAVID R. SMITH
*United States Geological Survey*

KARL F. NORDSTROM
*Rutgers University*

Sandy beaches in estuaries are recognized for their importance as habitat (Nordstrom 1992; Botton et al. 1994) but this habitat is decreasing due to beach erosion and human development. Along the east coast of North America, the horseshoe crab (*Limulus polyphemus*) is one of the most prominent species using sandy beaches and there is a need to preserve or restore beaches to enhance spawning and egg viability. Horseshoe crabs are found from Maine to the Yucatan, and closely related species are found in the Indo-Pacific region, but it is in the mid-Atlantic region of the eastern United States where their abundance and their role in the ecosystem are greatest. Each spring horseshoe crabs spawn and deposit great quantities of eggs on estuarine beaches, ensuring their own reproduction and providing critical nutrition for other aquatic and terrestrial species. Several species of migratory birds feed so heavily on horseshoe crab eggs during their springtime migration that it is thought that the ultimate fate of these birds may depend on spawning horseshoe crab populations (Tsipoura and Burger 1999). Horseshoe crabs also support multi-million dollar biopharmaceutical and fishing industries. Biopharmaceutical companies collect horseshoe crab blood to derive a diagnostic product used to test for bacterial contamination of injectable drugs and implantable medical devices. Recent increases in harvest of horseshoe crabs for bait in the eel and conch fisheries have raised concerns for the status of horseshoe crabs and the species that depend on them. Horseshoe crab populations also are threatened by loss of beach due to construction of shorefront buildings and shore protection structures (bulkheads and seawalls), and by inefficient spawning, as crabs are forced to use remaining less-desirable habitat, e.g., where beaches are underlain by peat layers. Policy action has focused on controlling crab harvesting to maintain sufficient numbers of

*D. G. Janelle et al. (eds.), WorldMinds: Geographical Perspectives on 100 Problems, 493–496.*
© 2004 *Springer. Printed in the Netherlands.*

the species, but interest in enhancing the suitability of available sand beach
is increasing. A multidisciplinary team, including geographers, is assessing the
following important questions.

## WHERE ARE THE LOCATIONS OF GREATEST HORSESHOE CRAB SPAWNING?

Horseshoe crabs spawn in greatest numbers on sandy beaches in estuaries.
Estuarine beaches are characterized by a steep foreshore, with little micro-
topographic variation, and a broad, flat low-tide terrace (Figure 80.1). Egg
development appears to be less on the lower foreshore where greater moisture
and less interstitial oxygen reduce development, and on the upper foreshore,
where less moisture causes eggs to desiccate (Penn and Brockman 1994).

   The World's largest population of spawning horseshoe crabs is found on
beaches in Delaware Bay, USA (Figure 80.1). At bay scale, the location of
heaviest spawning is a function of the salinity and the wave energy gradient.
In a funnel-shaped estuary, such as Delaware Bay, wave energies are greatest
near the mouth of the basin and decrease with distance toward its head (Figure
80.1). The viability of horseshoe crab embryos decreases with decreasing
salinity (Sekiguchi 1988), so conditions for egg development decline toward
the head of the estuary. Thus, the best sites for spawning in Delaware Bay
are on beaches in the mid-section of the estuary (Sullivan 1994).

## WHAT ARE THE MORPHODYNAMICS OF THE BEACHES WHERE GREATEST SPAWNING OCCURS?

Egg distribution in foreshore sediments is a function of tides and beach
morphology (width, slope) (Smith et al. 2002). Horseshoe crabs deposit their

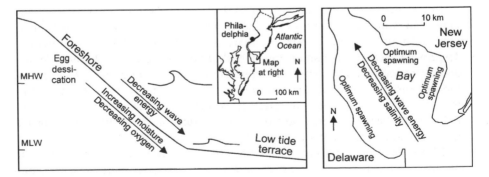

*Figure 80.1.* Profile and plan view of optimum spawning locations for horseshoe crabs, based
on conditions in Delaware Bay.

eggs 0.10–0.20 m below the sediment surface, initially out of reach of shore-birds (Botton et al. 1994). Eggs buried at depth are not exhumed by most non-storm waves unless burrowing of spawning females displaces the eggs upward in the sediment column (Botton et al. 1994). Swash uprush and backwash is the dominant process during low-wave energy conditions (breaking wave heights < 0.25 m) common on estuarine beaches (Jackson and Nordstrom 1993). Under these conditions, activation will not reach depths of 0.10 m and egg availability will be confined to the top few centimeters. A high-energy event (breaking wave heights > 0.60 m) can result in activation depths in excess of 0.10 m and erosion of sediment from the upper foreshore (Jackson 1999), bringing some eggs to the foreshore surface, but also eroding some eggs from the foreshore. Only eggs within the sediments near the surface are available to foraging shorebirds.

## HOW CAN BEACH MANAGEMENT AND RESTORATION ACTIVITIES MAINTAIN AND/OR ENHANCE THE QUALITY OF BEACH HABITAT FOR HORSESHOE CRABS?

Understanding beach morphodynamics is important to identifying relationships between biological productivity and beach change and predicting the effects of shore management strategies. Efforts are being devoted to evaluating ways that habitat may be restored and enhanced through beach nourishment projects and through manipulation of the beach profile. Nourishment operations in estuaries generally are undertaken to protect shorefront development or to dispose of sediments from dredging projects (Nordstrom 1992), but there is increasing interest in creating habitat where protection structures have eliminated the upper beach or along eroding shorelines where hard clay substratum or marsh peat have been exposed (Jackson et al. 2002).

Differences in the form and mobility of estuarine beaches can occur with differences in grain-size characteristics. Finer sizes of fill sediment can lead to flattening of the foreshore slope and burial of surface gravel. Silts and clays cause increased turbidity during placement, and they affect the structure of habitats after they settle and are incorporated into the beach matrix. The low-tide terrace is habitat for juvenile horseshoe crabs (Sekiguchi 1988), and effects of sediment deposition from nourishment operations are unknown. Fine-grained materials create different moisture-retention characteristics and may create substrate more resistant to waves and burrowing by organisms. All of these beach changes have potential impacts on the quality of the beach for horseshoe crab spawning and egg development.

Movement of the beach water table over the tidal cycle is influenced by beach morphology, sediment size, sorting and porosity, tidal elevation, and wave setup and runup. Well-sorted, medium-to-coarse sands are the dominant fraction on many sandy estuarine beaches, causing beaches to be well drained.

The hydraulic conductivity of a beach nourished with fine-grained sediment may decrease, causing lower rates of water table discharge. Beach water table movement affects viability of species by influencing erosion and deposition and flushing of oxygen and organic material (McLachlan and Turner 1994). Horseshoe crab egg viability depends on temperature, salinity, moisture, and oxygen gradients (Sekiguchi 1988, Penn and Brockman 1994). These considerations indicate that greater attention should be devoted to finding the optimum sediment sizes to use as fill materials to restore the habitat lost through shorefront development.

A challenge facing geographers is finding ways to restore or improve landscapes threatened by humans. On estuarine beaches, the task of designing ways of enhancing horseshoe crab spawning and egg availability involves identifying broad scale spatial controls on modification and use of the foreshore, determining the effect of the timing of high-energy events, determining the effects of shore protection structures, and determining the effects of changes in foreshore dimensions and sediment characteristics due to nourishment operations. There will be a continuing need to measure biological variables relative to specific geomorphic process regimes and sediment characteristics on sites where shoreline orientations and foreshore dimensions differ to understand the significance of spatial variability on the viability of species that use the shore as habitat.

## REFERENCES

Botton, M. L., R. E. Loveland, and T. R. Jacobsen. 1994. Site Selection by Migratory Shorebirds in Delaware Bay, and its Relationship to Beach Characteristics and Abundance of Horseshoe Crab (*Limulus Polyphemus*) Eggs. *The Auk* 111: 605–616.

Jackson, N. L. 1999. Evaluation of Criteria for Predicting Erosion and Accretion on an Estuarine Sand Beach, Delaware Bay, New Jersey. *Estuaries* 22: 215–223.

Jackson, N. L. and K. F. Nordstrom. 1993. Depth of Sediment Activation by Plunging Waves on a Steep Sand Beach. *Marine Geology* 115: 143–151.

Jackson, N. L., K. F. Nordstrom, and D. R. Smith. 2002.Geomorphic-Biotic Interactions on Sandy Estuarine Beaches. *Journal of Coastal Research* SI36: 414–424.

McLachlan, A. and I. Turner.1994. The Interstitial Environment of Sandy Beaches. *Marine Ecology* 15: 177–121.

Nordstrom, K. F. 1992. *Estuarine Beaches*. London: Elsevier Science Publishers.

Penn, D. and H. F. Brockman, 1994. Nest-Site Selection in the Horseshoe crab, *Limulus polyphemus*. *Biological Bulletin* 187: 373–384.

Sekiguchi, K. ed. 1988. *Biology of Horseshoe Crabs*. Tokyo: Science House Co., Ltd.

Smith, D. R., P. S. Pooler, R. E. Loveland, M. L. Botton, S. F. Michels, R. G. Weber, and D. B. Carter, 2002. Indices of Horseshoe Crab (*Limulus Polyphemus*) Reproductive Activity on Delaware Bay Beaches: Interactions With Beach Characteristics. *Journal of Coastal Research* 18: 730–740.

Sullivan, J. K. 1994. Habitat Status and Trends in the Delaware Estuary. *Coastal Management* 22: 49–79.

Tsipoura, N. and J. Burger. 1999. Shorebird Diet During Spring Migration Stopover on Delaware Bay. *Condor* 101: 635–644.

# 81

# RECONSTRUCTING PAST WATERSHED AND ECOSYSTEM DEVELOPMENT IN THE COAST MOUNTAINS, BRITISH COLUMBIA, CANADA

CATHERINE SOUCH

*Indiana University-Purdue University at Indianapolis*

Conventionally, records of glacial activity over the Holocene period (~ the last 10,000 years) have been reconstructed by dating the terminal and lateral moraines of alpine glaciers (Grove 1988). This approach has yielded extensive information on the timing of glacial activity. However, one of the major limitations of moraine-based chronologies is their lack of continuity; at best they provide "snapshots" on the timing of glacial positions, occasionally with information on retreat phases. This problem is compounded in western Canada, as in many regions, by the fact that the most recent phase of glacial activity, commonly referred to as the Little Ice Age, was the most extensive glacial advance of the Holocene. Hence, as glaciers expanded to their Little Ice Age maxima, they over-rode, reworked, and destroyed evidence of previous events. The example presented here demonstrates how lake sediments from Kokwaskey Lake, located within the Kwoiek Creek watershed, Coast Mountains of British Columbia (50° N, 122.8° W; elevation 1,050 m), provide a solution to this problem.

## RECORDS OF PAST ENVIRONMENTAL CONDITIONS FROM LAKE SEDIMENTS

Lakes with appropriate geometry and hydraulic conditions trap a significant fraction of inflowing sediment from their surrounding watersheds, thus recording a comprehensive integrated signal of drainage-basin scale geomorphic activity (Oldfield 1977). The location and size of a lake relative to its contributing area, in addition to the hydraulics of the system, dictate the sensitivity and resolution of the record. Analysis of rates of sediment accumulation and the physical, biological, and geochemical properties of the sediments with depth provides insight into watershed and ecosystem development through time.

Geographers have been particularly active in pursuing integrated lake

*D. G. Janelle et al. (eds.), WorldMinds: Geographical Perspectives on 100 Problems, 497–501.*
© 2004 *Springer. Printed in the Netherlands.*

sediment – drainage basin studies, often employing detailed sediment budgets in combination with analysis of sediment properties, using mineral magnetic and geochemical techniques, to consider explicitly the origin, transport, transformation, and accumulation of sediments in lake basins. This work has been conducted in a wide range of settings by a large number of investigators. Examples of the problems addressed using this methodology are diverse, including studies of past natural environmental change, the focus of this paper, but also human impacts on the environment, notably on rates of soil erosion (Foster et al. 1985) and anthropogenic contamination (Winter et al. 2001).

## THE LAKE-SEDIMENT DRAINAGE-BASIN APPROACH

The movement of sediment through a drainage basin can be considered as a cascade of material moved by a suite of transfer processes episodically between storages (Gregory and Walling 1973). Many geomorphological studies of alpine environments have been undertaken to quantify the rates of sediment transfer (fluvial, colluvial, eolian), their variability and controls (Gurnell and Clark 1997). Other studies have focused on rates of accumulation and changing sediment properties in the sediment storages, notably on slopes, floodplains, and in wetlands/lakes. In alpine environments, lake sediments are particularly well suited to recording drainage-basin scale changes. Alpine lakes often are oligotrophic, dominated by clastic sediment inputs from slopes and fluvial systems. Under these low-productivity conditions, bioturbation is low and effects of in lake-productivity do not complicate interpretation of the drainage-basin record. Moreover, if the alpine lake is deep relative to its surface area, wind-induced currents will not mix the sediments. Thus, sediment properties are fixed through time and provide an archive of past environmental conditions. With appropriate dating of the sediments (from annual layers – varves, tephras, radiometric techniques such, as $^{210}$Pb or $^{14}$C), detailed reconstructions of past drainage-basin scale conditions can be obtained (Karlen 1981; Leonard 1986).

## ALPINE ENVIRONMENTS

At the Holocene time scale, the dominant controls on sediment transfers in glacierized alpine environments relate to climate change, specifically to the effects of glacial advance and retreat. Periods of maximum glacial advance and the initial period of glacial retreat are associated with maximum sediment yield. This is due both to the greater availability of sediment, as large expanses of unvegetated, recently deposited glacial materials are exposed by retreating ice sheets, and greater meltwater discharges, hence increased capacity and competence of proglacial rivers.

The 4.2-m lake sediment core analyzed from Kokwaskey Lake is described in detail by Souch (1994). Chronology is provided by radiocarbon dates and tephra layers (Mazama and Bridge River). Approximately 40 percent of the contributing watershed (total area ~ 42 km$^2$) currently is above tree line and 10 percent is glacierized. The watershed has a high range of relief (150–2,944 m), with most summits and ridge crests above 2,000 m. Valley slopes are steep (average slope 31°), with vertical rock faces common. The lake sediments are composed almost exclusively of terrigenous material from upstream glaciers and surrounding slopes, with little material contributed by lake productivity.

Sedimentation rates in Kokwaskey Lake reflect the deglacial and neoglacial history of the region (Figure 81.1), verified by moraine-based studies in the

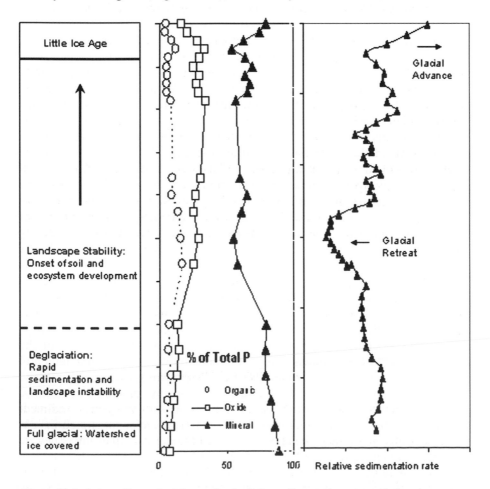

*Figure 81.1.* Lake-sediment based records of relative sedimentation rates and phosphorus geo-chemistry, with interpretations of glacial advances and retreats over the last ~10,000 years (based on Souch 1994 and Slaymaker et al. 2003).

region, and results of pollen and macrofossils in the lake sediments (Souch 1994). Sedimentation rates immediately following deglaciation were very high. Rates of sedimentation decreased and stabilized 10,000 to 7,000 years before present (YBP), as the landscape stabilized and soil development was initiated. Climatic deterioration in the mid-Holocene and renewed neoglacial activity subsequently (Ryder and Thompson 1986) increased glacial sediment supply, resulting in greater sedimentation rates 6,000–5,000, 3,500–2,900, and since 750 YBP.

Phosphorus geochemical analyses of lake sediments provide further insight into rates of erosion, soil development, and landscape stability over the postglacial period (Filippelli and Souch 1999). Results from this site show continual erosion from steep slopes and glacial sources, evidenced by the dominance of mineral forms of phosphorus in the lake sediments (Figure 81.1). The mineral phosphorus fraction does exhibit a decrease, however, to ~ 50 percent during the latter part of the mid-Holocene warmer, drier interval – indicative of landscape stabilization and soil development (~ 9,000 to 6,000 YBP). From the mid- to late-Holocene (6,000 to 1,000 YBP), a period of cooler/wetter conditions, each the phosphorus fractions vary slightly. The last 1,000 years of this record, marked by the most extensive Holocene glacial activity (Ryder and Thompson 1986), are characterized by a rapid return of phosphorus geochemistry to glacial/deglacial conditions (Figure 81.1).

Throughout the Holocene, the high-relief Coast Mountains watershed led to constant loss of surface sediments with poorly developed soils and relatively little organic phosphorus. Changes in the volume of material eroded, as recorded by material trapped in lakes, and the geochemical properties of those sediments provide information on glacial activity in the contributing watershed.

## FINAL COMMENTS

Direct observations of the environment, the atmosphere, hydrosphere, lithosphere, and biosphere, have been conducted for only a tiny fraction of the Earth's history. Lake sediments provide an opportunity to determine temporal variability of sediment movement in a catchment, in terms of volume and provenance, which in turn can be related to paleoenvironmental conditions. Explicit analysis of lake sediments in combination with analysis of sediment sources and transfer processes provides an opportunity for integrated insight into variations in drainage-basin scale processes over multiple timescales.

# REFERENCES

Filippelli, G. M. and C. Souch. 1999. Effects of Climate and Landscape Development on the Terrestrial Phosphorous Cycle. *Geology* 27: 171–174.

Foster, I. D., J. A. Dearing, A. D. Simpson, A. D. Carter, and P. G. Appleby. 1985. Lake Catchment Based Studies of Erosion and Denudation in the Merevale Catchment, Warwickshire, U.K. *Earth Surface Processes and Landforms* 10: 45–68.

Gregory, K. J. and D. E.Walling. 1973. *Drainage Basin Form and Process: A Geomorphological Approach.* New York: John Wiley.

Grove, J. M. 1988. *The Little Ice Age.* New York: Methuen.

Gurnell, A. M. and M. J. Clark. 1987. *Glacio-Fluvial Sediment Transfer: An Alpine Perspective.* Chicheste: John Wiley.

Karlen, W. 1981. Lacustrine Sediment Studies: A Technique to Obtain a Continuous Record of Holocene Glacier Fluctuations. *Geographiska Annaler* 63A: 273–281.

Leonard, E. 1986. Use of Lacustrine Sedimentary Sequences as Indicators of Holocene Glacial Activity, Banff National Park, Alberta, Canada. *Quaternary Research* 26: 18–231.

Oldfield, F. 1977. Lakes and Their Drainage Basins as Units of Sediment Based Ecological Study. *Progress in Physical Geography* 1: 460–454.

Ryder, J. M. and B. Thompson. 1986. Neoglaciation of the Southern Coast Mountains of British Columbia: Chronology Prior to the Late Neoglacial Maximum. *Canadian Journal of Earth Sciences* 23: 273–287.

Slaymaker, O., C. Souch, B. Menounos, and G. Filippelli. 2003. Advances in Holocene Mountain Geomorphology Inspired by Sediment Budget Methodology. *Geomorphology* (in press).

Souch, C. 1994. Downvalley Lake Sediments as Records of Neoglacial Activity, Kwoiek Creek Watershed, Coast Mountains, British Columbia. *Geografiska Annaler* 76A: 169–186.

Winter L. T., I. D. L. Foster, S. M. Charlesworth, and J. A. Lees. 2001. Floodplain Lakes as Sinks for Sediment-Associated Contaminants – A New Source of Proxy Hydrological Data? *The Science of the Total Environment* 266: 187–194.

# 82

# WATER RESOURCE DEVELOPMENT ON SMALL CARBONATE ISLANDS: SOLUTIONS OFFERED BY THE HYDROLOGIC LANDSCAPE CONCEPT

DOUGLAS W. GAMBLE
*University of North Carolina at Wilmington*

One of the greatest challenges facing small island states is the sustainable development of natural resources (Nurse et al. 1998). The small area of the small island states limits the resource base available to residents. Too high of a population can create pressure on a resource base, forcing the small island states to search "off island" to supplement limited resources and support an undue large population. Freshwater resources can be especially problematic given their importance to all facets of society: health, industry, tourism, and agriculture. Without sustainable development of water resources, the economic and social development of small island states can be delayed or diminished, lowering the overall quality of life on an island.

Freshwater can be found on small islands in three forms, precipitation, surface water, and groundwater. However, many islands exist in Tropical regions where evaporation exceeds total precipitation inputs. The result is few potable surface water sources. Lakes or ponds eventually become brackish or hyper-saline due to the incessant evaporation. Thus, citizens of small islands are forced to focus upon precipitation and groundwater for freshwater supply.

Knowledge of tropical island surface and groundwater hydrology has grown from traditional hydrologic theory that was developed for continental settings (Vacher 1997). The common approach to tropical island hydrology is to "downscale" traditional continental concepts and models to "fit" island-scale processes. One unexpected negative aspect of this "downscaling" is a simplification of physical processes. Particularly, it is assumed that little variability exists in physical processes, because area limits the potential range of different landscapes. The end result is models and concepts that assume homogeneity and isotropic properties for meteorological process and aquifer characteristics on small tropical islands.

For example, in regard to rainfall processes, small islands with little relief are perceived to be too small to sufficiently modify prevailing winds and lift moist maritime air to the condensation level (Granger 1985). The result is the inability of the island to "create" its own weather or create variability in

*D. G. Janelle et al. (eds.), WorldMinds: Geographical Perspectives on 100 Problems, 503–507.*

passing storms. So, it is assumed in many studies, and much government planning, that input to the tropical island hydrologic system (rainfall) is spatially consistent or homogenous across an island surface. The only exception is if significant topographic barriers exist that modify prevailing winds to produce orographic precipitation (Granger 1983).

In regard to island aquifers, the Ghyben-Herzburg Principle frames the description of groundwater resources (Vacher 1997). Island aquifers take the form of a fresh-water lens floating on the more dense salty water. According to the Ghyben-Herzburg Principle, if the lens is unconfined, the depth to which fresh water extends below sea level is approximately 40 times the height of the water table above sea level. Consequently, the deepest portion of the aquifer will be in the center of the island, while on the margins of the island, the aquifer is thin, with constant fresh-salt water mixing along the coast.

As such principles were applied to small-island states, development efforts, it was quickly discovered that hydrologic systems on tropical islands are not simple and homogeneous (Davis and Johnson 1988). In fact, most systems are just as complex as continental counterparts and require unique hydrologic models for a more accurate assessment of available water resources. Geography offers some of the greatest potential in providing such unique hydrologic models. Geography has a long tradition of incorporating spatial variability and location-specific physical processes into research and application in environmental sciences. In particular, the concept of the landscape has been particularly effective.

A landscape can be broadly defined as the appearance of an area, the assemblage of objects used to produce that appearance, and the area itself (Johnston et al. 1991). Assessment of a geographic landscape relies upon a holistic view that recognizes the unique aspects of a specific location. This general definition of a landscape has been modified in geographic analysis many times to meet specific needs, such as assessing the impact of society upon the physical environment, studying biogeography, or cultural ecology. In terms of hydrology, a landscape can be characterized by the land-surface form, hydraulic properties of their geologic framework, and their climatic setting (Winter 2001). Thus, geographers can use the hydrologic-landscape concept to characterize the physical processes and features distinct to an island landscape, allowing for characterization of complex, spatially heterogeneous hydrologic-landscapes, a task that was near impossible to complete through the use of traditional hydrologic theory.

This essay demonstrates the utility of the hydrologic landscape approach to the development of water resources on the small tropical carbonate island of San Salvador, Bahamas. In particular, the spatial variability of the climatic setting and its impact upon the hydrologic landscape of San Salvador will be discussed. This case study represents research completed by the author and collaborators to assist in the development of water resources on this small island.

## SAN SALVADOR

San Salvador is a small "family-island" of the Bahamian archipelago located approximately 600 kms east-southeast of Miami, Florida (Figure 82.1). The island is 155 km$^2$ and represents one of the small isolated carbonate platforms common to the southeastern Bahamas. The climate of San Salvador has been classified as the Koeppen Tropical Savanna (Aw) climate type (Shaklee 1996). This climate type is characterized by warm temperatures (monthly mean temperatures 22–27 °C) and distinct dry and wet seasons created by the seasonal oscillation of the Bermuda High. On San Salvador, the dry season extends from December to April, with a small wet season in May-June due to passage of frontal systems originating over continental North America. Dry conditions prevail again in July–August followed by a wet season in September–November due to hurricanes and tropical systems. Yearly rainfall ranges between 1,000 and 1,250 mm with potential evaporation between 1,250 and 1,375 mm, creating negative annual water balance for the island (Davis and Johnson 1988).

In January 2001, a network of five rain gages was established on San Salvador to determine the spatial variability of precipitation on the island. Up to this point in time, it has been assumed that precipitation totals are consistent across the island and the majority of annual rain total falls during the hurricane season. Rainfall observations for 2001 and 2002 indicate that the majority of rain fell during spring, not during the hurricane season. In particular, large frontal systems created deluges across the entire island in April

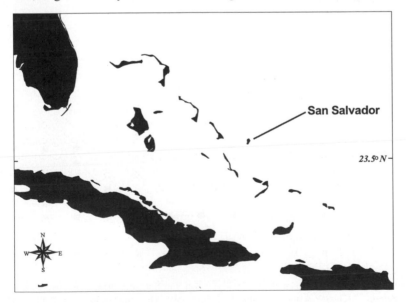

*Figure 82.1.* Location of San Salvador, Bahamas.

and May. Both of the months experienced near record rainfall. Such a year represents a climatic anomaly in regard to San Salvador's long-term climatology. However, literature regarding rainfall on the island does indicate great inter-annual variability is quite frequent, or, in other words, rain patterns are rarely the same from year to year.

To determine how much of this rainfall may reach the aquifer, daily and monthly Thornthwaite and Mather water budgets were calculated for 2001 (Thornthwaite and Mather 1957). The water budget determines the amount of runoff available for a given environment based upon daily temperature observation and estimated evaporation values. The water budget analysis indicated that on the monthly timescale, little freshwater is available for aquifer recharge due to cumulative evaporation effects. However, the daily water budget indicated that a water surplus exists on a short-term basis (days to weeks) directly after heavy rain events, allowing for aquifer recharge.

A spatial analysis of rain events indicated that only 32 percent of the rain events in 2001 covered the entire island. The majority of these events were the large frontal storms that created high precipitation amounts in the spring. The remaining 68 percent of the rain events were created by a mixture of low pressure cells, isolated convective thunderstorms, tropical waves, meso-

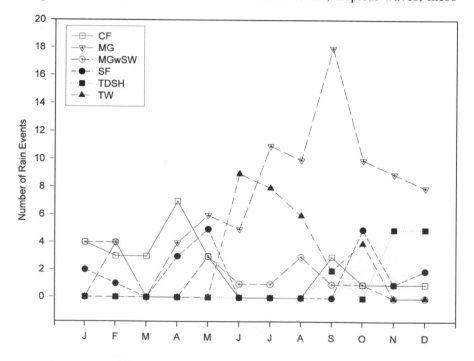

*Figure 82.2.* Monthly frequency of storm types on San Salvador 2001. CF = cold front, MS = meso-gamma, MGwSW = meso-gamma with shortwave, SF = stationary front, TDSH = tropical depression, storm or hurricane, and TW = tropical wave.

scale convective complexes, and tropical depressions or storms (Figure 82.2). This mixture of storms caused rain to occur on all scales below complete island coverage, from heavy localized rain (single rain gage) to partial coverage of the island (3–4 rain gages).

## CONCLUSIONS

This in-depth analysis of the climatic setting for San Salvador's hydrologic landscape indicates the utility of geographic analysis to water resource development. In particular, without this analysis, water resource managers may not realize that the majority of annual rainfall total is not guaranteed during the hurricane season. Also, water managers may not recognize the need to quickly harvest water surpluses after heavy rain events. Finally, water managers without an understanding of the climatic setting, may fall into the trap of assuming all rain events cover the entire island with precipitation. In fact, only about one third of storms produce rain over the entire island, and a large variety (in terms of both meteorological mechanism and spatial coverage) create rain throughout the year. With an understanding of the climatic setting, local officials can use more accurate, realistic, and successful approaches to water resource development. Such efforts will lead toward the sustainable development on small tropical islands.

## REFERENCES

Davis, R. L. and C. R. Johnson. 1988. Karst Hydrology of San Salvador. In J. E. Mylroie, ed. *Proceedings of the 4th Symposium on the Geology of the Bahamas*, 118–136. San Salvador: Bahamian Field Station Limited.

Granger, O. E. 1983. The Hydroclimatonomy of a Developing Tropical Island: A Water Resources Perspective. *Annals of the Association of American Geographers* 73(2): 183–205.

Granger, O. E. 1985. Caribbean Climates. *Progress in Physical Geography* 9(1): 16–43.

Johnston, R. J., D. Gregory, and D. M. Smith, eds. 1991. *The Dictionary of Human Geography*. Oxford, U.K.: Blackwell Publishers.

Nurse, L. A., R. F. McLean, and A. G. Suarez. 1998. Small Island States. In R. T Watson, M. C. Zinyowera, R. H. Moss, and D. J. Dokken, eds. *The Regional Impacts of Climate Change: An Assessment of Vulnerability*, 331–354. Cambridge, Britain: Cambridge University Press.

Shaklee, R. V. 1996. *Weather and Climate San Salvador Island, Bahamas*. San Salvador, Bahamas: Bahamian Field Station Limited.

Thornthwaite, C. W. and J. R. Mather. 1957. Instructions and Tables for Computing Potential Evapotranspiration and the Water Balance. *Publications in Climatology* 10(3): 185–311. Centerton, NJ: Drexel Institute of Technology, Laboratory of Climatology.

Vacher, H. L. 1997. Introduction: Varieties of Carbonate Islands and a Historical Perspective. In H. L. Vacher and T. M. Quinn, eds. *Geology and Hydrogeology of Carbonate Islands: Developments in Sedimentology* 54: 1–33. Amsterdam: Elsevier Science.

Winter, T. C. 2001. The Concept of Hydrologic Landscapes. *Journal of the American Water Resources Association* 37(2): 335–350.

# 83

# THE HYDRODYNAMIC EFFICIENCY OF NON-TRADITIONAL LEVEE PROTECTION METHODS IN THE SACRAMENTO RIVER DELTA

DOUGLAS SHERMAN and JEAN ELLIS
*Texas A&M University*

JEFFREY HART
*Habitat Assessment and Restoration Team, Inc.*

DAVID HANSEN
*University of Southern California*

Levee erosion and failure is a widespread environmental management problem. In many low-lying regions around the world, levees are the only line of protection against inundation. Erosion, as a result of natural and anthropogenic processes, reduces the structural integrity of levees and increases the threat of failure. Standard actions to mitigate levee erosion involve engineering responses, such as levee enlargement, bank armoring, or building flow-control structures. There are environments, however, where management objectives, such as recreation enhancement or habitat preservation, encourage non-traditional approaches to protection. This is the case for many of the waterways in the Delta region of California's Central Valley.

The Delta is formed by the confluence of the Sacramento and San Joaquin Rivers, approximately 80 km east of San Francisco. The region was once a marshland of about 140,000 hectares, but it now comprises some of the richest agricultural lands in the world. Approximately 500,000 residents live near or below sea level, mainly in the cities of Sacramento and Stockton (CALFED 2000). Communities and agricultural lands in the Delta are protected against flooding by more than 1,700 km of levees. This network began in the 1850s, and has required continued expansion and maintenance (Arreola 1975). Levee failure and flooding occurs frequently (more than 28 breaches between 1967 and 1992, CDWR 1995), and there are increased efforts to ensure the integrity of the Delta levees.

CALFED, established in 1994, is a cooperative venture that links 23 California and federal agencies with a mission that includes improving the health of the Delta and overseeing the integrity of the levee system (CALFED

*D. G. Janelle et al. (eds.), WorldMinds: Geographical Perspectives on 100 Problems*, 509–514.
© 2004 *Springer. Printed in the Netherlands.*

2000). The CALFED Levee Integrity Program Plan calls for the investment of $1.5 billion over 30 years to maintain the general levee system configuration (CALFED 2000). However, some aspects of levee improvement and maintenance, especially armoring banks with rock or broken concrete, reduce the quantity and quality of riparian habitat along channel margins. CALFED recognizes the potential conflict, and aims to coordinate efforts between its Levee Integrity and Ecosystem Restoration Programs (CALFED 2000). Part of the effort includes the encouragement of approaches to levee protection that minimizes or offsets deleterious impacts to habitat.

Several such alternatives have been developed by Habitat Assessment and Restoration Team (H.A.R.T.), Inc., a Delta-based company. One has been the design and installation of brush bundles along lengths of levee bank that have been scalloped by localized erosion. Brush bundles are coir-wrapped packets of orchard prunings, typically about 2 m long and a 0.5 m in diameter. Wooden pilings are used to anchor sets of bundles across the mouths of the scallops to create a relatively calm pool between the bundles and the levee bank. Sediment-laden water can flow through, or over, the bundles, especially during high tides or floods. Indeed, the bundles are designed to work best when overtopping occurs regularly, as that recharges the suspended sediments in the pool behind the structure. The intended result is the deposition of suspended sediments in the calmed waters and, ultimately, the infilling of the scallop, burial of the brush bundles, and additional bank protection. The calm water and sedimentation encourages revegetation, forming new habitat for aquatic and terrestrial organisms.

A second approach has been to plant tule stands along the banks of armored or unarmored levees, primarily to create new habitat. If the geometry and density of the new stand is appropriate, it will cause substantial flow dissipation, with concomitant sedimentation and enhanced bank protection. The stilling potential of vegetation stands is well known (e.g., Kobayashi 1992), and the principles underlying the functions of brush bundles and tule stands are understood. However, the hydrodynamic efficiencies of tule stands and brush bundles have not been demonstrated empirically, nor have they been tested from a design perspective.

Recreational boat wakes are one cause of bank erosion in the Delta (Bauer et al. 2002). We designed experiments to test the ability of brush bundles and tule stands to reduce boat wake energy. Boat wakes were chosen because we can control their characteristics by generating them in a prescribed manner with a standard boat. In August 2000, we conducted a field experiment to assess the performance of brush bundles (Ellis et al. 2002), and in January 2001, we conducted a similar evaluation of a tule stand (Hansen 2002).

## BRUSH BUNDLE EXPERIMENT

The experiments used arrays of pressure transducers inside and outside of a brush bundle installation that was about 4 m in length, and about 1 m high and wide. Sets of boat wakes were generated at different tidal stages to test for potential water-depth effects. The brush bundles were then removed to establish control conditions (with the cooperation of H.A.R.T., Inc), and the experiments repeated. Pressure transducers were sampled at 5 Hz, and the results calibrated and corrected for depth attenuation. Index wake heights and periods were calculated, and the former were used to estimate wake energy using linear wave theory. Our results indicated that the average energy reduction caused by the presence of the brush bundles was about 60 percent when compared to the control condition. Bundle efficiency was also found to be depth dependent.

Regression analysis was used to assess the degree to which energy reduction is controlled by depth dependency, where the absolute depth was normalized using the height of the brush bundle (Figure 83.1). Energy reduc-

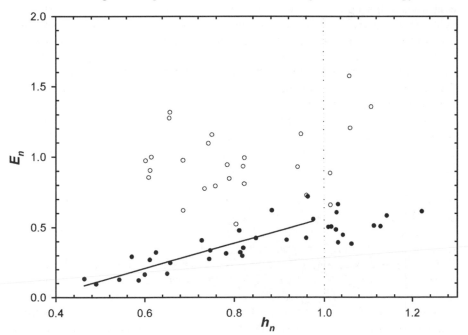

*Figure 83.1.* Depth control on energy attenuation by brush bundles. Wake energy is normalized ($E_n$) by input energy, water depths are normalized ($h_n$) by brush bundle and wake height. $E_n$ and $h_n$ are dimensionless. Open circles are measurements made without the bundles; closed circles are with the bundles in place. The dotted line at $h_n = 1.0$ represents the depth at which no water should pass over the top of the bundle. The only statistically significant relationship (indicated by the least squares line) occurs for low water conditions ($h_n < 1.0$) with the bundles in place ($R^2 = 0.73$, $n = 24$, $p < 0.001$).

tion was greatest when the height of the bundle exceeded the water level associated with the highest boat wake, and all water motion was through the bundle. However, economics and aesthetics favor a lower structure, and there is also evidence that periodic overtopping enhances sedimentation rates because it allows a more rapid exchange of suspended sediments from the river to the pool behind the bundles. Combining our results with knowledge of local tidal variability and flood stages allows economically and environmentally practical bundle geometry to be estimated. At present, more than 3,000 m of brush bundles have been installed in the Delta, with additional installations planned. In some locations, deposition behind the structures has exceeded 0.6 m in thickness after one winter season, and the revegetation of the sites has been rapid.

## TULE EXPERIMENT

The tule stand experiment was conducted using a linear array of six pressure transducers spaced at 2 m intervals through the stand. The instruments were sampled at 10 Hz, and the signals calibrated and corrected for water depth. The offshore instrument was outside of the stand, and was used to obtain

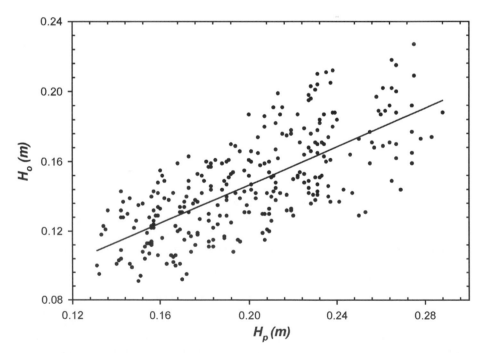

*Figure 83.2.* Comparison of measured wake heights ($H_o$) with heights predicted ($H_p$) by the independent variables in Dean's (1978) model ($R^2 = 0.51$, $n = 288$, $p < 0.001$).

the control data for the wakes that we generated. Our intention was to determine if Dean's (1978) model of wave attenuation through vegetation was applicable to boat wakes and tule stands. His model requires measurement of initial wave (wake) conditions, water depth, stem spacing (density), average stem diameter, and distance of wave propagation through the stand. To test the model, we measured wake-height changes along the instrument array. Over ten data-collection runs, individual tule plants were cut and removed to systematically reduce stalk density from a maximum of more than 80 stalks m$^{-2}$ to a state of complete clearance. The other measurements were obtained through the course of the experiment.

Linear regression was used to compare the predictions made with Dean's model with our measurements (Figure 83.2). Results indicate that the independent variables in Dean's model explain, statistically, about 50 percent of the variation in wake height through the tule stand. Further, when tule stalk densities were in excess of about 40 stalks m$^{-2}$, about 50 percent of the incident wake energy was dissipated over the 10 m length of the instrument array. The results of this experiment suggest that to accomplish significant attenuation of wakes, tule stands should be more than 10 m wide and have plant densities exceeding an average of about 40 stalks m$^{-2}$. These results also imply that tule stands may be tuned to obtain a particular degree of energy attenuation. However, the efficiency of a stand will vary through time as the plants mature and stalk diameters increase. Increased stalk diameter should increase energy attenuation directly, because of the larger surface area presented to the wakes, and indirectly, because the tule plants become less flexible to the wake motion.

## CONCLUSION

These two experiments provide valuable information concerning the hydrodynamic impacts of brush bundles and tule stands, and the results can be applied in environments where management policies require or encourage a balance between levee maintenance and habitat restoration. Understanding the hydrodynamic impacts of brush bundles and tule stands, and their effective geometries, improves our ability to efficiently design and install these systems, and to recognize appropriate sites for future restoration projects.

## REFERENCES

Arreola, D. D. 1975. The Chinese Role in the Making of the Early Cultural Landscape of the Sacramento-San Joaquin Delta. *The California Geographer* 15: 1–15.
Bauer, B. O., M. S. Lorang, and D. J. Sherman. 2002. Estimating Boat-Wake-Induced Levee Erosion Using Suspended Sediment Measurements. *Journal of Waterway, Port, Coastal and Ocean Engineering, ASCE* 128: 152–162.

CALFED. 2000. *Levee System Integrity Program Plan, Final Programmatic Environmental Impact Statement/Environmental Impact Report.* Sacramento, CA: CALFED Bay-Delta Program.

CDWR. 1995. *Sacramento-San Joaquin Delta Atlas.* Sacramento, CA: California Department of Water Resources.

Dean, R. G. 1978. Effects of Vegetation on Shoreline Erosional Processes. In P. E. Greeson and J. R. Clark, eds. *Wetland Functions and Values*, 415–426. Proceedings of the National Symposium on Wetlands, American Water Resource Association, Minneapolis, Minnesota.

Ellis, J. T., D. J. Sherman, B. O. Bauer, and J. Hart. 2002. Assessing the Impact of an Organic Restoration Structure on Boat Wake Energy. *Journal of Coastal Research SI* 36: 356–365.

Hansen, D. J. 2002. *An Evaluation of Boat-Wake Energy Attenuation by a Tule Stand on the Sacramento River.* Department of Geography, University of Southern California, unpublished M.S. thesis.

Kobayashi, N., A. W. Raichle, and T. Asano. 1992. Wave Attenuation by Vegetation. *Journal of Waterway, Port, Coastal and Ocean Engineering, ASCE* 119: 30–48.

# 84

# RECLAMATION OF SURFACE COAL MINED LANDS IN NORTHWEST COLORADO

RICHARD A. MARSTON
*Oklahoma State University*

DAVID M. FURIN
*University of Wyoming*

One of the greatest challenges, and most rewarding of professions, must certainly be restoring damaged lands to some productive status. Physical geographers are ideally positioned to contribute to the formation of reclamation plans because of the training, experience, and expertise they have developed in understanding the integration between biophysical phenomena on a variety of spatial scales. Moreover, physical geographers appreciate the synthesis between human systems and natural systems, necessary to fully comprehend the demand for resource use as well as the legal and socioeconomic forces that affect goals and procedures for reclamation of disturbed lands. This chapter evaluates the geomorphic adjustment of stream channels and hillslopes at the Colowyo Coal Mine, located just west of Colorado State Highway 13, between Craig and Meeker in northwest Colorado. The mine is operated by the Colowyo Coal Company, formerly a subsidiary of W.R. Grace and Co. and now under the ownership of Kennecott. This mine provides a unique opportunity to evaluate the success of reclamation measures that were adopted shortly after state and federal legislation had been enacted, at a time when reclamation science was relatively new (Grim and Hill 1974).

The Colowyo Mine is situated in the Danforth Hills of the Axial Basin at elevations between 2,130 and 2,320 m. Soils on the gently north-sloping mesa are clayey-loam to sandy-loam in texture, 50 to 100 cm thick, with organic matter content of 5 percent. The cool, continental, semi-arid climate supports communities of sagebrush steppe and pinyon-juniper woodland. No major aquifers are present at the site; groundwater occurs only in bedrock fractures. Fifteen separate seams of low-sulfur coal are found in the upper Cretaceous Williams Fork Formation. This presents a challenge to reclamation engineers who must store the soil and overburden for a considerable span of time before all coal is removed.

*D. G. Janelle et al. (eds.), WorldMinds: Geographical Perspectives on 100 Problems, 515–519.*
© 2004 *Springer. Printed in the Netherlands.*

## THE 1976 RECLAMATION PLAN

The 1974 Colorado Open Mine Land Reclamation Act, and the 1977 Federal Surface Mining and Reclamation Act both require that land be reshaped close to the original contour. In late 1974, W.R. Grace and Co. contracted with the Plant Sciences Department at Colorado State University to study the feasibility of using native species for revegetation. A need was soon recognized that this goal could openly be met by developing a complimentary reclamation plan involving geomorphology, soils, hydrology, climatology, wildlife, and aesthetics. The Colowyo Coal Company contracted with V.T.N. Engineers-Architects-Planners in 1975 to develop a reclamation plan that would include reclamation as an ongoing practice simultaneous with mining. In 1975–1976, the senior author of this paper served as a member of a multidisciplinary team of reclamation scientists who conducted initial baseline studies and formulated reclamation recommendations (V.T.N. Colorado 1976). After reviewing the literature and touring existing surface coal mines that represented both positive and negative examples of reclamation efforts, it was decided to pursue a goal for post-mining land use of wildlife habitat (for deer, elk, antelope, game birds) and cattle grazing. The Colowyo Mine Reclamation Plan was released in February 1976. Specific geomorphic recommendations were presented to minimize channel erosion, retain soil moisture, and prevent mass movement. Hypsometric profiles of land mass distribution were compiled, following the Strahler (1952) method to guide the redistribution of overburden. To minimize peak flows, it was recommended that the pre-mining surface drainage be restored: sub-parallel, ephemeral channels with an overall low drainage density and low bifurcation ratios. The channels were to incorporate sinuosity, boulder steps, and other roughness elements to prevent degradation. Additional recommendations were formulated to minimize land surface erosion by preventing overland flow, and where it might occur, create short slope lengths and high surface depression storage. To minimize slope failures and maximize soil moisture for revegetation, the maximum slope gradient on south and west aspects was set at 13 percent, and at 25 percent on north and east aspects. The design criteria for surface mine reclamation used in the Colowyo effort were based on general principles of physical geography, articulated by reclamation specialists in subsequent years (e.g., Lyle 1987; Toy and Hadley 1987).

## EVALUATION OF THREE DECADES OF RECLAMATION

Mining commenced in 1977 and may continue until 2040 and beyond. By the summer of 1999, 480 hectares had been reclaimed out of 5,700 hectares mined (Kiger 1999). To evaluate the success of the reclamation measures, a field survey was conducted in 1999 on three hillslopes and three drainage channels that ranged from 4 to 14 years of age since reclamation had commenced. A rainfall simulator was used to produce rainfall equivalent to the 100-year, 24-hour precipitation event at the mine site (Marston and Dolan 1999). Hillslope profiles were surveyed using a slope pantometer.

No runoff was observed in any experiment, nor could any rills, gullies, or mass movement be found throughout the reclaimed areas. The mean infiltration depth from the rainfall simulation experiments was 8 cm, with no significant difference between the three hillslopes. Pre-mining "S-shaped" slope profiles have been reconstructed, consistent with pre-mining profiles, even though overall relief has been reduced (Table 84.1). Benches and furrows were installed on the longest slope to dissipate the erosive energy of any overland flow. The A and B horizons of the soil were removed separately using a rubber-tire scrapper. The soil was stored in a location where weathering and erosion would not affect soil fertility. Once the overburden had been regraded to the approximate original topography, the soil was reapplied, shaped, disked, harrowed, and drill-seeded. The regraded soils were scarified to form 15–20 cm deep depressions. A fiber mulch was used to promote germination of seeds, prevent soil crusting, increase infiltration, and retain soil moisture. Revegetation was achieved using quick-growing and dense native grass species.

The total drainage density increased 19 percent from 1.33 to 1.58 km/km$^2$. Low stream bifurcation ratios have also been reproduced, although overall drainage density has increased. Ephemeral channels on the reclaimed property exhibit increased hydraulic roughness due to higher sinuosity, placement of boulder steps, and rip-rap placed along the length of the channels (Figure 84.1). Longitudinal profiles are more gradual than the pre-mining channels, exhibiting a smooth, concave-upward profile, joining channels at accordant junctions and avoiding knickpoints. The sinuosity of the reclaimed channels increased 11 percent over that of the pre-mining channels. Boulders have been properly placed for effective energy dissipation. Vegetation has been allowed to encroach on the channels to create roughness and to bind alluvium in the channels. Beneath the floor of channels, one meter of crushed sandstone was placed to enhance deep percolation and reduce runoff. Detention dams were constructed

*Table 84.1.* Comparison of Hillslopes and Stream Channels Pre-mining and Post-reclamation.

| Hillslope/Channel | 1 | 2 | 3 |
|---|---|---|---|
| Year Reclaimed | 1981 | 1983 | 1991 |
| Slope Aspect | NE | NW | N |
| Reclaimed Hillslope Gradient (%) | 13 | 18 | 17 |
| Pre-Mining Channel Gradient (%) | 21 | 7.6 | 16 |
| Pre-Mining Relief/Length (meters) | 30/610 | 380/5030 | 170/1070 |
| Reclaimed Channel Gradient (%) | 8.2 | 4.7 | 8.1 |
| Reclaimed Channel Relief/Length (meters) | 90/110 | 260/5490 | 80/990 |

*Figure 84.1.* Photograph of the north edge of the Colowyo property, showing reclaimed hillslope on the right and landscape undisturbed by mining on the left.

near the outlet of the mine to trap sediment, the only significant water quality parameter of concern.

## CONCLUSIONS

The reclaimed portion of the Colowyo Mine, developed over the past two decades, constitutes a stable geomorphic landscape that meets or exceeds legal requirements. All of the recommendations proposed in the 1976 reclamation plan have been followed with respect to the handling of overburden and soil, revegetation with native species, minimizing land-surface erosion, and conveying stream-channel flow with minimal erosion. Successful reclamation practices were built on lessons learned from nearby adjacent mines, especially with respect to removing topsoil and subsoil separately, and recontouring in a manner that minimizes erosion and water quality problems. Landscape complexity has been reduced, as measured by mesoscale topographic roughness and vegetation species diversity. Desired species of wildlife have returned to portions of the reclaimed lands farthest from ongoing mining. This study provided a rare opportunity to return to an area that had been mined at the same time that state and federal legislation appeared, requiring reclamation. The successful recommendations could not have been formulated without full appreciation of the ways in which phenomena of climate, water, topography, soils, and biota interact.

## REFERENCES

Grim, E. C. and R. D. Hill. 1974. *Environmental Protection in Surface Mining of Coal.* Cincinatti, OH: U.S. Environmental Protection Agency.

Kiger, J. A. Senior Environmental Engineer, Colowyo Coal Company, personal correspondence, August 1999.

Lyle, E. S., Jr. 1987. *Surface Mine Reclamation Manual.* New York: Elsevier.

Marston, R. A. and L. S. Dolan. 1999. Effectiveness of Sediment Control Structures Relative to Spatial Patterns of Upland Soil Loss in an Arid Watershed, Wyoming. *Geomorphology* 31: 313–323.

Strahler, A. N. 1952. Hypsometric (Area-Altitude) Analysis of Erosional Topography. *Bulletin of the Geological Society of America* 63: 1117–1142.

Toy, T. T. and R. F. Hadley. 1987. *Geomorphology and Reclamation of Disturbed Lands.* Orlando: Academic Press.

V. T. N. Colorado. 1976. *Preliminary Reclamation and Revegetation Design Criteria for the Colowyo Coal Mine.* Denver: V.T.N Colorado.

PART IX

# MOBILIZING GEOGRAPHIC TECHNOLOGIES

Maps, geographic information systems, remote sensing, spatial statistical analysis, and micro-simulation modeling are powerful tools for uncovering patterns of interaction and change of both human and physical phenomena. These technologies not only aid scientists and scholars, but also planners, community groups, and policy makers, shaping not only what problems we know but how we know them. The broad dissemination of these tools throughout industry and commerce and into neighborhoods is testimony on the extent to which knowledge of geographic patterns and processes aids problem identification and solutions.

*Don Janelle*

# 85

# COMMUNITY MAPPING AS A SOLUTION TO DIGITAL EQUITY

JEREMY W. CRAMPTON and DONA J. STEWART
*Georgia State University*

The *digital divide* is the gap between those who can effectively use information and communication tools, such as the Internet, and those who cannot (Benton Foundation 2003). The drive for *digital equity* is inspired by the need to close this gap and to improve access for those who are "digitally divorced" or excluded.

As Craig and Elwood state, "maps and geographic information can play an effective role in the success of a community" (1998: 95). Local communities are empowered by increasing their participation in external decision-making and by making the community more self-aware (Harris and Weiner 1998). Maps and GIS can build community networks, support collaborative planning and external coalition building, and can authoritatively present a community proposal to the outside world (Shiffer 1999).

Why is digital equity important? Today, the information economy is worth over 750 billion dollars, and the value of high-technology companies in the United States alone exceeds $3 trillion (despite recent poor performances in the stock market). But it's not just a question of dollars. Information technology is central to education (e.g., distance learning); it produces social capital (e.g., email and instant message networks), enables a truly participatory democracy (e.g., online voting), and stimulates economic development (e.g., infrastructure, fiber optics, and telecommunications). To sustain the information economy and to remain globally competitive, significant investment in the next generation of American knowledge workers is required. Will this next generation of American citizens have an equal opportunity to contribute and participate in the information economy?

Information technology use is correlated strongly with income. People in high-income neighborhoods more than in low-income neighborhoods use email and the Internet. Yes, Internet usage is increasing more rapidly for poorer people than richer people. This is good news – to an extent. It's always easier to have a high percentage increase of a small initial number. It's also more than offset by American incomes that have declined for the first time since 1991, and the gap between rich and poor is widening. In 2001, according to numbers released by the Census Bureau, the number of poor Americans rose

*D. G. Janelle et al. (eds.), WorldMinds: Geographical Perspectives on 100 Problems, 523–527.*
© 2004 *Springer. Printed in the Netherlands.*

to 32.9 million (the poverty level for a family of four was just over $18,000 in 2001).

The issue is not only that computer and Internet use varies among Americans, but that there may be deeply ingrained and worrying inequalities that spring from basic divisions in society, such as race, class, age, and gender. One of the misleading characterizations of innovations such as information technology is that it is a single innovation to which everybody gradually gains access. Instead, it's better to think of information technology occurring in "innovation waves;" which means that as soon as one wave passes another is hot on its heels. The first desktop computers appeared in the early 1980s, but we would hardly argue that they are adequate to cope with today's software needs. The famous 1984 Mac for example, ran at 8 megahertz and stored programs on 400 k floppy disks! By contrast, the latest version of a popular graphics program, Illustrator 10, is an 81-megabyte download. Computers with disk space adequate to store this program (let along run it fast enough without frustration) are not available except to the information elite.

## HOW COMMUNITY MAPPING CAN HELP

Community mapping is the idea that communities are empowered by the use of tools that help them cope with decisions they have to make about their own community. Many of these decisions are about the place in which they want to live, what kind of place it is, and what it could be. To make these decisions, communities need to know about themselves. Mapping is one very powerful way to gain this knowledge. This approach is also known as public participation GIS (PPGIS). Scholars in geography and geographic information science have been examining the role that PPGIS can play in community empowerment since the early 1990s (Craig et al. 2002). Although it is not a panacea, community mapping can enable coalition building across societal divides (e.g., between different neighborhoods) and help build local capacity for neighborhood improvement.

Mapping is also valuable because it generates new knowledge about the earth, its systems, and human activity across different scales. Community mapping is a very effective approach that stimulates access to and interest in information technology. By mapping their local area or neighborhood, local residents can engage in a practice with relevance to their lives, create valuable new information about the places in which they live, and understand how the local is integrated to the global. Using mapping technologies, residents can participate in planning and policy-making, understand sustainability and environmental impacts, and be introduced to mapping technologies across the life-span, from school to senior. The result is that communities are strengthened and enriched while residents access information technologies.

Communities and local neighborhoods are often swept up by decisions made elsewhere; their particular situation is passed over or forgotten. However, through community mapping, maps can be used as deliberate strategies of resistance or "countermapping" (Aberley 1993). Countermapping is the practice of recovering your own local knowledge to promote community interests and to resist powerful interests from outside the community.

## CABBAGETOWN, GEORGIA

How does a community take up its own "autobiography" as an active struggle over place? This is a question of "geography" in the original sense of "earth writing" (*geo-graphia*) or self-mapping. An example is offered by Cabbage-town, one of Atlanta's oldest industrial settlements, which was built for employees of the South's first cotton processing mill in 1881 during the heyday of southeastern textile mills. Today its identity and sense of place is highly contested. Originally a white working class community oriented around the mill, today it is undergoing gentrification – the mill has been converted into gated lofts, and there are few long-term residents still left. Remarkably, the neighborhood is still distinct and unusual. It's almost as if a sense of history has settled into the neighborhood's bones and imbued its new residents with a distinctive sense of place.

In Cabbagetown, the conversion of the mill into a gated loft building meant a radical disruption of both physical and psychic sense of place-identity. The mill, as old and new maps and photographs show, is still a physical presence on the landscape. The long border roads of Cabbagetown (Wylie and Carroll Streets) look down directly on the mill complex. Its tall stacks are visible even on the other side of the elevated MARTA rail tracks. But now, an ornate iron fence encloses the mill and supplements its physical (historic) wall. Many Cabbagetown residents outside the mill no longer see it as "in" Cabbagetown at all.

A group of graduate students recently performed some work in Cabbagetown to produce a community GIS, a record of life in the community (including qualitative data such as "porch furniture" and what people were doing in the community; running, walking the dog, carpentry, etc.) Although the process was highly educational for the students, it raises some interesting questions about what maps can and cannot do to empower communities.

## GIVING VOICE – BUT TO WHOM?

Though community mapping can be a powerful tool for creating agency and enabling local communities in their negotiations with larger governmental entities, the process of community mapping should not be viewed in exclusively positive terms. In creating a community map, we must question who plays a role in the mapping process and which constituencies in the community are better able to represent their interests or needs. Mapping and GIS are a "contradictory technology" that can simultaneously marginalize and empower (Harris and Weiner 1998: 68). Whose voices in the community will be represented? Income and education factors are important because the mapping process is likely to be dominated by members of the community with technical skills, or, as in Cabbagetown, outsiders. Without an intentional effort to represent all subgroups in the community, the map will likely reflect the concerns of the dominant. Unequal representation in community mapping is not only an issue of access to the mapping process itself, but also the choice of which characteristics will be mapped.

A harsh reality is that the process of map making is much better at representing differences within a community than representing the intangibles that bind a community together. Some kinds of quantitative maps (such as choropleth mapping of race and income) emphasize differences within a community. As a result, a community can appear more divided or fragmented than it may actually be. Qualitative characteristics that bind the community together either spiritually or functionally, such as shared historical experience or limited mobility, are less likely to be mapped. Community mapping must seek to express not only quantitative data, but also qualitative data, such as shared attitudes or concerns that cut across racial or income divides. More imaginative mapping techniques can also be used. Finally, the mapping process itself, especially as it relates to concretizing the community's identity, can be a contested process as divergent identities vie for representation.

## CONCLUSION

Community mapping and PPGIS can address unequal access to the knowledge economy by providing a set of tools for local neighborhoods to learn more about themselves and to make decisions that are more locally relevant. At the same time, mapping and GIS are themselves technologies of power (which is why they are useful to communities). Currently, maps and GIS strongly emphasize quantitative data over qualitative data: this should be challenged. Since there are many voices within communities, it is preferable to avoid making one "definitive" map and to enable multiple perspectives to find voice. Thus, mapping is not irrelevant to communities, but it is a geo-political process in the broadest most positive sense of a struggle over the meaning of place.

# REFERENCES

Aberley, D. ed. 1993. *Boundaries of Home. Mapping for Local Empowerment.* Gabriola Island, BC: New Society Publishers.

Benton Foundation. 2003. *Digital Divide Basics.* See http://www.digitaldividenetwork.org/

Craig, W. J. and S. A. Elwood. 1998. How and Why Community Groups Use Maps and Geographic Information. *Cartography & Geographic Information Systems* 25(2): 95–104.

Craig, W. J., T. M. Harris, and D. Weiner. 2002. *Community Participation and Geographic Information Systems.* London and NewYork: Taylor & Francis.

Harris, T. and D. Weiner. 1998. Empowerment, Marginalization, and "Community-integrated" GIS. *Cartography & Geographic Information Systems* 25(2): 67–76.

Shiffer, M. J. 1999. Planning Support Systems for Low-income Communities. In D.A. Schön, B., Sanyal, and W. J. Mitchell, eds. *High Technology and Low-income Communities: Prospects for Positive Use of Advanced Information Technology*, 191–211. Cambridge, MA: MIT Press.

# 86

# USING HISTORICAL GIS TO RESOLVE POLITICAL BOUNDARY DISPUTES ALONG RIVERS

WENDY BIGLER
*Arizona State University*

PATRICIA Q. DESCHAMPS
*JE Fuller Hydrology and Geomorphology, Inc.*

Take a look at any map and you are likely to see examples of political boundaries superimposed on river channels. This superimposition, however, leads to serious legal and political disputes because river channels are dynamic features of the landscape that can drastically change course over time. Geographers provide valuable insights into the resolution of these disputes through their understandings of cultural and physical landscapes. Geographical information systems (GIS) provide geographers the means to organize and analyze data that are fundamental to adjudicating river-defined boundaries.

Using rivers to demarcate political boundaries makes some sense – especially prior to dense colonization and modern surveying instruments. Rivers are conspicuous elements of the landscape and often the first features mapped in little-known territories. The alternative, using straight-line surveys, was often complicated by rugged terrain (Rebert 2001). The ubiquity of rivers serving as political boundaries indicates the utility of the approach, at least in the short term.

But in the long term, rivers move. Meander scars (see Gila and Colorado Rivers in Figure 86.1) provide physical evidence of past river channel locations. With each channel movement, legal questions concerning property ownership and political jurisdiction arise. These legal disputes can become contentious and can make obtaining clear title to river front property difficult.

Under United States law, the nature of the channel change determines whether the boundary moves with the river or remains in place (Simpson 1994). Accretion, barely perceptible changes through bank erosion and deposition, results in a boundary that moves with the river, potentially resulting in substantial changes in area included within a political boundary. Avulsion, rapid dramatic change in a river's course, results in the boundary remaining fixed in space. Avulsive channel change can be either natural or artificial. Flood-driven natural cutoffs of river meanders result in abandoned oxbow channels.

*D. G. Janelle et al. (eds.), WorldMinds: Geographical Perspectives on 100 Problems*, 529–534.

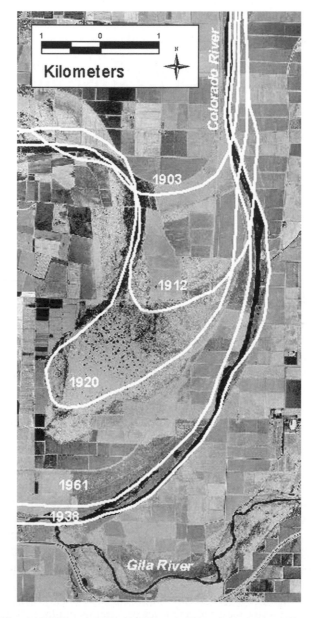

*Figure 86.1.* Historical GIS output showing Colorado River channel change for 1903–1961 along the California-Arizona border at Yuma Island.

Human-engineered cutoffs in the Palo Verde and Cibola Valleys of the Colorado River resulted in a net loss of 6,500 acres from Arizona to California (Lewis 1954).

When a boundary river is navigable, the situation is further complicated. When territories became states, navigable waterways were reserved as sovereign property. The bed of a navigable river belongs to the state. Since adjacent territories became states at different times, and channel change could occur at any time, the demarcation of river boundaries and riverbed ownership is legally complex.

Geographers, especially geomorphologists and historical geographers, are well positioned to help resolve these issues. Their broad knowledge base and specialized training, combined with the use of GIS, represent a powerful set of tools to provide the data to inform legal decisions. Historical GIS can help resolve these boundary and ownership questions by storing data, providing the means to analyze the data, and producing maps as an end product. The Colorado River presents an excellent example to illustrate the problems of a river interstate boundary and how geographers can contribute to solutions.

## THE PROBLEM

Under the Public Trust Doctrine, Equal Footing Doctrine, and Submerged Lands Act, the sovereign ownership of the beds of the Colorado River transferred from the Federal government to the State of California and to the State of Arizona upon admission to the United States in 1850 and 1912 respectively. Both states referenced the center of the river channel as their interstate boundary.

Arizona and California recognized the challenges that resulted from an interstate boundary based upon a dynamic river channel. Problems relating to conflicting land titles, rights or claims of private and public parties clouded title to properties lying adjacent to the river. Clouded titles hampered development in the area since financing depended upon the ability to obtain title insurance policies. Questions of judicial jurisdiction and taxation authority arose, along with confusion about the jurisdictions of public services, such as fire protection and law enforcement. Thus, in the 1950s and 1960s, the Colorado River Boundary Commission worked to establish a jurisdictional boundary independent of future natural and man-induced lateral movements of the river. In 1966, the Interstate Compact Defining the Boundary Between the States of Arizona and California established a jurisdictional boundary defining a fixed political boundary only. The jurisdictional boundary was comprised of 34 fixed points and 215 undocumented positions along the river. The jurisdictional boundary did not affect or modify the boundaries of upland ownership within either state. Further, the jurisdictional boundary established by the Compact did not address the following (State of California 1968):
• Settlement of land ownership title and boundary claims to sovereign lands under the river, either by agreement between the States and the upland owner, or by judicial decision in a quiet title action.

- Identification of lands owned by the State of Arizona that were physically located within the political jurisdiction of California, and vice versa.

Problems of jurisdiction arose because the location of the boundary between the states of California and Arizona was dynamic and, therefore, inexactly described as a result of continual lateral movement of the Colorado River between 1850 and 1912 and since that time. Further complications arose due to the differences in the definition of the limits of sovereign ownership in the two states. In Arizona, sovereign ownership extends to the limit of the ordinary high watermark (OHWM). However, the limit of sovereign title interest to the bed of navigable waterways in California is the ordinary low watermark (OLWM). California has certain public trust responsibilities in the land situated between the OLWM and the OHWM, but has no claim of title to these lands. This difference between the States' definitions of the limit of title sovereignty adds to the complexity of defining the channel centerline boundary.

## THE SOLUTION

In the mid-1980s, the Arizona State Land Department (ASLD) initiated a project to determine the extent of its sovereign landholdings along the Colorado River. Recognizing that Arizona holds title to land within California's jurisdiction, and vice versa, ASLD anticipates that this program will result in a series of land exchanges with California to eliminate these problematic inholdings along the border. ASLD contracted JE Fuller Hydrology and Geomorphology Incorporated to implement its Colorado River Boundary Determination program in a limited portion of the river near Yuma, Arizona (Fuller 2002).

The project consists of four phases: data collection, data management, data analysis, and boundary location recommendation. By breaking the Colorado River into reaches of similar geomorphic and hydrological characteristics, team members are able to more easily collect and organize detailed information. Data collection involves the following sources:
- Historic surveys and maps
- Aerial photographs
- Hydrologic records
- Geomorphologic records
- Engineering and other constructed works
- Legal and regulatory documents
- Land use and ownership records
- Field surveys to determine field evidence of OHWM and OLWM

GIS analysts enter these data into a GIS. In addition to the spatial attributes pertaining to these data sources, analysts record temporal attributes. This organization allows for concurrent spatial and temporal data analysis.

| | 1890s | 1900s | 1910s | 1920s | 1930s | 1940s |
|---|---|---|---|---|---|---|
| **MAPS/SURVEYS** | USRS Surveys (1874-1920) | | | | | |
| | 1895 BLM - right bank | 1903 USRS Lippincott; 1907 | 1912; 1913; 1916; 1917 | 1920 Before and after flood & avulsion; 1926 USACOE aerial photos; 1929 Warboys survey | 1938 USFS aerial | 1943 US War Dept. aerial photos; 1948 USGS aerial photos |
| **HYDROLOGY** | Peak discharge gage record (1878-1964) / Daily streamflowgage record (1904-1983) | | | | | |
| | 1894 Flow measurement begins; 1899 USGS collects first official data | 1905 Salton Sea flood | 1909 flood 150,000 cfs; 1912 flood 144,000 cfs; 1914 flood 137,000 cfs; 1916 flood 220,000 cfs; 1917 flood 142,900 cfs | 1920 flood 190,000 cfs major avulsion; 1921-1923 major floods; 1924-1925 lesser floods; 1926-1929 flood series | 1932 flood | 1948 IBWC, USACOE USBR set design flood |
| **ENGINEERING/ CONSTRUCTION** | | 1902 USRS established; 1903 USRS announces Colorado River water resources plan; 1907-08 Reservation Levee built; 1909 Laguna Dam completed | 1911 Congress approves Yuma Project; 1915 Reservation canal system completed; 1917 Arizona Levee built | 1920 Irrigation projects use all low flow of river | **1936 Hoover Dam completed**; 1938 Parker Dam completed; 1939 Imperial Dam completed | 1941 MWD diverts water; 1944 Headgate Rock Dam completed |
| **GEOMORPHOLOGY** | 1895 Channel breakout Flowed west of former channel | **1903-1917 Accretive southward progression of east peninsula loop** | | **June 8, 1920 Channel avulsion abandons oxbow channel Yuma Island formed** | 1935-1948 Channel degradation due to clear water releases from; 1926-1934 Gila R sediments deposited in Colorado R at confluence | **1929-1953 Channel movement northward at confluence** |
| **LEGAL/ REGULATORY** | 1884 Ft. Yuma Reservation | | 1912 Arizona statehood; 1914 City of Yuma incorporated | 1922-1927 Colorado R Compact; 1925 Colorado R Front Work & Levee System Act; 1929 Colorado R Compact ratified, except Arizona | | 1941 Arizona signs Colorado R Compact; 1944 US-Mexico water treaty |

*Figure 86.2.* Sample chronology showing relationships between historical events and geomorphic change near Yuma, Arizona. Arizona State Land Department, Colorado River boundary determination – Reach 1.

Chronologies of key events show how events and developments in hydrology, constructed works, geomorphology, and the legal system interacted to result in channel and landownership patterns we see today (Figure 86.2).

## CONCLUSIONS

Spatial data summaries show a synthesis of maps and coverages. As a subset of the data, composite maps show a time series of historical meander positions of the river (Figure 86.1). These maps serve to geo-rectify diverse sources of spatial data, displaying data at a uniform scale. In addition, analysts can use the spatial data in conjunction with the chronologies to suggest causal mechanisms for channel change. Because boundary recommendations depend on understanding the timing and nature of channel change, data interpretation requires a solid understanding of boundary law and geomorphological processes. These skills are rarely found in one individual, making a team approach valuable. Geographers can be excellent team members and leaders because of their broad backgrounds, specialized knowledge, and skills in spatial analysis.

Rivers are dynamic elements of the landscape. Because they are convenient boundary markers in the short term, they have commonly been used to delineate political boundaries. When river channels shift, disputes arise concerning the precise location of boundaries. Geographers make a powerful contribution to the resolution of these disputes by bringing to bear their understanding of cultural and physical landscapes and the use of GIS for data analysis.

## REFERENCES

JE Fuller Geomorphology and Hydrology, Inc. 2002. *Colorado River Boundary Determination Reach 1 Report.* Tempe, AZ: JE Fuller Geomorphology and Hydrology, Inc.
Lewis, P. M. 1954. *Report on the Colorado Reach Parker, Palo Verde, and Cibola Valleys of the Colorado River, November 1954.* Colorado River Boundary Commission State of Arizona.
Rebert, P. 2001. *La Gran Linea.* Austin: University of Texas Press.
Simpson, J. A. 1994. *Surveying Water Boundaries – A Manual.* Kingman, AZ: Plat Key Publishing Company.
State of California Colorado River Boundary Commission. 1968. *Report on Land Title Problems Along the Colorado River to the Legislature Pursuant to the Provisions of Chapter 124,* Statutes of 1967.

# 87

# TOWARD A PARTICIPATORY GEOGRAPHIC INFORMATION SCIENCE

TIMOTHY L. NYERGES
*University of Washington*

PIOTR JANKOWSKI
*San Diego State University*

GIS has long been touted as a "decision support system" (Cowen 1988). Consequently, research about participatory, spatial decision making is at the core of participatory geographic information science (PGIScience) and participatory geographic information systems (PGISystems). Two research questions underpin our overall research agenda. First, "How is GIS software with integrated decision support techniques used within group decision processes to address complex geographic problems?" Second, "How can we improve our research studies about participatory geographic decision support to contribute to geographic information science?" This overview provides a glimpse about both.

Research about participatory geographic decision support characterizes the dynamics of human-computer-human interaction, whether this concerns a single group session to address an immediate issue, e.g., allocation of funds for rural medical services, or whether concern is over multiple sessions with multiple groups working together, e.g., high capacity transit development in a metropolitan region. The research is about PGISystems design/development and evaluation on the impacts of those designs/developments on group use and society. Much research has been informal studies of application use. A gap between theory and applications creates a need to develop an understanding about how GIS software with integrated decision support techniques can be and is used within group decision processes. An effective way to close the gap is to outline a research agenda that intentionally and systematically emphasizes a different balance among the combination of theory, method, and substance research domains for different research projects (Brinberg and McGrath 1985). Intentionally choosing a different balance builds evidence in a systematic way as a basis for knowledge contributing toward a PGIScience.

*D. G. Janelle et al. (eds.), WorldMinds: Geographical Perspectives on 100 Problems, 535–539.*
© 2004 *Springer. Printed in the Netherlands.*

## A THEORETICAL FRAMEWORK FOR CHARACTERIZING
## PARTICIPATORY GEOGRAPHIC DECISION SUPPORT

Information needs and the associated decision support tool requirements can
be addressed by a good understanding of the decision situation at the time
and place (context) within which it occurs. As an example, we can make use
of the six-phase landscape modeling process elucidated by Steinitz (1990),
as used as a framework agenda in several large landscape planning projects.
That six-phase process involves: (1) database representation modeling, (2) land
development process modeling, (3) scenario evaluation modeling, (4) change
of landscape modeling, (5) impact on landscape modeling, and (6) decision
evaluation modeling. Each model result feeds to the next phase, but the entire
process is iterative to "catch" aspects overlooked. The flow of information
in participatory decision processes can be addressed by a two-level description of a process, what we call a macro-micro strategy (Jankowski and Nyerges
2001). To use a macro-micro strategy for characterizing participatory decision
situations, imagine a matrix comprised of six columns representing the
macro-phases defined by Steinitz's six-phase process, and four rows representing micro-activities: (1) gather, (2) organize, (3) select, (4) review. Those
micro-activities derive from Herbert Simon's (1979) work on management
decision-making. For any given decision task, Simon found that people perform
some amount of intelligence gathering, design some organization of a problem,
select a choice of options among designs, and review their work before
proceeding. Consequently, the six macro-phases together with four micro-
activities constitute twenty-four "phase-activity" steps in a macro-micro framework for systematic articulation of decision evaluation in landscape planning.

The significance of "phase-activity" labeling is that a *phase* speaks to the
issue of what is expected as an outcome in the overall strategy, while an *activity*
is an action (i.e., use of a GIS tool) that fosters creation of the outcome.
Thus, the beauty of the macro-micro approach is that a group could use any
GIS-supported planning process to articulate macro-phases, while asking
what information tools are needed to support the micro-activity processing.
The macro-micro strategy for analyzing decision situations is a normative
description of an expected decision process. Of course, decision processes
are not likely to proceed in a rational way, mostly because people's judgments often depart from normative rationality (Kahneman 1974). However,
if a group (or multiple groups) were dealing with rather complex geographic
decision situations, like transportation improvement or hazardous waste
cleanup, such a process could be used as a planning agenda to outline an
analytic-deliberative decision process as a recommended way to proceed
(National Research Council 1996). Whether groups follow their own
project/meeting agendas is up to them, but good plans are more often useful
than not.

Keeping these "actual" versus "normative" views in mind, it is important to understand how people undertake decision making while using geographic information technologies. That understanding, particularly if developed in a systematic way through social science research, provides an important contribution to GIScience. To provide an in-depth articulation of what can transpire during a participatory decision-making process involving the use of GIS and other decision support technologies, Nyerges and Jankowski (1997) developed Enhanced Adaptive Structuration Theory (EAST), which is now in a second version as EAST2 (Jankowski and Nyerges 2001). EAST2 is a network of eight constructs and seven premise relationships. The eight constructs of EAST2 together contain 25 *aspects* as basic elements for characterizing group decision-making. The seven premises describe the *relations* between the eight constructs (hence the aspects contained within those constructs). EAST2 as a framework is meant to be re-applied to every macro-phase of a planning agenda, e.g., to each of the six modeling phases in landscape planning elucidated by Steinitz (1990). Macro-micro EAST2 provides a way to unpack and understand decision-making complexity by articulating *each macro-phase* as eight constructs, hence 25 aspects. As EAST2 is re-applied, we can detect whether any one or more aspect(s) of the decision situation is changing from phase to phase. No wonder decision-making in many planning contexts seems to be so complex – when more aspects change as a process moves from one phase to the next phase, the greater the complexity of the decision situation. This is now testable in a systematic way. Despite the focus on theory, EAST2 has a rather practical value as well. Due to its comprehensive character, EAST2 provides the basis for developing PGISystems and for selecting tools appropriate for a given task. We can use its 25 aspects in different combinations to articulate systems design/critique for various decision situations.

## THE RESEARCH STRATEGIES FOR CONTRIBUTING TO PGISCIENCE

Having a framework that supports both research and practice brings us back to the reason for choosing a different intentional balance among theory, method, and substance domains in research (Jankowski and Nyerges 2001). Choosing which domain is emphasized in a particular study, and which ones follow that leading emphasis, establishes the research strategy and the potential contribution to knowledge. Choosing a leading domain imprints on a study as to whether it is basic (theory lead) research, methods (method lead) research, or applied (substance lead) research. For example, in one study (Jankowski and Nyerges 2001), we demonstrated an application of user needs (task) analysis guided by EAST2. A task analysis of primary health care funding allocation in Idaho proceeded by examining the convening process, decision process, and decision outcome constructs of EAST2's framework. A system was configured for and used by the Idaho Department of Health and Welfare staff and stakeholders to allocate funds for attracting doctors to rural areas across the state. This study focused on substance, method, and theory, in that order of emphasis; hence it was an applied research study.

In another study, we reported on transportation improvement decision making in the central Puget Sound region while making use of the EAST2 framework to guide a proposition analysis of that process. The propositions of EAST2 express relationships among pairs of constructs. Analyzing the relationships among constructs and construct-embedded variables of the transportation improvement process, we were able to partially reconstruct needs, process, and outcomes in a multi-participant setting that highlighted the state of PGISystems use. The state of GIS use in this decision situation indicates there is a lot of potential for decision support in the future. The emphasis here was on substance, then concept, then method, in social science, using proposition analysis to see what it could uncover about the substance of transportation improvement decision-making.

A third study explored opportunities for mixed-method approaches to research about a PGISystem used in habitat site (re)development in the Duwamish Waterway of Seattle, WA. We examined how two different social science data analysis strategies – traditional analysis techniques and lag sequential techniques – can address the same set of research questions and experimental data, but create different findings. We wanted to go beyond just a presentation of mixed-method analysis strategies, and explore how and why a mixed-method analysis could provide insight into how PGISystems are used by groups. Differences in the use of map decision aids and table decision aids can be detected by both strategies. Task complexity did not matter much in differentiating the use of maps and/or tables. In that study, the conceptual domain was emphasized, followed by the method of analysis, and then topic of site selection for habitat restoration.

Our studies over the past several years show that there is indeed considerable potential for how society might make use of geographic information technologies in participatory settings. There are many contexts for use, and that use is likely to make a difference in community and in quality of life, as suggested by a recent National Research Council (2002) report about *Community and Quality of Life*. We believe that formulating a systematic approach based on social-behavioral science research that links the development and evaluation of participatory geographic information technologies will move us closer toward a participatory geographic information science.

## REFERENCES

Brinberg, D. and J. McGrath. 1985. *Validity and the Research Process*. Thousand Oaks, CA: Sage Publishers.

Cowen, D. 1988. GIS Versus CAD Versus DBMS: What Are the Differences? *Photogrammetric Engineering and Remote Sensing* 54(11): 1551–1555.

Jankowski, P. and T. Nyerges. 2001. *Geographic Information Systems for Group Decision Making.* London: Taylor & Francis.

Kahneman, D. 1974. Cognitive Limitations and Public Decision Making. Science and Absolute Values. *Proceedings of the Third International Conference on the Unity of the Sciences*, 1261–1281. London: International Cultural Foundation.

National Research Council. 1996. *Understanding Risk: Informing Decisions in a Democratic Society.* Washington, D.C.: National Academy Press.

National Research Council. 2002. *Community and Quality of Life.* Washington, D.C.: National Academy Press.

Nyerges, T. and P. Jankowski. 1997. Enhanced Adaptive Structuration Theory: A Theory of GIS-Supported Collaborative Decision Making. *Geographical Systems* 4(3): 225–257.

Simon, H. 1979. Rational Decision Making in Business Organizations. *American Economic Review* 69: 493–513.

Steinitz, C. 1990. A Framework for Theory Applicable to the Education of Landscape Architects (and other design professionals). *Landscape Journal* 9(2): 136–143.

# 88

# EMPOWERING INDIGENOUS PEOPLES AND PROMOTING COLLABORATIVE NATURAL RESOURCE MANAGEMENT THROUGH MOBILE INTERACTIVE GIS (MIGIS)

JACK A. MCCONCHIE and JOHN M. MCKINNON
*Victoria University, New Zealand*

Deforestation caused by the felling of trees for firewood is a major cause of global environmental degradation. This is of particular concern in the headwaters of major catchments where increased runoff, erosion, and sediment yield impact on all those living down stream. While the link between forest clearance and the ability of farmers to sustain production is understood, the pressures on farmers in these marginal areas to survive from year to year are less well appreciated. The challenge, therefore, is to get these communities to recognize their situation and work out their own solution rather than suffer the imposition of well meaning, but too often inappropriate, top-down intervention.

It is widely accepted that community involvement is a critical component of effective environmental decision-making (Chambers 1997; Goodwin 1998; McKinnon et al. 2000). In fact, in many cases it is the local farmers who actually have the most profound knowledge of their environment and situation, and who know how best to manage their resources. An essential element of this decision-making process is to document and illustrate environmental change and its future impact. This information, however, must be portrayed in a manner that makes it accessible and comprehensible to all, including the preliterate members of the communities. It was the challenge of establishing better communication and dialogue with preliterate communities that drove the development of mobile interactive GIS (MIGIS) – an acronym for community based planning that integrates the techniques of PLA (participatory learning and action) and GIS (geographic information systems). MIGIS recognizes the considerable benefits of maps and graphics over textual and numerical analyses to preliterate peoples, and allows primary data obtained through PLA to be manipulated and explored within a wider context when augmented with existing secondary data. This approach brings the best of indigenous knowledge and scientific information together to provide common ground on which farmers, communities, government administrators, and

*D. G. Janelle et al. (eds.), WorldMinds: Geographical Perspectives on 100 Problems, 541–546.*
© 2004 *Springer. Printed in the Netherlands.*

planners can optimize their understanding of the environment and each other, and to work as a team to plan for a better future. This essay explores the use of MIGIS to facilitate a negotiated, bottom-up approach to addressing issues of deforestation confronting the Hani farmers of Luchun County, Yunnan, China.

## FIELD STUDIES

Following the success of a MIGIS feasibility trial in 1997 with the Karen of Ban Huai Hoi in northern Thailand, a full feasibility study was undertaken in Luchun County. Luchun is one of the fifty poorest counties in China, with almost half the population living below the poverty line. This study focused on the villages of Shapu (Xiashapu-Lower Shapu and Shangshapu-Upper Shapu), approximately 18km from the county town of Dashun. These villages, typical of those in this region, are located in an area characterized by steep slopes (mean > 35°) and rugged relief. Under natural forest cover, extreme runoff events were rare and precipitation stored on the slopes provided a reliable water supply throughout the year. Forest removal, particularly over the past 10 years, has led to a number of adverse environmental consequences, in particular increased erosion (over 30 percent of the ground is now exposed) and channel aggradation, and decreased reliability of stream flow and water supply. If these communities are to survive, they must adopt some form of sustainable agriculture and forestry and stop the present severe degradation of land resources (McConchie and McKinnon 2002).

A detailed study, combining both primary (via PLA) and secondary (via GIS) data, was therefore undertaken to quantify the resource base, estimate resource use and environmental constraints, and measure the rate of forest clearance. Village-focused primary data were collected from the communities via a range of PLA exercises. These data were encoded, manipulated, and analyzed in the field. Topographic and land use maps, at a scale of 1:25,000, were used to produce a terrain model to improve the quality of the presentation graphics and to quantify the physical constraints imposed by the landscape. The 1990 land use map was updated and, when combined with the village-based data, allowed both the quantification of current resources and changes over the previous 10 years. Having all the data in digital and graphic form allowed the rapid feedback and presentation of information to the entire village during an evening session on the day it was collected. Corrections, adjustments, and clarifications were made on-screen until the entire village was happy with the validity of the information. In this manner consensus was achieved on the accuracy and any interpretation of the data.

Although many people believe that villagers with little or no formal education cannot recognize and describe their environment on maps or aerial photographs, our experience with farmers in Thailand, Luchun, and more

*Figure 88.1.* Villagers with little or no education could readily recognize their environment on plan view, be it via maps or aerial photographs.

recently Kampong Spueu, in Cambodia, is quite the contrary (Figure 88.1). All of these groups, including men, women, and children, could identify their landscape, and locate themselves and their various resources, accurately. The resource and land use maps constructed by these preliterate farmers were field-checked and found to be accurate and reliable. The GIS and PLA were regarded as two interacting, inter-dependent tools used within an iterative process continually controlled, guided, and validated by the villagers themselves.

## USE OF SCENARIOS

Although the rapid destruction of forest cover is a major issue facing the Shapu villages, it was an awkward issue to address. Government policy was opposed to any forest clearance but a late 1980s boom in the market for lemon grass oil had resulted in a surge of deforestation via crop expansion and increased demand for timber to fuel the distilleries (Figure 88.2). By 1999 the price for the oil had plummeted but it remained the only instant cash crop available to the poor. Though farmers debated the environmental and economic cost of this excursion into a cash economy among themselves, they did not want to discuss the issue with personnel from the government's Environmental Protection Bureau. It was, therefore, decided to quantify the rate of forest clearance, and to model this into the near future to show what will happen if nothing is done to reverse the trend.

*Figure 88.2.* The introduction of lemon grass oil as a cash crop increased the rate of forest clearance dramatically.

Estimates of firewood usage and forest productivity were obtained through the PLA process with the local villagers:
- Approximately 36,700 baskets of firewood are used for domestic purposes each year.
- Approximately 50 baskets of firewood are used each year to distil the oil from 1 mu of lemon grass (a *mu* is the local unit of land measurement equivalent to 660 m$^2$).
- Based on an area of 0.95 km$^2$ under lemon grass (obtained from the land use and village resource maps) this requires 71,000 baskets of firewood a year for processing.
- The villages, therefore, burn about 108,000 baskets of firewood each year (McConchie and McKinnon 2002).
- A number of village women, responsible for gathering firewood (Figure 88.3), estimated that each mu of forest in the area yields approximately 86 baskets of firewood.

Based on this information, the current annual rate of forest destruction for firewood is approximately 1,250 mu (0.83 km$^2$). This compares to a loss of forest identified from change on the land use maps since 1990 of 0.81 km$^2$/yr. Since the two methods gave values within two percent, a high degree of confidence can be placed in the estimate of the current rate of forest clearance.

*Figure 88.3.* The gathering of fuel wood imposes a burden on both the time and labor of women, and the environment.

Using the existing forest area as the starting point, a model was developed to predict the pattern of forest clearance for the next one, two, and three years if it continues at a rate of 0.83 km²/yr. It was hypothesized that the areas most at risk were those closest to the villages and near access paths. While these assumptions may not be entirely valid, e.g., the *dragon forest* is likely to be protected; they are a reasonable first approximation. Each 10 m grid in the study area was, therefore, ranked on the basis of its distance from the villages and all access tracks. Areas most at risk were ranked with low values, while more remote areas received higher ranks. The "grids" were then removed at the appropriate rate of forest clearance in rank order. Modeling the disappearance of the forest was a shock for all. There was no way to avoid the fact that unless drastic action is taken, the last of the forest would be gone within three years (McConchie and McKinnon 2002).

## OUTCOMES

Foremost among a series of farmer-initiated action plans resulting from this exercise was a community reforestation project, and a decision to replant land cleared on dangerously steep slopes. The Shangshapu village headman and 14 other villagers formed a watershed protection committee. Land belonging to the collective considered unsuitable for cultivation was reclaimed and placed under protection. All remaining forest was declared part of a village reserve. A forest reserve agreement was negotiated, and the head of each household placed their fingerprint on the document as a commitment to abide by its principles. In addition, 10 mu was planted in Chinese fir, with the Department of Forestry providing the seedlings. The project was seen as a simple "first act" and was carried out *for this generation and the next* to mark the Half Year Ceremony. A levy of two yuan per household was collected to pay the honorarium of the person appointed as the forest guard.

Government administrators were amazed at the ability of farmers to work with scientists in the above manner and to use the resulting information to devise their own action plans. Because these officials now had a clear understanding of the farmer's needs, they went out of their way to provide assistance. MIGIS, therefore, added significantly to the standard PLA approach. It proved to be a very powerful set of tools to assist in defining, and refining development initiatives that are appropriate, are accepted and adopted by the local people, and have a high probability of success.

## ACKNOWLEDGEMENTS

The authors acknowledge the New Zealand Asia Development Assistance Facility, Ministry of Foreign Affairs for financial support to conduct the fieldwork during which the information on which this paper is based was collected. We also wish to acknowledge the contribution made to our understanding by the other members of the MIGIS team. The authors take sole responsibility for the opinions and interpretations offered.

## REFERENCES

Chambers, R. 1997. *Whose Reality Counts? Putting the First Last.* London: Intermediate Technology Publication.
Goodwin, P. 1998. 'Hired Hands' or 'Local Voice': Understandings and Experience of Local Participation in Conservation. *Transactions of the Institute of British Geographers* 23(4): 481–499.
McConchie, J. A. and J. M. McKinnon. 2002. MIGIS – Using GIS to Produce Community-Based Maps to Promote Collaborative Natural Resource Management. A Hani Case Study, Luchun County, Yunnan. *Asean biodiversity* 2(1): 27–34.
McKinnon, J. M., C. Kui, J. A. McConchie, H. C. Ma, and J. M. McKinnon. 2000. MIGIS – Mobile Interactive GIS [Online].

# 89

# TOWARDS STRUCTURED PUBLIC INVOLVEMENT: ENHANCING COMMUNITY INVOLVEMENT IN TRANSPORTATION DECISION MAKING

KEIRON BAILEY
*University of Arizona*

TED GROSSARDT
*Kentucky Transportation Center*

Although it is increasingly regarded as essential, public involvement in infra-structure decision-making has a highly problematic history. Public skepticism about the activities and motivations of planning and design professionals remains high. Arnstein's (1969) famous "Ladder of Citizen Participation" is still a useful way of characterizing levels of public involvement, ranging from the ideal of "citizen control" to creeping "manipulation" by officials and powerful interest groups. While infrastructure problems involve a range of stakeholder groups and span a variety of scales, from the neighborhood transit-station design to large-scale regional highway-corridor selection, in many cases the public involvement processes and associated problems are similar.

## PROBLEMS WITH PUBLIC INVOLVEMENT

In a typical public infrastructure process, a few design options are prepared in advance by design professionals and presented at public forums. Unstruc-tured feedback is then gathered using microphones and flip charts and is used in an unspecified way to determine which one should be selected. This limited-involvement and restricted-choice paradigm, termed Decide, Announce, and Defend (DAD), reinforces the distrust that many stakeholders hold toward public planning processes (Campbell-Jackson 2002). Loud voices, mobilized resistance groups, charges, and counter-charges typify many such public meetings. They are only half-jokingly referred to in some circles as "karaoke nights."

Besides open hostility, public distrust can manifest itself in non-participa-tion: people refuse to recognize the legitimacy of what appears to be a

*D. G. Janelle et al. (eds.), WorldMinds: Geographical Perspectives on 100 Problems*, 547–552.
© 2004 *Springer. Printed in the Netherlands.*

non-choice between equally distasteful, pre-given solutions (Lidskog and Soneryd 2000). Thoughtful citizens avoid the meetings, while those who expect dramatic impacts from the proposed options must attend. Thus, responsible authorities experience public input processes as painful exercises to be minimized rather than as opportunities to improve the design product (FHWA 1996). This self-fulfilling prophecy ensures that satisfaction with the planning process remains low on all sides. In short, discontent prevails with the type of public involvement; the tools that are used; the design process; and many other aspects (Maier 2001).

## STRUCTURED PUBLIC INVOLVEMENT: MAKING TECHNOLOGIES PART OF COMMUNITY DIALOG

The Policy and Systems Analysis Team at the University of Kentucky has been working with local governments, private organizations, and citizens' groups to improve public involvement in transportation decision-making. This research program is termed *Structured Public Involvement* or SPI. SPI treats technologies as *dialogic elements*; that is, as active elements in structuring the understanding and analysis of specific problems. SPI treats technologies of representations, such as geographic information systems and computer-generated visualizations, as shared languages and investigates how these can be used to build meaning among stakeholder groups. This requires a better appreciation of their strengths and weaknesses, and how they can be used to foster interchange in public forums.

Integrating visualization into SPI is more complex than showing computer-generated design scenarios, asking people how much they like them, and then telling planners that people prefer scenario "A" to scenario "B." People of different cultural backgrounds, or who live in different environments, or who have different occupations, do not "see" the landscape in the same way (Whitmore et al. 1994). Further, the particularities of this vision are influenced by different visualization methods. SPI calls for a fuller understanding of our tools before using them to investigate public preferences for prospective designs.

## CASEWISE VISUAL EVALUATION: A METHODOLOGY FOR INTEGRATING VISUALIZATION INTO SPI

One promising SPI methodology is Casewise Visual Evaluation (CAVE). Unlike traditional visual assessment methodologies, such as the widely used Visual Preference Survey® (or VPS; Nelessen 1994), CAVE allows preferences to be computed for individual design elements in a visualization scenario, even if these elements are not scored one by one. This is critical because peoples' tastes and preferences are non-additive: for example, consider a highway design problem. If people prefer narrow winding lanes and high-speed highways when asked separately, this does *not* mean that they consequently prefer a design featuring a dangerous combination of narrow, high-speed highways (otherwise known as the NASCAR racing circuit). Preferences respond in a complex, non-linear manner. This problem becomes particularly acute when there are many elements, all of which interact with each other. CAVE addresses this by asking people to score the complete visual scenario, comprised of all the design elements, and then by working to break down, or "decompose" the preference interrelationships.

Through a Transportation and Community Systems Preservation program grant, the team was charged to develop a way to involve the public in the evaluation of potential road improvements in a culturally and economically sensitive location in Kentucky's Bluegrass area. The tasks were to evaluate the usefulness of 2D (photos), 3D (rendering), and 4D (virtual reality) visualization modes as aids to public understanding and then to use the best mode in conjunction with CAVE to evaluate public preference for specific road design elements. An electronic scoring system allowed data to be gathered and processed rapidly, unobtrusively, and democratically from a diverse local focus group.

The research showed a public preference (in this case) for the 3D rendering mode, and that visualization complements, instead of replacing, traditional information, such as capacity, cost, or environmental impact. The 3D mode was used to present and solicit scoring feedback on a number of roadway design "samples." The groups' preference scores were processed using fuzzy-logic software to generate a preference knowledge base. Then, designers and planners could query this knowledge base to evaluate public preferences for specific design combinations.

Such an approach is possible because of CAVE's limited data requirement. Every possible combination of elements need not be scored to generate an accurate and useful understanding of preference response. Even the seemingly simple problem of roadway design can generate several hundred possible combinations: surely too many visual models to build or evaluate! This is typically an argument for professionals to "narrow" the set of possibilities. But CAVE does not require the building and scoring of hundreds of potential scenarios. Instead, as few as 12 or 15 can be scored and useful output can be generated. The fuzzy-logic based modeling system "fills in" the gaps in the knowledge base.

When embedded into an SPI protocol, this approach shows respect for the public by keeping demands on their time to a minimum. Seeing advanced methods used to translate their preferences directly and quickly into useful output gives people confidence in the public involvement process. This stimulates interest and encourages input. In the case study, the survey results showed that the focus group was very satisfied with the use of this tech-

*Figure 89.1.* Preferred composite highway design.

nology. Figure 89.1 shows a preferred composite design for this highway in Kentucky's Bluegrass Region, featuring a combination of three lanes, no fences, grass shoulders, and narrow lane width.

The team is now working with a coalition that includes Louisville's transit authority (TARC) and the Urban Design Studio (a joint Louisville Development Authority/University of Louisville/University of Kentucky College of Architecture venture) to generate and evaluate design options for a proposed transit-oriented development. Focus group meetings are underway in Louisville neighborhoods, using electronic scoring and the CAVE methodology to model community preferences.

## ANALYTIC MINIMUM IMPEDANCE SURFACE: A GEOSPATIAL SPI METHODOLOGY

The team is also implementing SPI in other contexts. For example, the Analytic Minimum Impedance Surface (AMIS) methodology is designed for participatory highway corridor planning (Grossardt et al. 2001). AMIS combines GIS and multicriteria decision methods to solicit and incorporate community values into a spatial decision support system (Jankowski and Nyerges 2001). In this case, a State highway agency required a group-input based quantitative method for comparing potential highway alignments. Stakeholder groups identified spatial decision criteria, including environmental features (such as rare species habitats), socioeconomic factors (such as poverty rates), and engineering factors

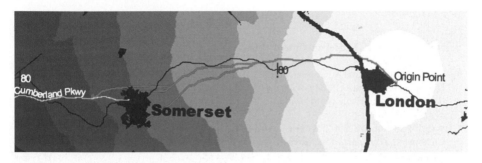

*Figure 89.2.* Sample AMIS output on a least-impedance highway alignment.

(such as slope). Stakeholders quantified their relative importance and then all of the chosen factors were integrated into a *decision surface*. This system allows the impedance of various alignments to be compared in several ways. An automatic software function can generate a "least-impedance" path between user-defined endpoints, or it can be used to "interrogate" the features of the path, showing how many meters of each feature will be affected by the chosen route. Figure 89.2 shows sample AMIS output.

## ADVANTAGES OF SPI

Ideally, by centering public involvement around the participants and fitting the technologies properly to the context and questions at hand, SPI offers several important advantages: (1) improved stakeholder buy-in, contribution, and satisfaction; (2) increased transparency; (3) clearer recommendations for planners, designers, and engineers; and (4) more efficient use of public and professional time.

In projects undertaken by the research team, SPI is seen to improve the quality of specific design decisions, but more importantly, it actively builds civic capacity (that is, willingness to participate in public decision-making; Docherty et al. 2001) by increasing confidence in the professionals and the processes undertaken (Bailey et al. 2001; Grossardt and Bailey 2002). This is particularly important since confidence in technologies and their properties cannot be separated from the mode of their deployment. After all, a "good design" is not one that maximizes use of technologies for their own sake nor is it one that is only responsive to a small cadre of professionals and their criteria; it is one in which as many participants as possible feel that they have been treated fairly and equitably by the professionals, and in which the final design is seen by all involved parties to be a direct product of such a process. Given the highly problematic history of public involvement in the transportation field, SPI is intended to move one step toward this desirable and highly challenging goal. For more information about the *Structured Public Involvement* program, see http://cvoz.uky.edu/psa.

# REFERENCES

Arnstein, S. 1969. A Ladder of Citizen Participation. *American Institute of Planners Journal* 35: 215–224.

Bailey, K., J. Brumm, and T. Grossardt. 2001, Towards Structured Public Involvement in Highway Design: A Comparative Study of Visualization Methods and Preference Modeling using CAVE (Casewise Visual Evaluation). *Journal of Geographic Information and Decision Analysis* 6(1): 1–15.

Campbell-Jackson, M. 2002. Public Involvement in Transportation: Collaborating With the Customers. *Transportation Research News* May–June 2002: 3. Washington D.C.: National Academies.

Docherty, I., R. Goodlad, and R. Paddison. 2001. Civic Culture, Community and Citizen Participation in Contrasting Neighborhoods. *Urban Studies* 38(12): 2225–2250.

Federal Highway Administration. 1996. *Public Involvement Techniques for Transportation Decisionmaking.* Washington D.C.: U.S. Department of Transportation.

Grossardt, T. and K. Bailey. 2002. *Transit-IDEA T-33 Expert Panel Review Meeting Interim Report.* Washington D.C.: Transportation Research Board, National Research Council.

Grossardt, T., K. Bailey, and J. Brumm. 2001. AMIS: Geographic Information System-Based Corridor Planning Methodology. *Transportation Research Record* 1768: 224–232.

Jankowski, P. and T. Nyerges. 2001. *Geographic Information Systems for Group Decision Making: Towards a Participatory Geographic Information Science.* London and New York: Taylor and Francis.

Lidskog, R. and L. Soneryd. 2000. Transport Infrastructure Investment and Environmental Impact Assessment in Sweden: Public Involvement or Exclusion? *Environment and Planning A* 32(8): 1465–1479.

Maier, K. 2001. Citizen Participation in Planning: Climbing a Ladder? *European Planning Studies* 9(6): 707–719.

Nelessen, A. 1994. *Visions for a New American Dream: Process, Principles and an Ordinance to Plan and Design Small Communities.* Chicago and Washington, D.C.: American Planning Association Press.

Whitmore, W., E. Cook, and F. Steiner. 1994. Public Involvement in Visual Assessment: The Verde River Corridor Study. *Landscape Journal*: 27–45.

# 90

# GIS EVALUATION FOR FIELD DISSIPATION IN THE NORTHERN WHEAT BELT

MARY BEE HALL-BROWN and ROY S. STINE
*University of North Carolina, Greensboro*

WARNER J. PHELPS
*Syngenta Crop Protection, Inc.*

Each year, tons of chemicals are sprayed on agricultural fields. It is the responsibility of the Environmental Protection Agency (EPA) to make sure that leaching, runoff, and carryover from these products do not present an environmental problem. To ensure safety, the EPA grants agricultural companies conditional registrations on specific products every year. Although registration requirements vary, one condition may demand the agricultural company to submit two or more *in-situ* dissipation studies, detailing all environmental concerns related to a specific chemical. A major problem for both the EPA and the agricultural company is how to select the location of these field tests.

To properly assess a chemical's impact on the environment, the entire region where that agricultural product is applied must be considered. For the EPA, the decision on appropriate locations can be hampered by lack of funding and human resources. For the agricultural company, field dissipation studies are time consuming and expensive; a single project can cost in excess of five hundred thousand dollars. This chapter presents a site selection process to identify a single area within the study region where the chemical has the highest probability of causing environmental damage. By conducting the field dissipation study at this worst-case scenario location, one well-sited test would serve to protect the environment more than multiple studies in random locations. A geographic information system (GIS) was developed to implement this strategy.

In this case study, Syngenta Crop Protection, Inc. is requesting a registration within the United States for an herbicide that has been used successfully in Canada. If the registration is granted, this chemical will be used on spring and durum wheat grown within Minnesota, Montana, North Dakota, and South Dakota. Using publicly available data, the spatial extent of the study area and the location where the herbicide has the greatest potential to leach into the ground water are defined. This site selection process, if adopted by the

*D. G. Janelle et al. (eds.), WorldMinds: Geographical Perspectives on 100 Problems*, 553–558.
© 2004 *Springer. Printed in the Netherlands.*

EPA, offers a better method to protect the environment and saves time and money for agricultural companies.


## METHODS

The location and extent of the study area is determined by county-level harvest and acreage data (USDA 1999). The number of acres harvested per crop in the county is averaged by the total acres within the county. Both the original harvest values and normalized harvest values are used to identify the spatial extent of the wheat, and to determine any overlap in wheat types. The most intense usage of county acreage for growing spring and/or durum wheat is located within North and South Dakota. Using the U.S. Department of Agriculture's State Soil Geographic (STATSGO) database, the characteristics of the underlying soil are analyzed.

The STATSGO database is structured in a hierarchical data schema, creating many-to-one relationships between layers of data and their corresponding soil polygons (USDA 1994). This type of relationship allows a STATSGO polygon to be characterized by particular soil chemistry, when in reality only a very small percentage of the entire polygon matches that chemistry. Previous analyses using STATSGO data had uncovered the difficulty of isolating individual polygons (Vogel 1999). Calculating weighted averages for each soil property reduces the many-to-one relationships to a one-to-one relationship, enabling a STATSGO soil polygon to be characterized by its major constituents. In addition, by giving each polygon a unique identifier, the location of each polygon could be ascertained (Allen et al. 2000). This study focuses on the uppermost soil profile.

The soil properties of texture, pH, percent clay, and percent organic matter are weighted according to the percent coverage of the specific polygon. Percent clay is identified within the STATSGO data set, and formulas taken from STATSGO are used to calculate the percent sand and silt content for each polygon (USDA 1994). The USGS Soil Triangle is used to classify soil textures, such as sandy-loam, clay-loam, or silt-loam.

Soil Texture is used as an important parameter for judging the chemical product's ability to leach into the groundwater. Soils containing greater percentages of clay or silt will help slow leaching and give more time for the chemical to degrade. Within the study area, most soils contain a relatively high percentage of silt and clay (Figure 90.1). The type of soil, either sand or silt-based, will be characteristic of the soil's absorption capabilities.

The water available within the system, through irrigation and precipitation, is taken into consideration by using irrigation data taken from the National Water Use Data Files (USGS 1995), and precipitation data taken from the Oregon Climate Service (Daly and Taylor 1998). The precipitation data is an average of 29 years and is used to offset drought or flood data of indi-

Percent Sand
☐ 0 - 18.77
☐ 18.77 - 31.2
☐ 31.2 - 43.8
■ 43.8 - 59.86
■ 59.86 - 100

Percent Silt
☐ 0 - 28.03
☐ 28.03 - 39.2
☐ 39.2 - 46.36
■ 46.36 - 54.39
■ 54.39 - 85

Percent Clay
☐ 0 - 13.45
☐ 13.45 - 21.33
☐ 21.33 - 29.16
■ 29.16 - 41.21
■ 41.21 - 67.81

Mary Hall-Brown
UNCG Syngenta Cooperative
Source: USDA STATSGO Pub. #1492, 1994

*Figure 90.1.* Soil texture.

vidual years. Total withdrawals for irrigation are converted to inches per year, per acre (Andrew Merritt, personal communication, Fall 2000). This conversion allows two things: (1) the average amount of yearly irrigation is normalized to the acreage within the county and (2) the average yearly irrigation values are added to the precipitation values, resulting in the average total water provided to the crops per year.

Areas that are considered vulnerable to the chemical product will contain relatively high sand and low organic matter content, and be more alkaline (pH value > 8) than most wheat-growing soils (Dr. Kimberly Winton, personal communication, Fall 2000). The higher levels of organic matter, from 3.0 to 18.2 percent, closely follow the spring and durum wheat growing areas. The pH levels for almost the entire four-state region are considered neutral (Figure 90.2).

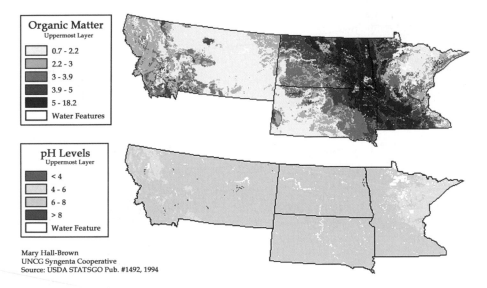

*Figure 90.2.* Organic matter and pH levels.


## LOCATION FOR WORST-CASE SCENARIO

The area with the highest potential for environmental damage within the spring and durum wheat region, as outlined above, is McHenry County, North Dakota (Figure 90.3). The soil within McHenry County has sand content between 43 and 100 percent and organic-matter content as low as 2.2 percent. This analysis could not be more specific on individual site locations within McHenry County, as the resolution of STATSGO data is 1:250,000. As the USDA Soil Survey Geographic (SSURGO) data sets become available for more areas, the selected location for the worst-case scenario will become even more precise. SSURGO data sets have mapping scales that generally range from 1:12,000 to 1:63,360, and are typically 1:20,000 or 1:24,000 (USDA 1995). If subsequent dissipation studies performed within McHenry County prove the chemical environmentally safe, application of the product throughout the four-state area in an environmentally prudent manner should be acceptable.


## CONCLUSION

Terrestrial field dissipation studies are extremely effective ways to determine how a chemical product will interact within the environment for which it is to be used. These projects "examine various routes and rates of pesticide dissipation, such as accumulation, degradation, transport by surface runoff and

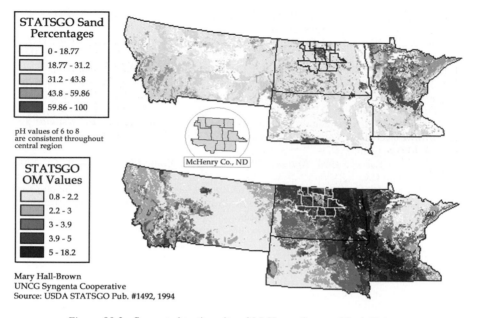

STATSGO Sand
Percentages

☐ 0 - 18.77
☐ 18.77 - 31.2
☐ 31.2 - 43.8
☐ 43.8 - 59.86
■ 59.86 - 100

pH values of 6 to 8
are consistent throughout
central region

STATSGO
OM Values

☐ 0.8 - 2.2
☐ 2.2 - 3
■ 3 - 3.9
■ 3.9 - 5
■ 5 - 18.2

Mary Hall-Brown
UNCG Syngenta Cooperative
Source: USDA STATSGO Pub. #1492, 1994

McHenry Co., ND

*Figure 90.3.* Suggested testing site of McHenry County, North Dakota.

leaching, and potential pesticide residue carryover" (Phelps et al. 2002: 8). Agricultural companies spend a tremendous amount of time and money to run field dissipation studies. The EPA must do the same to ensure that the location of the study is appropriate for the product and its crop. Due to the lack of funding and manpower for environmental departments within the U.S. government, GIS can help scientists make the decision process more reliable and less expensive. Within this study, the environmental concerns of runoff, carryover, and leaching are addressed by using GIS methods to analyze soil characteristics, climate, and agricultural data within the Northern Wheat Belt region. Locating one dissipation study based on a worst-case criterion within an area that has been selected through GIS methods would have greater potential to protect the environment than placing studies at two or more randomly selected sites.

## REFERENCES

Allen, D., M. Hall-Brown, and A. Meritt. 2000. *GIS Evaluations for Field Dissipation Work in North Dakota, South Dakota, Minnesota and Montana (The Northern Wheat Belt).* Syngenta Crop Protection, Inc.: In-house Report.

Daly, C. and G. Taylor. 1998. *United States Average Monthly or Annual Precipitation,* 1961–1990. See http://www.ocs.orst.edu/pub/maps/Precipitation/Total/U.S., 5 September 2000.

Merritt, A. 2001. Syngenta Crop Protection, Inc., personal communication.

National Climactic Data Center. 1998. *Cooperative Summary of the Day TD3200-Period of Record Through 1993.* Asheville, NC.

Phelps, W., K. Winton, and W. Effland. eds. 2002. *Pesticide Environmental Fate.* Oxford University Press.

U.S. Department of Agriculture. 1999. *1997 Census of Agriculture.* Washington, D.C.

U.S. Department of Agriculture's Natural Resources Conservation Service and National Soil Survey Center. 1995. *Soil Survey Geographic (SSURGO) Data Base.* USDA Publication No. 1527. Washington, D.C.

U.S. Department of Agriculture's Natural Resources Conservation Service and National Soil Survey Center. 1994. *State Soil Geographic (STATSGO) Data Base.* USDA Publication No. 1492. Washington, D.C.

U.S. Geologic Survey. 1995. *National Water Use Data Files.* USGS Circular No. 1200. Washington, D.C.

Vogel, J. 1999. *Utilizing GIS to Provide Site Selection of Agricultural Chemical Use on Crops Rotating With Cotton in the United States.* University of North Carolina at Greensboro, MA Applied Geography Internship Project.

Winton, K. 2000. Syngenta Crop Protection, Inc., personal communication.

# 91

# QUANTITATIVE SPATIAL ANALYSIS OF REMOTELY SENSED IMAGERY

NINA LAM
*Louisiana State University*

CHARLES EMERSON
*Western Michigan University*

DALE QUATTROCHI
*NASA Global Hydrology and Climate Center*

The rapid increase in digital remote sensing and GIS data raises a critical problem – how can such an enormous amount of data be handled and analyzed so that useful information can be derived quickly? Efficient handling and analysis of large spatial data sets is central to environmental research, particularly in global change studies that employ time series. Advances in large-scale environmental monitoring and modeling require not only high-quality data, but also reliable tools to analyze the various types of data.

## THE PROBLEM

A major difficulty facing geographers and environmental scientists in environmental assessment and monitoring is that spatial analytical tools are not easily accessible. Although many spatial techniques have been described recently in the literature, they are typically presented in an analytical form and are difficult to transform to a numerical algorithm. Moreover, these spatial techniques are not necessarily designed for remote sensing and GIS applications, and research must be conducted to examine their applicability and effectiveness in different types of environmental applications. For example, spatial techniques such as fractals, geostatistics, spatial autocorrelation, wavelets, and lacunarity analysis have been around for some time, but their uses in large-scale environmental research, such as land-cover change detection, have seldom been explored (Goodchild 1980; Mandelbrot 1983; Daubechies 1988; Lam and De Cola 1993; Cheng 1997; Atkinson and Lewis 2000). This poses a chicken-and-egg problem: we need more research to examine the usability of the newer techniques and tools; yet, this type

*D. G. Janelle et al. (eds.), WorldMinds: Geographical Perspectives on 100 Problems*, 559–564.
© 2004 *Springer. Printed in the Netherlands.*

of research is difficult to conduct if the tools to be explored are not accessible.

Another problem that is fundamental to environmental research is the scale issue (Quattrochi and Goodchild 1997). The scale issue is especially acute in the context of global change studies because of the need to integrate remote-sensing and other spatial data that are collected at different scales and resolutions. Extrapolation of results across broad spatial scales remains the most difficult problem in global environmental research (Turner et al. 1989). There is a need for basic characterization of the effects of scale on image data, and the techniques used to measure these effects must be developed and implemented to allow for a multiple-scale assessment of the data before any useful process-oriented modeling involving scale-dependent data can be conducted.

Through the support of research grants from NASA, we have developed a software module called ICAMS (Image Characterization and Modeling System) to address the need to develop innovative spatial techniques and make them available to the broader scientific communities. ICAMS provides new spatial techniques, such as fractal analysis, geostatistical functions, and multiscale analysis that are not easily available in commercial GIS/image processing software. By bundling newer spatial methods in a user-friendly software module, researchers can begin to test and experiment with the new spatial analysis methods and they can gauge scale effects using a variety of remote sensing imagery. In the following, we describe briefly the development of ICAMS and present application examples.

## THE SOLUTION

ICAMS originally runs on both the Intergraph-MGE and the Arc/Info Unix and Windows-NT platforms. However, the main disadvantage of designing ICAMS as a module on commercial software is that when the commercial software has a new version, ICAMS will need to be modified to make it compatible, and this is quite a tedious process. To overcome this disadvantage, standalone C++ based Windows and multi-platform Java applications are now being built to incorporate fractal, wavelet, spatial autocorrelation, and other GIS functions.

ICAMS has four subsystems: image input, image characterization, specialized functions, and image output (Figure 91.1). Detailed descriptions of the functions are not provided here but they can be found in Lam et al. (1998).

Image Characterization And Modeling System
(ICAMS)

| Image Input | Image Characterization | Specialized Functions | Image Display & Output |
|---|---|---|---|
| Format Transformation | Descriptive Statistics | NDVI | Two-dimensional Map |
| Geo-referencing | Histograms | Land/Water Interface | Three-dimensional Map |
| Co-registration | Fractal Analysis | Vegetated/Nonvegetated | Statistics Output |
| Noise removal/filtering | Variogram Analysis | Temperature | Digital Image Output |
| | Spatial Autocorrelation | Aggregation routines | |
| | Textural Measures | (Multiscale Analysis) | |

*Figure 91.1.* Main functions of ICAMS.

## APPLICATION EXAMPLES

Figure 91.2a shows a 1999 Landsat ETM+ image of the Muskegon, Michigan area. ICAMS was used to create a Normalized Difference Vegetation Index ((Band 4 − Band 3)/(Band 4 + Band 3)) image to highlight land/water differences and the presence or absence of green vegetation. Lighter areas in this image indicate the presence of green vegetation and black areas are bodies of water. Figure 91.2b represents the local fractal dimension of the NDVI image. The fractal dimension has a theoretical range between 2.0 and 3.0, with higher dimensions denoting higher spatial complexity of the surface. Lighter areas are therefore more "rough" or complex, and darker areas are smoother, with less local variation in brightness values. The local fractal dimension

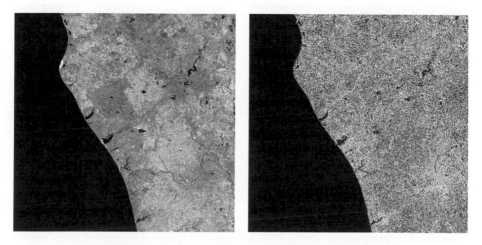

*Figure 91.2.* Landsat ETM+ image of Muskegon, Michigan.

measurement thus quantitatively measures image texture and also serves as an edge detector that highlights areas of steep brightness changes. By isolating areas of interest in these images and comparing them to other images obtained at different times, at different resolutions, or by different platforms, we are able to characterize the sensitivity of the imaged phenomenon to changes in scale and resolution, and can thus determine how much information is lost when newer, high resolution images are resampled to coarser resolutions and compared to older imagery with larger pixel sizes. Regions in an image that do not change their fractal dimension values significantly with changes in resolution are relatively scale invariant, while regions with different fractal dimension values at different resolutions depict features and surfaces that are most subject to scale effects.

In another application, Landsat-TM images acquired in 1984 and 1993 for the City of Lake Charles, Louisiana were used to demonstrate the utility of the fractal technique in characterizing landscapes and detecting changes (Figure 91.3). In addition to computing the fractal dimension, the coefficient of variation (the ratio between standard deviation and mean) of the pixel values for each band was also computed (Table 91.1). A comparison between the coefficients of variation and the fractal dimensions for the 1984 and 1993 images show a moderate correlation between these two sets of numbers. However, it is interesting to note that in the 1984 image, band 7 has the lowest fractal dimension (except band 6, which is a thermal band with large pixel size) with a value of 2.67, but it has the highest coefficient of variation with a value of 0.45. This demonstrates the utility of spatial indices: the coefficient of variation is a non-spatial index summarizing the variations of the pixel values regardless of their locations, and the fractal dimension, a spatial index, describes the spatial complexity of the pixel values. When the two

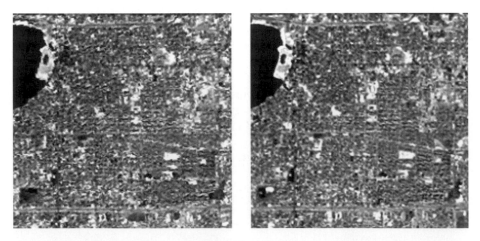

*Figure 91.3.* Comparison between the 1984 (left) and 1993 Landsat-TM images of Lake Charles, Louisiana using band 4 (infrared). Band 4 has the highest fractal dimension among all 7 bands.

*Table 91.1.* A Comparison Between the Fractal Dimension Values (*D*) and the Coefficients of Variation (*CV*) for the 1984 and 1993 Landsat™ Images of Lake Charles, Louisiana.

| Band | 1984 | | 1993 | |
|---|---|---|---|---|
| | D | CV | D | CV |
| 1 | 2.95 | 0.18 | 2.83 | 0.15 |
| 2 | 2.93 | 0.28 | 2.80 | 0.22 |
| 3 | 2.92 | 0.36 | 2.81 | 0.29 |
| 4 | 2.76 | 0.26 | 2.71 | 0.23 |
| 5 | 2.70 | 0.33 | 2.73 | 0.30 |
| 6 | 2.24 | 0.03 | 2.31 | 0.02 |
| 7 | 2.84 | 0.45 | 2.77 | 0.38 |

indices are used together, a broad but basic impression of an image can be formed, even without viewing the image. As such, these indices could be used as part of the metadata for the image. For example, when an image has a high coefficient of variation but relatively low fractal dimension, such as band 7 of the 1984 image, the surface would most likely exhibit a more spatially homogeneous pattern. On the contrary, if an image has a low coefficient of variation but high fractal dimension, the surface is much more fragmented and spatially varying. The fractal indices provide added information that is not contained in non-spatial statistics. The fractal indices, as well as other spatial indices contained in ICAMS, could serve as efficient tools for measuring and understanding the spatial dimension of the image.

## FUTURE DEVELOPMENT

To be truly useful to the broader scientific community, ICAMS must progress beyond its current pilot-study status. In our current project, also sponsored by NASA, more spatial methods, such as lacunarity measures and indices of spatial autocorrelation, will be included. Moreover, many more studies involving multiscale, multisensor, and multitemporal remote sensing imagery are needed for further testing and refinement of the methods. We argue that this new spatial approach is useful to a host of environmental applications, such as landscape measurement, spatial data mining, and land-cover change detection.

## REFERENCES

Atkinson, P. M. and P. Lewis. 2000. Geostatistical Classification for Remote Sensing: An Introduction. *Computers and Geosciences* 26: 361–371.

Cheng. Q. 1997. Multifractal Modeling and Lacunarity Analysis. *Mathematical Geology* 29(7): 919–932.

Daubechies, I. 1988. Orthonormal Bases of Compactly Supported Wavelets. *Communications on Pure Applied Mathematics* 41: 909–996.

Goodchild, M. F. 1980. Fractals and the Accuracy of Geographical Measures. *Mathematical Geology* 12: 85–98.

Lam, N. S.-N. and L. De Cola. 1993. *Fractals in Geography.* Englewood Cliffs, NJ: Prentice Hall.

Lam, N. S.-N., D. Quattrochi, H.-L. Qiu, and W. Zhao. 1998. Environmental Assessment and Monitoring With Image Characterization and Modeling System Using Multiscale Remote Sensing Data. *Applied Geographic Studies* 2(2): 77–93.

Mandelbrot, B. B. 1983. *The Fractal Geometry of Nature.* New York: W.H. Freeman and Co.

Quattrochi, D. A. and M. F. Goodchild, eds. 1997. *Scale in Remote Sensing and GIS.* Boca Raton, FL: CRC/Lewis Publishers.

Turner, M. G., V. H. Dale, and R. H. Gardner. 1989. *Predicting Across Scales: Theory Development and Testing. Landscape Ecology* 3(3/4): 245–252.

92

# GEOSPATIAL CONTRIBUTIONS TO WATERSHED-SCALE SURFACE WATER QUALITY MODELING

J. M. SHAWN HUTCHINSON and JOHN A. HARRINGTON, JR.
*Kansas State University*

LUKE J. MARZEN
*Auburn University*

First recognized in Section 208 of the 1972 Clean Water Act as a national issue, non-point source (NPS) pollution is the nation's largest water quality problem. NPS reduction is a major challenge facing society today. The U.S. Environmental Protection Agency (EPA) estimates that agriculture is by far the single largest contributor of NPS pollution (Table 92.1), responsible for the impairment of 25 percent and 19 percent, respectively, of the nation's surveyed river miles and lake acreage (EPA 2000).

## NON-POINT SOURCE POLLUTION MODELING

A common method in watershed management is use of distributed or lumped parameter models to simulate runoff, sediment, and nutrient transport. These models provide resource managers with the means to study hydrologic processes and assess current water quality conditions within a watershed. A major advantage associated with using simulation models is the ability to evaluate, via scenario-driven assessments, the effectiveness of different land management practices in meeting future water quality goals.

*Table 92.1.* Leading Sources of Impairment for Assessed Waters in the United States.

| Streams | Impoundments | Estuaries |
|---|---|---|
| Agriculture | Agriculture | Municipal Point Sources |
| Hydromodification | Hydromodification | Urban Runoff/Storm Sewers |
| Urban Runoff/Storm Sewers | Urban Runoff/Storm Sewers | Atmospheric Deposition |

Modified from Davenport and Kirschner (2002).

*D. G. Janelle et al. (eds.), WorldMinds: Geographical Perspectives on 100 Problems, 565–570.*
© 2004 *Springer. Printed in the Netherlands.*

## GEOGRAPHY AND WATER QUALITY MODELING

Geography's contributions to surface water quality modeling include (but are not limited to) the derivation, integration, visualization, and spatial modeling of hydrological, meteorological, and other environmental data. Geographic techniques, such as geographic information systems (GIS) and satellite remote sensing, have made possible the integration and modeling of spatial data at both large and small map scales, and the generation of spatially distributed model inputs, rather than relying on incomplete sets of point measurements (Mendoza et al. 2002).

Parameter estimation, using GIS and remote sensing techniques, is an important and active area of geographic research within the general field of hydrologic modeling (McDonnell 1996). However, a renewed post-modeling emphasis on the basic uses of GIS, such as mapping ("what" and "where") and querying ("how much/where") will emphasize the relationship between modeled results and possible solutions to the NPS problem (Nyerges 1991).

Maidment (1996) outlined a procedure to guide hydrologic modeling investigations using GIS. The first five steps describe the conceptual and structural framework of modeling environments, including study objectives, data requirements, "construction" of watershed units, and resolution issues (Table 92.2). Steps six through ten comprise computations and simulations describing water flow and balance within the system, human impacts on water, and generation of model results (including graphics or tables).

Step 10 ("Presentation of Results") amounts to dissemination of model outputs in forms ready for public consumption. We contend that an additional step (#11) be added where GIS is used to analyze spatial variability and spatial associations of model output(s) – a process of geographic visualization or "visual thinking." With this eleventh step, researchers examine spatial patterns associated with contaminant movement predicted by water quality models, attempting to make sense between the observed (estimated) NPS distribution (e.g., high levels of sediment transport) and causes (e.g., movement of land out of the Conservation Reserve Program).

## THE GEOGRAPHY OF WATER QUALITY

Key to watershed management is the identification of linkages between land use, land management practices, and pollutant transport. Establishment of these linkages increases prospects for achieving water quality management goals through determination of the most effective (considering both cost and result) options. Using the Agricultural Non-Point Source Pollution Model (AGNPS), Bhuyan et al. (2002) assessed runoff and sediment yields from watersheds draining into Cheney Reservoir, a primary source of drinking water for Wichita, Kansas. Key findings were a substantial increase in estimated depth of surface

*Table 92.2.* Steps for Hydrologic Modeling with GIS.

1. *Study Design:* Objectives; spatial/temporal requirements; necessary process models and inputs/outputs.
2. *Terrain Analysis:* Development of watershed boundaries, stream network.
3. *Land Surface:* Description of soils, land use/cover, permeability, chemical application, and other features or activities.
4. *Subsurface:* Description of aquifers.
5. *Hydrologic Data:* Locating/interpolating point weather data.
6. *Soil-Water Balance:* Partitioning precipitation into evapotranspiration, surface runoff, and ground/soilwater recharge.
7. *Water Flow:* Computing streamflow and groundwater flow rates.
8. *Constituent Transport:* Sediment and potential pollutant movement through watersheds and stream networks.
9. *Water Utilization:* Effect of impoundments and water withdrawals/discharge on flow and constituent transport.
10. *Presentation of Results:* Graphic and tabular dissemination of results.

Modified from Maidment (1996).

runoff (+0.3 mm) and soil erosion (+576 tonnes) between 1997 and 1998 using a one-year storm over the entire watershed. This "effect" was related back to changes in land use, specifically increased winter-wheat production in 1998.

Since Kansasís wheat fields are usually harvested in late spring, residual crop stubble greatly impacts sediment erosion associated with summer thunderstorms. Bhuyan et al. (2002) used remote sensing techniques to determine, in relative terms, the amount of biomass (crop stubble) remaining on agricultural fields after harvest – important information in model parameterization. Results suggest that fields with minimal crop stubble and near the Cheney Reservoir were major contributors of NPS pollution. It was concluded that implementation of agricultural best management practices (BMPs) on wheat lands would reduce sediment and nutrient transport and improve reservoir water quality.

Preliminary results from another AGNPS-based study, looking at sub-watersheds of Kanopolis Reservoir in central Kansas, illustrate the utility of surface water quality models in identifying locations of pollutant sources responsible for stream and lake impairment (Hutchinson et al., 2002). Spring and summer Landsat 5 satellite images were used to determine land cover following a modified USGS Level I classification scheme. Land cover in the Bluff Creek watershed (southwest Ellsworth County) is dominated by rangeland (80 percent) with much smaller fractions of agricultural land (15 percent) and forested land (5 percent) concentrated in bottomland sites (Figure 92.1a). The areal extents of the same land-cover classes in the Cow Creek watershed (northeast Ellsworth County) are very different, with agricultural land (75 percent) comprising most of the watershed area (Figure 92.1b), followed by rangeland (22 percent), and riparian forests (3 percent) (Figure 92.1b).

Comparing land cover to estimates of soil erosion for each watershed

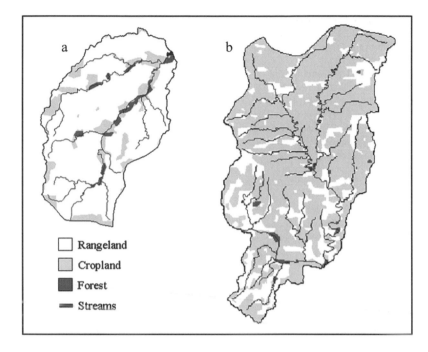

*Figure 92.1.* Land-cover classification for Bluff Creek (a) and Cow Creek (b) watersheds.

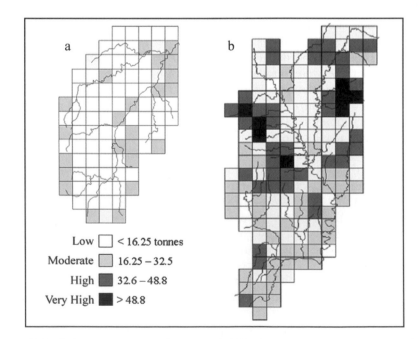

*Figure 92.2.* Estimated soil erosion from the Bluff Creek (a) and Cow Creek (b) watersheds.

permits conclusions to be drawn concerning the influence of land cover/land use on the extent and magnitude of soil erosion within the area. Based on a modeled 63.5 mm (2.5 in) rainfall event over a 24-hour period, Hutchinson et al. (2002) reported AGNPS-derived estimates of soil erosion per grid cell as "low" (< 16.25 tonnes per 64.75 hectare cell or 0.25 tonnes ha$^{-1}$) for 75 percent of Bluff Creek and 43 percent of the Cow Creek watersheds (Figure 92.2). Perhaps more indicative of land use within the watersheds, maximum estimated soil erosion within rangeland-dominated Bluff Creek was classified as "moderate," with less than 32.6 tonnes of sediment generated per cell (0.5 tonnes ha$^{-1}$). In agricultural Cow Creek, however, soil erosion was "very high" (> 48.8 tonnes per 64.75 hectare cell) for 5 percent, "high" (32.6–48.8 tonnes per cell) for 24 percent, and "moderate" for 28 percent of the watershed area. In both watersheds, distribution of moderate to very high erosion classes matched the cropland cell pattern.

## CONCLUSION

Addition of an eleventh "analysis" step to Maidment's (1996) hydrology-GIS process guidelines extends the simple communication of model results to a method for understanding the impact of natural processes and human activities on surface water quality. Practitioners can bridge the gap between academia and application by examining how complex landscapes influence the geography of watershed surface water quality. More effective use of surface water quality modeling output, supplemented with better data management, parameterization, and visualization capabilities of geospatial technologies, will be necessary to maximize impacts of scarce funding allocated NPS management programs, such as total maximum daily load (TMDL) regulations for surface waters.

## REFERENCES

Bhuyan, S. J., L. Marzen, J. K. Koelliker, J. A. Harrington, and P. L. Barnes. 2002. Assessment of Runoff and Sediment Yield Using Remote Sensing, GIS, and AGNPS. *Journal of Soil and Water Conservation* 57(5): 351–364.

Davenport, T. E., and L. Kirschner. 2002. Landscape Approach to TMDL Implementation Planning. *Proceedings of the Total Maximum Daily Load (TMDL) Environmental Regulations Conference*, 26–32. St. Joseph, MI: American Society of Agricultural Engineers (ASAE).

Hutchinson, J. M. S., J. A. Harrington, and L. J. Marzen. 2002. Geographic Applications in Nonpoint Source Pollution Assessment. *Papers and Proceedings of the Applied Geography Conference* 25: 18–26.

Maidment, D. R. 1996. GIS and Hydrologic Modeling – An Assessment of Progress. *Proceedings of the Third International Conference/Workshop on Integrating GIS and Environmental Modeling*. Santa Barbara, CA: NCGIA Publications, (http://www.ncgia.ucsb.edu/conf/SANTA_FE_CD-ROM/sf_papers/maidment_david/maidment.html).

McDonnell, R. A. 1996. Including The Spatial Dimension: Using Geographical Information Systems in Hydrology. *Progress in Physical Geography* 20(2): 159–177.

Mendoza, M., G. Bocco, and M. Bravo. 2002. Spatial Prediction in Hydrology: Status and Implications in the Estimation of Hydrological Processes for Applied Research. *Progress in Physical Geography* 26(3): 319–338.

Nyerges, T. L. 1991. Analytical Map Use. *Cartography and Geographic Information Systems* 18: 11–22.

U.S. Environmental Protection Agency. 2000. *National Water Quality Inventory: 1998 Report to Congress. U.S. Environmental Protection Agency* (EPA 841-F-00-006). Washington, D.C.: U.S. Government Printing Office.

# 93

# THE MODIFIABLE AREAL UNIT PROBLEM (MAUP)

DAVID W. S. WONG
*George Mason University*

Even though Gehlke and Biehl (1934) discovered certain aspects of the modifiable areal unit problem (MAUP), the term MAUP was not coined formally until Openshaw and Taylor (1979) evaluated systematically the variability of correlation values when different boundaries systems were used in the analysis. The problem is called "the modifiable areal unit" because the boundaries of many geographical units are often demarcated artificially, and thus can be changed. For example, administrative boundaries, political districts, and census enumeration units are all subject to be redrawn. When data are gathered according to different boundary definitions, different data sets are generated. Analyzing these data sets will likely provide inconsistent results. This is the essence of the MAUP.

"Modifiable areal units" can emerge from two spatial mechanisms in redefining boundaries. When the number of areal units is kept somewhat constant in a given region, new boundaries can be drawn to create new zoning systems or configurations. Data tabulated according to these different zoning systems will yield inconsistent analytical results. This is known as the "zoning effect," the first sub-problem of the MAUP. Another mechanism to create modifiable areal units is through spatial (dis)aggregation or by changing the spatial resolution of the data. Smaller areal units can be merged or aggregated into larger units, but fewer in number to cover the study area. Then, the spatial resolution of the data is lowered. Or, areal units can be subdivided into smaller areal units, provided that data pertaining to the larger areal units can be reasonably disaggregated. These two processes, which operate from opposite directions, create nested or hierarchical zonal systems. The problem of obtaining inconsistent analytical results from using data gathered at different spatial resolutions is known as the "scale effect," the second sub-problem of the MAUP.

In Figure 93.1, Washington, D.C. was divided into 188 census tracts for the 2000 Census. Each census tract was further subdivided into smaller block groups for census data tabulation and mapping. The District has 433 block groups in the 2000 Census. Because block groups are nested under the tract level, different results from these two levels reflect the scale effect. Aside from using census tracts, one may use postal zip code units to partition Washington,

*D. G. Janelle et al. (eds.), WorldMinds: Geographical Perspectives on 100 Problems*, 571–575.
© 2004 *Springer. Printed in the Netherlands.*

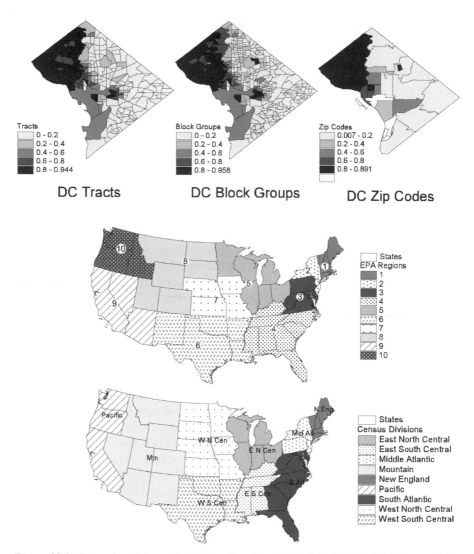

*Figure 93.1.* Examples of the scale effect and zoning effect of the MAUP: Washington, D.C., and the United States.

D.C. Figure 93.1 shows 27 zip code areas. Even though the number of zip code units and the number of census tracts are not identical, the two zoning systems yield different results that still can be regarded as the zoning effect. Figure 93.1 also shows two zonal systems for partitioning the United States. The Bureau of the Census divides the country into nine census divisions, while the U.S. Environmental Protection Agency uses ten administrative regions. Data gathered or tabulated according to these two zonal systems will provide different results, another example of the zoning effect.

*Table 93.1.* Selected Statistics for Washington, D.C.

|  | Proportion Non-white | Proportion Elderly ($\geq$ 65) |
|---|---|---|
| Entire D.C. | 0.6922 | 0.1222 |
| Tract | 0.7241 | 0.1200 |
| Block Group | 0.7181 | 0.1309 |
| Zip Code | 0.5613 | 0.1151 |

Table 93.1 reports on two sets of averages derived from several census variables of Washington, D.C. Proportions of non-white and proportions of elderly were computed for all census units at the census-tract and block-group levels, and averages were taken at the two levels. Results from the two census levels are not identical, even though they are not dramatically different. The same set of statistics was also derived for zip code areas, and these statistics are different from the results derived at the two census levels. The averages based on Washington, D.C. as a whole are also reported as references. The proportions of white, which are the reciprocal of the proportions of non-white, are also mapped at the two census levels in Figure 93.1 to illustrate the MAUP.

## WHY DOES THE MAUP ARISE?

In the spatial aggregation process, smaller areal units within a neighborhood are merged to form larger units. If all merged units have identical values, then the aggregated unit and the disaggregated units will have the same value, and, thus, there is no MAUP effect based on scale differences. Values at the aggregated level can preserve values at the disaggregated level. Although the "First Law" of Geography tells us that closer things are more similar, we can hardly find uniform geographical surfaces in the real world. Then, when slightly different neighboring units are aggregated to form larger units, the original values are averaged or smoothed at the aggregated level. For instance, standard deviations indicate the level of variation for the proportions of non-white in Washington, D.C. at the census-tract and block-group levels of 0.3270 and 0.3436, respectively. Thus the census tract data were smoothed and lost variation captured by the block-group data.

Although changing levels of variation pertain to the scale effect, the general concept of averaging or smoothing neighboring values also helps to explain the zoning effect. Nevertheless, we cannot predict the direction of changes in the variables between different spatial partitioning schemes.

## WHAT ARE THE SIGNIFICANCES OF THE PROBLEM?

The impacts of the MAUP effect are pervasive among various analytical techniques, including simple descriptive statistics (those used in the previous section), standard statistical procedures (such as various types of regression analysis), and spatial models (such as gravity-type spatial interaction models). Numerous studies have documented the impacts of the MAUP on other statistical and spatial techniques (multiple and logistic regressions, location-allocation models, and input-output models; for a review, see Fotheringham and Wong 1991). In addition, the MAUP is a concern in processing and analyzing remotely sensed data (Quattrochi and Goodchild 1997) and in handling spatial data in GIS (Tate and Aktinson 2001). In general, the MAUP is of significance in three major areas.

First, when smaller areal units are merged to form larger units, variable values are averaged. The correlations among variables for the aggregated units will likely be higher than that for the disaggregated level (Fotheringham and Wong 1991). For instance, the correlation coefficient, which has a theoretical range of $-1$ and $+1$, is 0.3247 at the census-tract level and 0.2800 at the block-group level when the number of white population counts and counts of elderly for Washington, D.C. were evaluated. But the most significant implication of the inconsistent correlation coefficient across scales is that correlations among variables are the bases of almost all statistical analyses involving more than one variable. Therefore, inconsistent correlations across scales imply that statistical results will also vary across scales.

Second, for any given study area, data with multiple resolutions or gathered according to different partitioning systems are likely available. Because of the presence of the MAUP, using different data sets for the same analysis will offer different results. Then, how should one decide which dataset to use? A related issue is to what extent the results are dependent upon the chosen dataset.

Third, at the conceptual level, most spatial data are aggregates of individuals, which can be persons or locations. Quite often, the goal of analysis is to identify patterns or systematic processes pertaining to individuals based on information derived from aggregated or ecological data. However, due to the MAUP, aggregated data for different scales or zonal systems cannot provide a consistent picture on the individual situation. It is also argued that it can be erroneous to infer individual situations based upon aggregated or ecological data. This is known as the ecological fallacy, and the MAUP is one of the sources of this fallacy.

## HOW TO HANDLE THE PROBLEM?

Some scholars argue that using data representing different spatial scales should yield different results because they reflect the different processes operating at different geographical levels. In other words, they do not recognize the MAUP as a geographical problem. On the other hand, some scholars are working diligently to identify solutions to the MAUP (e.g., King 1997). But, currently, no general solutions exist. Some scholars suggest that the zoning problem is simpler because it can be treated as a data interpolation or transformation problem (Fisher and Langford 1995). For scale effect, one approach is to develop relatively scale-insensitive analytical techniques. This approach has had limited success so far and the solutions are subject-dependent (e.g., Tobler 1989; Wong 2001).

Another approach to the MAUP is to acknowledge that there can be multiple results when different data sets of the same study area are used (Fotheringham 1989). This approach recognizes that a result based on one data set is only one of many possible results, and thus the range of possible results should be reported whenever possible. To implement this approach, multiple data sets are used and the same analysis is performed on each data set such that a range of outcomes can be reported. This approach is especially feasible when data for multiple scales and for different zoning systems are stored in a geographic information system (GIS), and the same analysis can be repeated for all data sets.

## REFERENCES

Fisher, P. F. and M. Langford, 1995. Modelling the Errors in Areal Interpolation Between Zonal Systems by Monte Carlo Simulation. *Environment and Planning A* 27 (2): 211–224.

Fotheringham, A. S. 1989. Scale-independent Spatial Analysis. In M. F. *Goodchild and S. Gopal, eds. Accuracy of Spatial Databases*, 221–228. London: Taylor and Francis.

Fotheringham, A. S. and D. W. S. Wong, 1991. The Modifiable Areal Unit Problem in Multivariate Statistical Analysis. *Environment and Planning A* 23: 1025–1044.

Gehlke, C. E., and K. Biehl, 1934. Certain Effects of Grouping Upon the Size of the Correlation Coefficient in Census Tract Material. *Journal of the American Statistical Association Supplement* 29: 169–170.

King, G. 1997. *A Solution to the Ecological Inference Problem*. Princeton, NJ: Princeton University Press.

Openshaw, S. and P. J. Taylor, 1979. A Million or so Correlation Coefficients: Three Experiments on the Modifiable Areal Unit Problem. In N. Wrigley, ed. *Statistical Applications in the Spatial Sciences*, 127–144. London: Pion.

Quattrochi, D. A. and M. F. Goodchild. eds. 1997. *Scale in Remote Sensing and GIS*. New York: CRC Press.

Tate, N. J. and P. M. Atkinson, eds. 2001. *Modelling Scale in Geographical Information Science*. West Sussex, England: Wiley.

Tobler, W. 1989. Frame Independent Spatial Analysis. In M. F. Goodchild and S. Gopal, eds. *Accuracy of Spatial Databases*, 115–122. London: Taylor and Francis.

Wong, D. W. S. 2001. Location-Specific Cumulative Distribution Function (LSCDF): An Alternative to Spatial Correlation Analysis. *Geographical Analysis* 33(1): 76–93.

## PART X

# (RE)ASSESSING CULTURE AND IDENTITY

Geography is more than analyses of tangible landscapes and the process that generate them; it is also entwined with the mental constructs and belief systems that structure the individual and collective identities of people and places. Culture and identity are fundamental to the ways in which individuals produce their biographies and societies reproduce themselves in time and space, to the manner in which differences are constructed, interpreted, and negotiated. Geographers must probe the ideologies that guide people and institutions if they are to succeed in helping to reinvent humankind as the effective steward for minding the world and its resources.

*Don Janelle and Barney Warf*

# 94

# PLACING CHILDREN AT THE HEART OF GLOBALIZATION

STUART C. AITKEN

*San Diego State University*

In the 1970s, Bill Bunge used the metaphor of a "canary in a coal-mine" to highlight a particularly important aspect of a child's place in society (Bunge and Bordessa 1975). Bunge was one of the first scholars to indicate a need to study children's geographies not just for their own sake, but also as a harbinger of larger social ills. Early work in the discipline focused on children's mapping and wayfinding abilities, and how they developed into spatially cognizant beings, but Bunge introduced the import of children's local geographies. The local geographies of children – their political and economic contexts – speak to larger societal problems, and I argue that those contexts are often hidden in adult moral panics over teenage pregnancies, schoolyard shootings, and the like. Thirty years after Bunge's prophetic words, the plight of young people is of increasing concern and geographers are at the forefront of elaborating that concern and suggesting some of its root causes. Contemporary geographic work places children closer to the center of our understanding of consumption, production, and reproduction, and at the heart of the inequities generated by globalization.

The understanding of children and their place in society has changed. A central thesis of early work in geography – including Bunge's – posits children as innocents, as passive victims of larger political, social, and economic forces. Most contemporary geographical studies do not portray children as passive; rather they articulate the ways children form critical and reflexive engagements with their worlds. My intent in this essay is not to synthesize recent geographic work on young people but, rather, to situate some of this work around a set of problems that encompass critical constructions of children's active engagement with the world through their identities (their bodies, ethnicities, sexualities, and politics) and through material social transformations (local/global). Some of these problems find accommodation in geographers' reworking of notions of reproduction and spatial justice. I weave these notions through an elaboration of problems with how young people are *placed* that relate to the *scale* at which they purportedly operate and the ways in which their identities are *fixed*.

*D. G. Janelle et al. (eds.), WorldMinds: Geographical Perspectives on 100 Problems*, 579–583.
© 2004 *Springer. Printed in the Netherlands.*

## PLACING CHILDREN

In poor neighborhoods in the global North, urban disinvestment, social disenfranchisement, the deskilling of children, and their policing on public streets
contribute to problems with the ways children are placed. Low-income children
in the United States are poorer than their counterparts in most other countries, and high-income children are seemingly better off. Of all poor children
in the United States, 47.6 percent are chronically poor and only 18 percent
are successful in exiting poverty (Mattson and Sanders 1998: 63–64). In more
affluent neighborhoods, children are missing from public streets, their lives
circumscribed by highly structured activities in safe havens (child care centers,
day camps, mini-theme parks) that also survey and discipline their bodies,
and behaviors. Put simply, it is another way of fixing children's laughter,
their bodies and their imaginations into hyper-commercial spaces. That all
this playful conviviality is based on carefully cultivated market principles is
more than ironic at a time when, in poorer neighborhoods, minority youth
are actively prevented from congregating. In the global South, economic
restructuring rather than disinvestment forces many children out of educational
institutions and onto streets, into bondage (e.g., the global sex industry), into
work (e.g., sweatshops), or into the practice of healthcare for ailing relatives
(Robson 1996).

These problems are not new, but I am writing in an era of profound change
in how we come to know them. Numerous geographical studies argue that
childhood is not only constructed in different ways at different times but also
varies depending upon where it is constructed (Holloway and Valentine 2000).
Interpretative lenses that focus simultaneously on local and global representations and lived experiences enable a more comprehensive analysis.
Traditional perspectives of childhood as a separate developmental and social
class are radically undermined, leaving views that valorize difference, producing work that is about the practices of young people (Hyams 2000), their
communities (Bauder 2001), and the institutions that shape (and are shaped
by) their lives (Fielding 2000).

## SCALING CHILDREN

A recent focus on the scaling as well as the placing of children highlights social
reproduction as an important, but as yet missing, aspect in globalization debates
(Katz 2001). Reproduction is not just about biological reproduction; it is also
about the daily health and welfare of people. And social reproduction is in
large part about the differentiation and skill of a labor force. Children's geographies are integrally linked to social reproduction but the places and practices
of children's everyday lives are rarely considered a dynamic context for
understanding historical change, geographic variation, and social differentia-

tion. While geographers have focused on important issues like child-care, child-rearing, and home rules (Wood and Beck 1994; Holloway 1998), these discussions speak only of the ways rules, ideals, and practices are transmitted to children; they say very little about the ways they are received, internalized, resisted, and mobilized. With a focus on the latter, young people provide a very different, and in many ways more illuminating, window upon social reproduction. It is during childhood and adolescence that the principles of society are mapped onto the consciousness and unconsciousness. This is also when some portion of social reproduction is contested and negotiated. Caught up in a web of flexible accumulation, with associated widespread economic uncertainty, unemployment, and decreasing public services, the lives of children are drastically affected. But unless it is through tokenism, children's voices are rarely included in the common political culture that defines public discourse.

## FIXING CHILDREN

Geographers argue that young people are *fixed* because their actions are often viewed as disorderly, chaotic, and needing discipline. This moral assault on children stems in part, first, from the perception that they need disciplining and, second, from the increasingly disembedded and disembodied contexts within which they find themselves. It is here that violence occurs, a wrenching assault that is in part quixotic but also terrifyingly real when translated into the body-mangling bullets that find their way onto schoolyards (Aitken and Marchant 2003). The morality, of course, is part of the violence because it appropriates and encompasses the notion of safe places (e.g., homes and school-yards) where children can be raised apart from seemingly immoral urban landscapes. The fact that schoolyard shootings are rare does not detract from fierce and unjust moral attacks on some young people and their families and communities. What is required is a new interdependent and connected form of justice and this kind of justice turns upon looking at identity in different ways.

There is an insidious irony in how we deal with children's rights to justice. The "canary in a coalmine" metaphor serves well the political claims of a broad constituency from the left to the right, but it usually revolves around the notion that childhood should be set aside as a time for play and education. The suggestion is that children are a separate culture, which nonetheless requires constant adult surveillance and tutelage. Geographers writing today refocus this problem by arguing that what is missing is a consideration of children as playful, active, interpretative subjects, which at the same time understands that they are not autonomous subjects who are always dealt with justly.

Keeping in mind that there is no universal form of play any more than there is a single monolithic children's culture, it is worthwhile considering a space of justice that elaborates play and takes it seriously. To suggest that children's play is simply acting out and learning aspects of adult life is, to me, a deeply disturbing notion that smacks of social engineering. It is the kind of notion that propelled nineteenth-century thinking behind the American Playground Association's creation of public parks where poor immigrant boys could march, drill, and play team games (in preparation for war), and girls met to sew and dance (in preparation for domestic and erotic support) (Gagen 2000). The same argument may be made for U.S. concerns about child obesity and taking DNA samples as a child safety precaution that are, at the moment, opening new niches for slick marketers.

Play is the active exploration of individual and social imaginaries, built up in the spaces of everyday life and unfettered from adult tutelage. The problem is to provide appropriate places for play. By "appropriate" I mean the protection of children's rights and bodies through places (and legal structures) that are robust enough to accommodate children's creativity, but flexible enough to guard against play's objectification and commodification. Put simply and geographically, a child's play, knowledge, maturation, and morality originates and develops through their active participation in the world around them, in all of its contingencies.

The separation of children from adults, the taking on of responsibilities and the enactment of rites of passage – the seeming nature of childhood – and all the problems associated with that separation are simultaneously tied to local contexts and larger global transformations. Geographers working at the forefront of these issues challenge myths that evoke childhood as a natural stage in the development towards adulthood. Rather, they provide less structured perspectives that value the embodiment and embeddedness of children, and their propensity to resist discipline and enervating notions of justice.

## REFERENCES

Aitken, S. and R. Marchant. 2003. Memories and Miscreants: Teenage Tales of Terror. *Children's Geographies.* In Press.

Bauder, H. 2001. Agency, Place, Scale: Representations of Inner-City Youth Identities. *Tijdschrift voor Economische en Sociale Geografie* 92: 279–290.

Bunge, W. and R. Bordessa. 1975. *The Canadian Alternative: Survival, Expeditions and Urban Change.* Geographical Monographs, No. 2. Toronto: York University.

Fielding, S. 2000. Walking on the Left!: Children's Geographies and the Primary School. In S. Holloway and G. Valentine, eds. *Children's Geographies: Playing, Living and Learning*, 230–244. London and New York: Routledge.

Gagen, E. 2000. An Example to Us All: Child Development and Identity Construction in Early 20th Century Playgrounds. *Environment and Planning A* 32: 599–616.

Holloway, S. 1998. Local Childcare Cultures: Moral Geographies of Mothering and the Social Organization of Pre-school Education. *Gender, Place and Culture* 5: 29–53.

Holloway, S. and G. Valentine. 2000. Children's Geographies and the New Social Studies of Childhood. In S. Holloway and G. Valentine, eds. *Children's Geographies: Playing, Living, Learning*, 1–28. London and New York: Routledge.

Hyams, M. 2000. Pay Attention in Class . . . [and] Don't Get Pregnant: A Discourse of Academic Success Among Adolescent Latinas. *Environment and Planning A* 32: 617–635.

Katz, C. 2001. Vagabond Capitalism and the Necessity of Social Reproduction. *Antipode* 33: 709–728.

Mattson, M. and R. Sanders. 1998. *Growing Up in America: An Atlas of Youth in the USA.* New York: Simon and Schuster Macmillan.

Robson, E. 1996. Working Girls and Boys: Children's Contributions to Household Survival in West Africa. *Geography* 81: 403–407.

Wood, D. and R. Beck. 1994. *The Home Rules.* Baltimore and London: Johns Hopkins University Press.

95

# HOW DO CHILDREN USE REPRESENTATIONS
# OF SPACE?

SCOTT BELL
*University of Saskatchewan*

Understanding how children use representations of space and how the ability to use representations of space develops are fundamental problems for which geographers are ideally suited. Representations of space include maps, diagrams, verbal and written descriptions, and virtual environments, among many others. Representations of space are unique in the way they deliver spatial information, and can take many forms; some may use space as a device for communication (in the case of maps, diagrams, and virtual displays), while others use less spatial devices, such as text, symbols, and verbal communication. All representations of space share in common the objective of communicating information concerning the spatial relationships between and among phenomena in the real world (MacEachren 1995). There is a great deal for active research in geography concerned with representations of space (visualization, human-computer interfaces, cognitive models of space, spatio-linguistic devices, etc.) (Landau 1996; Mark et al. 1999). One of the most fundamental areas of this research examines how humans first develop an understanding of the relationship between a symbolic representation of space (map-like representations) and the representation's reality-based referents. Interestingly, much of this research is being done by non-geographers, and as a result lacks some of the understanding of how important maps and other spatial representations are in how we look at the world around us (DeLoache 1989; Uttal et al. 1998; Uttal and Wellman 1989). At the same time, it is important to acknowledge the benefits of an interdisciplinary approach to such an important problem.

Children's use and comprehension of spatial representations has been explored by numerous researchers representing diverse disciplines (DeLoache 1989; Liben and Downs 1989; Plester et al. 2002; Stea et al. 1996). This research has examined a variety of representation types, including air photos (Plester et al. 2002), maps (Liben and Downs 1989), and other models of space (DeLoache 1989). Despite the breadth of research on the topic, there is still a great deal to learn with respect to a child's ability to use representations of space and how they are used to solve spatial problems. These questions include (but are not limited to), at what age can children start using maps

585

*D. G. Janelle et al. (eds.), WorldMinds: Geographical Perspectives on 100 Problems, 585–588.*
© 2004 *Springer. Printed in the Netherlands.*

and other representations of space? How does the ability to use representations of space develop? What tasks can be performed using representations and at what age? What characteristics of the representation influence its utility in different situations? Each of these questions has implications far beyond that of a child's ability to use a map; research can help elucidate the nature of how maps and other representations of space are used to solve problems at all ages, and how the nature of maps can influence spatial problem solving.

Representations of space are used by geographers and non-geographers in many different ways; they may be used to catalog and store spatial information, indicate locations, express spatial relationships, identify features and processes, and present wayfinding information. The development of our ability to use representations of space is a complex multivariate problem. Building on basic research involving object search, developmental psychologists and geographers have contributed to our understanding of how young children first use models and representations. Children as young as 33 months of age can use scale models to successfully find a hidden toy (DeLoache 1989). Further research by DeLoache (1991) and by others has identified additional stages in children's development of the ability to understand the complicated relationships that underlie representations of space.

Current results seem to indicate that the earliest stages of developing an understanding of spatial representations involve one-to-one mappings of individual locations with individual and unique symbols in three-dimensional models of space. These models have only a limited likeness to complex reference maps, such as United States Geological Survey and Natural Resources Canada topographical maps, with which most geographers are familiar. In fact, evidence suggests that it is not until about 4 years of age that children start to have the capacity to use more map-like (two-dimensional) representations. What makes this problem so complicated are the many variables that can be manipulated when creating representations of space. Symbols, scale, location, orientation, text, colors, perspective, among many other cartographic variables, are manipulated to create maps and other representations.

A critical component of cartographic representation of space is the use of symbology and systems of symbols for representing different types of features and phenomena across space. While the ability to find a single location as represented on a map or in a model of a larger space is also a component of these same cartographic representations, it is not necessarily based on the structure of a map. While the ability to find a location represented by a unique (single) symbol or icon is related to the map as a representation of space, it does not tackle the issue of map structure and the fact that maps are made up of a system of symbols, many of them repeated across the map to show the locations of many occurrences of the same type (Liben and Downs 1989). This typifies a more complex component of the map as an integrated system for communicating spatial information. In familiar spaces, a child's ability to understand these complex relationships develops around five years of age, but there are many individual differences (Liben and Downs 1989). To increase the complexity of this type of task, researchers have related a map to a second (three-dimensional) model of space. Liben and Downs (1989) have shown that children between the ages of five and six had great difficulty making map judgments based on referent locations in a model of the mapped space. As argued by several developmental researchers, it is often as important to report and examine what children are incapable of doing as it is to examine what they are capable of doing.

While researchers from a variety of disciplines have examined the nature of children's mapping ability, a vibrant discussion, held partially in the *Annals of the Association of American Geographers*, has exposed some of the principal questions in this field. The psychological tradition, exemplified in research by Liben and Downs (1989, 1993), Deloache (1989), Blades and Cooke (1994), and Uttal and Wellman (1989), has provided understanding through rigorous control of both subject and experimental variables, but has been limited by its narrow definition of what constitutes a map. Rarely do the maps in these studies bear much resemblance to the complicated representations with which most geographers are familiar. The geographer's approach is to sample a much broader range of maps, more representative of the instruments used in examining problems in large, environmental spaces (Blaut and Stea 1971). At the same time, the astute geographer would recognize that some of the representations that are used in this research are also not true maps but aerial photographs, and do not contain the standard components of a cartographic representation (orientation, scale, symbology, title, legend, etc.). Each approach has its characteristic weaknesses that leads to a reduction in the strength of the researchers' conclusions, at least to the extent that they add to our understanding of how complicated, multivariate representations of space are used to solve complex spatial problems (Liben and Downs 1989).

When and how children develop the ability to understand representations of space is an important question for geography because it deals with a way of communicating that is uniquely geographic. That so many researchers from different disciplines also believe it is a compelling problem for science should only bolster geography's resolve to explore it further. Implications for cartography, GIScience, education, and cognitive science should serve to sharpen that resolve. Exploring and understanding space are two reasons why maps have held human fascination for so long. Developing a clearer picture of how our ability to use maps emerges should do the same for a discipline in which the map is so clearly an integral part of its tradition.

# REFERENCES

Blades, M. and Cooke, Z. 1994. Young Children's Ability to Understand a Model as a Spatial Representation. *Journal of Genetic Psychology* 155(2): 201–218.

Blaut, J. M. and D. Stea. 1971. Studies of Geographic Learning. *Annals of the Association of American Geographers* 61(2): 387–393.

DeLoache, J. S. 1989. Young Children's Understanding of the Correspondence Between a Scale Model and a Larger Space. *Cognitive Development* 4(2): 121–139.

DeLoache, J. S. 1991. Symbolic Functioning in Very Young Children: Understanding of Pictures and Models. *Child Development* 62(4): 736–752.

Landau, B. 1996. Multiple Geometric Representations of Objects in Languages and Language Learners. In P. Bloom, M. A. Peterson, L. Nadel, and M. F. Garrett, eds. *Language and Space: Language, Speech, and Communication*, 317–363. Cambridge, MA: MIT Press.

Liben, L. S. and R. Downs. 1989. Understanding Maps as Symbols: The Development of Map Concepts in Children. In H. W. Reese, ed. *Advances in Child Development and Behavior*, 145–201. San Diego: Academic Press.

MacEachren, A. M. 1995. *How Maps Work: Representation, Visualization, and Design*. New York: Guilford Press.

Mark, D., C. Freksa, S. C. Hirtle, R. Lloyd, and B. Tversky. 1999. Cognitive Models of Geographic Space. *International Journal of Geographical Information Science* 13: 747–774.

Plester, B., J. Richards, M. Blades, and C. Spencer. 2002. Young Children's Ability to Use Aerial Photographs as Maps. *Journal of Environmental Psychology* 22(1–2): 29–47.

Stea, D., J. M. Blaut, and J. Stephens. 1996. Mapping as a Cultural Universal. In J. Portugali, ed. *The Construction of Cognitive Maps*, 345–360. The Hague: Kluwer Academic Publishers.

Uttal, D. H., D. P. Marzolf, S. L. Pierroutsakos, C. M. Smith, G. L. Troseth, K. V. Scudder, and J. S. DeLoache. 1998. Seeing Through Symbols: The Development of Children's Understanding of Symbolic Relations. In O. N. Saracho and B. Spodek, eds. *Multiple Perspecitives on Play in Early Childhood Education*, 59–79. Albany: SUNY Press.

Uttal, D. H. and H. M. Wellman. 1989. Young Children's Representation of Spatial Information Acquired From Maps. *Developmental Psychology* 25(1): 128–138.

# 96

# UNDOCUMENTED IMMIGRANTS IN THE 21ST CENTURY: PERCEPTIONS OF SPATIAL LEGITIMACY

GWEN GUSTAFSON SCOTT

*University of Oregon*

Geographers have addressed various aspects of migration for at least a century. They have typically focused on immigrant settlement patterns, adaptation, and acculturation. However, as the "face" of the American immigrant has changed dramatically since the 1960s, geography has been slow to provide more nuanced and theoretical assessments of the issues specific to these contemporary immigrants. Because of the recent increase in the number of undocumented immigrants, refugees, asylum-seekers, and potential immigrants (those entering on nonimmigrant visas who may later opt to immigrate), and because today's immigrants are far more ethnically and racially diverse than their predecessors, the contemporary period creates a far more challenging scenario.

In particular, geographic perspectives directed toward the problems and implications associated with undocumented immigrants have been minimal. The subject has been touched on by geographers studying domestic service and transnational processes (Pratt 1999; Mattingly 2001). However, a review of major North American geographic journals identified only a few articles pertaining specifically to undocumented and/or illegal immigrants (e.g., Jones 1995; McIntyre and Weeks 2002). By far the most extensive published work on the subject by a geographer is Nevins' (2002) book, *Operation Gatekeeper: The Rise of the 'Illegal Alien' and the Making of the U.S.-Mexico Boundary*, in which he challenges the geopolitical construction of the "illegal" alien as a threat to national boundaries and territorial identities. He writes, "While yesterday's undesirables were distinguished by racial factors, today's unwanted immigrants are marked by their legal status – or lack thereof. [Hence,] the 'illegal' is guilty of a geographical transgression for the putative crime of being in a particular space without state authorization" (121).

The notion that a person and/or group of people can commit a "geographical transgression" implies that certain occupants have more legitimate claims to space than others within the same geographic location. These spaces, be they social, cultural, or physical, are complex and often overlapping. They vary from place to place. Therefore, geographers have much to contribute to the

589

*D. G. Janelle et al. (eds.), WorldMinds: Geographical Perspectives on 100 Problems*, 589–593.
© 2004 *Springer. Printed in the Netherlands.*

sorting out, understanding of, and possible "solution" to conflicts over, around, and about the immigration of undocumented persons.

## DEFINING THE PROBLEM

Reports by the Immigration and Naturalization Service (U.S. Department of Justice 1999) estimate that in 1996, 5 to 6.5 million illegal immigrants lived in the United States. Forty-one percent of these people entered the United States legitimately (typically as visitors or students) and have over-stayed their non-immigrant visas. Fifty-four percent of all undocumented residents and workers are Mexican. However, only 16 percent of undocumented people from Mexico are overstayers, compared to 26 percent from other Central American countries. Of the undocumented immigrants from all other countries, 91 percent are nonimmigrant visa overstayers. With the exception of Canadians, who represent 2.4 percent of aliens, most illegal immigrants living and working in the United States are non-White (U.S. Department of Justice 1999). These data indicate that the undocumented population in the United States is diverse and that issues associated with "illegal" immigrants are spatially variable, and, undoubtedly, race-related.

Problems frequently associated with a high influx of undocumented immigrants include: welfare use by "illegals"; job loss by minorities and low-skilled American workers; ethnic conflicts; increased educational and environmental pressures; and increased crime. In contrast to claims that accuse illegal immigrants of causing economic hardship and social conflicts, some argue that undocumented laborers stabilize the U.S. economy as these workers typically take jobs that "Americans take for granted but won't do themselves," in agricultural, hospitality, manufacturing, health care, and construction sectors (U.S. Chamber of Commerce 2002: 2).

Nevins (2002) moves the debate regarding illegal immigration away from economic arguments toward issues of race. He states, "The focus on unauthorized immigration is part and parcel of a more general restrictionist sentiment, one rooted historically in notions of undesirable 'others' who have typically been non-white, non-English-speaking people from relatively poor countries" (118). He continues, claiming: "It is no mere coincidence that many of today's 'undesirables' are from groups previously excluded or marginalized for overtly racist criteria" (121).

Most undocumented immigrants who come to the United States settle in California, New York, Texas, and Florida (U.S. Department of Justice 1999; Fernandez and Robinson 1994). Unlike states with fewer undocumented immigrants, these states, especially California, have been proactive in their response. California alone is likely home to between 1,321,000 to 1,784,000 people classified as illegal aliens (Fernandez and Robinson 1994). In the last two decades, Californians passed three anti-immigration referendums.

Proposition 63, passed in 1986, laid the groundwork for other policies of exclusion by making English the official language of the State of California. Proposition 187, passed in 1994 and sponsored by the group Save Our State (SOS), required that state, local, and federal agencies be alerted regarding the presence of illegal aliens in order to prevent their receipt of benefits and services in California. Proposition 209, passed in 1996 and known as the California "Civil Rights" Initiative, sought to eliminate affirmative action laws affecting employment, education, and contracting. No other state has passed such elaborate legislation specific to immigrants, whether documented or undocumented.

## GEOGRAPHICAL CONTRIBUTIONS

Geography's contribution to the above discussion derives from the discipline's longstanding tradition of integrating social and spatial phenomena, including documenting flows and patterns of human mobility and settlement, and influencing policy decisions regarding borders and border crossings. However, geography's potential to further understanding about "illegal" immigration lies in the acknowledgement and critique of its role in creating the circumstances that generated the very conditions that produce such movements. The development of European navigational techniques, the subsequent cartographic centering of Europe and North America, and the solidification of state boundaries are parts of a highly racialized, Western, and largely academic, project that geography as a discipline has helped create and reinforce (Blaut 1993; Spate 1979). Thus, it is essential that geographers participate in disentangling the elaborate geopolitical and socio-cultural webs of spatial legitimacy endowed through such Eurocentrism at both theoretical and empirical levels, and also in the methodological approaches used to collect and analyze information and ideas relevant to "illegal" migration and border crossings. By redirecting thinking about this issue, geography may, in turn, redirect outdated immigration policies that serve to reinforce racism, and intellectual and technological elitism, and Judeo-Christian extremism.

The theoretical implications of such study include the exploration of how geographers construct, think about, and employ terms and concepts regarding the categorization of peoples and human mobility and settlement. Defining and refining concepts, such as permanent and temporary migration, along with return and circular migration, transnationalism, and others, as they apply to contemporary immigration, is crucial if we are to develop understanding of these complex and inherently spatial processes, especially as they relate to undocumented immigrants. Though Nevins (2002) began the discussion about how and why the terms "illegal" and "alien" have been applied to non-white immigrants, there is much more to be said about the political ramifications of such labeling. Likewise, the connotations of the more 'politically correct' label of "undocumented immigrant" should also be explored within the context of borders and other barriers.

Another important theoretical element is the categorization of immigrants and non-immigrant workers. Currently, people with advanced intellectual and technological skills are given immigration priority over "unskilled" laborers who typically work in agriculture, food service, the garment industry, and health care – industries that are integral to the basic functioning of the United States and American society. By questioning and examining from whence and how these value judgments were formed and spatially executed, geography will provide a platform through which these practices can be reconsidered.

The development, testing, and application of these theoretical ideas rely on geographers engaging in empirical research specific to the undocumented. Public responses to undocumented immigrants are also opportune grounds for empirical investigation. However, the sensitive nature of the research, the respondents' reluctance to participate, and the lack of availability of actual numbers, calls for integrated and innovative methodological approaches that include both qualitative and quantitative methods. Quantitative information, such as census data, Immigration and Naturalization Service (INS) figures, and election results, reveal part of the untold story of undocumented immigration from a geographic perspective. However, the story is incomplete without the inclusion of qualitative methods, including the use of life histories, personal interviews, and informed participant observation. In combination, these data will present a clearer scenario about the problems of undocumented immigration to and within the United States.

By documenting and differentiating the experiences of various immigrant groups, as they negotiate spaces and places "illegitimately," as well as by studying the responses to undocumented immigrants by specific receiving communities and/or regions, geographers can significantly contribute to sorting out the complexities and deepening understandings about spatial legitimacy as it is perceived in specific places at specific times. Further, as we divulge the ways through which certain groups gain and maintain control over particular space, geographers can influence new ways of thinking about more appropriate immigration policies that are designed to promote fairness and recognize the contributions of those who under-gird American society.

## REFERENCES

Blaut, J. 1993. *The Colonizer's Model of the World: Geographical Diffusionism and Eurocentric History*. NY: The Guilford Press.

Fernandez, E. and J. G. Robinson. 1994. Illustrative Ranges of the Distribution of Undocumented Immigrants by State. *Population Division Technical Working Paper No. 8*. Washington, D.C.: U.S. Bureau of the Census.

Jones, R. 1995. Immigration Reform and Migrant Flows: Compositional and Spatial Changes in Mexican Migration. *Annals of the Association of American Geographers* 85: 715–731.

Mattingly, D. 2001. The Home and the World: Domestic Service and International Networks of Caring Labor. *Annals of the Association of American Geographers* 91: 370–386.

McIntyre, D. and J. Weeks. 2002. Environmental Impacts of Illegal Immigration of the Cleveland National Forest in California. *The Professional Geographer* 54: 392–405.

Nevins, J. 2002. *Operation Gatekeeper: The Rise of the 'Illegal Alien' and the Making of the U.S.-Mexico Boundary.* New York, NY: Routledge.

Pratt, G. 1999. From Registered Nurse to Registered Nanny: Discursive Geographies of Filipina Domestic Workers in Vancouver, B.C. *Economic Geography* 75: 215–236.

Spate, O. 1979. *The Spanish Lake.* Minneapolis, MN: University of Minnesota Press.

U.S. Chamber of Commerce. 2002. *Essential Workers, Issues Index.* http://uschamber.com/Political+Acvocacy/Issues+Index/ [cited on 3 September 2002.

U.S. Department of Justice, Immigration and Naturalization Service. 1999. *Illegal Immigrants: Illegal Alien Resident Population.* http://www.ins.usdj.gov/publicaffairs/newsreels/top25.pdf. [cited on 8 August 2002].

# 97

# 'NOURISHING THE SOUL': GEOGRAPHY AND MATTERS OF MEANING

DANIEL H. OLSEN and JEANNE KAY GUELKE
*University of Waterloo*

Religion and spirituality matter to geographers because they concern so many people. Most humans experience sufficient uncertainty at some point in their lives to ask the basic question of human existence: "What is the meaning of life?" or, more specifically, "What is the meaning of *my* life?" In a recent Gallup Poll (2002), almost 40 percent of Americans said they seek answers to these questions through attendance at weekly worship services. Another 45 percent reported religion as important to their lives. But even atheists recognize that religious fundamentalism is a crucial political issue today, whether the fundamentalists in question are Christians who oppose abortion or the teaching of evolution in public schools, or Al Qaeda terrorists.

Geographical concepts and observations indicate a great deal about religion as a form of spatial behavior. Beyond the simple mapping and description of societies based on religious memberships, geographers have examined how religious believers modify landscapes to express their precepts. Geographers also study pilgrimages to holy sites, the role of the religious built environment within the global heritage tourism industry, and the political expression of religions in disputed places (e.g., Kong 1990; Park 1994). Geographers have also used their cultural and ecological expertise to develop links between belief systems and environmental sustainability. We propose two ways in which an understanding of geography informs the human quest for meaning through religion. We first discuss the phenomenon of religious tourism, which has grown in the past decade alongside the global tourism economy. Our second example discusses geography's contributions to environmental ethics and sustainability.

## RELIGIOUS TOURISM

Millions of people travel every year to major pilgrimage destinations, both of ancient and modern origin. The global resurgence of pilgrimage, due in part to the recent turn of the Millennium (Olsen and Timothy 1999), indicates that increasing numbers of people, in a world of shifting values and

*D. G. Janelle et al. (eds.), WorldMinds: Geographical Perspectives on 100 Problems*, 595–599.
© 2004 *Springer. Printed in the Netherlands.*

norms, attempt to leave behind their everyday lives and to travel in search of some authentic reality elsewhere. Motivations for undertaking transformational pilgrimages are complex. Some travel in search of or in order to maintain an identity, to satisfy feelings of nostalgia, to experience the transcendent or numinous; or to fulfill the teachings of particular faiths, such as journeys to Mecca for devout Muslims.

Members of religious faiths travel to places where they believe there is a possibility of communing with the divine. Here believers renew themselves through rediscovering and re-rooting themselves in their belief system. These sites of worship range from the great cathedrals of Europe and the Taj Mahal in India to the simple wooden churches of the Amish in North America to outdoor destinations like the Ganges River in India and Zen gardens in Japan.

The maintenance of such religious sites is costly, raising issues of whether people should have to "pay to pray" (Shackley 2002). The management of these sites has also become politicized, particularly where the tourism industry views religious sites as raw cultural materials to be commodified as part of local, regional, and national economic development strategies. Questions over who controls, manages, and maintains these places, as well as who interprets the meanings, can raise heated debates, particularly in zones of inter-faith strife (e.g., Jerusalem, Ireland) or where religious leaders differ with government and tourism officials over the function of religious sites.

Geographers, with their focus of the importance of places, contribute to the management of pilgrimage sites through understanding both the development of sacred places and the motivations and expectations of those who travel to these sites. This approach can help in the understanding of conflict between pilgrims and tourists, in turn helping religious site managers manage and meet the expectations of both parties while maintaining a sense of place. However, factual information is necessary to understand the complex relationships between religious authorities, government and tourism officials, pilgrims, and tourists regarding the nature and function of sacred sites and how these sites, along with appropriate infrastructures and derivative service industries, are best to be managed (Shackley 2001).

## THE ENVIRONMENT AND RELIGION

The earth's overall environmental quality remains precarious, despite the search for solutions to environmental problems through improved scientific understanding, gentler technologies, and more effective policies. Some environmentalists have argued that the root explanation for the environmental deterioration is not really scientific or economic, but is that society has not yet accepted ecological sustainability as a problem of morality or ethics. If they are correct, then religious approaches to sustainability must be considered.

Oeschlager (1994: 1–12) argued that the Bible is the root of western civilization from which current secular scientific, economic, moral, and other discourses have gradually emerged. Even secular environmentalists, he reasoned, must realize that religion in the United States today is the main form of non-market discourse concerning the collective good, and thus the principal institution capable of addressing ethical issues, including environmental deterioration. Reed and Slaymaker (1993; Pacione 1999) saw a critical engagement with ethics as geography's greatest need if it is to contribute to environmental sustainability, and they noted that religious beliefs cannot be overlooked as primary sources of ethics.

Such arguments have not always been persuasive, given standard criticisms of the moral lapses of religious institutions at various points in their history, or the Darwinian challenge to the Biblical account of the earth's origins. Some scholars view the Genesis mandate for humanity to procreate and to dominate nature as itself being the root cause of the environmental crisis, a western conceit that Tuan (1970) was quick to critique. Christian environmentalists responded to such criticisms primarily by espousing an ethic of stewardship based on caring for God's creation.

If Oelschlager's (1994) assessment is correct, however, secular science and policy, together with religious environmentalism, must collaborate in forging public environmental ethics simply because there are no other secular institutions of sufficient critical mass to address the challenge. Religious environmentalists in any event are already using organized religions as sites from which to launch a variety of political action and practical conservation strategies. These are summarized by the Harvard Forum on Religion and Ecology (http://environment.harvard.edu/religion/main.html). For example, the National Council of Churches, together with heads of major Jewish organizations, recently lobbied Congress "to adopt energy policies 'that embody and promote justice, stewardship, and intergenerational responsibility,'" including advice from major spiritual leaders on such specific matters as fuel cells and proposed Arctic oil drilling (COEJL 2002: 3). Religious educational programs increasingly include environmental units, just as many congregations include committees on greening their buildings and community.

Religious educators recognize, however, that their expertise seldom extends to environmental science. Neither can fundamentalist denominations quickly adopt environmental platforms developed at ecumenical conventions if their foundational principles derive from codified scriptures or divine revelations to centralized authorities.

Despite these constraints, geographers can analyze particular scriptures or theologies to identify points of compatibility between religious precepts and environmental management and thus offer believers an interpretive basis for practicing sustainability through their core religious beliefs. For example, Kay (1988) anthologized environmental volumes designed for Jewish and

Christian study groups. Landscapes, moreover, often figure prominently in religious texts like the Old Testament, and geographers are well positioned to explain how these relate to the ethical systems that developed (Guelke 2003). Most religious ethics are not explicitly environmental, yet environmental geographers have the expertise to explain how adherence to religious teachings on greed and anti-materialism would have collateral benefits of reducing resource consumption and pollution.


## CONCLUSIONS

One of the most inspirational leaders in American higher education today is Dr. Ruth Simmons, President of Brown University. She is the daughter of African-American sharecroppers and a former inner-city resident who succeeded as a scholar of French literature. When asked why, as a disadvantaged student, she did not pursue more career-oriented subjects, she responded that an education is not about getting a job, but is about nourishing one's soul. Simmons's point is that "relevance" for seekers of wisdom may be more metaphysical than physical. While discussions of religion and spirituality may seem rather remote from the other papers in this volume, there is no doubt that religion relates to some very practical problems as well. Our purpose here is to identify some "big" questions that engage most human beings and to demonstrate that even if geography is not specifically about the search for spiritual nourishment or meaning, it nevertheless has much to contribute to those who seek such sustenance.


## REFERENCES

Coalition on the Environment and Jewish Life. 2002. Energy Conservation and God's Creation: An Open Letter to the President, the Congress, and the America People. *Newsletter of the Coalition on the Environment and Jewish Life.*

Gallup, Jr., G. 2002. Religion/Church Attendance. *The Gallup Pool: Public Opinion 2001.*

Guelke, J. 2003. Judaism, Israel, and Natural Resources: Models and Practices. In H. Selin, ed. *Nature Across Cultures: Non-Western Views of Nature and Environment,* 433–456. Dordrecht: Kluwer Academic Publisher.

Kay, J. 1988. Concepts of Nature in the Hebrew Bible. *Environmental Ethics* 10: 309–327.

Kong, L. 1990. Geography and Religion: Trends and Prospects. *Progress in Human Geography* 14: 355–371.

Oelschlager, M. 1994. *Caring for Creation: An Ecumenical Approach to the Environmental Crisis.* New Haven: Yale University Press.

Olsen, D. H. and D. J. Timothy. 2002. Differing Views of Contested Mormon Heritage. *Tourism Recreation Research* 27(2): 7–15.

Pacione, M. 1999. The Relevance of Religion for a Relevant Human Geography. *Scottish Geographical Journal* 115(2): 117–131.

Park, C. 1994. *Sacred Worlds: Introduction to Geography and Religion.* London and New York: Routledge.

Reed, M. and O. Slaymaker. 1993. Ethics and Sustainability: A Preliminary Perspective. *Environment and Planning A* 25: 723–739.

Shackley, M. 2001. *Managing Sacred Sites: Service Provision and Visitor Experience.* London: Continuum.

Shackley, M. 2002. Space, Sanctity and Service: The English Cathedral as *Heterotopia.* *International Journal of Tourism Research* 4: 345–352.

Tuan, Y.-F. 1970. Treatment of the Environment in Ideal and Actuality. *American Scientist* 58: 2.

# 98

# THE PROFOUND PROBLEM OF LOCATING HUMANITY: THE SIGNIFICANCE AND IMPLICATIONS OF *FENG-SHUI*

DAVID NEMETH
*University of Toledo*

In 1862, Thomas Huxley published his classic *Man's Place in Nature*. In justifying the title and contents of this book, he wrote of a recurring "question of questions" for humankind; a problem "which underlies all others, and is more deeply interesting than any other." He identified this problem as "the ascertainment of the place which man occupies in nature, and of his relationship to the universe of things." An important corollary of this problem was "What are the limits of our power over nature?"

Throughout history and across the face of the earth, humans searching for guidance typically approached this problem by contemplating nature itself. Traditional Chinese civilization celebrated Nature by imitating it, and its story of creation claimed that the first emperor grew wise by looking upward and contemplating the images in the heavens, then looking downward and contemplating the markings of birds and beasts and their adaptation to the regions. Recognition of natural patterns was and is essential to human survival. Humans contemplated visible natural pattern and then generated cosmologies containing stories about hidden creative and destructive forces behind pattern formation. Geography everywhere began as a profession specializing in the recognition and interpretation of natural pattern in order to discover hidden knowledge and predict the future.

## ARE THE "BIG QUESTIONS" IN GEOGRAPHY ALSO PROFOUND?

Attempts to ascertain the limits of human power over nature attracted the attention of many geographers in the West during the 20th century, culminating with the famous international, interdisciplinary symposium and proceedings *Man's Role in Changing the Face of the Earth* (Thomas 1956). By that time, professional geographers were already rapidly losing interest in the problem of locating humanity with respect to nature. It is revealing that Lösch's influential *The Economics of Location* (1954) begins by announcing "Our existence

*D. G. Janelle et al. (eds.), WorldMinds: Geographical Perspectives on 100 Problems*, 601–605.
© 2004 *Springer. Printed in the Netherlands.*

in time is determined for us, *but we are largely free to select our location*"
[emphasis mine]. Lösch solved the problem of human placement by ascertaining without proof that humankind is an autonomous agent in nature and
therefore free of natural constraints to act in its own interests.

It comes as no surprise that Huxley's "question of questions" is nowhere
found among those ten "Big Questions in Geography" recently recommended
for scientific geographic study by Cutter, et al. (2002). Yet it may be prema-
ture and dangerous to dismiss so profound a question that by definition means
"deep" and, perhaps, unfathomable. Traditional Chinese civilization contemplated pattern in nature while considering the profound problem of human
placement and concluded that everything that grows has limits, and that
everything changes but is restrained. Is the human condition part of "everything"? If so, how does that square with what Lösch's asserts about human
freedom?

## *FENG-SHUI*: AN INDIGENOUS EAST ASIAN GEOGRAPHICAL TRADITION

This essay briefly introduces *feng-shui* (various spellings), an indigenous
geographical tradition from East Asian neo-Confucian civilizations, as a way
to locate humanity's position in the natural order. *Feng-shui* has been described
as "the Chinese art of placement" (Rossbach 1983). A more comprehensive
and accurate definition is provided in Chatley (1917: 175): "The art of adapting
the residences of the living and dead so as to co-operate and harmonize with
the local currents of the cosmic breath." *Feng-shui* was thoroughly integrated
into the great consolidation of Chinese neo-Confucian cosmological thinking
about eight centuries ago. The neo-Confucians perceived the world as a finite
and hierarchical order and whole; a self-sufficient system in which humanity
had inherent tendencies to move toward certain places.

There has not been much scholarly study of *feng-shui* in the West, where
scholars and scientists typically dismiss it as a "pseudo-science." The logic
of the *feng-shui* symbolic system and how it operates in the context of neo-
Confucian cosmology is, however, internally consistent. Knowledge of the
*feng-shui* system in terms of its own internal logic is a prerequisite to understanding how it helps people find their place in nature and their relation to
the universe of things. However, there may be a shortcut to accessing this
knowledge. Access to this knowledge is valuable to humankind because the
*feng-shui* response to the profound problem of human placement seems to have
benefited neo-Confucian society: "In the 18th century China possibly had
the highest standard of living in the world" (Spencer 1954: 315).

## THE *FENG-SHUI* MAP AS A DIDACTIC AND MNEMONIC MODEL OF NEO-CONFUCIAN COSMOLOGY

*Feng-shui* experts are skilled cartographers, and their maps are neo-Confucian cosmological models. The drafting of maps and diagrams that describe the locations of *feng-shui* sites constitutes the oldest cartographic tradition in East Asia. *Feng-shui* maps are designed to be powerful didactic (teaching) and mnemonic (memory) devices in the service of reproducing the ideology of neo-Confucianism, so their hidden knowledge is intended to be accessible.

Figure 98.1 is an example of a typical *feng-shui* map from an unknown cartographer's 18th-century manuscript atlas. The map invests with cosmological symbolism an earth surface pattern on Cheju Island covering several square miles surrounding the *feng-shui* site. Patterns in nature represented on the map consist of important topographic and man-made features illustrated in black ink against a white background. Generally, black represents positive terrain and white represents negative terrain, as observed or intuited from the cartographer's perspective at the selected site (which is usually represented by a small circle at the center of the completed map).

*Figure 98.1.* An 18th century Cheju Island Feng-shui map.

## A MAP-TRACING EXERCISE: RATIONALE AND RESULTS

During five years of residence on Cheju Island over the past thirty years, I learned quite a bit about the *feng-shui* response to the problem of human placement in nature, including the power of *feng-shui* maps as neo-Confucian didactic and mnemonic devices. A typical *feng-shui* map offers "morphological" insight into the universe of neo-Confucian cosmology. The map invites exploration of, and speculation about, the creative patterns, processes, and underlying principles of map formation. The *feng-shui* place-finder and cartographer is not the creator of the map, but is the medium by which nature's self-organizing principles in physical space create the map. Contemplating map pattern is a potential gateway for individuals to access hidden knowledge about creative forces of growth and decay in nature.

Contemplating patterns in nature is not the only way to access appreciation of the forces that create natural pattern. To the extent that a *feng-shui* map represents a local response to the general problem of ascertaining human placement in nature, it may be possible for seekers of neo-Confucian cosmological knowledge to participate in reproducing the neo-Confucian world-view in order to learn about it's theory of human placement. They can attempt to do so by tracing or re-drawing the *feng-shui* map pattern.

I conduct *feng-shui* map-tracing exercises as part of my "Advanced Cultural Geography" class. With a few simple brush strokes, the students reiterate the natural and unobstructed flow of the universe that produced the map in the first (creative) instance. The original cartographer and the student both become mediums for the release of creative forces beyond their understanding. This experience of being "possessed by nature" through map tracing is spiritually and intellectually satisfying.

These are what my students and I conclude to be nature's hidden principles of self-organization in physical space:
– centrality
– connectivity
– periodicity
– symmetry (proportion)
– hierarchy
– similarity at different scale
– completeness

They are familiar concepts. Students had encountered these principles in isolation of one another, primarily in the context of their exposure to positivist geography, and rarely expected them to be part of a conversation about humanity's place in nature. Yet, the map-tracing exercise revealed these notions to us in a refreshing, new way. Each principle was no longer isolated from the others, but was revealed as an essential part of the unified and creative forces behind patterns in nature.

## CONCLUSIONS

Huxley's profound problem of humankind's place in nature and its relation to the universe of things is inherently geographical, universal, recurrent, and unsolved. The significance and implications of addressing the problem at the present time bears on the collective destiny of the human species. Neo-Confucian cosmology, through its association with *feng-shui* practices, directs attention to the importance of the study of patterns in nature as they relate to the problem of human placement.

Cheju Island's symbolically charged *feng-shui* maps were intended to be didactic and mnemonic devices through which knowledge about nature's self-organizing principles could be accessed and identified. *Feng-shui* map-tracing exercises revealed these principles to my students, and their significance and implications became the focus of our discussions.

Tentative conclusions based on the findings of advanced Western scientific studies of pattern recognition, network analysis, chaos theory, and fractals seem to be converging on the same view of humanity's place in nature articulated by the ancient Chinese. For example, "the immense variety that nature creates emerges from the working and reworking of only a few formal themes," and "space prohibits so much and permits so little" (Stevens 1974: 3, 222).

If *feng-shui* wisdom continues to be validated, then Lösch was very wrong and the profound problem is still unanswered. Humans are *not* largely free of natural constraints to select our location, but must seek our place amidst nature's own limiting principles of self-organization in physical space.

## REFERENCES

Chatley, H. 1917. Feng-Shui. In S. Couling, ed. *Encyclopaedia Sinica*, 75. Shanghai: Kelly and Walsh.

Cutter, S., R. Golledge, and W. Graf. 2002. The Big Questions in Geography. *The Professional Geographer* 54: 305–317.

Huxley, T. 1906. *Man's Place In Nature and Other Essays by Thomas Henry Huxley.* London and Toronto: J.M. Dent & Sons.

Lösch, A. 1954. *The Economics of Location.* Trans. by W. Woglom. New Haven: Yale University Press.

Rossbach, S. 1983. *Feng Shui: The Chinese Art of Placement.* New York: E. P. Dutton.

Spencer, J. 1954. *Asia East By South.* New York: John Wiley.

Stevens, P. 1974. *Patterns in Nature.* Boston: Little, Brown and Company.

Thomas, W. 1956. *Man's Role in Changing the Face of the Earth.* Chicago: University of Chicago Press.

# 99

# ALEXANDER VON HUMBOLDT AND THE ORIGINS OF OUR MODERN GEOGRAPHICAL VIEW OF EARTH

ALEJANDRO GUARÍN

*Pennsylvania State University*

> Throughout, the simple and scientifically descriptive must be incorporated with the rhetorical. It is so in nature herself. The glittering stars delight the senses and inspire the mind, and yet everything beneath the vault of heaven moves in a path of mathematical precision.
>
> (Alexander von Humboldt 1860: 70)

Alexander von Humboldt (1769–1859) was probably the most important and certainly the most famous scientist in 19th-century Europe before Darwin. Many have pointed to his unparalleled contribution to the modern geographical view. Cannon (1979) and Buttimer (2001) have characterized the key features of "Humboldtian science," as distinguished by synthetic thinking, special attention to scale, and innovative use of maps and diagrams. Why, then, talk about Humboldt – *again*? I argue that understanding Humboldt's fundamental contributions to science in general, and to geography in particular, provide elements to solve two concrete problems in the way we study our world today: over-specialization and over-rationalization. The broad geographical view of earth, owed in great part to Humboldt, constitutes a distinctive character of our discipline, and a key feature to address today's relevant issues. In this essay, I trace the origins of Humboldt's worldview, underscoring his contributions in the form of "Enlightened Romanticism" and an integration of mechanistic and holistic philosophies. This integration, I believe, is a defining characteristic – and a strength – of geography.

## ROMANTICISM VERSUS ENLIGHTENMENT? ROMANTICISM *AND* ENLIGHTENMENT!

During the later part of the 18th century, several artists and scholars (outstandingly, Goethe, Schelling, and Schiller, in Germany) turned away from what they viewed as a soul-less mechanical conception of nature derived

*D. G. Janelle et al. (eds.), WorldMinds: Geographical Perspectives on 100 Problems, 607–611.*
© 2004 *Springer. Printed in the Netherlands.*

from the "Enlightenment," and started to seek a more spiritual and integral explanation of the world (Cunningham and Jardine 1990), referred to today as *Romanticism*. Romantics were convinced that human's relation to nature had been perverted by extreme rationalism, and they longed for a new spiritual communion with the natural world. They were reacting against the host of scientists that saw in the "clear light of reason" the road to human freedom. In the rational perspective of the Enlightenment, understanding was provided by direct, factual experience and/or experimentation.

Humboldt's ties to German Romanticism date to the late 18th century, when he befriended such renowned figures as Goethe, Schelling, and Schiller. From them Humboldt inherited a holistic conception of the world, a desire to

> . . . arrive at a point of view from which all organisms and forces of nature may be seen as one living, acting whole, animated by one sole impulse. "Nature", as Schelling remarks in his poetic discourse on art, "is not an inert mass; and to him that can comprehend her vast sublimity, she reveals herself as the creative force of the universe" (Humboldt 1845: 55).

Humboldt's interest in disclosing and understanding the complex and holistic interactions found in the natural world are particularly evident in his study of vegetation. In his *Essay on the Geography of Plants*, he departed from the description of individual plant species and attempted to look at the *physiognomy* of vegetation. By relating plant life forms to climatic and other physical conditions, Humboldt was among the first to advance a scheme to explain how plants are distributed on a planetary scale. To add to his romantic conception, Humboldt was also very concerned about the aesthetic impression that vegetation had on a viewer. He believed that the emotional sensations produced by the plants were as important as the rationale that explained them (Nicolson 1987). In his other great book, *Cosmos*, Humboldt sought to provide a complete view of nature, from the stars to the mosses. The very title indicates his belief in an ordered, harmonic universe that could be understood as a whole, and within which every constituent part was affected by common underlying processes and driving forces (Rupke 1997).

Despite his intimacy with the Romantics, Humboldt was formally trained as an empirical scientist. He studied botany, mineralogy, geography, and geology. His science was exact. He made major contributions to measurement and scientific tool development. In 1799, on a journey to the American tropics, Humboldt set out to *measure* nature armed with compasses, barometers, sextants, quadrants, and other state-of-the-art equipment. He calculated the height of mountains, recorded data on air pressure, temperature, rainfall, magnetism, and electricity, and made rigorous astronomical observations. The bulk of his fieldwork was largely a painstaking and systematic collection of facts and measurements. Humboldt's disdain for what he regarded as a shallow Linnaean taxonomy did not prevent him from collecting thousands of botanical and zoological specimens. Although he abhorred empty description, Humboldt spent almost five years doing precisely that: describing.

Humboldt's science integrated elements of Romanticism and Enlightenment. Both of these perspectives were indispensable to the development of his broader, richer view of the world. The Holism and mechanism are logically and necessarily interwoven into his discourse, allowing him to "analyze the individual parts of the natural phenomena without succumbing beneath the weight of the whole" (Humboldt 1845: 2–3). Pure empiricism lacked a metaphysical frame that could provide a complete and true view of nature, while pure Romanticism was nonsensical in rejecting experimentation and measurement as a way of knowing the world. The program for "his" science is put forward, albeit timidly, in the preface to the *Essay on the Geography of Plants* (Humboldt and Bonpland 1804: ii/translation from Spanish by the author):

> Loyal to the traditions of empirical research in the field of Natural Sciences to which I have devoted my entire life, I want in this work to present in an orderly fashion the various phenomena, side by side, instead of explaining, inquiring deep into the nature of things and their internal relations. This confession, that states my position under which I hope it will be commented, also shows the possibility that some time it will be possible to present a picture of nature in a different way, both broader and in a more Natural-Philosophical level.

The particularities of Humboldt's methods and ideas make him a central figure in the epistemological transformation leading from a descriptive to an explanatory science at the turn of the 19th century. Humboldt's departure from standard 18th century natural history may be seen as a change of *episteme*, in Foucault's words (although he did not refer particularly to Humboldt), as he "opposed historical knowledge of the visible to *philosophical* knowledge of the invisible, of what is hidden and of the causes" (Foucault 1973). Through the merging of apparently irreconcilable physics and metaphysics, Humboldt's innovative and synthetic interpretation of phenomena in the context of their interrelations is a landmark in the development of modern geography and of science in general.

In the late 19th century, the age of great discoveries and explorations was coming to an end, and geographers were struggling to define their discipline and to stop being regarded as mere "map makers" (Bowen 1981). Humboldt was not a proper "geographer" to his contemporaries' standards, nor was his name tied intellectually to the foremost geographers of his time (Godlewska 1999). Nonetheless, his novel insights into the relation of the physical, the biological, and the human are outstandingly geographical.

## HUMBOLDT'S SOLUTION TO OUR "PROBLEM"

Still showing the enduring marks of the Enlightenment project, our science is pervaded by reductionism, mechanism, and specialization. While the romantic/holistic perspectives were fundamental in bringing us into the modernity of explanatory science, we still regard forms of understanding that are not completely rational or specialized as primitive and non-rigorous. I argue that geography's importance as a discipline at the dawn of the 21st century lies not (exclusively) on the production of specialized knowledge, but in its synthetic approach that bridges so many different interests and methods. Those who run computer simulations to model the population dynamics of a country and those who write about how the sense of place is constructed by an individual, feel equally at home in geography.

But geographers themselves seem to be trapped in the "human versus physical" or "qualitative versus quantitative" dilemma, as if they had inherited the intellectual conflict between Romanticism and Enlightenment. I suggest that we recognize that geography nurtures from the existence of such opposing views, and that the relevance of the discipline lies on its intellectual and epistemological diversity. The world – a world of people and landforms, of money and language, of technology and individual sense of place – should "fit" into geography's broad view of the world.

I have argued that Humboldt's work brought geography to modernity, as he shifted from a descriptive to an explanatory interpretation of the world. This was possible because of a geographical vision that was broad enough to accommodate large and small, human and physical, but also intellectuality and spirituality. To solve the problems of planet earth, we must first deal with the problem of our own narrow-mindedness. I propose that we recognize the exceptional contributions of Alexander von Humboldt as a place to start.

## ACKNOWLEDGEMENTS

This paper started as part of a graduate seminar with Dr. Amy Glasmeier at The Pennsylvania State University. Her support and advice made all this possible. I thank Dr. Malcom Nicolson for insightful comments on a previous draft of my work.

## REFERENCES

Bowen, M. 1981. *Empiricism and Geographical Thought. Cambridge.* U.K.: Cambridge University Press.
Buttimer, A. 2001. Beyond Humboldtian Science and Goethe's Way of Science: Challenges of Alexander von Humboldt's Geography. *Erdkunde* 55(2): 105–120.
Cannon, S. 1978. *Science in Culture.* New York: Dawson and Science History Publications.
Cunningham, A. and N. Jardine, eds. 1990. *Romanticism and the Sciences.* Cambridge, U.K.: Cambridge University Press.
Foucault, M. 1973. *The Order of Things: An Archaeology of the Human Sciences.* New York: Vintage books.
Godlewska, A. 1999. From Enlightenment Vision to Modern Science? Humboldt's Visual

Thinking. In D. Livingstone and C. Withers, eds. *Geography and Enlightenment*, 236–275. Chicago: The University of Chicago Press.

Humboldt, A. [1845] 1997. *Cosmos, a Sketch of the Physical Description of the Universe*, Volume 1. Baltimore: The Johns Hopkins University Press.

Humboldt, A. and A. Bonpland. [1804] 1985. *Ideas para una geografía de las plantas más un cuadro de los países tropicales*. Bogotá: Jardín Botánico de Bogotá.

Humboldt, A. and K. Varnhagen. 1860. *Letters of Alexander von Humboldt Written Between the Years 1827 and 1858 to Varnhagen von Ense, Together with Extracts From Varnhagen's Diaries, and Letters of Varnhagen and Others to Humboldt*. London: Trübner.

Nicolson, M. 1987. Alexander von Humboldt, Humboldtian Science and the Origins of the Study of Vegetation. *History of Science* 25: 167–194.

Rupke, N. R. 1997. Introduction. In A. Humboldt, *Cosmos, a Sketch of the Physical Description of the Universe*, Volume 1. Baltimore: The Johns Hopkins University Press.

# 100

# NATURAL AND CULTURAL LANDSCAPES: ONE HERITAGE OR TWO?

## THOMAS F. MCILWRAITH
*University of Toronto at Mississauga*

One humid summer day I was sitting beside a sprightly little waterfall where Ancaster Creek tumbles over the Niagara Escarpment in southern Ontario. I was on the dry side of a huge glass pane in the Old Ancaster Mill restaurant, savoring a salad in air-conditioned elegance. It struck me that the predicament of which I write here had been largely resolved. Here was a waterfall in a post-Pleistocene setting, and beside it a rusty iron pipe as big as an embrace, leading by a number of kinks and bends to an 1862 gristmill turning out stone-ground flour. Waterpower had long since yielded to electrical, and the product once exported internationally was now being picked up in small packages for home baking locally by discerning, health-conscious urban gourmands driving SUVs. The artifacts told the story. My musings led me easily to conclude that human and physical geography processes were inseparably intertwined in this bucolic, gentle setting, much as they had been since the start of recorded history in this place some two hundred years earlier.

## TWO SOLITUDES?

The professional geographer side of me retorts that the human and physical heritage encompassed in this scene commonly have stood poles apart for scholars, and that this dichotomy has spread to the wider public who, nevertheless, do not practice it. Hence the predicament. How is it that these two inseparables – natural and cultural – have indeed stood among scholars as two solitudes throughout much of the 20th century, and are geographers really to be recognized as having provided, or currently providing, reconciliation?

The predicament might never have occurred had George Perkins Marsh received in 1870 the attention bestowed on him (by geographers, among others) a century later. It was his fate to have encouraged ecological synthesis at a time when the intellectual world was moving steadily towards compartmentalizing the swelling body of knowledge as a tactic for bringing it under control. Theology, ethics, and mathematics drifted apart; sociology spawned anthro-

*D. G. Janelle et al. (eds.), WorldMinds: Geographical Perspectives on 100 Problems*, 613–617.

pology; geology spawned geomorphology; political economy fostered human geography. Scientists experimented, humanists reflected, and fieldwork declined. The breach between physical and social studies deepened through the middle of the 20th century, rubbing off on educational systems and an urban society broadly detached from living off the land. For most citizens the science of the landscape moved beyond them, while the humanity of its space and symbolism continually imbued them.

## GEOGRAPHERS TO THE RESCUE

"Rescue" is perhaps self-serving, for geographers have taken considerable time to understand such divergence, even lagging behind public inclination for synthesis. The cultural landscape has seemed to scholars a lumpy mechanical mixture, while the natural environment more of a chemical brew, smooth and viscous; research has approached them from different angles. Yet both worlds are eternally current, and that is a tremendously important similarity that geographers have recognized. It has allowed us the useful role of refocusing on the intimacy of land and people. Geographers stir thinking about context and processes that demonstrate physical and cultural attributes intruding on each other. Agriculturalists may look at crops and architects at farmhouses, but geographers contemplate the farmstead on the farmlot along the road running through the county neighborhood. An urban river is as important as a mountain stream. Torrance Barrens, an Ontario park that "preserves" the nighttime sky, is both a cultural and a natural place. The more one stands back, the more is included.

Fieldwork is the key to reconvergence. It immerses scholars in living and inanimate objects all bumping into each other. It cannot be done in fragments, or in shifts as in a laboratory; it is holistic. Fieldwork invites questions on a limitless range of subjects lying at hand, and no one of them can be presumed to be irrelevant. Synthesis follows, and geographers have heard the call. Fieldwork also connects professional and personal lives, so that resolving the predicament thus becomes a personal task, and not just the geographer's day job.

For the historian "the field" is the library or archive shelf, where interaction with words occurs, and too seldom even a whisper of recognition of the trees from which all that paper comes. For biologists "the field" is botanic, irritated by instances, such as air pollution, of human imposition. Even geographers' physiographic tradition long positioned human beings as detached and of little influence. The landscape architect is rather closer to bridging the physical-cultural gap, but it is human geographers, looking at cultural landscapes, who have been particularly active in reconnecting the two themes over recent decades. "Natural" landscapes and environments no longer exist, climate is being manipulated, and many botanic features are themselves manmade. Such truths are pushing the physical and cultural worlds together faster than any provostial edict. I find that natural scientists are well inclined to acknowledge the cultural heritage theme, yet often constrained by labs and granting structures from being active participants. The initiative for overcoming the gap may be more in the hands of human geographers. Many are creatively involved in heritage activities, and few subjects attract volunteers to heritage agencies and boards as readily.

Lobby groups are regularly attentive to threatened landscapes. Again, consider a farm. Cultural heritage activists focused on fences and buildings are apt to neglect physical attributes. Natural heritage activists see a woodlot and its adjacent springline and boulder pasture, and find cultural elements intrusive and irrelevant. Biologists seldom serve on heritage (meaning "cultural heritage") committees; it is the rare historian who has been accepted into the structure of a conservation (meaning "natural heritage") authority. Geographers are found in both groups, bridging between, say, field naturalists and local history enthusiasts, and their presence adds strength to fragmented defenses. Industrial heritage, offering a humanistic celebration of science, is a particularly inviting way to blend the two solitudes. Celebrate the river the Howe truss bridge spans, and the need for a community to get to the other side, and not just the bridge itself. Celebrate a quarry both for the geology revealed and the infrastructure material liberated. See in the rehabilitation of brownfields an opportunity for cultural and physical heritage interests to converge.

The world's distinguished physical attributes are commonly sprawling and pervasive: a barrier reef, or the Precambrian Shield. Distinguished cultural artifacts are commonly confined and discrete. One step towards convergence would be to generalize the cultural, by recognizing neighborhoods, say, and particularize the natural as, for instance, marking a drumlin field or an urban heat island. It is all one heritage, one environment.


## CONCLUDING

Environmentalism is rooted in "saving" the natural world; heritage started by taking on the built world. Growing recognition that natural heritage and cultural environments belong in the same breath leads me to believe that in the not-too-distant future the adjectives may vanish from common usage. The landscape of trees and houses will stand recognized for what it is, a human concoction at once botanic, zoologic, and inanimate. All landscapes are the current expression of what is natural – that is, normal – in the behavior of living species. The world is moving towards one solitude. Predicaments resolved may, of course, breed further ones. It is a large challenge to organize research support for the best pursuit of an integrated mix of natural and human activity, one that puts mankind back into the natural world, rather than stand beyond it. Connecting natural and human heritage in our professional world is itself natural, and a worthy goal.

My eye scans a list of websites in a new heritage glossy under headings including Resources, Canadian Heritage – General, Natural Heritage, Conservation, International Heritage, and For Kids. Such overlap is symptomatic of the struggle to sort scattered interests into outmoded pigeon-holes. Thank goodness for hyperlink, and I rejoice in technology easing my predicament. But so too can a simple drawing, such as done some sixty years ago by the artist, naturalist, and heritage conservationist Thoreau MacDonald (1901–1989) (Figure 100.1). With a name like that he was destined to recognize the unity of nature and culture, and he has done so with bold simplicity.

*Figure 100.1.* Natural and cultural elements converge beneath a wintry sun setting over rural Ontario landscape, probably about 1950. Copied with permission of Susan M. MacDonald (copyright), *House and Barn: Drawings by Thoreau MacDonald* (Toronto: D.M. Press, n.d.), frontispiece.

## REFERENCES

Heritage links. 2002. *Heritage Matters: The Newsletter of the Ontario Heritage Foundation* 1 (No. 1, Summer): 11.

James, P. E. and C. F. Jones. 1954. *American Geography, Inventory and Prospect.* Syracuse: Syracuse University Press.

Lowenthal, D. 1998. *The Heritage Crusade and the Spoils of History.* Cambridge: Cambridge University Press.

Marsh, G. P. 1864 [1965]. *Man and Nature, or Physical Geography as Modified by Human Action.* Cambridge: Harvard University Press.

Meinig, D., ed. 1979. *The Interpretation of Ordinary Landscapes.* New York: Oxford University Press.

Raitz, K. B. 2001. Field Observation, Archives, and Explanation. *The Geographical Review* 91: 121–131.

Tuan, Y.-F. 2001. Life as a Field Trip. *The Geographical Review* 91: 41–45.

Wright, J. B. 2001. Conservation's Radical Center in Geography. *The Geographical Review* 91: 9–18.

Zelinsky, W. 2001. The Geographer as Voyeur. *The Geographical Review* 91: 1–8.

# INDEX

# GEOJOURNAL LIBRARY
# Book Series

## International Handbook on Geographical Education

Edited by
**Rod Gerber**

The *International Handbook on Geographical Education* is the first truly
international publication in the field of geographical education for several
decades ... It is a publication for the thinking geographer and educator who
appreciate where international education is traveling to and how its challenges
can be met.

BOOK SERIES:
GEOJOURNAL LIBRARY 73
Hardbound, ISBN 1-4020-1019-2
November 2002, 360 pp.
€ 135.00 / $ 149.00 / £ 93.00

*FREE online preview available*

## Modelling Geographical Systems
*Statistical and Computational Applications*

Edited by
**Barry Boots, Atsuyuki Okabe and Richard Thomas**

This book presents a representative selection of innovative ideas currently
shaping the development and testing of geographical systems models by means
of statistical and computational approaches ... this volume would provide a useful
supplementary text for courses on quantitative geography and geographical
systems modelling in both human and physical geography, and GIS and
geocomputation.

BOOK SERIES:
GEOJOURNAL LIBRARY 70
Hardbound, ISBN 1-4020-0821-X
August 2003, 368 pp.
€ 130.00 /$ 130.00 / £ 82.00

## Managing Intermediate Size Cities
*Sustainable Development in a Growth Region of Thailand*

Edited by
**Michael Romanos and Christopher Auffrey**

This book applies a sustainable development framework to the planning and
managing of an intermediate size city in a developing region of a developing
nation, and assesses the potential of such a framework to effectively guide the
city's development.

BOOK SERIES:
GEOJOURNAL LIBRARY 69
Hardbound, ISBN 1-4020-0818-X
October 2002, 356 pp.
€ 130.00 / $ 143.00 / £ 90.00

For more information visit: www.wkap.nl/prod/s/GEJL

Contact Information:
**Customers in Europe, Middle East, Africa, Asia and Australasia:**
Kluwer Academic Publishers, Order Dept, P.O. Box 322, 3300 AH Dordrecht, The Netherlands
F +31-78-6576476 T +31-78-6576050 E orderdept@wkap.nl W www.wkap.nl
**Customers in the Americas:**
Kluwer Academic Publishers, Order Dept, P.O. Box 358, Accord Station, Hingham MA 02018-0358, USA
F +1-781-681-9045 T +1-781-871-6600 or (toll free within US) +-1-866-269-wkap E kluwer@wkap.com W www.wkap.com

kluwer
the language of science